Statistical Issues

A Reader for the Behavioral Sciences

Statistical Issues

A Reader for the Behavioral Sciences

Edited by

Roger E. Kirk
Baylor University

Brooks/Cole Publishing Company
Monterey, California

A Division of Wadsworth Publishing Company, Inc.
Belmont, California

ISBN: 0-8185-0005-0
L.C. Catalog Card No.: 73-150744
Printed in the United States of America

2 3 4 5 6 7 8 9 10

Preface

This book of readings has been compiled for introductory- and intermediate-level statistics courses in the behavioral sciences; it is an outgrowth of my efforts over the years to enrich my introductory statistics course with selected articles from the statistical literature. All too often students see statistics as a series of cookbook techniques to be slavishly applied to data. Hopefully this book will dispel that notion and help the reader catch a glimpse of the excitement of statistics.

The selection of articles has been guided by a number of criteria, the most important of which is whether an article will help broaden the student's understanding of important concepts and issues in statistics. Preference has been given to articles dealing with conceptual issues as opposed to those that are technique-oriented. A number of articles have been selected because they trace the development of such controversial statistical issues as the relevance of levels of measurement for the selection of a statistic, the use of one- versus two-tailed tests, the logic of hypothesis testing, and the choice of an error rate for multiple comparisons. The articles are organized into chapters that generally parallel the contents of contemporary textbooks.

Editorial commentaries preceding the articles point up key issues to be discussed, provide background information, and in many cases summarize the conclusions of the articles. Such overviews can provide a useful conceptual framework for the integration of new ideas. An additional aid is a glossary, in the appendix, containing 160 definitions of statistical terms.

Most of the articles in this book are appropriate for students whose backgrounds include only college algebra; however, several selections require some knowledge of calculus. Chapter 10 and portions of Chapters 3 and 7 are included primarily for use in intermediate-level courses. The teacher who is familiar with the mathematical preparation of his students can best judge which articles are appropriate.

"Classic" articles in the statistical literature are always stimulating, often interestingly written, and sometimes even intelligible to college students. A number of these articles appear in the book, although many could not be included because of their length or because they assume a mathematical sophistication that the typical student does not possess. Annotated bibliographies are provided for the student who wants to pursue in depth a particular topic or statistical issue.

The preparation of a book of this type always involves a number of compromises for an editor, particularly regarding breadth of coverage and length. Nevertheless, I think this book provides a balanced presentation of contemporary thinking about statistical issues in the behavioral sciences.

Many people have contributed to the preparation of this book. Although I cannot acknowledge all of them individually, I do want to express my appreciation to the authors and publishers who gave permission to use copyrighted material and to James V. Bradley of New Mexico State University, Arthur L. and Linda W. Dudycha of Purdue University, and John C. Flynn of Baylor University, who contributed original papers. Professors J. Barnard Gilmore of the University of Toronto and William L. Sawrey of California State College at Hayward reviewed the manuscript; their comments were especially helpful. I am indebted to hundreds of undergraduate and graduate students at Baylor University, whose reactions to the proposed selections were invaluable to me in making my final choices. And I am grateful to my wife for her editorial assistance and to the staff of Brooks/Cole for their splendid assistance and cooperation.

Roger E. Kirk

Contents

Statistical Issues

A Reader for the
Behavioral Sciences

Behavioral Statistics: Historical Perspective and Applications

A mastery of the current knowledge in many scientific areas does not require an appreciation of, or even familiarity with, the historical backgrounds of those areas. This is especially true of statistics, which has a short but colorful history. In today's statistics courses, emphasis on theoretical and methodological topics to the exclusion of history and the men who shaped that history is a natural consequence of the ever-increasing volume of information that must be taught. After covering theorems, derivations, and formulas, the teacher has little time to detail the interesting events that led to break-throughs in statistics. Thus it is easy for the student to lose sight of the fact that the body of knowledge in statistics is the product of people whose contributions were shaped by their personal attributes as well as the times in which they lived. This depersonalization of statistics may explain why introductory students rarely develop a real interest in their statistics course.

Clearly what is needed is an interestingly written overview of the development of statistics, with particular emphasis on the personal attributes of the giants in the field. The first article in this chapter provides such an overview. The authors, Arthur L. and Linda W. Dudycha, begin with the earliest work on probability and conclude with the contributions of Ronald A. Fisher, Jerzy Neyman, and Egon Pearson to statistical theory and practice. Their paper has three sections: (I) Probability Theory and the Normal Curve, (II) Descriptive and National Statistics, and (III) Statistical Inference and Experimental Design. These sections are largely self-contained and can be assigned in the sequence in which they are taught in class. Subsequent articles by L. McMullen and William G. Cochran provide personal reminiscences of two figures prominent in the development of modern statistics, W. S. Gosset and R. A. Fisher.

1.1 Behavioral Statistics: An Historical Perspective

Arthur L. Dudycha and Linda W. Dudycha
Department of Psychology, Purdue University

Prologue

Contemporary behavioral statistics, so impersonally discussed in often forbidding textbooks, has its primary roots—both subtle and impassioned—in the post-Renaissance period. The statistics used and revered in the behavioral sciences today descended from a varied and colorful ancestry: from the greed of ancient monarchs and gamblers to the quest for knowledge by the intellectually elite, ranging from pure mathematicians to clergymen to a brewer.

Forty years ago Helen Walker noted the rapidity with which the use of statistics was advancing and argued for an understanding of its origin and development by its students. Her prophecy concerning the salience of statistics was indeed accurate—perhaps understated. Statistics has become one of the major modes of communication among behavioral scientists—sometimes to the detriment of sound deliberate thought on the behavioral phenomenon in question. For example, "statistical significance" to many has become almost sacred and itself a goal, all too often at the sacrifice of practical or meaningful significance. Nonetheless, social scientists must continue to measure behavioral phenomena and submit these data to rigorous statistical analysis, but a close relationship must be maintained cognitively between the measured phenomenon and the statistical analysis.

Students of behavior, through exposure to the historical perspectives of statistics, are provided with the opportunity to gain a deeper appreciation for the necessity of this partnership. However, prior to Walker's book *Studies in the History of Statistical Method* (1929), there was, and still is, a paucity of definitive writing in the area.[1] This is not to imply that statistical history is not well documented. Various writers do take cognizance of statistics' heritage, but the literature is not replete with chronologies of statistical development.

This article will show some of the currents in the stream of statistical development through the thoughts, contributions, and personalities of its forefathers. Ideally, to best understand their contributions one should discuss these men through the prevailing social, political, economical, and religious frameworks of their times. We cannot accomplish completely so laudable a mission in one article, but we can provide a sample (nonrandom) of the contributions and polemics of these great men.

[1] Koren, John, *The History of Statistics*, 1918; Westergaard, Harald, *Contributions to the History of Statistics*, 1932.

A knowledge of the backgrounds of the giants in the evolution of statistics will give the serious student a better perspective from which to understand and evaluate current statistical practices in his science. For those yet needing reassurance that the perhaps not-so-palatable subject of statistics is both important and necessary, Sir Francis Galton's words surely must suffice:

General impressions are never to be trusted. Unfortunately when they are of long standing they become fixed rules of life, and assume a prescriptive right not to be questioned. Consequently, those who are not accustomed to original inquiry entertain a hatred and a horror of statistics. They cannot endure the idea of submitting their sacred impressions to cold-blooded verification. But it is the triumph of scientific men to rise superior to such superstitions, to desire tests by which the value of beliefs may be ascertained, and to feel sufficiently masters of themselves to discard contemptuously whatever may be found untrue [1908].

I. Probability Theory and the Normal Curve

Since the origin of most statistical concepts is rooted in the mathematical theory of probability, it seems appropriate to begin with the developments in that field.

The earliest faint traces of probability, found in the Orient around 200 B.C., were concerned with whether an expected child would be a male or a female. However, the first real cornerstone of the calculus of probability seems to have been laid in Italy when a commentary (Venice, 1477) on Dante's *Divine Comedy* referenced the different throws which could be made with three dice[2] in the game of Hazard. The first mathematical treatment of gambling problems was *Suma* (1494) written by Luca Paccioli (1445–1509). This work gives the

first version of the celebrated "problem of points," which concerns the "equitable division of the stakes between two players of unequal skill when the game is interrupted before its conclusion" (Walker, 1929, p. 5). This problem was destined to occupy the minds of probability theorists for two centuries. However, neither Paccioli nor the later Cardano (1501–1576), who wrote what is considered a "gambler's handbook" (1663), offered any general principles of probability, and they were often incorrect in solutions to the simple problems they did consider.

Though the conception of probability occurred in Italy, French mathematicians deserve credit for the first concerted efforts to master problems in probability. Blaise Pascal (1623–1662), who gained renown as a mathematician and physicist (but became a religious recluse at age 25), and Pierre de Fermat (1601–1665), a distinguished mathematician, exchanged a volley of letters during 1654 on problems suggested to Pascal by a gambler, Chevalier de Méré. Among these was the problem of points. Through the year, Pascal and Fermat gradually chiseled out of this "gambler's perplexity" an extremely important foundation stone in mathematical concepts—later to be known as the "theory of probability." Unfortunately, this correspondence was temporarily obscured by the then highly visible writings of Newton and Leibnitz; thus, the calculus of probabilities was not yet placed on a sound footing.

According to David, Christianus Huygens was "the scientist who first put forward in a systematic way the new propositions evoked by the problems set to Pascal and Fermat, who gave the rules and who first made definitive the idea of mathematical expectation" (1962, p. 110). Lord Huygens (1629–1695) was a Dutch astronomer and natural scientist who came from a family of wealth and position (as did many of the pioneers whom this article will mention). His mathematical treatise on dice games, *De Ratiociniis in Aleae Ludo* (1657), stood for a half century as the "unique" introduction to the theory of probability, and was only superceded when it

---

[2] The modern day die probably originated from the *astragalus*, used especially in the gaming of the Middle Ages. It is a knuckle bone having four sides on which it can rest (scored 1, 3, 4, 6); the other two sides are rounded (David, 1962).

inspired the major works of James Bernoulli, Pierre Montmort, and Abraham de Moivre.

James Bernoulli (1654–1705) was the eldest son of a family of Swiss merchant bankers in Basel, and the first of nine distinguished mathematicians of that famed name. He first took a degree in theology because his parents expected him to become a minister of the Reformed Church. What first set James on the path of astronomy and mathematics is speculative, but early in his career he became interested in the calculus of probability, as evidenced by a number of papers on this subject which were obviously inspired by Huygens. The single work for which James Bernoulli is best known is *Ars Conjectandi*, which was written during the latter part of his life but not published until eight years after his death. Nicholas Bernoulli, though only 18 years of age at the time, a nephew and pupil of James and already a probability theorist of some stature himself, was asked to edit James' all-but-complete manuscript for publication. Nicholas felt reticent and incompetent to do so (possibly because of Leibnitz's criticism of it), but finally capitulated when the pressure of public opinion allowed no further delay.

Ars Conjectandi is divided into four parts. In the preface Nicholas states:

... the first contains the treatise of the illustrious Huygens, "Reasoning on Games of Chance," with notes, in which one finds the first elements of the art of conjecture. The second part is comprised of the theory of permutations and combinations, theory so necessary for the calculation of probabilities and the use of which he explains in the third part for solution of games of chance. In the fourth part he undertook to apply the principles previously developed to civil, moral and economic affairs. But held back for a long time by ill-health, and at last prevented by death itself, he was obliged to leave it imperfect [from David, 1962, p. 134].

The first three parts alone would have established James Bernoulli as a probability theorist. It was, however, in the fourth part, *Pars Quarta*, that he introduced his celebrated but controversial "golden theorem"—his solution of the problem of "assigning the limits within which, by the repetition of experiments, the probability of an event may approach indefinitely to a given probability"—about which he wrote:

This is therefore the problem that I now wish to publish here, having considered it closely for a period of twenty years, and it is a problem of which the novelty, as well as the high utility, together with its grave difficulty, exceed in value all the remaining chapters of my doctrine. Before I treat of this "Golden Theorem" I will show that a few objections, which certain learned men [e.g., Leibnitz] have raised against my propositions, are not valid [*Ars Conjectandi*, 1713, p. 327; from K. Pearson, 1925, p. 206].

Bernoulli's Golden Theorem, which has since become the well-known "Bernoulli Theorem," was usually incorrectly phrased by textbook writers until the early 1900's as: "Accuracy increases with the square root of the number of observations." Further, early writers repeatedly stated that an illustration of Bernoulli's principle was "the fact that the constants [statistics] of frequency distributions in the case of large samples have standard deviations varying inversely as the square root of the size of the sample ...." This principle—currently often called the Law of Large Numbers—although admittedly closely allied to both Bernoulli's Theorem and the Tchebycheff Inequality, should be rightfully attributed to de Moivre (K. Pearson, 1925, p. 201). Pearson, taking note of the manner in which history often manages to misrepresent authorships, set the record straight as to what Bernoulli "really did achieve," which was "... to prove that by increasing sufficiently the number of observations he can cause the probability—i.e., that the ratio of observed successful to unsuccessful occurrences will differ from the true ratio within certain small limits—to diverge from certainty by an assignable limit" (K. Pearson, 1925, pp. 201–202). In modern terminology, the Bernoulli Theorem states that the "probability that a frequency v/n

differs from its mean value p by a quantity of modulus at least equal to ε tends to zero as $n \to \infty$, however small $\varepsilon > 0$ is chosen" (Cramér, 1946, p. 196; see also Hays, 1963).

Also in *Pars Quarta* Bernoulli proceeded to turn his argument for the Golden Theorem around, which resulted in the first definite suggestion of inverse and fiducial probability that is so essential to modern statistical theory in the now familiar form of confidence intervals. It is to this reverse principle, which Bernoulli simply stated without offering proof, that Leibnitz raised strong objections. The Golden Theorem thus sparked the beginning of the controversy on inverse probability and foreshadowed further developments of it in the writings of de Moivre, Bayes, Laplace, and Gauss.

Bernoulli, through *Ars Conjectandi*, is responsible for many other of our "modern" ideas. He developed the binomial theorem which is the basis for many distribution-free tests. Binomial trials often go by his name— Bernoulli trials. He also can be accounted responsible for inspiring de Moivre's derivation of the "normal curve" limit to the sum of a number of binomial probabilities when he (de Moivre) refined the Golden Theorem. If it were not for James Bernoulli, who had the mathematical prowess to digest the then modern analysis of Leibnitz (1646–1716) and Newton (1642–1727) and apply it to the analysis of games of chance, it is doubtful whether Montmort or de Moivre would have contributed what they did to the development of probability theory—to which we shall now turn.

Pierre Rémond de Montmort (1678–1719) was born in Paris of nobility. Contemptuous of parental control, young Pierre left home to avoid having to study law as his father had intended and instead traveled throughout Europe. In 1699, however, Pierre returned home and made peace with his father, who died shortly afterwards, leaving him a large fortune. Though having received this large inheritance, he did not plunge into the dissolute life of wine, women, and song which was thought natural for a young

nobleman of his time. Instead, he purchased the estate of Montmort, resigned his stall as a canon of Notre Dame in order to marry, and settled down in relative seclusion on his country estate to work on problems of probability—though he was no gambler.

The results of his efforts were first published in *Essai d'Analyse sur les Jeux de Hasard* (1708). A much more comprehensive second edition, which also included the extensive Montmort-Bernoulli (John and Nicholas) correspondence, was published in 1714. In the first edition Montmort began by finding the chances involved in various card games. He exhibited great agility and insight in his use of the principle of conditional probability, often attributed to de Moivre but probably dating back to Huygens or James Bernoulli. He also decided that while the rules of probability could be applied to the game of life, the chances in this game were too difficult to compute! The second edition, which contained much new material, reflected Montmort's maturity of thought on the subject and the influence of Nicholas Bernoulli. Here he presented generalized solutions for many games of chance discussed in the first edition, made the first but rather maladroit attempts toward questions of annuities, and solved the problem of points in full generality with two players of unequal skill. Montmort's contributions to probability theory probably lie not in the novel ideas he introduced but in his algebraic methods of attack (David, 1962).

Montmort, however, felt a strong compunction for having spent most of his life working on gambling problems and so apologized:

It is particularly in games of chance that the weakness of the human mind appears and its leaning towards superstition. ... There are those who will play only with packs of cards with which they have won, with the thought that good luck is attached to them. Others on the contrary prefer packs with which they have lost, with the idea that having lost a few times with them it is less likely that they will go on losing, as if the past can decide something for the future.... Others refuse to shuffle the cards and believe they must infallibly lose if they deviate from their rules. Finally there are those who look for advantage where there is none, or at least so small as to be negligible. Nearly the *same thing can be said of the conduct of men in all situations of life where chance plays a part*. It is the *same* superstitions which govern them, the *same* imagination which rules their method of procedure and which blinds their fears and hopes. ... The general principle of these superstitions and errors is that most men attribute the distribution of good and evil and generally all the happenings in this world to a fatal power which works without order or rule.... I think therefore it would be useful, not only to gamesters *but to all men* in general, to know that *chance has rules which can be known*, and that through not knowing these rules they make faults every day, the results of which with more reason may be imputed to themselves than to the destiny which they accuse.... It is certain that men do not work honestly as hard to obtain what they want as they do in the pursuit of Fortune or Destiny.... The conduct of men usually makes their good fortune, and wise men leave as little to chance as possible [from David, 1962, pp. 143–144; present authors' italics].

His prophecy concerning the applicability of "chance rules" to all walks of life seems to have been accurate, and his advice seems as necessary in the twentieth century as in the eighteenth.

Abraham de Moivre (1667–1754), an extremely influential but often underrated probabilitist of this period, was born in Champagne—then a province of eastern France (and still famous for its dry white wine). He, like Bernoulli, was a Huguenot, but unlike many other early probability theorists was neither wealthy nor of noble birth. At the age of eleven de Moivre began studying the humanities at a Protestant college. It was not until his family moved to Paris and he began attending classes at the Sorbonne, where he came into contact with the great teacher of mathematics Ozanam, that de Moivre became interested in mathematics. His education, however, was interrupted when the Edit of Nantes[3] was revoked by Louis XIV in 1685

and de Moivre, then eighteen, was imprisoned. He was released in 1688 and fled from France immediately, never to return and never to publish in his native tongue. The embittered young de Moivre landed in England devoid of money, friends, and influence and soon realized he had been disillusioned even about his "profound" knowledge of mathematics.

De Moivre spent his early years in London as a visiting tutor to the sons of noblemen. Finally he broke into the charmed circle of English mathematicians and was elected a Fellow of the Royal Society in 1697. Even so, de Moivre was destined to tramp about the London streets from pupil to pupil. In an effort to escape these humble circumstances, he begged John Bernoulli (James' brother) to intercede with Leibnitz and use his influence to get him a university post somewhere, but to no avail. Nevertheless, while augmenting his income by calculating odds for gamblers at a coffee-house that he frequented after his long days of tutoring, de Moivre unexpectedly found and developed a lasting friendship with Newton.

The early part of the eighteenth century saw de Moivre grow rapidly in mathematical stature. In 1711 he published *De Mensura Sortis, seu de Probabilitate Eventuum in Ludis a Casu Fortuito Pendentibus*, which led to a flurry of charges and counter-charges between Montmort and de Moivre. De Moivre insinuated that Montmort had done no more than slightly improve on Huygens, while Montmort insinuated that de Moivre had taken his ideas solely from *Essai d'Analyse*. De Moivre wrote in the preface of his memoir:

Huygens, first, as I know, set down rules for the solution of the same kind of problem as those which the new French author[4] illustrates freely with diverse examples. But these famous men do not seem to have been accustomed to that simplicity and generality which the nature of the thing demands [from David, 1962, p. 152].

It is particularly surprising that Montmort, who usually avoided getting embroiled in

---

[3] A law promulgated by Henry IV of France in 1598, granting considerable religious and civil liberty to the Huguenots.

[4] Referring to *L'Analyse des Jeux de Hasard* by Montmort.

arguments, replied so vigorously to the *De Mensura Sortis* in the "Avertissement" of the second edition of *Essai d'Analyse* (1714), writing:

The author did me the honour of sending me a copy.... M. Moivre[5] was right to think I would need his book to reply to the criticism he made of mine in his introduction. His praise-worthy intention of boosting and increasing the value of his work has led him to disparage mine and to deny my methods the merit of novelty. As he imagined he could attack me without giving me reason for complaint against him, I think I can reply to him without giving him cause to complain against me.... [from David, 1962, p. 153].

And indeed he did reply—with both length and uncharacteristic adamance—expounding on the history of probability theory from the Pascal-Fermat debate up to and including his own contributions. A partial precipitating factor to this entire controversy may also have been that Montmort felt impelled to react against the émigré Frenchman in London, especially since a few years earlier English troops had been knocking at the gates of France. Apparently the quarrel had resolved itself by 1718, because in that year de Moivre acted as Montmort's guide and interpreter during a visit he made to London (David, 1962).

In 1718, de Moivre published and dedicated to Newton *The Doctrine of Chances*, which was an expanded version, in English, of *De Mensura Sortis*. In the preface of this first edition de Moivre was obviously trying to gain favor with Montmort and the Bernoullis when he wrote apologetically for his remarks about Montmort's memoir:

As for the French book, I had run it over but cursorily, by reason I had observed that the Author chiefly insisted on the Method of Huygens, which I was absolutely resolved to reject.... However, had I allowed myself a little more time to consider it, I had certainly done the Justice to its Author, to have owned that he had not only

illustrated Huygen's Method by a great variety of well chosen examples, but that he had added to it several curious things of his own Invention.... [He] published a Second Edition of that Book, in which he has particularly given many proofs of his singular Genius and extraordinary Capacity; which Testimony I give both to Truth, and to the Friendship with which he is pleased to Honour me. ... [from David, 1962, p. 166].

As David notes, it is revealing that the words after "reject" were omitted from the second edition (1738) and the posthumous third edition (1756)—both of which were published well after the death of Montmort in 1719!

The Preface of de Moivre's *Doctrine* contains a lengthy summary of the contents of the book, discusses the problem of points for two gamblers of unequal skill, and relates the author's general ideas about chance. He begins by giving the definitions of probability, the addition of probabilities, expectation, independence of events, and joint and conditional probabilities. The remainder of the book is divided into discussions of specific problems. It is of note that in Problem V, de Moivre reaches what has been commonly called Poisson's (1781–1840) binomial exponential limit, and in Problem VII he gives us the multinomial distribution, which was also arrived at independently by Montmort and Nicholas Bernoulli in their correspondence. There is no doubt that the first edition was written by " ... a man who was already superior to Montmort and the Bernoullis in his mathematical powers, and who, when he came to maturity, was to produce in this third edition the first modern book on probability theory" (David, 1962, p. 171).

De Moivre also had a life-long interest in the theory of annuities, an interest that undoubtedly grew from his early and lasting acquaintance with Edmund Halley,[6] who had

---

[5] Note that Montmort did not use the noble prefix "de," which de Moivre himself is believed to have added while crossing the English Channel.

[6] It was Halley (1656–1742) who in 1692, while Secretary to the Royal Society, met de Moivre and introduced him to the mathematical elite of Britain and tried to interest him in astronomy. Halley, an astronomer, was the first to predict the return of the comet which now bears his name.

constructed one of the early life-expectancy tables (1693). In 1724 de Moivre published the first edition of his *Annuities on Lives*, with subsequent editions in 1743 and 1750. The second edition was probably published in protest against Thomas Simpson[7] (1710–1761), who not only published *The Nature and the Laws of Chance* in 1740, which was very similar in content to the 1738 edition of the *Doctrine*, but in 1742 published a treatise on annuities which was an out-right plagiarism of de Moivre's earlier work on annuities. De Moivre was justifiably disturbed over Simpson's behavior—especially since he was then 76, still lived a humble existence, and depended heavily on his book profits. It was probably poetic justice that Simpson's single real contribution to probability theory, *An Attempt to Show the Advantage Arising by Taking the Mean of a Number of Observations in Practical Astronomy* (1757), was plagiarized by Count Joseph LaGrange in 1773 (Walker, 1934).

The discovery of the normal curve of errors, a distribution of extreme salience to both probability and statistics, was for over a century primarily attributed to Gauss (1777–1855) and much less often to the earlier Laplace (1749–1827). Karl Pearson, true to his keen interest for historical accuracy, wrote: "But in studying de Moivre I have come across a work which long antedates both Laplace and Gauss" (1924, p. 402). In 1730 de Moivre published his *Miscellanea Analytica de Seriebus et Quadraturis*. It was in this book that de Moivre first gave the expansion of factorials that is now known as Stirling's theorem but would be more appropriately called the de Moivre-Stirling theorem. Many copies of this work printed in 1730 contain a *Supplementum* in which de Moivre had further thoughts on the expansion of factorials and presented a table of fourteen-figure logarithms of factorials.

Of paramount importance, however, is the extremely rare second supplement discovered by Pearson. He wrote: "But only a *very few* copies have a second supplement also with separate pagination (pp. 1–7) and dated Nov. 12, 1733. This second supplement could only be added to copies sold three years after the issue of the original book, and this accounts for its rarity" (1924, p. 402).[8] This "landmark" second supplement, written in Latin as was the rest of the book, is entitled *Approximatio ad Summam Terminorum Binomii $(a + b)^n$ in Seriem Expansi*. Walker observed that "In this obscure treatise on abstract mathematics . . . supposed by its author to have no practical implications outside the realm of games of chance . . . we have the first formulation of the momentous concept of a law of errors [normal curve]" (1929, p. 14).

In the second and third editions of the *Doctrine*, de Moivre printed his own English translation of the *Approximatio*.[9] He introduced it at the end of a corollary to Problem LXXXVII with the remarks:

---

[7] Thomas Simpson is described as a somewhat erratic and irascible genius who wrote on calculus, algebra, and probability. He was, however, one of the distinguished English mathematicians in the relatively sterile period following the death of Newton (Walker, 1929).

[8] There are only two original copies of this second supplement known to exist—one at University College, London, and another in the Preussische Staatsbibliothek, Berlin.

[9] There appears to have been some confusion about when de Moivre first printed his translation of the *Approximatio*. David writes in a footnote (1962, p. 174): "I find it curious that de Moivre did not incorporate it in the Second Edition of *The Doctrine of Chances* in 1738"; and Karl Pearson writes: "The same matter is dealt with twenty-three years later in the 1756 edition of the *Doctrine of Chances*, pp. 243–250" (1924, p. 404). Helen Walker, however, in her book (1929) correctly states that de Moivre first translated this momentous document in the 1738 edition of the *Doctrine*. Thus, it appears as though Pearson and David must have overlooked the 1738 edition, since the 1967 edition of de Moivre's *Doctrine*, which is supposedly an exact impression of the 1738 edition except for the addition of an analytic table of contents and an index, does indeed contain on pages 235–243 the translation of the *Approximatio*. It is even more "curious" that the discovery of the normal curve had not been rightfully attributed to de Moivre long before Pearson's 1924 discovery of the obscure Latin document, especially since it was contained in the 1738 and the 1756 editions of the *Doctrine*, which were in English and which apparently enjoyed widespread acclaim.

...I'll take the liberty to say, that this is the hardest Problem that can be proposed on the Subject of Chance, for which reason I have reserved it for the last. ... I shall derive from it some Conclusions that may be of use to every body: in order thereto, I shall here translate a Paper of mine which was printed November 12, 1733, and communicated to some friends, but never yet made public, reserving to myself the right of enlarging my own Thoughts, as occasion shall require [1967, pp. 234–235].[10]

Its English title is "A Method of approximating the Sum of the Terms of the Binomial $(a + b)^n$ expanded into a Series, from whence are deduced some practical Rules to estimate the Degree of Assent which is to be given to Experiments." It begins with the following introductory remarks:

Altho' the Solution of Problems of Chance often require that several Terms of the Binomial $(a + b)^n$ be added together, nevertheless in very high Powers the thing appears so laborious, and of so great a difficulty, that few people have undertaken that Task [1967, p. 235].

De Moivre's method of approximating to the "sum of the terms of the binomial" yielded the first "normal" limit to be derived in probability, and many of the modern treatises on probability (and sampling) approach this problem similarly.

Upon reading this work, one is immediately struck by de Moivre's remarkable accomplishments and accuracy. For example, de Moivre gave very close approximations (relative to present tabled values) to specific areas under the "normal" curve, as:

	de Moivre's value		Present tabled values
$\pm 1\sigma$	.682688	(Corollary 3)	.682689
$\pm 2\sigma$	.95428	(Corollary 6)	.95450
$\pm 3\sigma$	.99874	(Corollary 6)	.99730

The small discrepancies result from the slight inaccuracies in his quadrature formula,

[10] Toward the end of his long and arduous life his memory was undoubtedly beginning to fail since this paper was in fact included in some copies of *Miscellanea Analytica* as a second supplement.

but his reasoning was correct. In addition, he knew that the maximum ordinate (probability density) of the curve of error is $1/(\sigma\sqrt{2\pi})$ (current notation).

As a close to the final chapter of de Moivre's life, it is interesting to note that in the 1738 edition of the *Doctrine*, and to a greater extent in the posthumous 1756 edition, he relates his theory of probability inextricably with theology, finally remarking:

... as it is thus demonstratable that there are, in the constitution of things, certain Laws according to which Events happen, it is no less evident from Observation, that those Laws serve to wise, useful and beneficient purposes; to preserve the stedfast Order of the Universe, to propagate the several Species of Beings, and furnish to the sentient Kind such degrees of happiness as are suited to their State.

But such Laws, as well as the original Design and Purpose of their Establishment, must all be *from without*: the *Inertia* of matter, and the nature of all created Beings, rendering it impossible that any thing should modify its own essence, or give to itself, or to any thing else, an original determination or propensity. And hence, if we blind not ourselves with metaphysical dust, we shall be led, by a short and obvious way, to the acknowledgement of the great MAKER and GOUVERNOUR of all; *Himself all-wise, all-powerful* and *good* [*The Doctrine of Chances*, 1756].

And so de Moivre died—old, feeble, and still indigent—little suspecting that his work on the difficulty of the "thing" would attain stature commensurate with the brilliance that gave it birth.

A now well-known contemporary of de Moivre was the Reverend Thomas Bayes (1702–1761). He was born in London, and since his family were Nonconformists, Bayes was forced to seek a private education. It is believed that de Moivre might have been his mentor for a short period when he was 12. He was eventually ordained and first assisted his father at a Presbyterian meeting house in London. Later he became the minister at the Presbyterian Chapel in Tunbridge Wells. In 1731 and 1736 Bayes, under the pseudonym of John Noon, published two treatises that probably led to his election in 1742 as a Fellow of the Royal Society. In 1752 he either retired from or was forced to abdicate his ministerial position. He did remain, however, in Tunbridge Wells until his death, when the details of his will revealed that he was certainly other than a "poor church mouse" (Barnard, 1958).

The single most referenced and celebrated essay written by Bayes is one published posthumously in 1763—"An Essay Towards Solving a Problem in the Doctrine of Chances."[11] In a letter to John Canton dated November 10, 1763, Richard Price wrote: "I now send you an essay which I have found among the papers of our deceased friend Mr. Bayes, and which, in my opinion, has great merit, and well deserves to be preserved" (Barnard, 1958, p. 296). The "problem" on which Bayes had worked was, in his words:

Given the number of times in which an unknown event has happened and failed: *Required* the chance that the probability of its happening in a single trial lies somewhere between any two degrees of probability that can be named [from Barnard, p. 298].

Through the years the theorem and the inferential process that now bear his name evolved from considerations of this problem. The "Essay," however, has been the focus of what may be one of the more heated controversies in the history of probability and statistics, extending to the present. Many current authors feel that the contents of Bayes' paper are mathematically unsophisticated and that the scope of the applicability of "Bayesian" thinking is severely limited (e.g., David, 1951; Feller, 1957). In fact, Bayes himself may have had doubts about the validity of the application of his theorem, since he had himself withheld its publication.

The period of the eighteenth century following the death of de Moivre was relatively

[11] This essay was published in the *Philosophical Transactions* of the Royal Society of London, 1763, **53**, pp. 370–418.

dormant, and there were few important advances made in the theory of probability until near the close of the century. The first two decades of the nineteenth century, however, saw renewed activity. The law of errors was established on scientific principles and the foundation was laid for a theory of errors of observations. These new discoveries were widely disseminated, especially among European astronomers. Laplace and Gauss, the two greatest mathematical astronomers then living, were primarily responsible for this renewed enthusiasm for probability.

Pierre-Simon de Laplace (1749–1827) was born in Beaumont-en-Auge (Normandie, France) to parents of very modest means—his father was in some manner concerned with the production of cider. Laplace was destined for the church and at age six was sent to a Benedictine school where his uncle was a priest and teacher of mathematics. In 1765 Laplace entered the Jesuit College of Arts at Caen with the idea of continuing study in the humanities and taking the clerical robe, but instead he found the path of his life-work in celestial mechanics and probability. In 1768 at the age of 19 he left for the big city (Paris) with a letter of introduction to D'Alembert (1717–1783), who was then one of the most influential men on the French scientific scene. D'Alembert, finally convinced of Laplace's abilities, secured for him a post with the Military Academy of Paris, first as a teacher of mathematics and later as an "examiner" for the finals of the cadets. It is recorded that Napoleon Bonaparte was examined for his commission of Second Lieutenant by Laplace in September, 1785.

Laplace, an ambitious young man with a precocious intellect and prodigious memory, was a prolific writer during his early years in Paris—only to have most of his memoirs rejected. Finally in 1774 the Paris Academy chose and published two of his many submitted papers. They dealt with the theory of chance, the first being a rehash of problems discussed exhaustively by de Moivre and the second being applications of Bayes' Theorem.

During his early years in Paris as well as in

later years Laplace was a social climber—a "cultivator of social position" (David, 1965, p. 34). One of his early acquaintances was Condorcet (1743–1794), who was a Marquis of the old French nobility and a friend and disciple of the then elderly Voltaire (1694–1778). Condorcet and Laplace, like many of the French mathematicians and philosophers of that period, discussed at length social justice, the credibility of witnesses, and the probability that a tribunal will reach a correct decision—all from the point of view that a mathematical probability can be considered as a degree of belief. Later Laplace cultivated the friendship of Lavoisier (1743–1794), who was a nobleman, a noted chemist, and a hated tax collector. He had, however, considerable patronage at his disposal. Laplace and he worked together on the generation of electricity (1781) and specific heat (1782–1784). Also about this time Laplace wrote on probability-generating functions, worked on the speed of sound and on capillary action, and improved Legendre's[12] and Lagrange's[13] methods for approximating to integrals by series expansion.

In 1788 Laplace married into minor nobility. The following year heard the first rumblings of the French Revolution, and saw the old academies dissolved and their archives divided among their members, probably as a result of Robespierre's thundering that the Revolution had no use for scientists. Louis XVI and Marie-Antoinette were guillotined in 1793. By now the Revolution was at full throttle, the bloodbath had begun, and the streets of Paris were unsafe. In 1794 Condorcet, Lavoisier, and many other close friends of Laplace were arrested as "men not sufficiently trustworthy either in their republican sympathies or in their hatred of royalty" and lost their heads. By the end of the year the terror had subsided and somehow Laplace had survived. He became assistant to Lagrange[14] at the Ecole Normale, and in 1795 both were appointed to the Bureau of Longitudes, where Laplace calculated artillery tables. This was by no means the end of his "cultivating of social positions." In 1799 he published the first two books of the *Celestial Mechanics* (Book 3 in 1802; Book 4, 1805; Book 5, 1825) and sent copies to Napoleon Bonaparte, who replied with an invitation to dinner. Within a month Napoleon seized power and named Laplace minister of the interior. During the short six-week tenure as minister Laplace " . . . sent troops of comedians and dancing girls to the army in Egypt" and named July 14 a national holiday—Bastille Day. He was then moved to the senate and by 1803 had become chancellor. Although Napoleon bestowed many honors upon Laplace, he was not without enemies—both political and scientific. It is claimed (probably unjustly) that it was Laplace who had ill-advised Napoleon on the Russian weather, thus causing the debacle of the French army. After the Russian campaign of 1812 Frenchmen began to turn against Napoleon, and in 1814 Laplace, as chancellor of the senate, signed the document deposing Napoleon. He soon resigned as chancellor but remained in the Bureau of Longitudes during the Bourbon Restoration. His final

[12] Although little is known about the French mathematician Adrien Marie Legendre (1752–1833), he did make a significant contribution to statistics in his book entitled *New Methods for Determining the Orbits of Comets* (1805). In an appendix "On the Method of Least Squares" he introduced the technique of least sum of squared residuals and proceeded to deduce his now well-known rules for forming the normal equations.

[13] Joseph Louis Lagrange (1736–1813) was born, reared, and educated in Turin, Italy, where his family had settled. He was, however, of French extraction (Réné Descartes, 1596–1650, was his ancestor) but always thought of himself as Italian. Among the many honors bestowed on Lagrange was the Grand Prize of the French Academy of Sciences in 1764 for his solution of an astronomical problem: "The libration of the moon—why does the moon always present the same face to the earth?" In 1766 Lagrange succeeded Euler (1707–1783, Venn-*Euler* diagrams) in Berlin where his interest was diverted to the calculus of hazard. He published *Miscellanea Taurinensia* in 1773, which he had plagiarized from Simpson. He did, however, introduce the terminology *curve of errors*.

[14] Lagrange had returned to Paris in 1787 at the request of Louis XVI. Although he was a friend and protégé of Marie Antoinette, he was "tolerated" by the revolutionaries since he was Italian.

honor was being made a Marquis by Louis XVIII (David, 1965).

Throughout his life Laplace had written many memoirs on probability, on the method of least squares, and on the curve of errors. Many of these were brought together and reprinted along with some new material in a book entitled *Théorie Analytique des Probabilitiés* (1812). Some consider this 500-page book the greatest single work on the subject of probability. Later editions in 1814 and 1820 contained a long introduction that had been published separately under the title " Essai philosophique sur les probabilitiés."

Laplace's contributions to the development of statistics are best summed in the words of David:

It seems unquestionable . . . that Laplace needed, very strongly, the stimulus of other men's creative ideas. He, himself, was not an originator like Fermat or Newton, but a synthesizer and improver, a mathematician unrivalled in beating out and extending the paths pointed out by other men's intuition. Once he got the idea, he was magnificent in generalization and application [1965, p. 36].

While the influence of Laplace was reaching its zenith, the keen mathematical abilities of Carl Friedrich Gauss (1777–1855) were beginning to be recognized in Germany. Gauss, often referred to as the "Prince of Mathematics," is ranked with Archimedes and Newton as one of the three greatest mathematicians of all time (Bell, 1937). He was born in Brunswick, Germany, under very humble circumstances. Had his father prevailed, this gifted boy probably would have become a gardener or bricklayer. However, Gauss's mother (who could barely read and write but was intelligent and strong-willed) and his uncle quickly recognized his unusual talents, which they continually cultivated and sharpened. Like Laplace, Gauss's mathematical precocity was evidenced time and time again throughout his life. Even before he was three he is reported to have detected an error in his father's payroll calculations and announced the correct "reckoning" (Bell,

1937). In school he repeatedly astonished his teachers with his mental perspicacity and celerity. He soon came to the attention of Carl Wilhelm Ferdinand, Duke of Brunswick, who became Gauss's patron and paid for his education at Caroline College (1792–1795) and the University of Göttingen (1795–1798). Not until 1796 did Gauss decide to make mathematics, rather than philology, his career. It was then that he solved a 2,000-year-old problem by proving that a regular seventeen-sided polygon is amenable to straight-edge-and-compass construction. The same year he developed the first rigorous proof of the fundamental theorem of algebra (i.e., that every nonconstant polynomial has a root), for which he was awarded a Ph.D. *in absentia* by the University of Helmstedt in 1799 (Eisenhart, 1968). In 1798, while still a student, Gauss completed his masterpiece *Disquisitiones Arithmeticae*, which was finally published in 1801. This great work is a rigorous reformulation and blending of Gauss's own original contributions to the theory of integral numbers and rational fractions with those of his predecessors.

Gauss's achievements and the honors he received are numerous. He made significant contributions to every branch of pure and applied mathematics that existed in his day, some of which he had founded. He also made major contributions in astronomy, geodesy, physics, and metrology. Among his many honors, the ones he cherished above all else were his honorary citizenships of Brunswick and Göttingen.

From the many contributions of Gauss, those used most widely in the biological, physical, and social sciences today relate to the method of least squares. His first formulation of the method was made between 1795 and 1798, when he was still a student. It is often referred to as his minimum-sum-of-squared-residuals (mssr) technique. This technique was a simpler and more objective procedure than the method of averages, which had been independently devised in 1748 by Leonard Euler and Johann Tobias Mayer (1723–1762). Gauss first attempted to justify his mssr technique by means of probability theory but was soon aware that the "determination of the most probable value of an unknown quantity is impossible unless the probability distribution of errors of observation is known explicitly" (1821, in Gauss, 1880, 4, p. 98). Therefore he proceeded to seek the probability distribution of errors using inverse probability. He found $f(v) = (h/\sqrt{\pi}) \exp(-h^2 v^2)$ to be the required probability distribution of errors[15] and noted that h can be considered as the measure of precision of the observations (in modern notation h equals $1/\sigma\sqrt{2}$ where σ is the root-mean-square-error, or standard deviation, of the distribution). Gauss then proceeded to derive and justify his mssr technique from his law of errors by what would now be called the method of maximum likelihood. According to Eisenhart (1964, p. 31), "The Method of Least Squares was, therefore, regarded as firmly established, not merely on grounds of algebraic and arithmetical convenience, but also via the calculus of probabilities—at least when the number of independent observations is large!" Concurrently, Gauss deduced the rule of formation of the normal equations that are used to determine optimal estimators for unknown quantities in a linear function and outlined his method of elimination for solving the normal equations, which is most familiar today through a modification of M. H. Doolittle in 1878 (Eisenhart, 1964).

In 1801, when Gauss began to apply his principles of least squares and made, from very limited data, remarkably accurate predictions of the orbits of some newly discovered planets, he gained immediate recognition as an astronomer. Yet nearly a decade elapsed before Gauss published anything on least squares. In fact, Gauss originally "did not

attach great importance to the method of least squares; he felt it was so natural that it must have been used by many who were engaged in numerical calculations" (Dunnington, 1955, p. 113). However, in 1809 Gauss finally published his second great work, *Theoria Motus Corporum Coelestium in Sectionibus Coniciis Solem Ambientium* (Theory of Motion of the Heavenly Bodies Moving about the Sun in Conic Sections), and devoted the third section to a detailed exposition of the theory and application of his method of least squares. Gauss had offered a German version of this great work for publication in 1807, but the publisher would accept it only if Gauss translated it into Latin. This was due to the shaky political scene in Germany following the decisive defeat of the Prussian Army (led by Duke Ferdinand) by the Napoleonic forces in 1806.

In 1821 Gauss presented a second and completely new formulation of the least-squares method that was independent of any particular probability distribution of errors. This was prompted by some earlier work in which Laplace (1774) had adopted the principle of least-mean absolute error of estimation and had shown that in the case of independent observations Gauss's mssr technique led to the same estimators as his own procedure, but only when the errors of observations behaved according to the Gaussian law of errors. Gauss instead adopted the principle of least-mean squared error of estimation to find the "best mean" of a number of independent observations. He showed that the mssr technique yields minimum mean-square error estimators of the unknown parameters, regardless of the error distribution of observations, but only when the mean values of a set of n independent observations are linear functions of the unknown parameters and the variances (modern terminology) of the observations are all finite. In a second memoir, appearing in 1823, he extended the preceding result to estimators of linear functions of the m unknown parameters and showed that the resultant minimum sum of squared residuals is equal to the sum of

---

[15] When Gauss began to deduce his law of errors, many other shapes had already been suggested by earlier authors: a normal law of error by de Moivre (1733); a discrete rectangular or a discrete triangular law of error by Simpson (1755); a continuous triangular distribution by Simpson (1757); rectangular by Lagrange (1774); double exponential by Laplace (1774); semicircular by Daniel Bernoulli (1778); and double-logarithmic by Laplace (1781).

$n - m$ independent errors. These results often have been attributed erroneously to Andrey Andreevich Markov (1856–1922).[16] Some authors (such as Scheffé, 1959), however, have derived and discussed them under the rubric "Gauss-Markov Theorem." Today statisticians say that this method of least squares yields minimum-variance linear unbiased estimators, which are found in techniques such as linear regression and analyses of variance and covariance.

In his later years Gauss took great pride in his contributions to the development of the least-squares method and preferred his second formulation above all others. In an 1839 letter to F. W. Bessel, " ... he remarked that he had never made a public statement of his reasons for abandoning the metaphysical approach of his first formulation, but that a decisive reason was his belief that *maximizing* the probability of a *zero* error is less important than *minimizing* the probability of committing *large* errors" (Eisenhart, 1968, p. 78; italics added). Also in these later years Gauss, like James Bernoulli, Laplace, and others, turned his attention to the emerging social sciences. He was particularly active in applying probability theory to social laws, concerning himself with such things as mortality and the construction of mortality tables.

II. Descriptive and National Statistics

At the same time that mathematical astronomers and other scientists and philosophers were contributing to the development of probability theory, there was a concurrent and yet almost entirely unrelated growth in two other movements vital to modern statistical theory and practice: descriptive statistics and national statistics (census). A little more than a century ago these three movements were

merged in the work of Adolphe Quetelet, who is sometimes considered the father of social science. Prior to discussing the influence of Quetelet and those following in his footsteps, let us look briefly at the course of developments in national and descriptive statistics.[17]

National statistics go as far back as recorded history—David "numbered" his people, and the Egyptians, Romans, and Greeks recorded many types of state resources. However, one of the first systematic attempts to compile statistics on national resources was made by Sebastian Münster (1489–1552), professor at Heidelberg and Basel, in the mid-1500s. He tried to collect statistics on the geography, history, political organizations, social institutions, commerce, and military strength of all the principal states then existing—a Herculean task indeed (Walker, 1929). Since the Middle Ages data were collected on such topics as land, serfs, births, marriages, deaths, and industrial production, and later statistics began to play an important role in determining state policies. The first work to shed light on the regularity of social phenomena was *Observations on the London Bills of Mortality, 1662* by John Graunt (1620–1674), a shopkeeper. This and similar works marked the beginning of a theory of annuities and led to the founding of insurance societies. Further advances in census taking were spurred by the social upheavals that marked the close of the eighteenth century—movements that emphasized the salience of the proletariat or masses. The U.S. Federal Constitution (1787) provided for the taking of a decennial census, and the first was taken in

---

[16] A. A. Markov was a prominent Russian mathematician. He was a student of Tchebycheff (1821–1894), who is known for the inequality which bears his name (see Tchebycheff's Inequality, Hays, 1963, p. 188). Markov is best known for what today are called Markov chains, which he introduced in 1907.

[17] The term *statistik* probably occurred for the first time in the writings of Gottfried Achenwall (1719–1772), who lectured at Göttingen and is sometimes hailed as the father of statistical science. For an account of the early use of the term, see G. U. Yule's "Introduction of the words 'statistics,' 'statistical' into the English language," *J. of Roy. Stat. Soc.*, 1905, **68**, 391–396. The term *statistic* and its modern usage is attributed to R. A. Fisher: " ... a *statistic* is a value calculated from an observed sample with a view to characterising the population from which it is drawn," *Statistical Methods for Research Workers*, 1925, p. 44.

1790. Most countries had provided for some form of enumeration of their people by the 1850s.

Also growing out of this movement were numerous statistical societies. In 1834 the Statistical Society of London was organized; it was renamed in 1877 and survives today as the Royal Statistical Society. In 1839 the American Statistical Association was founded in Boston. During the early meetings of this association, the addresses covered such topics as statistics of pauperism, crime, and immigration. A member once presented the statistics of the number and kind of carriages in Massachusetts in 1756, and another deemed it worthy to read a paper on the statistics of the funeral charges in the interment of Governor Winslow in 1680![18] Today both of these societies enjoy memberships spanning the globe, but their founding fathers would have difficulty recognizing their progeny (Koren, 1918; Walker, 1929).

The developments in national statistics (also called "political arithmetic"), however, were closely related to and relied heavily upon those in descriptive statistics. The modern indices of central tendency and variability were earlier referred to as "frequency constants." Undoubtedly, the oldest index was the arithmetic mean, which was known to Pythagoras in the sixth century B.C., along with the geometric mean and the "subcontrary" (harmonic) mean. Fechner (1874) outlined extensive procedures for arriving at the value of the "middlemost ordinate," an index that had long been known and used. It was dubbed "median" by Galton (1883) while he was working on percentile formulations. Another already familiar descriptor was called the "mode" by K. Pearson (1894); it referred to the abscissa value of the ordinate of maximum frequency.

Conceptually, many indicants of variability or dispersion had been in existence long before J. F. Enke (1791–1865, a student of Gauss)

formally set down in 1832 a gross-score formula for the standard deviation and formulae for the standard errors of the mean and standard deviation, all of which he learned or modified from Gauss (Walker, 1929). Gauss (1816) also had used the mean deviation around the middlemost value, which today is the median deviation. The term "standard deviation" was originated by K. Pearson (1894), whereas "variance" was originated by R. A. Fisher (1918). Another index of variability, the semi-interquartile range, is attributable to Galton (1889). It evolved from his development of percentiles or "scales of merit" (Walker, 1929).

These three movements were then integrated by Lambert Adolphe Jacques Quetelet (1796–1874), who was born and schooled in Ghent, Belgium. Upon completion of his secondary schooling he became a teacher of mathematics in a Ghent secondary school, but his true interests at that time were in the arts—he painted, wrote poetry, even collaborated on an opera with an old school chum. Later he was persuaded to turn from his artistic endeavors to the study of advanced mathematics at the newly founded University of Ghent, from which he received his doctorate in 1819. At age 23 he accepted a chair in mathematics at the Athenaeum in Brussels, where he immediately began publishing a prodigious array of memoirs, mostly in mathematics and physics. In 1824 he assumed the additional task of delivering public lectures. A number of elementary treatises sprang from these lectures designed to popularize the fields of astronomy (1826, 1827), physics (1827), and probability (1828). His "Instructions populaires sur le calcul des probabilité" marked the beginning of his emerging interest in statistics and its applications to social phenomena. However, from 1823 to 1832 he was preoccupied with establishing an observatory at Brussels. It was finally completed in 1832, after having been used as a fortress during the Belgian Revolution of 1830. It was this endeavor that led him to Paris in 1823 to become acquainted with the latest astronomical techniques and instru-

---

[18] Not to be outdone, the latter half of the twentieth century saw, for example, the "statistical" report of the "First 1,945 British Steamships" (*Amer. Stat. Assoc.*, 1958, **53**, p. 360).

ments; but more importantly, it afforded him the opportunity to meet Poisson, Fourier, and, in particular, Laplace, from whom he received valuable instruction and stimulation. In 1829 he also visited with Gauss, and they collaborated on geomagnetic measurement experiments in Gauss's yard.

Quetelet wore many hats simultaneously. He was a mathematician, physicist, astronomer, anthropometrist, supervisor of official statistics for his country, prime mover in the organization of a central commission on statistics, instigator of the first nationwide census, ardent and systematic collector of statistics on a wide variety of phenomena, university teacher, public lecturer, author of numerous books and articles, prolific correspondent with many of the scholars of his day (approximately 2,500 have been identified!), and in his leisure time a poet.

That Quetelet may merit being called the father of social science is evidenced by the large amount of his writing (sometimes controversial) on social theory and research. The concept of "social physics" emerged in his 1835 publication of *Sur l'homme et le dévelopment de ses faculties*, subtitled *Physique sociale*. This and later publications contained his two basic principles in "moral statistics" (the regularities of moral characteristics, such as crime and suicide rates). These principles are (1) ". . . social phenomena in general are extremely regular and . . . [their] empirical regularities can be discovered through the application of statistical techniques," and (2) " . . . these regularities have causes . . . [which are] the social conditions at different times and . . . places" (Landau and Lazarsfeld, 1968, p. 250). In offering support of his views, Quetelet made use of what are now known as multivariate techniques, though of course in a less sophisticated form.

Quetelet vehemently believed in the concept of an "average man" or *homme moyen*, seen both in earlier memoirs and in *Du système social et des lois qui le régissent* (1848). That characteristics of the average man (i.e., the man representing a nation) must be estab-

lished through knowledge of a mean value plus knowledge of upper and lower limits of variation within a group was something of a novel idea introduced by Quetelet. He noted that various characteristics distribute themselves systematically around their means according to the "law of accidental causes" which is "general . . . [and] applies to *individuals* as well as *peoples* and which governs our *moral* and *intellectual* qualities . . . [as well as] physical" (Landau and Lazarsfeld, 1968, p. 250, italics added). He had indeed taken note of the fact that many of his sets of empirical observations were distributed according to the binomial and normal theoretical distributions. In the process of comparing these kinds of distributions (theoretical versus empirical), Quetelet unexpectedly "discovered" an interesting bit of trivia—an alarming amount of draft evasion in the French army. It seemed on the basis of his calculations that some 2,000 men had managed to escape conscription by "somehow shortening" themselves below the minimum height (Landau and Lazarsfeld).

III. Statistical Inference and Experimental Design

A man whose contributions to the evolution of behavioral statistics rank alongside Quetelet's was his contemporary Sir Francis Galton (1822–1911). While Galton's life work was not the development of statistics *per se*, he, more than anyone before him, applied the knowledge gained from the mathematical astronomers to an almost endless variety of problems—from tea making to eugenics.

Galton was born near Sparkbrook, Birmingham, England. Both of his grandfathers were Fellows of the Royal Society. His father came from a line of prosperous Quaker businessmen and had a strong interest in the natural sciences and statistical inquiry. His grandfather on his mother's side, the British naturalist and poet Erasmus Darwin (1731–1802), was also the grandfather of Charles Darwin (1809–1882) by a previous marriage;

thus Francis and Charles were half-cousins. It was, in fact, Darwin's *Origin of Species* (1859) which was later to influence greatly Galton's thinking and his achievements in genetics.

Like Gauss, Galton was a precocious child (Terman estimated his IQ at about 200). His early training came largely from his invalid sister, Millicent Adéle Galton (1810–1883), and it was "Sister Delly" who encouraged his literary and scientific tastes. At age eight he was sent to a series of private boarding schools, which he remembered as fine places for bullying and birchings, but where he claimed to have learned nothing— "I had craved for what was denied, namely an abundance of good English reading, well-taught mathematics, and solid science. Grammar and the dry rudiments of Latin and Greek were abhorrent to me" (Galton, 1908). In 1838 Galton, at his parents' wish, began the study of medicine in the Birmingham General Hospital. Later in London he studied anatomy and botany at King's College and St. Georges' Hospital, and took his degree at Trinity College, Cambridge, in 1843. While at Cambridge he devoted much time to the study of mathematics.

Due to his poor health after strenuous university work and his father's death (1844), which left him financially independent, Galton abandoned medicine and took up travel. For many years he was a traveler and geographer in Egypt and unfrequented parts of Africa. He wrote numerous accounts of these adventures, such as *Narrative of an Explorer in South Africa* (1853), *Art of Travel* (1855), and *Vacation Tourists* (1860). Galton then turned to meteorology, authoring *Meteorographica* (1863), the first attempt to chart weather patterns systematically. By 1865 his interests and labors shifted to his life's work in heredity. The probable catalyst, *Origin of Species*, converted Galton to the study of human characteristics, especially with respect to heredity and its implications for the emerging science of eugenics. In 1865 he published two articles in *Macmillan's Magazine* which contained the basic but undeveloped premises of his later work on heredity and eugenics. He suggested that systematic efforts should be made to improve the breed of mankind by controlling the birth rate of the unfit and furthering the productivity of the fit. These articles contain the first "statistical demonstration" of the inheritance of intellectual and moral qualities, for which his biographer, Francis Darwin, labeled Galton a "passionate statistician, pre-eminently a lover of statistics" (Darwin, 1917, p. 28). Galton continued the study of heredity, writing *Herediturx Genius* (1869), *Natural Inheritance* (1889), *Noteworthy Families* (1906), and *Inquiries into Human Faculty and its Development* (1907). In 1904 he endowed a research fellowship at the University of London for promoting the knowledge of eugenics. One of Galton's later interests, on which he wrote three books, was fingerprints and their use in criminal detection. It was largely through his efforts that the fingerprint identification system became widely established. In 1908 he published *Memories of My Life*, and in 1909 he was knighted. He died in 1911.

This thumbnail sketch evidences the wide diversity of Galton's interests and talents, but thus far only hints at his contributions to behavioral statistics. Galton always possessed a passion to count and measure, and he applied this urge to a great many aspects of human existence: heads, noses, eye color, breathing power, strength, reaction time, frequency of yawns, number of fidgets per minute among persons attending lectures, and almost anything that captured his attention. For example, he devised a means of marking, via a hidden recorder, the beauty of women he observed on the streets of different towns, classifying them as pretty, ugly, or indifferent—an entertaining endeavor indeed! He also spent a number of months trying to brew the perfect cup of tea.

It was, however, Galton's momentous discovery of regression (or reversion), from which the concept of correlation developed, for which he has been heralded in the evolution of statistics. Galton, in his *Memories*

(p. 300), related that he first recognized regression as the key to the problems of heredity when he sought refuge from a shower "in a reddish recess in the rock. . . . There the idea flashed across me, and I forgot everything else for a moment in my great delight." The question that had aroused Galton's curiosity around 1875 and that led to his law of reversion was "How is it possible for a whole population to remain alike in its features, as a whole, during many successive generations, if the *average* produce of each couple resemble their parents?" In order to obtain data to answer this question, he raised sweet-peas, experimented with pedigreed moths, studied hounds, and finally offered prizes for the records of human families. In 1877 Galton delivered a lecture in which he related that the crude data on sweet-peas had suggested to him the law of regression as well as the formula for the standard error of estimate, and that the various arrays in a correlation table had equal variability. However, it was not until he published *Natural Inheritance* in 1889 that the concepts of regression and correlation[19] became generally known. However it was K. Pearson, Yule, Edgeworth, and others who developed and refined the mathematical theory of regression and correlation in the 1890s. And they, as did Galton, acknowledged their debt to Gauss for his least-squares methods, which were immediately applicable in correlation analysis in spite of the fact that the aims of correlation are the very antithesis of those of the theory of errors. Galton remarked (1908, p. 305): "The primary objects of the Gaussian Law of Error were exactly opposed, in one sense, to those to which I applied them. They were to get rid of or to provide a just allowance for errors. But these errors or deviations were the very things I wanted to preserve and to know about." Our debt to

Sir Francis Galton is that it was he, the last of the great Victorian scientists, who enormously widened the range of topics susceptible to statistical treatment through his firm belief that "until the phenomena of any branch of knowledge have been submitted to measurement and number it cannot assume the dignity of a science."

The stage was now set. Powerful mathematical tools had been hammered and molded from the rich ore of probability and least-squares considerations. Quetelet and Galton had realized the efficacy of applying these tools to social phenomena. The behavioral sciences stood at the brink of the twentieth century, awaiting the refinement of correlation-regression principles. It was at this time that classical parametric tests emerged. Among the principal actors were Karl Pearson, Gosset, Fisher, Egon Pearson, and Neyman.

Karl Pearson (1857–1936), considered by some biographers as the "founder of the science of statistics," was born in London of Quaker ancestry, the son of a Queen's Counselor (barrister). He received his early education at University College School in London. At age 18 he received a scholarship to King's College, Cambridge, and was graduated with mathematical honors in 1879. During his school years Pearson's interests revolved about mathematics, theory of elasticity, poetry, philosophy, religion, and a search for a palatable concept of the Deity. Late in life he wrote "Til this day I think Spinoza the sole philosopher who provides a conception of the Diety in the least compatible with scientific knowledge" (E. Pearson, 1938, p. 4).

The decade of the eighties represents a period of great importance in Pearson's development: "It was a period for much reading, of intense intellectual activity with an urge for self-expression, whether in letters, poems, critical reviews, essays or lectures" (p. 5). After graduation he spent a number of years in Germany and embarked on an extensive course of studies in Reformation history, German folklore, and the characters

---

[19] Galton first used the term *correlation* in 1888; previously *co-relation* had been used. The symbol *r*, familiar today as the Pearson product-moment correlation coefficient, was originally used by Galton (1877) to denote the slope of regression lines. He actually called the slope *r* for *reversion*.

of the German Humanists and of Luther. He returned to England, set up headquarters in London, and was called to the Bar in 1881; but he often returned to the peace and quiet of the Black Forest to read and write. Reading Pearson's writings of this period, one can see gradually taking shape that faith of a scientist which formed the basic philosophy of his life.

In 1884 at age 27 Pearson finally received an appointment to the Goldsmid Chair of Applied Mathematics and Mechanics at University College, after several unsuccessful applications for professorships elsewhere. Some of his University College lectures, which were published along with earlier essays in *The Ethic of Freethought, a Selection of Essays and Lectures* (1888), reflect the maturation of Pearson's philosophy. In 1885 Pearson became an active member of a newly formed club "for the free and unreserved discussion of all matters in any way connected with the mutual position and relation of men and women." This small group broke down the puritanical barrier that then obstructed the free discussion of the problems of morality and sex. He read two essays before this group—"The Woman's Question" and "Socialism and Sex." At this club Pearson met Maria Sharpe, whom he married in 1890 and who was to be his companion for 38 very productive years.

That same year W. F. R. Weldon accepted the Jodrell Professorship of Zoology at University College. A friendship developed, and Weldon's interests in evolution became a profound influence on the direction of Pearson's career—which was increasingly leaning toward the application of statistics to heredity and evolution. It was Weldon's personality "teeming with vigour" and his eagerness "to enthuse a mathematician with his project of demonstrating Darwinian evolution by mathematical enquiries" that played a major role in drawing Pearson from more orthodox fields of applied mathematics (E. Pearson, 1938, p. 19). Pearson also acknowledged the influence of Galton when he interpreted part of the introduction of *Natural Inheritance* as meaning:

... that there was a category broader than causation, namely correlation, of which causation was only the limit, and that this new conception of correlation brought psychology, anthropology, medicine and sociology in large parts into the field of mathematical treatment. It was Galton who first freed me from the prejudice that sound mathematics could only be applied to natural phenomena under the category of causation. Here for the first time was a possibility, I will not say a certainty, of reaching knowledge—as valid as physical knowledge was then thought to be—in the field of living forms and above all in the field of human conduct [K. Pearson, 1934: from E. Pearson, 1938, p. 19].

From 1891 to 1894 Pearson assumed an additional post as lecturer in geometry at Gresham College. The gamut of these lectures revealed by the syllabi reflects Pearson's emerging thoughts and labors in statistics. In 1894 he "popularly" lectured on the scope, methods and concepts of science, the scientific "law" and scientific "fact," the classification of the sciences, the geometry of motion, and matter and force. The Gresham Lectures from late 1891 through 1894 are noticeably different. In them one can trace the rapid advancements of statistical theory that Pearson was then achieving with Weldon's encouragement. Pearson's outlook on statistics passed in three short years from a formal handling of descriptive methods, through a discussion of the theory of probability illustrated on games of chance, to the treatment of skew and compound frequency curves and biometry. These lectures "show how the author of *The Grammar of Science* (1892)[20] was led to realize the essential place of the theory of probability among the tools of the scientific investigator

---

[20] Many of Pearson's Gresham Lectures were enlarged and reproduced in *The Grammar of Science*, along with some of his earlier essays. This work was also later to have a profound effect on the scientific thinking of Jerzy Neyman—"its teaching remained to influence our outlook."

. . . [and] they mark the start of Pearson's career as a teacher of theoretical and applied statistics" (E. Pearson, 1938, p. 22). It is interesting to note that Pearson's developments of statistical theory during this period were closely related to the problem of regression and correlation described in Galton's *Natural Inheritance* and that his approach to the problem of statistical inference was a generalization of Bayes' Theorem.

The ensuing mathematical and experimental labors of Pearson during the years 1893–1901 are indeed astonishing. They were directed toward the refinement and testing of Galton's Law of Ancestral Heredity, which particularly involved the development of the theory of multiple correlation. The fruits of his labor are revealed in a series of papers afterward entitled "Mathematical Contributions to the Theory of Evolution." The first of these (1894) introduced the method of moments for fitting a theoretical curve to observational data, gave definitions of the normal curve and standard deviation, and indicated the latter for the first time by σ. In 1895 the second paper appeared, introducing Pearson's comprehensive system of frequency curves. Their importance in statistical theory has been increased by the later discovery that, provided one is dealing with normal variation, they represent the sampling distributions of many criteria used in statistical tests— for example, χ^2, t, r (when $\rho = 0$), and F. In 1895 he published the third paper, "Regression, Heredity, and Panmixia." Here Pearson summarized the work of Bravais, Galton, and Edgeworth on correlation and advanced ever closer to the problems in multiple correlation. He also proved, by what is now called the method of maximum likelihood, that the product-moment r, which now bears his name, is the best estimate of the population correlation ρ. In 1897 the fourth paper appeared, coauthored by his student L. N. G. Filon and entitled "On the Probable Errors of Frequency Constants and on the Influence of Random Selection on Variation and Correlation." This work was later greatly extended by Fisher in "Theory of Estimation" (1922). The numerous theoretical contributions of

this period befittingly culminated with the introduction of the χ^2 test for goodness of fit (1900), which his son Egon considered to be "one of Pearson's greatest single contributions to statistical theory" (1938, p. 29). It was the precursor of many similar tests. Pearson, however, did not fully understand the appropriate degrees of freedom to be used with his test. It is noteworthy that in this paper χ^2 values were calculated to seven decimal places, which was typical of the "carefree extravagance of the owner of a Brunsviga"—one of the first calculators.

Continuing even this cursory a survey of Pearson's contributions in the field of statistics would require many more pages, and would not acknowledge his contributions to biometry and eugenics. Therefore a reckoning of the debt modern behavioral statistics owes to Pearson is in order. To paraphrase and amend Stouffer (1958), Pearson was the perfecter of simple linear correlation; the inventor of multiple, partial, biserial, and tetrachoric correlation coefficients; the discoverer of curvilinear regression and correlation (Eta), the correlation ratio and contingency coefficient, and the chi-square goodness of fit test; the originator of the terms homo- and heteroscedasticity, kurtosis, standard deviation (standard error is due to Yule, Pearson's student), and mode; and developer of numerous variations of the preceding. Along with Weldon and Galton he was founder of *Biometrika* (1901) and was its editor from 1901 to 1936. He was founder and editor of the *Annals of Eugenics* (1925). He was elected a member of the Royal Society in 1896 and was awarded its Darwin Medal in 1898 (but did not belong to the Royal Statistical Society). He served as director of his Biometric Lab and of Galton's Eugenics Lab and was editor of their respective publications. In 1911 the two labs merged to form Pearson's Department of Applied Statistics, but they split again upon Pearson's retirement in 1933 to form the Department of Statistics (headed by Egon) and the Department of Eugenics (headed by R. A. Fisher), much to Pearson's chagrin.

Finally, with a note of levity, the key to

how Pearson was able to accomplish his staggering achievements is revealed in a comment to Stouffer: "You Americans would not understand. I never answer a telephone or attend a committee meeting" (1958, p. 25). What a perfect world!

The characters remaining on the stage are to a greater or lesser degree discussed in subsequent papers in this chapter. So as to not upstage those papers, the following comments will be brief.

William Sealy Gosset (1876–1937), a student of Karl Pearson, was an unassuming man who wrote under the pseudonym of "Student." The concern in this article is to indicate succinctly some of Gosset's statistical accomplishments that were born out of his concern with the reliability of small samples. His major work, "The Probable Error of a Mean" (1908), led to the t statistic and its sampling distribution. Other noteworthy features introduced in this paper were the first use of random sampling experimentation; the recognition of difficulties in applying χ^2, which later led to Fisher's modification of the degrees of freedom; the prophecy, later confirmed, that departure from population normality would little affect the sampling distribution of z (or t); and the introduction of separate notation for sample and population characteristics (such as s and σ). In E. Pearson's words:

It is probably true to say that this investigation published in 1908 has done more than any other single paper to bring [chemical, biological, and agricultural] subjects within the range of statistical inquiry; as it stands it has provided an essential tool for the practical worker, while on the theoretical side it has proved to contain the seed of new ideas which have since grown and multiplied in hundredfold [1939, p. 224].

An immense figure in the evolution of both theoretical and applied statistics was Sir Ronald Aylmer Fisher (1890–1962), who was undoubtedly indebted to Gosset and K. Pearson, though he would in all certainty not acknowledge the latter. Fisher precipitated what is now commonly referred to as the "Fisherian revolution" in statistical thinking

and experimental design. He was born the youngest of eight children (his twin brother died in infancy) in East Finchley, a northern suburb of London. His father was a partner in a prominent firm of auctioneers. From early childhood Fisher showed great promise but had extremely myopic eyesight, which probably forced the development of his clear spatial reasoning and mental problem-solving abilities. This might account for the sometimes incompleteness of proofs and lack of mathematical rigor in his numerous papers. He graduated from Gonville and Caius College, Cambridge, in 1912 with a BA in astronomy and was among those in the first grade of honors in mathematics. He remained an additional year in Cambridge on a studentship in physics, emphasizing statistical mechanics, quantum theory, and theory of errors.

Fisher's first job was as a statistician for a London bonding company, after which he taught math and physics in public schools. He then was persuaded to accept a post as chief statistician at the Rothamsted Agricultural Station, although he had already rejected a similar position under K. Pearson at the Galton Laboratory. However, Fisher did accept the headship of the newly formed Department of Eugenics at University College, London, in 1933 (after Pearson's retirement). Ten years later he returned to his alma mater to assume the Balfour Chair of Genetics and remained there until his retirement in 1957.

Fisher's profound impact on the behavioral sciences has usually been regarded as naturally falling into three streams: contributions to mathematical statistics; design and analysis of experiments; and genetics and eugenics.

A number of Fisher's papers were concerned with the determination of exact sampling distributions derived from small samples—for example, of the correlation coefficient (1915), of the partial (1924) and multiple (1928) correlations, of the correlation ratio (1924), and of the F statistic (1924)—and also were concerned with their implications for testing statistical hypotheses. Two of his more conceptual papers were "On the Mathematical Foundations of Theoretical Statistics" (1922) and "Theory of Statistical Estimation" (1925).

These papers stand as some of his greatest works. They were the beginning of a new theory of estimation, as opposed to the established Pearsonian procedures. Fisher defined salient properties of point estimators —*consistency*, *efficiency*, and *sufficiency*— and demonstrated that the method of maximum likelihood yields a sufficient statistic when one exists and that it always provides an efficient statistic. The method of maximum likelihood had been suggested by Fisher in 1912 as a basis for estimation (extending Edgeworth's 1908–1909 work), and it is worth noting that Gauss's method of least squares is in fact a special case of maximum likelihood. The 1922 paper also contained the exact distribution of regression coefficients.

Knowledge of exact probability distributions of statistics and of the emerging theory of estimation combined to provide the foundations for most of what were then the principal tests of significance. Fisher's impact on the general area of inferential statistics cannot be overstated. Also among his mathematical accomplishments are found the successful solution to the degrees of freedom problem associated with the χ^2 test in contingency tables (1922) and the introduction of *fiducial probability* as a basis for inductive inference (1930, 1932, 1934; Yates and Mather, 1963). Controversy still reigns over some of the positions Fisher either advocated or disputed, and the foregoing is but a minute representation of his contributions to mathematical statistics.

Since Rothamsted was an agricultural experiment station, much of Fisher's experimental design work was related to crop and greenhouse data. In fact, his first published (1923) application of the analysis of variance technique was of "an experiment . . . on the effect of potash (sulphate, cloride, and none) on twelve varieties of potatoes, with three replicates on each of two series, dunged and undunged respectively" (Yates and Mather, 1963, p. 107).

The basic elements of the very powerful analysis of variance were set down by Fisher in "The Correlation between Relatives on the Supposition of Mendelian Inheritance" (1918) and in the first edition of *Statistical Methods for Research Workers* (1925; subsequent editions totaled 12 in 7 languages). In *Statistical Methods*, Fisher summarized the state of analysis of variance, indicating an improved knowledge of experimental error partitioning and associated degrees of freedom. He elaborated on the principle of randomization and also included a table of the z-distribution for $P = .05$, thus providing for exact tests of significance. The F statistic associated with analysis of variance was first called F by Snedecor in 1934 in honor of Fisher, who had himself used a form of the z-variable:

$$z = 1/2 \log_e(n_2\chi_1^2)/(n_1\chi_2^2) = 1/2 \log_e F.$$

Over the next few editions of *Statistical Methods* Fisher's ideas on factorial design, blocking, replications, and confounding were being groomed. The fourth edition (1932) demonstrates the utility of the analysis of covariance in increasing the precision of an experiment. And so it was up to the early 1960s; each new publication by Fisher or each new test added to the knowledge of the procurement and analysis of data. The resultant influence of Fisher's thinking on the behavioral sciences is eloquently acknowledged by G. A. Miller (1963):

. . . for it was Fisher more than any other person . . . who placed the powerful tools of mathematics at the disposal of even the humblest psychologist.

Of his many contributions to statistical theory, the most important to psychologists were analysis of variance and the design of multivariate experiments. So ubiquitous today is analysis of variance, so densely now do t-test and F-ratios fill our journals, that it is difficult, especially for younger psychologists, to realize how recently we discovered Fisher's work. Only twenty-five years ago the design of experiments was still an intuitive art and there was almost no trace of small-sample statistics. . . . Few psychologists have educated us as rapidly, or have influenced our work as pervasively, as did this fervent, clear-headed statistician.

Jerzy Neyman (1894-) and Egon Sharpe Pearson (1895-) were Fisher's contemporaries, and many of their individual and joint contributions were extensions of or reactions to Fisher's theory of statistical inference and estimation.

Neyman was born in Bendery, Bessarabia, formerly a province of Rumania but now the Moldavian Republic in the Soviet Union. He was educated at the University of Kharkov in Russia (1912–1916) and then held a lectureship at the Institute of Technology in Kharkov until 1921. For the next two years Neyman worked as an agricultural statistician in Bydgoszcz, Poland. In 1923 he received his Ph.D. from the University of Warsaw. His early life in Russia, the disruptive forces of war in Central Europe, and the Russian Revolution all left an indelible impression on the young statistician, who was later to become a naturalized citizen of the United States.

From 1923 to 1934 Neyman was associated with the Nencki Institute for Experimental Biology in Warsaw. During this time he was awarded a Polish Government Fellowship for post-doctoral studies at the University College, London, (1925–26) and at the University of Paris (1926–27). When he returned to Poland, Neyman established and directed the Biometrics Lab of the Nencki Institute. In 1938 he accepted an appointment as professor of mathematics at the University of California, Berkeley. Three years later he was appointed professor of statistics and director of the Berkeley Statistics Lab, a position he still holds (E. Pearson, in David, 1966).

Egon Pearson was born in London and educated at Trinity College in Cambridge (1917–1921). He received his D.Sc. degree from the University of London in 1926. The influence during these years of his father Karl Pearson was profound and enduring. E. Pearson was a member of the faculty of the Department of Statistics at University College, London, from 1921 to 1960, when he became professor emeritus. He succeeded his father as managing editor of *Biometrika* and held that position with distinction for nearly thirty years (1936–1965).

Neyman and E. Pearson first met in 1925 when Neyman entered Karl Pearson's "Gower Street world" of post-graduate students as a research fellow. Karl Pearson fondly referred to this group at the 1926 annual Galton dinner (E. Pearson, in David, 1966, p. 3):

No workers in the laboratories are more welcome here than the post-graduates. It is they who keep us alive, bring new ideas and new questions.... If undergraduates teach a teacher to teach, it is equally certain that post-graduates teach a teacher to think.

During this period new ideas were stirring among members of the group. It was late in the spring of 1926 that E. Pearson first suggested that he and Neyman collaborate on some unsolved puzzles concerning a general principle for choosing the most stringent statistical test. Since Neyman was leaving for a year in Paris and then returning to Warsaw, the ensuing years of collaboration had to be conducted mostly by mail and during intermittent short holiday work-sessions, when one or the other would cross the Channel to meet in England, France, or Poland (E. Pearson, in David, 1966).

The biometric school founded by K. Pearson and Weldon emphasized the collection of large samples from natural populations rather than the use of controlled experiments. Their theoretical orientation greatly influenced the current outlook on tests of significance. However, in the mid-1920s the development of small-sample distribution theory, stemming largely from the work of Student and Fisher, called for a rethinking of the philosophy of inference. Although Neyman and Pearson believed that Fisher's theory of estimation did not provide the complete answer, it was an incitement for further research. E. Pearson feels that two works particularly influenced the direction that the Neyman-Pearson theory was to follow. One was a 1924 *Biometrika* paper by E. C. Rhodes entitled "On the problem of whether two given samples can be supposed to have been drawn from the same population"; the other was a letter that he received from Student in May 1926.

During the autumn of 1926 Neyman and Pearson began to crystallize such ideas as the necessity of specifying the class of alternative hypotheses that should be accepted as admissible for formal treatment; the difference between simple and composite hypotheses; the rejection region in the sample space; and the two sources of error in statistical inference. By November of that year Pearson was using the likelihood ratio criterion as a principle for choosing among possible contours in the sample space in such as way that the hypothesis tested became less likely and alternatives became more likely as a sample point moved outward across them.

Gradually Neyman and Pearson expanded and systematized the rationale for significance testing and interval estimation which today is routinely taught in the behavioral sciences but rarely used effectively. They reasoned that a significance test is a decision procedure to be used as an aid in deciding which one of two or possibly three actions should be taken in the face of uncertainty. Their approach focused attention on (1) the necessity of carefully considering the alternatives to the hypothesis tested; (2) the notions of errors of the first and second kinds, the former being more serious but fortunately under the control of the experimenter; (3) the power of a test in discriminating between the tested hypothesis and the alternative hypothesis; (4) the comparison of the power functions of different tests as an aid in selecting the most powerful statistical test; and (5) the importance of determining the sample size prior to experimentation. They also expanded Fisher's maximum likelihood method (Neyman-Pearson lemma) and developed likelihood ratio tests for both univariate and multivariate cases (E. Pearson, in David, 1966).

The Neyman-Pearson theory also considered the related problem of gauging the reliability of an estimate—that is, surrounding it with a band of error. However, two very disparate lines of development were evolving between 1925 and 1930: the "confidence intervals" of Neyman and Pearson, and the "fiducial intervals" of Fisher. For a short time the two procedures amiably existed side by side, and it seemed that they were equivalent; they certainly led to the same results in simple cases. In fact, Neyman and Pearson at first thought they were simply expanding Fisher's notions. By 1935, however, it was clear that the two procedures were conceptually very different, and an embittered argument ensued that continued until Fisher's death in 1962. The principle issues are still unresolved (Neyman, 1941).

Since the "Fisherian Revolution" and "Neyman-Pearson era" many developments have occurred in behavioral statistics; only the mellowing and solidifying effects of time can permit these developments and their originators to be placed in proper perspective. When a future historical account is written, however, the ancestry of the newer developments will surely include a generous acknowledgment of the contributions of the men described in this article.

Epilogue

This article has presented some of the highlights in the development of statistics. They are offered in the belief that an acquaintance with the origins of statistics will make its study and use more meaningful, less arbitrary, and less mechanical. Behavioral statistics is more complex and less well understood than many introductory texts would have the student believe. Its formulations have sometimes been the product of years of arduous thinking, as was the concept of probability; sometimes a eureka experience, as was Galton's concept of reversion; sometimes the product of applied needs, as was Student's t; and sometimes the product of years of shared thinking, as was Neyman-Pearson hypothesis testing. It is difficult to separate the men from their statistics and to understand statistics fully without knowing how it has developed. We hope this backward glance into history gives the student a greater appreciation of present-day statistics.

References

Barnard, G. A. Thomas Bayes—A biographical note. *Biometrika*, 1958, **45**, 293–295.

Bayes, T. Essay towards solving a problem in the *Doctrine of Chances. The Philosophical Transactions (Royal Society)*, 1763, **53**, 370–418. Reprinted in *Biometrika*, 1958, **45**, 296–315.

Bell, E. T. Gauss: The prince of mathematics. In J. R. Newman (Ed.), *The world of mathematics*, Vol. 1. New York: Simon and Schuster, 1937. Pp. 295–299.

Cramér, H. *Mathematical methods of statistics.* Princeton: Princeton University Press, 1946.

Darwin, F. Francis Galton. In *Rustic sounds.* London: John Murray, 1917.

David, F. N. *Probability theory for statistical methods.* Cambridge: The University Press, 1951.

David, F. N. *Games, gods and gambling.* New York: Hafner, 1962.

David, F. N. Some notes on La Place. In J. Neyman and L. M. LeCam, *Bernoulli 1713— Bayes 1763—La Place 1813.* New York: Springer Verlag, 1965.

David, F. N. *Research papers in statistics.* New York: John Wiley, 1966.

de Moivre, A. *The doctrine of chances* (2nd edition, 1738). London: Exact impression by Frank Cass and Son, 1967.

Dunnington, G. W. *Carl Friedrich Gauss, titan of science: A study of his life and work.* New York: Hafner, 1955.

Eisenhart, C. The background and evolution of the method of least squares. Paper presented for 34th session of the International Statistical Institute, Ottawa, Canada, August 1963.

Eisenhart, C. The meaning of "least" in least squares. *Journal of the Washington Academy of Sciences*, 1964, **54**, 24–33.

Eisenhart, C. Gauss, Carl Friedrich. *International Encyclopedia of the Social Sciences*, 1968, **6**, 74–81.

Feller, W. *An introduction to probability theory and its applications*, Vol. 1, 2nd ed. New York: John Wiley, 1957.

Fisher, R. A. "Student." *Annals of Eugenics*, 1938, **9**, 1–9.

Galton, F. *Memories of my life.* New York: E. P. Dutton, 1908.

Gauss, C. F. *Carl Friedrich Gauss Werke* (collected works, 12 vols.). Göttingen: Dieterichsche Universitäts–Druckerei, 1870–1933.

Hays, W. L. *Statistics for psychologists.* New York: Holt, Rinehart & Winston, 1963.

In memoriam: Ronald Aylmer Fisher, 1890–1962. *Biometrics*, 1964, **20**, Whole No. 2.

Koren, J. *The history of statistics: Their development and progress in many countries.* New York: Macmillan, 1918.

Landau, D., and Lazarsfeld, P. F. Quetelet, Adolphe. *International Encyclopedia of the Social Sciences*, 1968, **13**, 247–257.

Miller, G. A. Ronald A. Fisher: 1890-1962. *American Journal of Psychology*, 1963, **76**, 157–158.

Neyman, J. Fiducial argument and the theory of confidence intervals. *Biometrika*, 1941, **32**, 128–150.

Pearson, E. S. *Karl Pearson: An appreciation of some aspects of his life and work.* Cambridge: The University Press, 1938.

Pearson, E. S. "Student" as a statistician. *Biometrika*, 1939, **30**, 210–250.

Pearson, K. *The life, letters and labours of Francis Galton.* Cambridge: The University Press, Vol. I, 1914; Vol. II, 1924; Vols. IIIA and IIIB, 1930.

Pearson, K. Notes on the history of correlation. *Biometrika*, 1920, **13**, 25–45.

Pearson, K. Historical note on the origin of the normal curve of errors. *Biometrika*, 1924, **16**, 402–404.

Pearson, K. James Bernoulli's theorem. *Biometrika*, 1925, **17**, 201–210.

Scheffé, H. *The analysis of variance.* New York: John Wiley, 1959.

Snedecor, G. W. *Analysis of variance and covariance.* Ames, Iowa: Collegiate Press, 1934.

Stouffer, S. A. Karl Pearson—an appreciation on the 100th anniversary of his birth. *American Statistical Association Journal*, March 1958, 23–27.

Walker, H. M. *Studies in the history of statistical method.* Baltimore: Williams and Wilkins, 1929.

Walker, H. M. Abraham de Moivre. *Scripta Mathematica*, 1934, **2**, 316–333.

Walker, H. M. The contributions of Karl Pearson. *American Statistical Association Journal*, 1958, **53**, 11–22.

Westergaard, H. *Contributions to the history of statistics.* Westminster: P. S. King and Son, Ltd., 1932.

Yates, F., and Mather, K. Ronald Alymer Fisher. *Biographical Memoirs of Fellows of the Royal Society*, 1963, **9**, 91–129.

Editor's comments: Most introductory statistics students learn to associate the name Student with the *t* test statistic. The *t* ratio, as it is sometimes called, was derived by William Sealy Gosset, an unassuming man who wrote under the pseudonym of Student.

Gosset is credited with starting a trend toward the development of exact statistical tests—that is, tests independent of knowledge of or assumptions concerning population parameters. This development marked the beginning of the modern era in mathematical statistics.

The following article by L. McMullen, Gosset's colleague and friend, provides a rare glimpse of Student as a man.

1.2 "Student" as a Man

L. McMullen

William Sealy Gosset was the eldest son of Colonel Frederic Gosset, R.E., and was born at Canterbury in 1876. In 1906 he married Marjory Surtees Phillpotts, daughter of the late headmaster of Bedford School, and they had one son and two daughters. He died on 16 October 1937, and was survived by both his parents, his wife and children, and one grandson.

He was educated at Winchester, where he was a scholar, and New College, Oxford, where he studied chemistry and mathematics.

He entered the service of Messrs Guinness as a brewer in 1899.

It is not known exactly how or when "Student's" interest in statistics was first aroused, but at this period scientific methods and laboratory determinations were beginning to be seriously applied to brewing, and it is obvious that some knowledge of error functions would be necessary. A number of

Abridged from *Biometrika*, 1939, **30**, 205–210, by permission of the publisher.

university men with science degrees had been taken on, and it is probable that "Student," who was the most mathematical of them, was appealed to by the others with various questions and so began to study the subject. It is known that he could calculate a probable error in 1903. The circumstances of brewing work, with its variable materials and susceptibility to temperature change and necessarily short series of experiments, are all such as to show up most rapidly the limitations of large sample theory and emphasize the necessity for a correct method of treating small samples. It was thus no accident, but the circumstances of his work, that directed "Student's" attention to this problem, and so led to his discovery of the distribution of the sample standard deviation, which gave rise to what in its modern form is known as the *t*-test. For a long time after its discovery and publication the use of this test hardly spread outside Guinness's brewery, where it has been very extensively used ever since. In the Biometric school at University College the problems investigated were almost all concerned with much larger samples than those in which "studentizing," as it was sometimes called, made any difference. Nevertheless, although their lines of research diverged somewhat rapidly, the close statistical contact and personal friendship between Karl Pearson and "Student," which began during his year at University College, were only terminated by death.

The purpose of this note is not however to give an account of "Student's" statistical work, but to try to give a more general impression of the man himself. Although his public reputation was entirely as a statistician, and he was acknowledged to be one of the leading investigators in that subject, his time was never wholly and rarely even mainly occupied with statistical matters. For one who saw enough of him to know roughly how his time was spent both at work and at home, it was very difficult to understand how he managed to get so much activity into the day. At work he got through an enormous amount of the ordinary routine of the brewery, as well as his statistics. Until 1922 he had no regular statistical assistant, and did all the statistics and most of the arithmetic himself; later there was a definite department, of which he was in charge till 1934, but throughout he did a great deal of arithmetic and spadework himself. It might be supposed from the amount he did in the time that he was unusually good at arithmetic and the arrangement of work; such, however, was not the case, for his arithmetic frequently contained minor errors. In one of his obituary notices a tendency to do work on the backs of envelopes in trains was mentioned, but this tendency was not confined to trains; even in his office much work was done on random scraps of paper. He also had a great dislike of the tabulation of results, and preferred to do everything from first principles whenever possible. This preference led in certain instances to waste of time in routine work, but was of assistance in maintaining that flexibility and speed of attack on new problems which was so characteristic of him. . . .

His method was, of course, not necessarily the most suitable for others not aspiring to the same degree of versatility. Perhaps it is not altogether fanciful to compare the two methods with the organic evolution of, say, the human hand, the most versatile object known, and the construction of some highly efficient but absolutely specialized piece of machinery. I do not mean to imply that he gave this explanation, or was even altogether conscious of it. When he handed over to me a routine calculation which he had done for many years, I was astonished to find that he had written out every week an almost unvarying form of words with different figures. To my question, "Why ever don't you get a printed form?" he did not reply, "Doing it from first principles every time preserves mental flexibility." He would have considered such a remark unbearably pompous. He said, "Because I'm too lazy," to which I replied, "Well, I'm too lazy not to."

To many in the statistical world "Student" was regarded as a statistical adviser to Guinness's brewery; to others he appeared to be a brewer devoting his spare time to statistics. I have tried to show that though there is some truth in both of these ideas they miss the central point, which was the intimate connexion between his statistical research and the practical problems on which he was engaged. I can imagine that many think it wasteful that a man of his undoubted genius should have been engaged in industry, yet I am sure that it is just that association with immediate practical problems which gives "Student's" work its unique character and importance relative to its small volume. On at least one occasion he was offered an academic appointment, but it is almost certain that he would not have been a successful lecturer, though perhaps a good individual teacher; nor is it likely that his research work would have flourished in more academic circumstances; his mind worked in a different way.

The work in connexion with barley breeding carried out by the Department of Agriculture in Ireland, in which Messrs Guinness took a prominent part, enabled "Student" to get that first-hand experience of yield trials and agricultural experiments generally which contributed so largely to his great knowledge of the subject. He did not merely sit in his office and calculate the results, but discussed all the details and difficulties with the Department officials, and went round all the experiments before harvest, when a "grand tour" is annually carried out by the Department, the brewery, and sometimes statisticians or others interested from England or abroad. As well as the work carried out at the actual cereal station near Cork, three or four varieties of barley are grown in $\frac{3}{4}$ or 1 acre plots at ten farms representing all the principal barley-growing districts of Ireland, so a visit to all of them entails a fairly comprehensive inspection of the crops.

"Student" took a great deal of interest in this work from the beginning and correspondence shows that he discussed the results of these tests with Karl Pearson at great length when he went to study with him at University College in 1906.

In the last ten years or so of his time in Ireland he played a leading part in these investigations, and thus had a perhaps unique opportunity of following experimental varieties from sowing through growing and harvest to malting and brewing results, and also of carrying out or supervising all the relevant mathematical work. At one time he also made some barley crosses in his own garden, and accelerated their multiplication by having one generation grown in New Zealand during our winter. These crosses were known as Student I and II, and have now been discarded as failures, the inevitable fate of the large majority. With characteristic self-effacement he was the first to point out that they were not worth going on with.

He also made frequent visits to Dr. E. S. Beaven, whose work on barley breeding is well known, and discussed every aspect of yield trials with him. These visits were undoubtedly very useful, and although Dr. Beaven is never tired of protesting that he is no mathematician and does not understand "magic squares" or "birds of freedom," which he prefers to the more orthodox expressions, he has a vast experience of agricultural trials and is very quick to see the weak point of any experiment.

In spite of the quantity of work "Student" did he was never in a hurry or fussed; this was largely due to the absence of lag when he turned his mind to a new subject; unfortunately others were not always equal to this. He would ring one up on the phone and plunge straight into some subject which might have been discussed some days previously. The slower-witted listener would probably lose the thread of his discourse before realizing what it was about and would ignominiously have to ask him to begin again. I have many times seen him hard at it on a Monday morning, but at first meeting it was always "How did the sailing go?" "Well, did you catch any fish?," and he would recount any notable event of his own weekend before plunging into the very middle of some subject. I never heard him say "I'm busy."

"Student" had many correspondents, mostly agricultural and other experimenters, in different parts of the world. He took immense pains with these and often explained points to them at great length when he could easily have given a reference. His letters contain some of his clearest writing, and the more difficult points are often better elucidated than in his published papers.

Karl Pearson emphasized the fact that a statistician must advise others on their own subject, and so may incur the accusation of butting in without adequate knowledge. "Student" was particularly expert at avoiding any such disagreement; usually he was such an enthusiastic learner of the other's subject that the fact that he was giving advice escaped notice.

The reader will by now have realized that "Student" did a very large quantity of ordinary routine as well as his statistical work in the brewery, and all that in addition to consultative statistical work and to preparing his various published papers. It might thus be thought that he could have done nothing else but eat and sleep when at home; this, however, was far from being the case, and he had a great many domestic and sporting interests. He was a keen fruit-grower and specialized in pears. He was also a good carpenter, and built a number of boats; the last, which was completed in 1932, and on whose maiden voyage I had the honour to be nearly frozen to death, was equipped with a rudder at each end by means of which the direction and speed of drift could be adjusted—an advantage which will be readily appreciated by fly-fishermen. This boat with its arrangement of rudders was described in the *Field* of 28 March 1936. In his carpentry he showed preferences analogous to his mathematical ones previously mentioned; he disliked complicated or specific tools, and liked to do anything possible with a pen-knife. On one occasion, seeing him countersinking screwholes with a pocket-knife, I offered him a proper countersink bit which I had with me, but he declined it with some embarrassment, as he would not have liked to explain or perhaps could not have explained why he preferred using the pen-knife. Out of doors he was an energetic walker and also cycled extensively in the pre-war period. He did a lot of sailing and fishing. For his last boat he had a most unconventional sail, which cannot be exactly described under any of the usual categories; it was illustrated in the *Field* article referred to above.

In fishing he was an efficient performer; he used to hold that only the size and general lightness or darkness of a fly were important; the blue wings, red tails and so on being only to attract the fisherman to the shop. This view was more revolutionary when I first heard it than it is now. He was a sound though not spectacular shot, and was well above the average on skates. Until the accident to his leg in 1934 he was quite a regular golfer, and once went round a fairly difficult course in 85 strokes and $1\frac{1}{4}$ hours by himself. He used a remarkable collection of old clubs dating at

least from the beginning of the century. In the last few years since his accident he took up bowls with great keenness, and induced many other people to play as well. One of his last visits to Ireland was with a team which he had organized at the new brewery at Park Royal.

On top of all this he knew as much as most people of the affairs of the world in general and of what was going on about him. It became very difficult to imagine how he found 24 hours in any way a sufficient length for the day. His wife certainly organized things so that the minimum amount of time was wasted, but even so few people could approach such activity in quantity or diversity.

In personal relationships he was very kindly and tolerant and absolutely devoid of malice. He rarely spoke about personal matters but when he did his opinion was well worth listening to and not in the least superficial.

In the summer of 1934 he had a motor accident and broke the neck of his femur. He had to lie up for three months, of course working at statistics, and was a semi-cripple for a year. This was particularly irksome for such an active man, as was the sheer unnecessariness of the accident, for he ran into a lamp-post on a straight road, through looking down to adjust some stuff he was carrying; but with great hard work and persistence he eventually reduced the disability to a slight limp.

At the end of 1935 he left Ireland to take charge of the new Guinness brewery in London, and I saw comparatively little of him after that. The departure from Ireland of "Student" and his family was a great loss to many who had experienced their hospitality.

His work in London was necessarily very hard and accompanied by all the vexations inevitably associated with a big undertaking in its first stages, before any settled routine has been established; nevertheless, he still found time to continue his statistical work and wrote several papers.

His death at the comparatively early age of

61 was not only a heavy blow to his family and friends, but a great loss to statistics, as his mind retained its full vigour, and he would undoubtedly have continued to work for many more years.

I am painfully conscious of the inadequacy of this sketch, which cannot hope to convey more than a faint impression of his unique personal quality to those who did not know him, but it will have served its purpose if it helps any readers to grasp the essential unity and directness of the personality which lay behind such widely varied manifestations.

Editor's comments: The contributions of Sir Ronald A. Fisher to statistics are legion. He determined the sampling distribution of the ratio $(\chi^2_{v_1}/v_1)/(\chi^2_{v_2}/v_2)$, which was subsequently named F in his honor by G. W. Snedecor. Fisher is credited with introducing, among other principles that form the basis of modern experimental research, the principle of random assignment of treatment levels to experimental units.

Some personal reminiscences of Fisher are given by another statistician, William G. Cochran, in the following article.

1.3 Footnote

William G. Cochran

In adding a few notes to Neyman's summary and appraisal of Fisher's contributions, I would like to present an impression of my own about Fisher's outlook, and to give some personal reminiscences of Fisher.

The subject matter of statistics has been defined in various ways. I believe that Fisher thought of statistics as essentially an important part of the mainstream of research in the experimental sciences. His major books, *Statistical Methods for Research Workers*, *Design of Experiments*, and the Fisher-Yates *Statistical Tables* (1), were addressed not to statisticians but to workers in the experimental sciences. The 1925 preface to the first edition of *Statistical Methods* opens as follows (1): "For several years the author has been working in somewhat intimate co-operation with a number of biological

This is the text of an address delivered at the December 1966 meeting of the AAAS in Washington.

Abridged from *Science*, 1967, **156**, 1460–1462, by permission of the publisher and author. Copyright 1967 by the American Association for the Advancement of Science.

research departments; the present book is in every sense the product of this circumstance. Daily contact with the statistical problems which present themselves to the laboratory worker has stimulated the purely mathematical researches upon which are based the methods here presented." In those days the principal departments at Rothamsted, from which Fisher was writing, were soil chemistry, soil physics, bacteriology, microbiology, entomology, insecticides, botany, and plant pathology. Except possibly for botany, he contributed to the work of every one of these. His series of papers on distribution theory, which Neyman has described, were undertaken to provide working scientists with a battery of new tools to guide them in analyzing their data.

Neyman has reminded us of the occasion when Fisher was invited to present his ideas before the Royal Statistical Society. He entitled his paper "The logic of inductive inference," but hastened to tell his readers that the title might just as well have been "On making sense of figures"—inserting this homely alternative, I believe, in case his main title might suggest a rather rarified discussion remote from the real task of handling scientific data.

Consistent with this view was his assertion that decision theory was in no sense a generalization of his ideas. He writes (2): " . . . the Natural Sciences can only be successfully conducted by responsible and independent thinkers applying their minds and their imaginations to the detailed interpretation of verifiable observations. The idea that this responsibility can be delegated to a giant computer programmed with Decision Functions belongs to the phantasy of circles rather remote from scientific research." These fighting words might suggest a blanket disapproval of decision theory, but elsewhere he writes of decision theory as the correct approach to a different kind of problem, acceptance sampling (3): "The procedure as a whole is arrived at by minimizing the losses due to wrong decisions, or to unnecessary testing, and to frame such a procedure successfully the cost of such faulty decisions must be assessed in advance; equally, also, prior knowledge is required of the expected distribution of the material in supply."

In his own researches he took little interest in the problem of collecting non-experimental data, as in sample surveys, or in the more perplexing problem of making sound inferences from uncontrolled studies, although such data abound even in the experimental sciences. It is true that his late pamphlet *Smoking—The Cancer Controversy* dealt with the pitfalls in drawing conclusions about cause and effect from nonexperimental data (4). However, his main contribution in that pamphlet, as I see it, was to claim that there were two alternative hypotheses, both reasonable and neither implicating smoking as the culprit, that might explain the available data on the relation between cigarette smoking and lung cancer death rates. Until data had been gathered that disproved both these alternatives, there was no justification in his view for diatribes or action against cigarette smoking. But as indicative of a rather half-hearted interest in the matter, he presented no detailed analysis to support the claim that these alternatives were in fact consistent with all the observed data, nor did he attempt to outline the types of data that would be needed for a crucial comparison among these hypotheses.

His concept of statistics may explain some of the things that irritated him—for instance, the teaching of a test of significance as a rule for "rejecting" or "accepting" a hypothesis. Like others, I have difficulty in understanding exactly what Fisher meant by a test of significance: he seems to imply different things in different parts of his writings. My general impression is that he regarded it as a piece of evidence that the scientist would somehow weigh, along with all other relevant pieces, in summarizing his current opinion about a hypothesis or in thinking about the nature of the next experiment. A passage in the seventh (1960) edition of *Design of Experiments*, inserted in order to clarify this point, reads as follows (3, p. 25):

In "The Improvement of Natural Knowledge," that is, learning by experience, or by planned chains of experiments, conclusions are always provisional and in the nature of progress reports, interpreting and embodying the evidence so far accrued. Convenient as it is to note that a hypothesis is contradicted at some familiar level of significance such as 5% or 2% or 1% we do not, in Inductive Inference, ever need to lose sight of the exact strength which the evidence has in fact reached, or to ignore the fact that with further trial it might come to be stronger, or weaker.

In this connection I wish that Fisher had given more advice on how to appraise "the exact strength of the evidence." I have often wondered, as I suppose does Neyman, why Fisher seems not to have regarded the power of the test as relevant, although he developed the power functions of most of the common tests of significance.

He was also unhappy, particularly later in life, at seeing statistics taught essentially as mathematics by professors who overelaborated their notation (in order to make their theorems seem difficult, in his opinion) and who gave the impression that they had never seen any data and would hastily leave the room if someone appeared with data.

Although not disagreeing, I perhaps rate Fisher's positive contribution to estimation theory more highly than does Neyman. Given the probabilistic model by which the data were generated, the concept that a specific sample contains a measurable amount of information about a parameter, the delineation of cases in which a sufficient statistic exists, the notion of the efficiency of a estimate and the development of a technique for measuring efficiency—these, although not all original with Fisher, were great steps forward. It was soon evident that these concepts were oversimplifications, applicable to only a limited range of problems, and Fisher, like others after him, struggled hard to find ways of extending their range. But despite the mass of solid and difficult research that has been done since his work, it is noteworthy how often the methods actually used nowadays in data analysis are Fisherian, or are fairly straightforward extensions of his methods.

To turn to some reminiscences . . .

A first reminiscence might be entitled "Fisher explaining a proof." In one of his lecture courses, he quoted without proof a neat result for what appeared a complex problem. Since all my attempts to prove the result foundered in a maze of algebra, I asked him one day if he would show me how to do the proof. He stated that he had written out a proof, but after opening several file drawers haphazardly, all apparently full of a miscellaneous jumble of papers, he decided that it would be quicker to develop the proof anew. We sat down and he wrote the same equation from which I had started. "The obvious development is in this direction," he said, and wrote an expression two lines in length. After "then I suppose we have to expand this," he produced a three-line equation. I nodded—I had been there too. He scrutinized this expression with obvious distaste and began to stroke his beard. "The only course seems to be this" led to an expression four

and one-half lines long. His frown was now thunderous. There was silence, apart from beard stroking, for about 45 seconds. "Well," he said, "the result must come out something like this" and wrote down the compact expression which I had asked him to prove. Class dismissed.

My second experience concerns our joint project. When I went to Cambridge as a student in 1931 my supervisor, Wishart, instructed me, at the request of Fisher, to compute a table of the 1 percent levels of $z = \log_e F$ to seven decimal places for a large panel of different pairs of degrees of freedom. Fisher was doing a corresponding table of the 5 percent levels. For those who think that graduate students nowadays are exploited by their professors, I might mention that Wishart told me he expected me to be working on this table 3 hours a day, 6 days a week, and that the labor was unpaid.

My contacts with Fisher on this project went through three stages. At first, when we met, he would ask about my progress: he had started sooner and was well ahead. Then came a period when he didn't ask, so I would ask him how he was coming along: I was gaining and towards the end of this period I was ahead. The third stage is easily foreseen. I would ask him, and he would hastily change the subject. I can take a hint as well as the next man. I believe that we last mentioned the project sometime in 1936. The third incident exemplifies Fisher as the outraged professor. When he was professor of genetics at Cambridge I called on him one spring morning at his working quarters, Whittinghame Lodge. I was told that he had just received some upsetting news, and was walking in the garden to calm himself. The news was a report from the university committee that was to approve Fisher's proposed teaching program in genetics, to the effect that they had not yet completed their study of his proposal, and there would be a further postponement of a decision, for the seventh time as I recall it, until their next meeting in October. "Cambridge University," said Fisher, "should never appoint a professor who is older than

39. If they do, then by the time his proposal for his teaching program has been approved by the university, he will have reached retirement age."

Finally, there is Fisher, the applied geneticist. We were standing at the corner of Euston Road and Gower Street in London, waiting to cross the road on our way to St. Pancras Station. Traffic was almost continuous and I was worried, because Fisher could scarcely see and I would have to steer him safely across the road. Finally there was a gap, but clearly not large enough to get us across. Before I could stop him he stepped into the stream, crying over his left shoulder "Oh, come on, Cochran. A spot of natural selection won't hurt us."

The experience of a period of association with a genius is so exhilarating that I wish every young scientist could have it. I don't know that it helps one to become a better scientist, because relationships and results that we can discern only with great effort, if at all, seem to come in a flash to someone like Fisher. But a glimpse of what the human brain at its best can do encourages a spirit of optimism for the future of *Homo sapiens*.

References

1. R. A. Fisher, *Statistical Methods for Research Workers* (Oliver and Boyd, Edinburgh, ed. 13, 1963); ———, *Design of Experiments* (Oliver and Boyd, Edinburgh, ed. 7, 1960); R. A. Fisher and F. Yates, *Statistical Tables for Biological, Agricultural and Medical Research Workers* (Hafner, New York, ed. 6, 1963).
2. R. A. Fisher, *Statistical Methods and Scientific Inference* (Hafner, New York, 1956), p. 100.
3. ———, *Design of Experiments*, Sec. 12–1.
4. ———, *Smoking—The Cancer Controversy* (Oliver and Boyd, Edinburgh, 1959).

Editor's comments: Students who are required to do original research as a part of their academic program often find that the selection of the problem is their most difficult hurdle. It is unlikely that rules can be formulated that will lead the student unerringly to choose a problem that will merit his teacher's approval. However, when a student is considering a particular problem it should be helpful to have a list of criteria against which to judge the merits of the problem. Wilse B. Webb presents just such a list in the following article.

1.4   The Choice of the Problem

Wilse B. Webb

The matter of how to judge the goodness or badness of a result, particularly when this result is a theoretical formulation, has received considerable attention in recent years. Today in psychology, such decisions are increasingly necessary. Our subject matter has become quite boundless: muscle twitches and wars, the sound of porpoises and problems of space, the aesthetic qualities of tones and sick minds, psychophysics and labor turnover. The range of organisms involved in the studies of these problems extends from pigeons to people, from amoeba to social groups. Further the techniques of measurement have been honed and sharpened by electronic tubes, computing machines, mathematical niceties, and imaginative testing procedures. Too often, in this lush environment, we as researchers may find after some

An abridged version of the Presidential Address to the Southern Society for Philosophy and Psychology given at the fifty-second Annual Meeting in Biloxi, Mississippi, April 15, 1960.

months of toil and research that our findings, although in accord with nature and beautifully simple, are utterly petty and we ourselves are no longer interested in them much less anyone else being interested.

This problem of carefully selecting and evaluating a problem does not involve the researcher alone. In our complex beehive of today, this is a question for the teacher of research, the thesis and dissertation director, the research director of laboratories or programs, and the dispensers of research funds— in a small way, department chairmen and deans, and in a large way, the guardians of the coffers of foundations and government agencies.

We are not without criteria, of course. Either implicitly or explicitly we seek justification for what we do. Certainly when grants are involved we seek for some reasons to justify the getting or the giving of money. For our consideration, I have rummaged around and turned up six widely used bases for doing an experiment: curiosity, confirmability, compassion, cost, cupidity, and conformability—or, more simply, "Am I interested," "Can I get the answer," "Will it help," "How much will it cost," "What's the payola," "Is everyone else doing it?"

I believe you will find that these are the things that enter our minds when we evaluate a student's problem, dispense a sum of research money, or decide to put ourselves to work. To anticipate myself, however, I will try to establish the fact that these bases, used alone or in combination although perhaps correlated positively with a "successful" piece of research, will probably have a zero or even negative correlation with a "valuable" piece of research.

Before proceeding to examine these criteria, however, let me introduce a clarifying footnote. Although I am concerned about selecting a problem beyond the routine and "successful" experiment—and here I shall use completely indiscriminately (with apologies to the philosophers) such terms as "good," "valued," "enduring," "worthwhile"—I do not wish to disparage the necessary place of routine experiments, i.e., well conducted experiments which fill in and extend those more creative ones. I think that even the most cursory consideration of the history of science reveals that the "original" or the "important" experiment almost inevitably pushes out from routine work that has preceded it and is further dependent upon the supportive routine experiments for their fruition into the field. Most definitely, I would contend that it is far more important to do a routine experiment than no experiment at all.

But back to the problem before us: Are our reasons for experimenting sufficient guidelines to decide about experiments? The first of these, *curiosity*, is the grand old man of reasons for experimentation and hence, justification for our experimentation. In the days when knighthood was in flower, this was the most familiar emblem on the scientist's shield. It was enough to seek an answer to "I wonder what would happen if . . . . " This was sometimes formalized with the dignified phrase, "knowledge for knowledge's sake."

Today this is not a strong base of operation. Perhaps costs have outmoded whimsey; perhaps the glare of the public stage has made us too self-conscious for such a charming urge. Or, perhaps more forebodingly, we are less curious—or, perhaps a combination of these has made curiosity less defensible. More critically, when we look more closely at this justification for doing, or proposing to do an experiment, it does not turn out to be one. Clearly a person can be curious about valuable things, trivial things, absurd things, or evil things. I think we would all find it a little difficult to judge the relative merits of two pieces of completed research by trying to decide which of the two experimenters was the most curious. Perhaps wisely, then, it seems more and more difficult to convince deans or directors of research, or dispensers of funds that a problem is worth investigating because we, personally, happened to be puzzled by it.

The criterion of *confirmability* as a criterion of worthiness of the pursuit of a topic has two sides: a philosophic one and a pragmatic one. Philosophically, this criterion reached its glory in the '30s and '40s when the voice of a logical positivist was heard throughout the land. In no uncertain terms, they told us that the criterion for a problem was "that the question asked could be answered." On the pragmatic side, this criterion is interpreted to mean: "Pick variables which are likely to be statistically significant." Undoubtedly, the philosophical point of view has done much to clear up our experimental work by hacking through a jungle of undefined and ambiguous terms. From the pragmatic point of view, this has been a much valued criterion for the graduate student with several kids and who must finish his thesis or dissertation to get out and start earning a living. However, it is just this criterion that may be voted the most likely to result in a pedestrian problem. It demands problems which have easily measurable variables and clearly stated influences. It discourages the exploration of new, complex, or mysterious areas. To exercise this criterion alone would force one to choose an experiment of measuring age as related to strength of grip on the handle of the dynamometer against exploring variables associated with happiness.

Both may be quite worthwhile, but the latter has less of a chance of being approached so long as we exercise the criterion of confirmability alone.

The problem of *costs* must enter into considerations of undertaking an experiment. In the real, live world, determining the value of a thing is very simple. Find out how much it costs. Clearly, anything that costs a lot is very valuable. A car or house which costs more than another car or another house is naturally worth more. One pays for what one gets and one gets what one pays for. This thinking carries over into the world of scientific affairs. Space probes are obviously important because they cost a lot; a project which can get a large grant must be a good one or else it would not cost so much.

Certainly this is very faulty reasoning. That methodologies differ in their costs is quite obvious; the more expensive the methodology, the more valuable and important the activity is not a direct derivative. Einstein undoubtedly used less equipment than his dentist but we may suspect that Einstein attacked the more valuable problem. Departments of Philosophy are far less expensive than Departments of Veterinary Medicine but I do not believe them to be necessarily less valuable.

It is quite true that when a large sum of money is expended on a particular project that some decisions have been made that the desirability or the value of that project justifies the expenditure of this large sum. Money may serve as a crude index of where to look for decision bases that justify large expenditures but cannot serve as judgment bases themselves. If a person says, "To do this piece of research will cost x amount of money and will occupy x amount of my life," he has merely brought his problem into focus and has raised the critical question more clearly, i.e., is such an expenditure worth it? He has not solved the problem of how valuable his experiment is, but raised the question —some other criterion must be sought for an answer to that question.

A somewhat new criterion has entered into thinking today: *compassion*. As we have moved into the applied world, this rather new criterion has come into increasing use—at least this seems true of psychology. A person asks himself as he begins an experiment, "Will the results make things better?" and implicitly assumes that an affirmative answer will make his experiment or his project more valuable. The problem then is assessed in terms of its solutions or answers resulting in a patient's improvement, a reduction in prejudice, a happier or a healthier world, etc. A variation on this question, as it is asked in the market place and in some of the other sciences directly, is in a slightly cruder form: "Will this be useful; what service will this finding perform?"

H. G. Wells has a comment on this guideline for performance in his book *Meanwhile*:

The disease of cancer will be banished from life by calm, unhurrying, persistent men and women working with every shiver of feeling controlled and suppressed in hospitals and in laboratories. ... Pity never made a good doctor, love never made a good poet, desire for service never made a discovery.

In one sense of the word, this criterion is a form of the old, applied vs. basic issue that plagues all sciences that live with one foot in technology and the other foot in theory. I cannot begin to resolve this issue here. I can, however, I believe, say this: it is quite possible for a piece of research undertaken in compassion or for utility to be quite valuable, enduring, well thought of, etc. It is also quite possible that it be trivial, superficial, limited, useless, etc. The same may be said for any given piece of basic research in which utility or compassion was never an issue. I am not one who believes that because a piece of research has no relevant use, it then by definition is valuable; or that a finding because it is useful is worthless. If these statements may be granted, it would appear then that some more fundamental criteria must be applied.

Cupidity is a variation on the criterion of compassion. Here, however, the "pay-off" is not for others but for oneself. Very simply the research is evaluated in terms of whether it will get one a promotion, favorable publicity, peer applause. Well, sometimes, undoubtedly what is good for you is good for others, and hence of general value.

However, the contrary is just as likely to be true: namely, what is good for you is not necessarily good for others at all. For example, making the natural assumptions that deans cannot read and department chairmen do not, the greater the number of papers, then the greater the probability of promotion. This results in a whirling mass of fragmented, little, anything printables to be constantly turned out instead of mature, integrated, programmatic articles. In being good to yourself you have done little or no good for others. The most casual recall would suggest that the impelling motives behind most significant advances in thought have not been cupidity, rather to the contrary such advances seemed to be more " selfless " than " selfish."

The last of my " useful " criteria is that of *conformity*. In these days of togetherness, it is not at all surprising to find conformity and its cousin, comfortableness, serving as guidelines for determining the should or should not of experimentation. We mean, by conformity, the choice of the currently popular problem, i.e., one within an ongoing and popular system, for example, operant conditioning, statistical learning theory; or a currently popular area of investigation, for example, sensory deprivation, or the Taylor Manifest Anxiety Scale.

As with all of our preceding criteria for deciding about a research project, conformity clearly has its merits. It would be foolish to turn one's back on new methods or a recent breakthrough of ideas which have been developed and certainly the interactive stimulation of mutual efforts in the same area are helpful factors in research. These, however, seem to be more means than ends to be sought for. One must be cautioned against becoming overenamored by the availability of a method at the expense of thoughtfulness or being charmed by the social benefits of

working in an active area at the expense of the scientific implications of such work. More simply, some things that a lot of people are doing are quite worthwhile, and some are quite ridiculous.

Unfortunately, however, the assiduous cultivation of all these virtuous goals may still, it would seem, even in combination result in the most pedestrian of problems. We must then search further for means of assuring ourselves that the problem is a good one. The usual criteria do not seem to answer.

I am going to say that there are three fundamentals which form the basis for good experiments or good problems. None of these are new. By disclaiming originality, however, I can claim that they are profound and that their presence or absence makes a significant effect in the value of the problem to an individual. Two of these are characteristics of the person, and one of the problem. The most common tags for my trio are: knowledge, dissatisfaction, and generalizability. The first two of these, of course, refer to the individual himself and the third to the problem itself.

I think that there is very general agreement that one can only work effectively in an area when he has a thorough understanding of this general area of concern. It is quite often that the significant finding comes from a fusion of quite a number of simple studies or a perception of gaps in the detailed findings or the methodologies and procedures of others. I am quite sure that the vaunted, creative insight of the scientist occurs more frequently within a thorough knowledge of one's area than as a bolt from the blue.

In a quite mechanical sense, a failure to obtain all knowledge possible about one's problem area is to fail to profit from the errors of thought of the past, if they be errors, or the knowledges obtained, if such knowledges are correct. Either represents an arrogance of a witless kind. Moreover, in this very practical sense, unless these backgrounds are well assessed by the worker, one may find oneself both discovering a most well-known discovery and be in the embarrassing position of arriving at the party dressed in full regalia a day late. Or perhaps, more tragic for the world of ideas, such a person without knowledge may remain unheard, being incapable of gaining the attention of competent workers through an ineptitude of expression or lack of relation to the field.

Secondly, however, to avoid the sins of conformity, suggested previously—for specialized knowledge groups can become more ingroupish than a band of teenagers—a healthy opposition must be present. I have designated this as dissatisfaction. Other terms to be used are skepticism, negativism, or perhaps more charmingly, inconoclasm.

It is, of course, quite possible that I am very wrong in emphasizing the necessity of an opposing set to the existent knowledges and methodologies in one's time. Clearly the most convenient position would be to invoke the concept of "genius" or "insight," leading to important problems as a result of broad surveys of the literature, or efficiency in employing procedures. This, however, would hardly be useful as a guideline. One can hardly suggest to a person that they strain and have an insight, or try hard and be a genius.

There is, on the other hand, considerable empirical evidence, or at least examples, to substantiate the fact that original discoveries contain an element of active revolution. Skipping lightly through psychological history, we can point to Hemholtz' classical rate of nerve conduction experiments which flew in the face of established knowledge about immediate conductivity, Freud defined the reign of conscious thought, Watson negated mentalism, Köhler set against the tide of trial and error learning, and in recent times Harlow has spoken not so much for positive adient motivation as against avoidant motives as the prime mover of man.

Logically, and psychologically (and happily they conjoin occasionally) this seems to make good sense. A significant research problem is a creative act. One can hardly be creative if one is avidly listening to the voice of others. We have good evidence from such experimentations as the Luchin's jar experi-

ments that developed sets can clearly block solutions to problems. More simply, if one agrees with everything that everyone else says, one's role is automatically limited to feeding the fires, applauding the words or, at best, carrying the word. None of these are actions that lead to truly important research activities.

We may have, however, great knowledge and object quite violently to the items of this knowledge and proceed to conduct small experiments to substantiate our objections and still be doing little more than picking at a pimple on the face of one's science, to use a vulgar analogy. A further critical requirement must be recognized—a critical requirement that is most difficult to capture in words: to be an important result, one's findings require "extensity." Another word used here is that one's results must be generalizable. Poincaré, in his *Methods of Science*, states most clearly the reasoning underlying this requirement:

What, then, is a good experiment? It is that which informs us of something besides an isolated fact. It is that which enables us to foresee, i.e., that which enables us to generalize. ... Circumstances under which one has worked will never reproduce themselves all at once. The observed action then will never recur. The only thing that can be affirmed is that under analogous circumstances, analogous action is produced.

A further quotation from the same book amplifies this point of view:

... it is needful that each of our thoughts be as useful as possible and this is why a law will be the more precious the more general it is. This shows us how we should choose. The most interesting facts are those which may serve many times.

Very simply, this boils down to being able to evaluate the probable consequence of your findings with the question which goes something like this: "In how many and what kind of specific circumstances will the relationships or rules that hold in this experiment hold in such other instances?" If the answer to this is only in instances almost exactly replicable of this particular circumstance, the rules that we obtain are likely to be of little consequence. If, however, the rule applies to what apparently is a vast heterogeneity of events in time and space, in varieties of species and surrounds, this rule is likely to have great value. Stated otherwise, the extent to which our variables and situations are unique and rare in contrast to universal and common largely determines the extent to which the findings are likely to be considered trivial or tremendous in their implications.

My summary can be quite simple:

Research today is both complex and costly. Guidelines are needed to sort among these complexities to enhance our chances of a sound investment, be this personal, financial, or temporal.

Six criteria may be, and often are, applied to judge a project's "success" potential: curiosity, confirmability, cost, compassion, cupidity, and conformability. There is probably a good probability that studies meeting the guidelines will "pay off" in some form of coinage—perhaps small change.

However, for a study to be an enduring and critical one for the history of ideas or to enter into that stream, three further items seem involved:

1. You must know thoroughly the body of research and the techniques of experimentation which are related to a given problem area. Naivete may be a source of joy in an artistic field but is not the case in valued research efforts.

2. You should be able to disbelieve, be dissatisfied with, or deny the knowledge that you have. (This is no paradox in relation to our first statement. Recognize that the first requirement is propaedeutic to this one. This is an active, not a passive state; this is to know and then know differently, rather than a know-nothing state.) Valued research seems to grow from dissatisfactions with the way things are, rather than agreeable perpetuation of present ways of proceeding.

3. You should, very simply, look for the forest beyond the tree, test the generality of your proposed finding. If your finding is referent to a rat in a particular maze, a patient on a particular couch, or a refined statistical difference, then that rat, that patient, or a captive statistician may listen to you. This would be a skimpy and disappointing audience to my way of thinking.

It is quite likely that we cannot all become geniuses. We can at least try to be less trivial. Learn as much as we can, believe in new ways, seek as great extensity in our variables as we can.

1.5   Criteria for Criteria

Joseph Weitz

When one selects a criterion to evaluate the effectiveness of some independent variable, the choice is frequently made on one of several bases. Forgetting reliability for the moment, these might be summarized under the rubrics of relevance, expedience, or precedence.

In the industrial area, for example, where studies of selection or training are under consideration, we try to use criterional measures which management feels are relevant to the situation, such as termination-survival, production, etc.

In learning studies we frequently use criteria which historically have been used by other investigators: number of trials necessary to a certain level of performance, latency of response, number of errors, etc.

In many instances we use criteria which are expedient. In these cases, certain behavioral measures are found readily available (or more available than those we would prefer to have), so we use them.

Reprinted from *American Psychologist*, 1961, **16**, 228–231, by permission of the publisher and author. Copyright 1961 by the American Psychological Association.

Since the criterion is a representation of the dependent variable, does the choice of the criterion have any effect on the results (or lack of them) attributed to the independent variable? If, for example, we are evaluating the effectiveness of two simulators as training devices, would our conclusions concerning the utility of these simulators vary depending upon whether we chose as our criterion: the length of time required on each simulator to reach some level of performance on the actual instrument itself, the number of errors on the actual device after N trials on each simulator, the speed with which a certain level of performance on the actual device can be reached, or the performance of individuals on the actual device 6 months after simulator training? All of these criteria can perhaps be thought of as "reasonable" yet our conclusions concerning the value of the simulator might vary, depending upon which one was chosen. If we are to evaluate our conclusions, do we not need to understand the effect of choosing a particular criterion?

I believe that certain dimensions of criteria should be studied. Given enough time and data, it should be possible to determine the effect of using criteria having certain characteristics as well as the relationships of these characteristics with those of independent variables.

Some of the criterional dimensions which might be studied are:

Time: When do you decide to take a measurement? If, for example, you are working in the field of training and are trying to determine whether training Procedure A is superior to Procedure B, when do you measure the effects of these two procedures: immediately following the training, during the training, 6 months later, or when? Far different results may occur depending on when the measurement is taken. And if this is so, conclusions reached may have to be modified depending on the time at which the measure is taken. This can also be true in areas other than training. When should we attempt to take measures of attitude change, effect of therapy, extinction of a response, etc.?

Type: What performance measure do you select and why that particular measure? If we consider a training study again, we might use as a criterion such variables as accidents, attitudinal measures, output measures, or a host of others. Or in learning we might use the number of errors, latency of response, or any of a variety of measures commonly used. Again, how dependent are our conclusions on the type of criterion we have selected? Usually our choice here is determined either by history or precedent but in some cases expediency and availability are the criteria for choosing the criteria.

Level: What level of the performance measure chosen is considered success or failure? Why that particular level? We frequently find reports which say an animal has "learned" if he responds correctly in 19 out of 20 successive trials. What is sacred about this figure? Why not 2 out of 3 or 50 out of 51? More important, what effect does the chosen level of performance have on the conclusions we reach? In selection testing, for example, we might say our criterion of success is producing 250 "grumbles" per hour or we might use as our criterion, producing at or

above the average number of "grumbles" produced in one hour by a certain group of "grumble" makers. Does it matter in evaluating the test?

By studying the effect of these dimensions, and probably others, across a variety of areas, it should be possible to obtain some lawful generalizations. The sort of generalizations to which I am referring might be such things as these:

1. The shorter the time period between the introduction of the independent variable and the measurement of the dependent variable (criterion), the greater/less the likelihood of showing that the independent variable is effective.

2. The more clearly the criterion measure used resembles what might be thought of as "ultimate" criteria, the greater/less the likelihood of showing the independent variable is effective.

3. The easier the task to be performed, the more/less difficult the level of the criterion must be in order to show that the independent variable is effective.

These are some types of general hypotheses and are meant only to be illustrative. Actually the entire function relating these variables should be studied. The best guess is that the function describing any of these relationships may well be curvilinear.

Let us now examine some data to see how we would approach this problem. In studying these data we shall interest ourselves in two variables: the type of criterion selected and the level of the criterion.

The study to be discussed was performed by Cramer and Cofer (1960)[1] and was reported, in part, at the 1960 APA meeting. The purpose of the investigation was to determine the role of verbal associations in mediating transfer of learning from one list of paired associates to another.

The procedure was first to learn a set of paired associates composed of zero-association value nonsense syllables and words, then

[1] The author wishes to express his appreciation to Phebe Cramer and Charles Cofer for providing him with their original data.

another list composed of the same nonsense syllables and other words. In the second learning half of the words (experimental) had forward association value (Kent-Rosanoff) of 30% and backward association of 0%. The other half of the words in the second learning were control words of equal Thorndike-Lorge frequency with the experimental words but nonassociated. There were eight words in all, four experimental and four control.

Various paradigms were used so that the original learning consisted of nonsense-syllable–word, or word–nonsense-syllable. In certain of the paradigms, foward association was possible and in others backward associations were possible. Examples of some of these paradigms are as follows:

Original learning	New learning	
GEX–Justice	GEX–Peace	(experimental word, forward association)
GEX–Peace	GEX–Justice	(experimental word, backward association)
Peace–GEX	Justice–GEX	(experimental word–nonsense-syllable)
WUB–Tobacco	WUB–Butter	(nonsense syllable–control-words)

Eighteen trials were given in the original learning, followed by a one-minute rest period. The new list learning then began and continued for eight trials.

The measure of the effectiveness of mediation on the new list learning used by Cramer and Cofer was the number of correct responses to the first members of the experimental and control pairs in the test runs on Trials 1 and 2 of the second list learning. They report that the remaining six trials showed few differences of a significant nature.

They found facilitation of second list learning as measured by the average difference in number of correct responses between experimental and control words on Trials 1 and 2.

According to the findings of Cramer and Cofer, facilitation of second list learning occurred in all paradigms whether the direc-

tion of the association was forward or backward.

What occurs if we now examine the data, using different criteria? It will be recalled that the criterion for evaluation used by Cramer and Cofer was the difference between the number of correct responses for the control words and the experimental words on both Trials 1 and 2 of the new list learning. Now let us take a different criterion of success: the number of trials to reach a certain percentage of correct response. That is, the number of trials required to reach 50%, 75%, 100% correct on any one trial, and a fourth criterion of 100% correct on two successive trials. Notice here we are not only using a slightly different type of criterion, but also various levels of this criterion. Now our analysis consists of determining the number of trials needed to reach each of the criteria for the control words and experimental words. If there is a mediational effect with both forward and backward association, will the experimental words, as opposed to the control words, show fewer trials required to meet the various criteria?

Figure 1 shows the average difference between the control and experimental words in the number of trials required to reach criterion.

It can be seen from these curves that for Paradigm B we find a relatively large difference between the control and experimental words if we chose as our criterion 75% correct.

For Paradigms A and C we reach a relative maximum difference at the 100% criterion and for Paradigm D we find that 100% twice in succession gives us the maximal effect.

Notice that if we used certain criteria as opposed to others, we would come to quite different conclusions. For example, using 100% correct on two successive trials for Paradigm A, we would have concluded that there was no mediational effect. But for other criteria (100% correct on one trial) there is a significant difference between the control and experimental words.

In examining the paradigms we find differ-

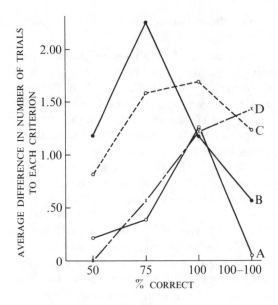

Fig. 1. Average difference between control and experimental words in trials to reach criteria.

For Paradigm A the experimental words are of the forward association type:

        Original learning: GEX–Justice
        New learning: GEX–Peace

In Paradigm B the experimental words are of the reverse association type:

        Original learning: GEX–Peace
        New learning: GEX–Justice

Paradigm C consists of forward associations, but in this manner:

        Original learning: Justice–GEX
        New learning: Peace–GEX

Paradigm D is similar to C, but with the reverse associations:

        Original learning: Peace–GEX
        New learning: Justice–GEX

ences in ease of learning. The order of difficulty turns out to be D easiest, C next, A next, and B hardest. This fits, in part, with other studies which indicate that S_1–R_1 then S_1–R_2 type of learning is more difficult than S_1–R_1 then S_2–R_1 type of learning. That is, learning two different responses to the same stimulus is more difficult than learning the same response to two different stimuli.

Let us look at Paradigms D and B. B is more difficult to learn than D. Before the data were analyzed (honestly!) the hypothesis was made that the more difficult the task, the easier the criterion should be in order to show an effect of the independent variable; con-

versely, the easier the task, the more difficult the criterion should be in order to show an effect. In examining these data we see that this is precisely what we find. For Paradigm B a criterion of 75% is most sensitive in showing an effect and for Paradigm D, two successive trials of 100% shows the maximal effect. This same finding has been substantiated on data from other similar learning studies performed by Cramer and Cofer. That is, a more difficult criterion showed maximal effects for easier tasks and easier criteria showed maximal effects when difficult tasks were involved.

The importance of this finding is that, depending upon the criterion chosen, we might radically change the conclusions reached in this type of investigation.

The foregoing discussion is presented only to show the type of analysis which I believe should be made across a variety of experimental areas. It would be helpful, of course, if studies were carried out so that data could be observed for longer time periods (or more trials). This would enable one to evaluate not only the importance of criterion level, but also the effect of time at which measurements are taken. Time may be an especially important variable when attempting to obtain criteria for selection tests or in such areas as evaluating the effectiveness of certain kinds of therapy.

This approach raises the question of how frequently do we say a given procedure was ineffective in producing the hypothesized result because the criterion was of such a nature that it was "inappropriate?" For example, Crawford and Vanderplas (1959) find little evidence for mediation in a learning study not too dissimilar from Cramer and Cofer. If you examine the criteria used in these two investigations you find them quite different. At what conclusions should we then arrive? Is there or is there not mediation in paired associate learning? If the criterion chosen determines the conclusions of a study, is this not an important parameter in and of itself?

An example of an instance where the criterion might be an important parameter is in the work reported by Bitterman (1960). He was interested in a "comparative psychology of learning" and in his article he describes a number of interesting studies on a variety of animals. The investigator was apparently not as much concerned with whether fish, for example, learn faster or better than rats, but whether they learn differently. In other words, how does learning theory, built in large part from observations of the rat, fit the findings obtained from other species? This is an interesting approach. Where differences between species occur, say, in the acquisition or extinction of a habit, can the neurological organization of the animal give new insights into learning theory, and can we get greater understanding from this approach concerning brain functions?

In a series of ingenious experiments the author showed differences among species in such behavior as acquisition of habits, reversal learning, and extinction. Care was taken to manipulate such variables as motivational states, and an attempt was made to take into account various sensory differences. The one thing which disturbed me was the fact that frequently the criterion used, let us say for extinction, was the same for a fish as for a rat. Now if differences in extinction to a certain criterion *do* occur, can we say our theoretical framework must be altered, or is it possible that a criterion of performance like behaving in a particular way on 17 out of 20 trials is a different level of difficulty for a rat and a fish? If the level of difficulty is different, then can we conclude necessarily that the functions obtained are in fact dissimilar? Perhaps criterional measures should be one of the parameters in investigations of this sort.

The criterion, if properly understood, could give us further insights into the effect of the independent variable, and perhaps even help identify some of the intervening variables.

Once we know something of the nature of criterional variables and their effects on the outcomes and results attributed to the independent variable and after discovering some of the "laws" of criteria, we should be able to make clearer statements of our hypothesis. For example, should we ever set up an hypothesis stating something like: "Anxiety leads to a decrement in performance?" If we know something about criteria, we should be able to clarify our hypothesis, allowing us to state something of the nature of the performance we expect to be affected.

In substance, then, the measure of the dependent variable (the criteria) should give us more insight into the operation of the independent variable if we know the rules governing the operation of criterional measures.

Being able to make "cleaner" hypotheses, perhaps in the future we will be able to restate our anxiety hypothesis to read something like: "Anxiety leads to a decrement of A, B, C types of performance but not x, y, z kinds of performance." With this type of analysis of criteria, we should be able to determine not only the effect of anxiety on performance but also more about the nature of the relationship and in fact more about the nature of anxiety.

If we knew more about the functioning of criterional variables, we should be able to predict which criteria are relevant for assessing effects of independent variables and with this knowledge, be able to state more concerning the operation of the independent and the intervening variables.

References

Bitterman, M. E. Toward a comparative psychology of learning. *Amer. Psychologist*, 1960, **15**, 704–712.

Cramer, P., & Cofer, C. N. The role of forward and reverse association in transfer of training. *Amer. Psychologist*, 1960, **15**, 463. (Abstract)

Crawford, J. L., & Vanderplas, J. M. An experiment on the mediation of transfer in paired-associate learning. *J. Psychol.*, 1959, **47**, 87–98.

1.6 Suggestions for Further Reading

I. *History of Statistics*

Neyman, J. R. A. Fisher (1890–1962): An appreciation. *Science*, 1967, **156**, 1456–1460. The principal contributions of R. A. Fisher to statistics are reviewed.

Pearson, E. S. Studies in the history of probability and statistics: Some early correspondence between W. A. Gosset, R. A. Fisher, and Karl Pearson, with notes and comments. *Biometrika*, 1968, **55**, 445–457. The embryonic development of statistics comes to life in this series of letters between giants in the field.

II. *Applications of Statistics*

McNemar, Q. At random: Sense and nonsense. *The American Psychologist*, 1960, **15**, 295–300. A potpourri of incisive and humorous comments on the state of psychology in general and research techniques in particular.

Selvin, H. C. Mathematics and sociology. *American Sociological Review*, 1965, **30**, 264–265. The importance of mathematical literacy for sociologists is discussed.

Wallis, W. A., and H. V. Roberts. *The Nature of Statistics*. New York: Free Press, 1962, Chapter 4. How to mislead with statistics is illustrated with a variety of real-life examples.

2

Measurement

S. S. Stevens[1] in 1946 published an article that set the stage for two decades of controversy about "levels of measurement." Measurement can be defined as the process of assigning numerals to objects or events according to a set of rules. The rules one uses in assigning numerals to the objects or events define the level of measurement obtained. Stevens classified scales of measurement into four types: nominal, ordinal, interval, and ratio. He listed both the mathematical transformations that would leave the scale-form invariant and the appropriate statistical operations for each of the four levels of measurement.

A number of Stevens' views have been subjected to criticism, particularly his assertion that a precise relationship exists between level of measurement and the permissible statistical operations that should be performed (1946, 1968). The following articles by John Gaito, C. Alan Boneau, Frederic M. Lord, and Norman H. Anderson take issue with this point of view. Stevens' justification for his position is set forth in the last article in the chapter.

[1] S. S. Stevens. On the theory of scales of measurement. *Science*, 1946, **103**, 677–680.

2.1   Scale Classification and Statistics

John Gaito

The classification of psychological scales into nominal, ordinal, interval, and ratio categories by Stevens (1951) has been a significant contribution to psychophysical and measurement theory. This classification has been widely accepted and applied by psychologists in conversation, teaching, and in publication (e.g., Edwards, 1958; Guilford, 1954; Senders, 1958; Siegel, 1956). In considering these scales Stevens has specified the appropriate statistical measures for use with each. This latter aspect has had a tremendous impact on writers of statistical texts. For example, these categories form the basis for the texts by Senders (1958) and by Siegel (1956). There is no doubt that this classification has great practical and theoretical value. However, we feel that the application of this classification has led to some misunderstanding in regard to the use of various statistical techniques and to an overemphasis on the utility of nonparametric techniques in psychological research.

As an example of this tendency Siegel (1956, p. 26) maintains that nonparametric tests of significance should be used with subinterval type data. He lists the requirement of interval scale as one of the assumptions for the use of the analysis of variance. However, this assumption cannot be found if one looks to the mathematical bases of

assumptions (Eisenhart, 1947). In fact a noted statistician, Kempthorne (1955), after showing mathematically that the normal theory analysis of variance test can approximate the randomization test, states that

This serves as some theoretical basis for the fact which has been noticed by most statisticians, that the level of significance of the analysis of variance test for differences between treatments is *little affected by the choice of a scale of measurement for analysis* [italics added].

The important consideration for the use of the analysis of variance is not that the data have certain scale properties but that *the data can be related to the normal distribution*, plus approximating the other assumptions of independence and homogeneity of errors. Furthermore, it must be emphasized that the normal distribution refers to the distribution of those portions of the data which are used as the appropriate estimates of error and not necessarily to the distribution of the observations. This is an important consideration because in multivariable designs the variation within each group may not distribute normally but the portions of certain interaction terms may. In many of these designs some of the interaction effects provide the valid estimate of error in testing certain effects.

It would seem that the comments concerning the relation between the various scales and the appropriate statistical procedures should serve merely as a guide.

In some cases empirical data may indicate that rigid adherence to such statistical methods is not required and is wasteful of the data. The author has propounded elsewhere this viewpoint and cited relevant data concerning the analysis of variance technique (Gaito, 1959).

Furthermore, the same data may be considered to have the properties of two or more scales, depending on the context in which it is considered. For example, if we look at the response of one subject (*S*) to a single item, the properties of the data are those of a nominal scale, i.e., right or wrong. However, if we concentrate on the total score for one *S* or the total scores for a group of *S*s, we have at least an ordinal scale. This is similar to the situation of approximating the binomial distribution (a distribution having nominal, or possibly ordinal, scale properties) by the normal distribution (a distribution having at least ordinal, possibly interval, scale properties) when *n* increases and *p* is close to .5. Another example is the relation between the sign test and the binomial distribution test. For a given set of data these two methods will give the same result; yet Siegel lists the former as of an ordinal nature, the latter as nominal data.

It appears that in dealing with descriptive statistics the comments made about the use of nonparametric statistics for subinterval-type data are most appropriate. However, in dealing with tests of significance such statements are not as appropriate, even though descriptive statistics are involved in the procedure. For each test the mathematical requirements are expressed in the assumptions, and nothing concerning specific scale properties is required. Supplementing the requirements will be evidence to indicate the effects on the data of failure to meet the assumptions (Gaito, 1959; Lindquist, 1953). Thus if empirical research indicates little effect on the data when assumptions are not met or are merely approximated, the investigator may use a parametric technique and allow for the effects of the deviations from assumptions in his interpretation.

It is encouraging to note that some individuals have been reluctant to embrace wholeheartedly the nonparametric technique (e.g., Guilford, 1954; McNemar, 1957, 1958; Savage, 1957). In a review of the statistical literature in the *Annual Review of Psychology*, Grant (1959, p. 137) appropriately summarizes this position by the comment: "*Some much-needed negative thinking has recently appeared on nonparametric techniques*" [italics added].

In conclusion we wish to restate that the statements concerning scale properties and statistical procedures should be guides and that context, mathematical assumptions of statistical procedures, and the results of research concerned with failure to satisfy assumptions should be the ultimate determiners of the choice of statistical techniques.

References

Edwards, A. E. *Statistical analysis*. New York: Rinehart, 1958.

Eisenhart, C. The assumptions underlying the analysis of variance. *Biometrics*, 1947, 3, 1–21.

Gaito, J. Nonparametric methods in psychological research. *Psychol. Rep.*, 1959, 5, 115–125.

Guilford, J. P. *Psychometric methods*. New York: Wiley, 1954.

Kempthorne, O. The randomization theory of experimental inference. *J. Amer. Statist. Ass.*, 1955, 50, 946–967.

Lindquist, E. F. *Design and analysis of experiments in psychology and education*. New York: Houghton Mifflin, 1953.

McNemar, Q. On Wilson's distribution-free test of analysis of variance hypotheses. *Psychol. Bull.*, 1957, 54, 361–362.

McNemar, Q. More on the Wilson test. *Psychol. Bull.*, 1958, 55, 334–335.

Savage, I. R. Nonparametric statistics. *J. Amer. Statist. Ass.*, 1957, 52, 331–344.

Senders, V. L. *Measurement and statistics*. New York: Oxford Univer. Press, 1958.

Siegel, S. *Nonparametric statistics for the behavioral sciences*. New York: McGraw-Hill, 1956.

Stevens, S. S. Mathematics, measurement, and psychophysics. In S. S. Stevens (Ed.), *Handbook of experimental psychology*. New York: Wiley, 1951.

2.2 A Note on Measurement Scales and Statistical Tests

C. Alan Boneau

In the past several months, the author, having once risen to the defense of parametric tests (Boneau, 1960), has been challenged on more than one occasion to justify the use of the *t* test in many typical psychological situations where there are measurement considerations. Research involving the concept of intelligence is often given as an instance, the point being that intelligence is actually measured by an ordinal scale, that equal differences between scores (on, say, a test) represent different magnitudes at different places on the underlying continuum. This is seen as somehow invalidating the use of the *t* test with such scores or as producing queasy feelings in those individuals who bravely resort to it in the face of uncertainty.

Burke (1953) has presented an argument which should have ended further discussion, but, in view of the present concern, a restatement of the argument and the addition of a few comments would seem indicated. The present concern seems to have been stimulated by the publication by psychologists of two recent texts in the field of statistics (Senders, 1958; Siegel, 1956) both of which are organized around Stevens' (1951) system of classifying measurement scales. Siegel and Senders belabor the point that parametric statistics, specifically the *t* and *F* tests, should be avoided when the measurement scales are

no stronger than ordinal, a state of affairs purportedly typical in psychology. To quote Siegel (1956):

Probability statements derived from the application of parametric statistical tests to ordinal data are in error to the extent that the structure of the method of collecting data is not isomorphic to arithmetic. Inasmuch as most of the measurements made by behavioral scientists culminate in ordinal scales (this seems to be the case except in the field of psychophysics, and possibly in the use of a few carefully standardized tests), this point deserves strong emphasis (p. 26).[1]

If one were to take Siegel seriously on this point one would hesitate to use parametric tests with practically any attitude measure in social psychology, with practically any performance measure in experimental psychology, and with practically any rating method or personality scale in clinical psychology.

A more realistic attitude is that parametric tests are useful whenever a measurement operation exists such that one of several possible numbers (scores) can be assigned unambiguously to an item of behavior without considering the relation of that item of behavior to other similar items, i.e., without ranking. This is typically the case with attitude scales, performance measures, and rating methods. If such numbers can be assigned, then conceptually there exists a population of these numbers having a specific distribution function, mean, and variance.

When we perform a test of a hypothesis we draw samples of numbers from two or more such populations each of which we can conceive of as being the totality of all possible measurements under the specific appropriate conditions. From the samples we can make estimates of the population parameters. Typically, we are interested in deciding whether or not the samples could have arisen from populations having the same parameters. Quite frequently we are concerned with statements about the means of the populations, but sometimes with variances.

In rejecting the null hypothesis, if we may do so, we are implying that the probability is high that if we draw further samples from the same populations (i.e., under the same conditions and using the same measuring instrument), we will get significant differences on each occasion. In other words we are asserting that a difference exists between the means (or variances, etc.) of the populations of assigned numbers or scores.

Note that we make absolutely no assumptions about any underlying dimension and we deal only with the numbers assigned by the measurement operation. We, and statistical tests, are concerned only with this operationally defined manifestation of the underlying dimension. The statistical test cares not whether a Social Desirability scale measures social desirability, or number of trials to extinction is an indicator of habit strength. It does not even care whether the measuring scale is monotonically related (or unrelated) to the underlying dimension. Given unending piles of numbers from which to draw small samples, the t test and the F test will methodically decide for us whether the means of the piles are different. If the distributions of the numbers in the piles are normal and have equal variances we can make exact statements as to the probability that the t or F test is mistaken. Even if this is not so, the probability statements are generally not greatly in error (see Boneau, 1960, for a discussion of this point). A further point: when the measurement operation is used to generate scores rather than ranks, inferences based upon nonparametric tests will be inferences about the populations of scores, the same sorts of inferences which we make when we use parametric methods. The nonparametric methods in no sense get closer to reality or peer at the underlying continuum through the screen of numbers which we have constructed to measure it, although some would have us think that these methods have such a magical property. Since parametric as well as nonparametric methods are called upon in most instances in psychological research to make decisions about differences in population parameters, the issue of relative sensitivity to such differences must be faced. Present evidence seems to point to the superiority of the parametric methods, specifically the t and F tests, in a large number of instances, even though the assumptions underlying these tests are rather drastically violated.

Certainly one cannot ignore the problem of measurement. It would seem to make a difference to psychology whether or not (and in what sense) the numbers we assign by means of an intelligence test are related to the underlying concept, but the problem is a measurement problem, not a statistical one. No matter how one assigns these numbers (even at random), however, he can expect to get the same objective, impartial, and neutral judgment whenever he resorts to the use of the parametric or nonparametric tests of significance.

References

Boneau, C. A. The effects of violations of assumptions underlying the t test. *Psychol. Bull.*, 1960, **57**, 49–64.

Burke, C. J. Additive scales and statistics. *Psychol. Rev.*, 1953, **60**, 73–75.

Senders, V. L. *Measurement and statistics.* New York: Oxford Univer. Press, 1958.

Siegel, S. *Nonparametric statistics for the behavioral sciences.* New York: McGraw-Hill, 1956.

Stevens, S. S. Mathematics, measurement, and psychophysics. In S. S. Stevens (Ed.), *Handbook of experimental psychology.* New York: Wiley, 1951. Pp. 1–49.

2.3   On the Statistical Treatment of Football Numbers

Frederic M. Lord

Professor X sold "football numbers." The television audience had to have some way to tell which player it was who caught the forward pass. So each player had to wear a number on his football uniform. It didn't matter what number, just so long as it wasn't more than a two-digit number.

Professor X loved numbers. Before retiring from teaching, Professor X had been chairman of the Department of Psychometrics. He would administer tests to all his students at every possible opportunity. He could hardly wait until the tests were scored. He would quickly stuff the scores in his pockets and hurry back to his office where he would lock the door, take the scores out again, add them up, and then calculate means and standard deviations for hours on end.

Professor X locked his door so that none of his students would catch him in his folly. He taught his students very carefully: "Test scores are ordinal numbers, not cardinal numbers. Ordinal numbers cannot be added. *A fortiori*, test scores cannot be multiplied or squared." The professor required his students to read the most up-to-date references on the theory of measurement (e.g., 1, 2, 3). Even the poorest student would quickly explain that it was wrong to compute means or standard deviations of test scores.

Reprinted from *American Psychologist*, 1953, **8**, 750–751, by permission of the publisher and author. Copyright 1953 by the American Psychological Association.

When the continual reproaches of conscience finally brought about a nervous breakdown, Professor X retired. In appreciation of his careful teaching, the university gave him the "football numbers" concession, together with a large supply of cloth numbers and a vending machine to sell them.

The first thing the professor did was to make a list of all the numbers given to him. The University had been generous and he found that he had exactly 100,000,000,000,000,000 two-digit cloth numbers to start out with. When he had listed them all on sheets of tabulating paper, he shuffled the pieces of cloth for two whole weeks. Then he put them in the vending machine.

If the numbers had been ordinal numbers, the Professor would have been sorely tempted to add them up, to square them, and to compute means and standard deviations. But these were not even serial numbers; they were only "football numbers"—they might as well have been letters of the alphabet. For instance, there were 2,681,793,401,686,191 pieces of cloth bearing the number "69," but there were only six pieces of cloth bearing the number "68," etc., etc. The numbers were for designation purposes only; there was no sense to them.

The first week, while the sophomore team bought its numbers, everything went fine. The second week the freshman team bought its numbers. By the end of the week there was trouble. Information secretly reached the professor that the numbers in the machine had been tampered with in some unspecified fashion.

The professor had barely had time to decide to investigate when the freshman team appeared in a body to complain. They said they had bought 1,600 numbers from the machine, and they complained that the numbers were too low. The sophomore team was laughing at them because they had such low numbers. The freshmen were all for routing the sophomores out of their beds one by one and throwing them in the river.

Alarmed at this possibility, the professor temporized and persuaded the freshmen to wait while he consulted the statistician who lived across the street. Perhaps, after all, the freshmen had gotten low numbers just by chance. Hastily he put on his bowler hat, took his tabulating sheets, and knocked on the door of the statistician.

Now the statistician knew the story of the poor professor's resignation from his teaching. So, when the problem had been explained to him, the statistician chose not to use the elegant nonparametric methods of modern statistical analysis. Instead he took the professor's list of the 100 quadrillion "football numbers" that had been put into the machine. He added them all together and divided by 100 quadrillion.

"The population mean," he said, "is 54.3."

"But these numbers are not cardinal numbers," the professor expostulated. "You can't add them."

"Oh, can't I?" said the statistician. "I just did. Furthermore, after squaring each number, adding the squares, and proceeding in the usual fashion, I find the population standard deviation to be exactly 16.0."

"But you can't multiply 'football numbers,'" the professor wailed. "Why, they aren't even ordinal numbers, like test scores."

"The numbers don't know that," said the statistician. "Since the numbers don't remember where they came from, they always behave just the same way, regardless."

The professor gasped.

"Now the 1,600 'football numbers' the freshmen bought have a mean of 50.3," the statistician continued. "When I divide the difference between population and sample means by the population standard deviation. . . ."

"Divide!" moaned the professor.

". . . And then multiply by $\sqrt{1,600}$, I find a critical ratio of 10," the statistician went on, ignoring the interruption. "Now, if your population of 'football numbers' had happened to have a normal frequency distribution, I would be able rigorously to assure you that the sample of 1,600 obtained by the freshmen could have arisen from random sampling only once in 65,618,050,000,000,000,000,000 times; for in this case these numbers obviously would obey all the rules that apply to sampling from any normal population."

"You cannot . . ." began the professor.

"Since the population is obviously not normal, it will in this case suffice to use Tchebycheff's inequality,"[1] the statistician continued calmly. "The probability of obtaining a value of 10 for such a critical ratio in random sampling from any population whatsoever is always less than .01. It is therefore highly implausible that the numbers obtained by the freshmen were actually a random sample of all numbers put into the machine."

"You cannot add and multiply any numbers except cardinal numbers," said the professor.

"If you doubt my conclusions," the statistician said coldly as he showed the professor to the door, "I suggest you try and see how often you can get a sample of 1,600 numbers from your machine with a mean below 50.3 or above 58.3. Good night."

To date, after reshuffling the numbers, the professor has drawn (with replacement) a little over 1,000,000,000 samples of 1,600 from his machine. Of these, only two samples have

---

[1] Tchebycheff's inequality, in a convenient variant, states that in random sampling the probability that a critical ratio of the type calculated here will exceed any chosen constant, c, is always less than $1/c^2$, irrespective of the shape of the population distribution. It is impossible to devise a set of numbers for which this inequality will not hold.

had means below 50.3 or above 58.3. He is continuing his sampling, since he enjoys the computations. But he has put a lock on his machine so that the sophomores cannot tamper with the numbers again. He is happy because, when he has added together a sample of 1,600 "football numbers," he finds that the resulting sum obeys the same laws of sampling as they would if they were real honest-to-God cardinal numbers.

Next year, he thinks, he will arrange things so that the population distribution of his "football numbers" is approximately normal. Then the means and standard deviations that he calculates from these numbers will obey the usual mathematical relations that have been proven to be applicable to random samples from any normal population.

The following year, recovering from his nervous breakdown, Professor X will give up the "football numbers" concession and resume his teaching. He will no longer lock his door when he computes the means and standard deviations of test scores.

References

1. Coombs, C. H. Mathematical models in psychological scaling. *J. Amer. stat. Ass.*, 1951, **46**, 480–489.
2. Stevens, S. S. Mathematics, measurement, and psychophysics. In S. S. Stevens (Ed.), *Handbook of experimental psychology*. New York: Wiley, 1951. Pp. 1–49.
3. Weitzenhoffer, A. M. Mathematical structures and psychological measurements. *Psychometrika*, 1951, **16**, 387–406.

2.4 Scales and Statistics: Parametric and Nonparametric

Norman H. Anderson

The recent rise of interest in the use of nonparametric tests stems from two main sources. One is the concern about the use of parametric tests when the underlying assumptions are not met. The other is the problem of whether or not the measurement scale is suitable for application of parametric procedures. On both counts parametric tests are generally more in danger than nonparametric tests. Because of this, and because of a natural enthusiasm for a new technique, there has been a sometimes uncritical acceptance of nonparametric procedures. By now a certain degree of agreement concerning the more practical aspects involved in the choice of tests appears to have been reached. However, the measurement theoretical issue has been less clearly resolved. The principal purpose of this article is to discuss this latter issue further. For the sake of completeness, a brief overview of practical statistical considerations will also be included.

A few preliminary comments are needed in order to circumscribe the subsequent discussion. In the first place, it is assumed throughout that the data at hand arise from some sort of measuring scale which gives numerical results. This restriction is implicit in the proposal to compare parametric and nonparametric tests since the former do not apply to strictly categorical data (but see Cochran, 1954). Second, parametric tests will mean tests of significance which assume equinormality, i.e., normality and some form of homogeneity of variance. For convenience, parametric test, F test, and analysis of variance will be used synonymously. Although this usage is not strictly correct, it should be noted that the t test and regression analysis may be considered as special applications of F. Nonparametric tests will refer to significance tests which make considerably weaker distributional assumptions as exemplified by rank order tests such as the Wilcoxon T, the Kruskal-Wallis H, and by the various median-type tests. Third, the main focus of the article is on tests of significance with a lesser emphasis on descriptive statistics. Problems of estimation are touched on only slightly although such problems are becoming increasingly important.

Finally, a word of caution is in order. It will be concluded that parametric procedures constitute the everyday tools of psychological statistics, but it should be realized that any area of investigation has its own statistical peculiarities and that general statements must always be adapted to the prevailing practical situation. In many cases, as in pilot

An earlier version of this paper was presented at the April 1959 meetings of the Western Psychological Association. The author's thanks are due F. N. Jones and J. B. Sidowski for their helpful comments.

work, for instance, or in situations in which data are cheap and plentiful, nonparametric tests, shortcut parametric tests (Tate & Clelland, 1957), or tests by visual inspection may well be the most efficient.

Practical Statistical Considerations

The three main points of comparison between parametric and nonparametric tests are significance level, power, and versatility. Most of the relevant considerations have been treated adequately by others and only a brief summary will be given here. For more detailed discussion, the articles of Cochran (1947), Savage (1957), Sawrey (1958), Gaito (1959), and Boneau (1960) are especially recommended.

Significance level. The effects of lack of equinormality on the significance level of parametric tests have received considerable study. The two handiest sources for the psychologist are Lindquist's (1953) citation of Norton's work, and the recent article of Boneau (1960) which summarizes much of the earlier work. The main conclusion of the various investigators is that lack of equinormality has remarkably little effect although two exceptions are noted: one-tailed tests and tests with considerably disparate cell n's may be rather severely affected by unequal variances.[1]

A somewhat different source of perturbation of significance level should also be mentioned. An overall test of several conditions may show that something is significant but will not localize the effects. As is well known, the common practice of t testing pairs of means tends to inflate the significance level even when the over-all F is significant. An

analogous inflation occurs with nonparametric tests. There are parametric multiple comparison procedures which are rigorously applicable in many such situations (Duncan, 1955; Federer, 1955) but analogous nonparametric techniques have as yet been developed in only a few cases.

Power. As Dixon and Massey (1957) note, rank order tests are nearly as powerful as parametric tests under equinormality. Consequently, there would seem to be no pressing reason in most investigations to use parametric techniques for reasons of power *if* an appropriate rank order test is available (but see Snedecor, 1956, p. 120). Of course, the loss of power involved in dichotomizing the data for a median-type test is considerable.

Although it might thus be argued that rank order tests should be generally used where applicable, it is to be suspected that such a practice would produce negative transfer to the use of the more incisive experimental designs which need parametric analyses. The logic and computing rules for the analysis of variance, however, follow a uniform pattern in all situations and thus provide maximal positive transfer from the simple to the more complex experiments.

There is also another aspect of power which needs mention. Not infrequently, it is possible to use existing data to get a rough idea of the chances of success in a further related experiment, or to estimate the N required for a given desired probability of success (Dixon & Massey, 1957, Ch. 14). Routine methods are available for these purposes when parametric statistics are employed but similar procedures are available only for certain nonparametric tests such as chi square.

Versatility. One of the most remarkable features of the analysis of variance is the breadth of its applicability, a point which has been emphasized by Gaito (1959). For present purposes, the ordinary factorial design will serve to exemplify the issue. Although factorial designs are widely employed, their uses in the investigation and control of minor variables have not been fully

---

[1] The split-plot designs (e.g., Lindquist, 1953) commonly used for the analysis of repeated or correlated observations have been subject to some criticism (Cotton, 1959; Greenhouse & Geisser, 1959) because of the additional assumption of equal correlation which is made. However, tests are available which do not require this assumption (Cotton, 1959; Greenhouse & Geisser, 1959; Rao, 1952).

exploited. Thus, Feldt (1958) has noted the general superiority of the factorial design in matching or equating groups, an important problem which is but poorly handled in current research (Anderson, 1959). Similarly, the use of replications as a factor in the design makes it possible to test and partially control for drift or shift in apparatus, procedure, or subject population during the course of an experiment. In the same way, taking experimenters or stimulus materials as a factor allows tests which bear on the adequacy of standardization of the experimental procedures and on the generalizability of the results.

An analogous argument could be given for latin squares, largely rehabilitated by the work of Wilk and Kempthorne (1955), which are useful when subjects are given successive treatments; for orthogonal polynomials and trend tests for correlated scores (Grant, 1956) which give the most sensitive tests when the independent variable is scaled; as well as for the multivariate analysis of variance (Rao, 1952) which is applicable to correlated dependent variables measured on incommensurable scales.

The point to these examples and to the more extensive treatment by Gaito is straightforward. Their analysis is more or less routine when parametric procedures are used. However, they are handled inadequately or not at all by current nonparametric methods.

It thus seems fair to conclude that parametric tests constitute the standard tools of psychological statistics. In respect of significance level and power, one might claim a fairly even match. However, the versatility of parametric procedures is quite unmatched and this is decisive. Unless and until nonparametric tests are developed to the point where they meet the routine needs of the researcher as exemplified by the above designs, they cannot realistically be considered as competitors to parametric tests. Until that day, nonparametric tests may best be considered as useful minor techniques in the analysis of numerical data.

Too promiscuous a use of F is, of course, not to be condoned since there will be many situations in which the data are distributed quite wildly. Although there is no easy rule with which to draw the line, a frame of reference can be developed by studying the results of Norton (Lindquist, 1953) and of Boneau (1960). It is also quite instructive to compare p values for parametric and nonparametric tests of the same data.

It may be worth noting that one of the reasons for the popularity of nonparametric tests is probably the current obsession with questions of statistical significance to the neglect of the often more important questions of design and power. Certainly some minimal degree of reliability is generally a necessary justification for asking others to spend time in assessing the importance of one's data. However, the question of statistical significance is only a first step, and a relatively minor one at that, in the over-all process of evaluating a set of results. To say that a result is statistically significant simply gives reasonable ground for believing that some nonchance effect was obtained. The meaning of a nonchance effect rests on an assessment of the design of the investigation. Even with judicious design, however, phenomena are seldom pinned down in a single study so that the question of replicability in further work often arises also. The statistical aspects of these two questions are not without importance but tend to be neglected when too heavy an emphasis is placed on p values. As has been noted, it is the parametric procedures which are the more useful in both respects.

Measurement Scale Considerations

The second and principal part of the article is concerned with the relations between types of measurement scales and statistical tests. For convenience, therefore, it will be assumed that lack of equinormality presents no serious problem. Since the F ratio remains constant with changes in unit or zero point of the measuring scale, we may ignore ratio scales and consider only ordinal and interval scales. These scales are defined following Stevens (1951). Briefly, an ordinal scale is one in which the events measured are, in some empirical sense, ordered in the same way as the arithmetic order of the numbers assigned to them. An interval scale has, in addition, an equality of unit over different parts of the scale. Stevens goes on to characterize scale types in terms of permissible transformations. For an ordinal scale, the permissible transformations are monotone since they leave rank order unchanged. For an interval scale, only the linear transformations are permissible since only these leave relative distance unchanged. Some workers (e.g., Coombs, 1952) have considered various scales which lie between the ordinal and interval scales. However, it will not be necessary to take this further refinement of the scale typology into account here.

As before, we suppose that we have a measuring scale which assigns numbers to events of a certain class. It is assumed that this measuring scale is an ordinal scale but not necessarily an interval scale. In order to fix ideas, consider the following example. Suppose that we are interested in studying attitude toward the church. Subjects are randomly assigned to two groups, one of which reads Communication A, while the other reads Communication B. The subjects' attitudes towards the church are then measured by asking them to check a seven category pro-con rating scale. Our problem is whether the data give adequate reason to conclude that the two communications had different effects.

To ascertain whether the communications had different effects, some statistical test must be applied. In some cases, to be sure, the effects may be so strong that the test can be made by inspection. In most cases, however, some more objective method is necessary. An obvious procedure would be to assign the numbers 1 to 7, say, to the rating scale categories and apply the F test, at least if the data presented some semblance of equinormality. However, some writers on statistics

(e.g., Siegel, 1956; Senders, 1958) would object to this on the ground that the rating scale is only an ordinal scale, the data are therefore not "truly numerical," and hence that the operations of addition and multiplication which are used in computing F cannot meaningfully be applied to the scores. There are three different questions involved in this objection, and much of the controversy over scales and statistics has arisen from a failure to keep them separate. Accordingly, these three questions will be taken up in turn.

Question 1. Can the F test be applied to data from an ordinal scale? It is convenient to consider two cases of this question according as the assumption of equinormality is satisfied or not. Suppose first that equinormality obtains. The caveat against parametric statistics has been stated most explicitly by Siegel (1956) who says:

The conditions which must be satisfied . . . before any confidence can be placed in any probability statement obtained by the use of the t test are at least these: . . . 4. The variables involved must have been measured in *at least* an interval scale . . . (p. 19). (By permission, from *Nonparametric Statistics*, by S. Siegel. Copyright, 1956. McGraw-Hill Book Company, Inc.)

This statement of Siegel's is completely incorrect. This particular question admits of no doubt whatsoever. The F (or t) test may be applied without qualm. It will then answer the question which it was designed to answer: can we reasonably conclude that the difference between the means of the two groups is real rather than due to chance? The justification for using F is purely statistical and quite straightforward; there is no need to waste space on it here. The reader who has doubts on the matter should postpone them to the discussion of the two subsequent questions, or read the elegant and entertaining article by Lord (1953). As Lord points out, the statistical test can hardly be cognizant of the empirical meaning of the numbers with which it deals. Consequently, the validity of a statistical inference cannot depend on the type of measuring scale used.

The case in which equinormality does not hold remains to be considered. We may still use *F*, of course, and as has been seen in the first part, we would still have about the same significance level in most cases. The *F* test might have less power than a rank order test so that the latter might be preferable in this simple two-group experiment. However, insofar as we wish to inquire into the reliability of the difference between the measured behavior of the two groups in our particular experiment, the choice of statistical test would be governed by purely statistical considerations and have nothing to do with scale type.

Question 2. Will statistical results be invariant under change of scale? The problem of invariance of result stems from the work of Stevens (1951) who observes that a statistic computed on data from a given scale will be invariant when the scale is changed according to any given permissible transformation. It is important to be precise about this usage of invariance. It means that if a statistic is computed from a set of scale values and this statistic is then transformed, the identical result will be obtained as when the separate scale values are transformed and the statistic is computed from these transformed scale values.

Now our scale of attitude toward the church is admittedly only an ordinal scale. Consequently, we would expect it to change in the direction of an interval scale in future work. Any such scale change would correspond to a monotone transformation of our original scale since only such transformations are permissible with an ordinal scale. Suppose then that a monotone transformation of the scale has been made subsequent to the experiment on attitude change. We would then have two sets of data: the responses as measured on the original scale used in the experiment, and the transformed values of these responses as measured on the new, transformed scale. (Presumably, these transformed scale values would be the same as the

subjects would have made had the new scale been used in the original experiment, although this will no doubt depend on the experimental basis of the new scale.) The question at issue then becomes whether the same significance results will be obtained from the two sets of data. If rank order tests are used, the same significance results will be found in either case because any permissible transformation leaves rank order unchanged. However, if parametric tests are employed, then different significance statements may be obtained from the two sets of data. It is possible to get a significant *F* from the original data and not from the transformed data, and vice versa. Worse yet, it is even logically possible that the means of the two groups will lie in reverse order on the two scales.

The state of affairs just described is clearly undesirable. If taken uncritically, it would constitute a strong argument for using only rank order tests on ordinal scale data and restricting the use of *F* to data obtained from interval scales. It is the purpose of this section to show that this conclusion is unwarranted. The basis of the argument is that the naming of the scales has begged the psychological question.

Consider interval scales first, and imagine that two students, P and Q, in an elementary lab course are assigned to investigate some process. This process might be a ball rolling on a plane, a rat running an alley, or a child doing sums. The students cooperate in the experimental work, making the same observations, except that they use different measuring scales. P decides to measure time intervals. He reasons that it makes sense to speak of one time interval as being twice another, that time intervals therefore form a ratio scale, and hence a fortiori an interval scale. Q decides to measure the speed of the process (feet per second, problems per minute). By the same reasoning as used by P, Q concludes that he has an interval scale also. Both P and Q are aware of current strictures about scales and statistics. However, since each believes (and rightly so) that he has an interval scale,

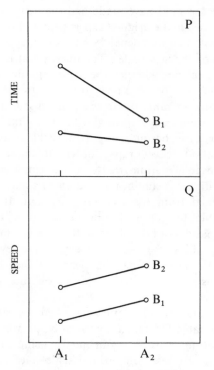

Fig. 1. Temporal aspects of some process obtained from a 2 × 2 design. (The data are plotted as a function of Variable A with Variable B as a parameter. Subscripts denote the two levels of each variable. Note that Panel P shows an interaction, but that Panel Q does not.)

each uses means and applies parametric tests in writing his lab report. Nevertheless, when they compare their reports they find considerable difference in their descriptive statistics and graphs (Figure 1), and in their F ratios as well. Consultation with a statistician shows that these differences are direct consequences of the difference in the measuring scales. Evidently then, possession of an interval scale does not guarantee invariance of interval scale statistics.

For ordinal scales, we would expect to obtain invariance of result by using ordinal scale statistics such as the median (Stevens, 1951). Let us suppose that some future investigator finds that attitude toward the church is multidimensional in nature and has, in fact, obtained interval scales for each of the dimensions. In some of his work he chanced to use our original ordinal scale so that he was

able to find the relation between this ordinal scale and the multidimensional representation of the attitude. His results are shown in Figure 2. Our ordinal scale is represented by the curved line in the plane of the two dimensions. Thus, a greater distance from the origin as measured along the line stands for a higher value on our ordinal scale. Points A and B on the curve represent the medians of Groups A and B in our experiment, and it is seen that Group A is more pro-church than Group B on our ordinal scale. The median scores for these two groups on the two dimensions are obtained simply by projecting Points A and B onto the two dimensions. All is well on Dimension 2 since there Group A is greater than Group B. On Dimension 1, however, a reversal is found: Group A is less than Group B, contrary to our ordinal scale results. Evidently then, possession of an ordinal scale does not guarantee invariance of ordinal scale statistics.

A rather more drastic loss of invariance would occur if the ordinal scale were measuring the resultant effect of two or more underlying processes. This could happen, for instance, in the study of approach-avoidance

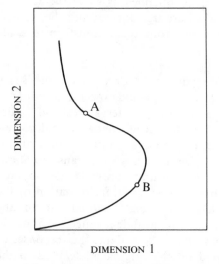

DIMENSION 1

Fig. 2. The curved line represents the ordinal scale of attitude toward the church plotted in the two-dimensional space underlying the attitude. (Points A and B denote the medians of two experimental groups. The graph is hypothetical, of course.)

conflict, or ambivalent behavior, as might be the case with attitude toward the church. In such situations, two people could give identical responses on the one-dimensional scale and yet be quite different as regards the two underlying processes. For instance, the same resultant could occur with two equal opposing tendencies of any given strength. Representing such data in the space formed by the underlying dimensions would yield a smear of points over an entire region rather than a simple curve as in Figure 2.

Although it may be reasonable to think that simple sensory phenomena are one-dimensional, it would seem that a considerable number of psychological variables must be conceived of as multidimensional in nature as, for instance, with "IQ" and other personality variables. Accordingly, as the two cited examples show, there is no logical guarantee that the use of ordinal scale statistics will yield invariant results under scale changes.

It is simple to construct analogous examples for nominal scales. However, their only relevance would be to show that a reduction of all results to categorical data does not avoid the difficulty with invariance.

It will be objected, of course, that the argument of the examples has violated the initial assumption that only "permissible" transformations would be used in changing the measuring scales. Thus, speed and time are not linearly related, but rather the one is a reciprocal transformation of the other. Similarly, Dimension 1 of Figure 2 is no monotone transformation of the original ordinal scale. This objection is correct, to be sure, but it simply shows that the problem of invariance of result with which one is actually faced in science has no particular connection with the invariance of "permissible" statistics. The examples which have been cited show that knowing the scale type, as determined by the commonly accepted criteria, does not imply that future scales measuring the same phenomena will be "permissible" transformations of the original scale. Hence the use of "permissible" statistics, although guaran-

teeing invariance of result over the class of "permissible" transformations, says little about invariance of result over the class of scale changes which must actually be considered by the investigator in his work.

This point is no doubt pretty obvious, and it should not be thought that those who have taken up the scaletype ideas are unaware of the problem. Stevens, at least, seems to appreciate the difficulty when, in the concluding section of his 1951 article, he distinguishes between psychological dimensions and indicants. The former may be considered as intervening variables whereas the latter are effects or correlates of these variables. However, it is evident that an indicant may be an interval scale in the customary sense and yet bear a complicated relation to the underlying psychological dimensions. In such cases, no procedure of descriptive or inferential statistics can guarantee invariance over the class of scale changes which may become necessary.

It should also be realized that only a partial list of practical problems of invariance has been considered. Effects on invariance of improvements in experimental technique would also have to be taken into account since such improvements would be expected to purify or change the dependent variable as well as decrease variability. There is, in addition, a problem of invariance over subject population. Most researches are based on some handy sample of subjects and leave more or less doubt about the generality of the results. Although this becomes in large part an extrastatistical problem (Wilk & Kempthorne, 1955), it is one which assumes added importance in view of Cronbach's (1957) emphasis on the interaction of experimental and subject variables. In the face of these assorted difficulties, it is not easy to see what utility the scale typology has for the practical problems of the investigator.

The preceding remarks have been intended to put into broader perspective that sort of invariance which is involved in the use of permissible statistics. They do not, however, solve the immediate problem of whether to use rank order tests or F in case only permissible transformations need be considered. Although invariance under permissible scale transformations may be of relatively minor importance, there is no point in taking unnecessary risks without the possibility of compensation.

On this basis, one would perhaps expect to find the greatest use of rank order tests in the initial stages of inquiry since it is then that measuring scales will be poorest. However, it is in these initial stages that the possibly relevant variables are not well-known so that the stronger experimental designs, and hence parametric procedures, are most needed. Thus, it may well be most efficient to use parametric tests, balancing any risk due to possible permissible scale changes against the greater power and versatility of such tests. In the later stages of investigation, we would be generally more sure of the scales and the use of rank order procedures would waste information which the scales by then embody.

At the same time, it should be realized that even with a relatively crude scale such as the rating scale of attitude toward the church, the possible permissible transformations which are relevant to the present discussion are somewhat restricted. Since the F ratio is invariant under change of zero and unit, it is no restriction to assume that any transformed scale also runs from 1 to 7. This imposes a considerable limitation on the permissible scale transformations which must be considered. In addition, whatever psychological worth the original rating scale possesses will limit still further the transformations which will occur in practice.

Although rank order tests do possess some logical advantage over parametric tests when only permissible transformations are considered, this advantage is, in the writer's opinion, very slight in practice and does not begin to balance the greater versatility of parametric procedures. The problem is, however, an empirical one and it would seem that some historical analysis is needed to provide an objective frame of reference. To quote an after-lunch remark of K. MacCorquodale, "Measurement theory should be descriptive, not proscriptive, nor prescriptive." Such an inquiry could not fail to be fascinating because of the light it would throw on the actual progress of measurement in psychology. One investigation of this sort would probably be more useful than all the speculation which has been written on the topic of measurement.

Question 3. Will the use of parametric as opposed to nonparametric statistics affect inferences about underlying psychological processes? In a narrow sense, Question 3 is irrelevant to this article since the inferences in question are substantive, relating to psychological meaning, rather than formal, relating to data reliability. Nevertheless, it is appropriate to discuss the matter briefly in order to make explicit some of the considerations involved because they are often confused with problems arising under the two previous questions. With no pretense of covering all aspects of this question, the following two examples will at least touch some of the problems.

The first example concerns the two students, P and Q, mentioned above, who had used time and speed as dependent variables. We suppose that their experiment was based on a 2×2 design and yielded means as plotted in Figure 1. This graph portrays main effects of both variables, which are seen to be similar in nature in both panels. However, our principal concern is with the interaction which may be visualized as measuring the degree of nonparallelism of the two lines in either panel. Panel P shows an interaction. The reciprocals of these same data, plotted in Panel Q, show no interaction. It is thus evident in the example, and true in general, that interaction effects will depend strongly on the measuring scales used.

Assessing an interaction does not always cause trouble, of course. Had the lines in Panel P, say, crossed each other, it would not be likely that any change of scale would yield uncrossed lines. In many cases also, the scale used is sufficient for the purposes at hand and future scale changes need not be considered. Nevertheless, it is clear that a measure of caution will often be needed in making inferences from interaction to psychological process. If the investigator envisages the possibility of future changes in the scale, he should also realize that a present inference based on significant interaction may lose credibility in the light of the rescaled data.

It is certainly true that the interpretation of interactions has sometimes led to error. It may also be noted that the usual factorial design analysis is sometimes incongruent with the phenomena. In a 2 × 2 design it might happen, for example, that three of the four cell means are equal. The usual analysis is not optimally sensitive to this one real difference since it is distributed over three degrees of freedom. In such cases, there will often be other parametric tests involving specific comparisons (Snedecor, 1956) or multiple comparisons (Ducan, 1955) which are more appropriate. Occasionally also, an analysis of variance based on a multiplicative model (Williams, 1952) will be useful (Jones & Marcus, 1961). A judicious choice of test may be of great help in dissecting the results. However, the test only answers set questions concerning the reliability of the results; only the research worker can say which questions are appropriate and meaningful.

Inferences based on nonparametric tests of interaction would presumably be less sensitive to certain types of scale changes. However, caution would still be needed in the interpretation as has been seen in Question 2. The problem is largely academic, however, since few nonparametric tests of interaction exist.[2] It might be suggested that the question of interaction cannot arise when only the ordinal properties of the data are considered since the interaction involves a comparison of differ-

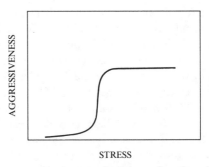

Fig. 3. Aggressiveness plotted as a function of stress. (The curve is hypothetical. Note the hypothetical threshold effect.)

ences and such a comparison is illegitimate with ordinal data. To the extent that this suggestion is correct, a parametric test can be used to the same purposes equally well if not better; to the extent that it is not correct, nonparametric tests will waste information.

One final comment on the first example deserves emphasis. Since both time and speed are interval scales, it cannot be argued that the difficulty in interpretation arises because we had only ordinal scales.

The second example, suggested by J. Kaswan, is shown in Figure 3. The graph, which is hypothetical, plots amount of aggressiveness as a function of amount of stress. A glance at the graph leads immediately to the inference that some sort of threshold effect is present. Under increasing stress, the organism remains quiescent until the stress passes a certain threshold value, whereupon the organism leaps into full scale aggressive behavior.

Confidence in this interpretation is shaken when we stop to consider that the scales for stress and aggression may not be very good. Perhaps, when future work has given us improved scales, these same data would yield a quite different function such as a straight line.

One extreme position regarding the threshold effect would be to say that the scales give rank order information and no more. The threshold inference, or any inference based on characteristics of the curve shape other than the uniform upward trend, would then be

[2] There is a nomenclatural difficulty here. Strictly speaking, nonparametric tests should be called more-or-less distribution free tests. For example, the Mood-Brown generalized median test (Mood, 1950) is distribution free, but is based on a parametric model of the same sort as in the analysis of variance. As noted in the introduction, the usual terminology is used in this article.

completely disallowed. At the other extreme, there would be complete faith in the scales and all inferences based on curve shape, including the threshold effect, would be made without fear that they would be undermined by future changes in the scales. In practice, one would probably adopt a position between these two extremes, believing, with Mosteller (1958), that our scales generally have some degree of numerical information worked into them, and realizing that to consider only the rank order character of the data would be to ignore the information that gives the strongest hold on the behavior.

From this ill-defined middle ground, inferences such as the threshold effect would be entertained as guides to future work. Such inferences, however, are made at the judgment of the investigator. Statistical techniques may be helpful in evaluating the reliability of various features of the data, but only the investigator can endow them with psychological meaning.

Summary

This article has compared parametric and nonparametric statistics under two general headings: practical statistical problems, and measurement theoretical considerations. The scope of the article is restricted to situations in which the dependent variable is numerical, thus excluding strictly categorical data.

Regarding practical problems, it was noted that the difference between parametric and rank order tests was not great insofar as significance level and power were concerned. However, only the versatility of parametric statistics meets the everyday needs of psychological research. It was concluded that parametric procedures are the standard tools of psychological statistics although nonparametric procedures are useful minor techniques.

Under the heading of measurement theoretical considerations, three questions were distinguished. The well-known fact that an interval scale is not prerequisite to making a statistical inference based on a parametric test was first pointed out. The second question took up the important problem of invariance. It was noted that the practical problems of invariance or generality of result far transcend measurement scale typology. In addition, the cited example of time and speed showed that interval scales of a given phenomenon are not unique. The discussion of the third question noted that the problem of psychological meaning is not basically a statistical matter. It was thus concluded that the type of measuring scale used had little relevance to the question of whether to use parametric or nonparametric tests.

References

Anderson, N. H. Education for research in psychology. *Amer. Psychologist*, 1959, **14**, 695–696.

Boneau, C. A. The effects of violations of assumptions underlying the *t* test. *Psychol. Bull.*, 1960, **57**, 49–64.

Cochran, W. G. Some consequences when the assumptions for the analysis of variance are not satisfied. *Biometrics*, 1947, **3**, 22–38.

Cochran, W. G. Some methods for strengthening the common χ^2 tests. *Biometrics*, 1954, **10**, 417–451.

Coombs, C. H. A theory of psychological scaling. *Bull. Engrg. Res. Inst. U. Mich.*, 1952, No. 34.

Cotton, J. W. A re-examination of the repeated measurements problem. Paper read at American Statistical Association, Chicago, December 1959.

Cronbach, L. J. The two disciplines of scientific psychology. *Amer. Psychologist*, 1957, **11**, 671–684.

Dixon, W. J., & Massey, F. J., Jr. *Introduction to statistical analysis.* (2nd ed.) New York: McGraw-Hill, 1957.

Duncan, D. B. Multiple range and multiple *F* tests. *Biometrics*, 1955, **11**, 1–41.

Federer, W. T. *Experimental design.* New York: Macmillan, 1955.

Feldt, L. S. A comparison of the precision of three experimental designs employing a concomitant variable. *Psychometrika*, 1958, **23**, 335–354.

Gaito, J. Nonparametric methods in psychological research. *Psychol. Rep.*, 1959, **5**, 115–125.

Grant, D. A. Analysis-of-variance tests in the analysis and comparison of curves. *Psychol. Bull.*, 1956, **53**, 141–154.

Greenhouse, S. W., & Geisser, S. On methods in the analysis of profile data. *Psychometrika*, 1959, **24**, 95–112.

Jones, F. N., & Marcus, M. J. The subject effect in judgments of subjective magnitude. *J. exp. Psychol.*, 1961, **61**, 40–44.

Lindquist, E. F. *Design and analysis of experiments.* Boston: Houghton Mifflin, 1953.

Lord, F. M. On the statistical treatment of football numbers. *Amer. Psychologist*, 1953, **8**, 750–751.

Mood, A. M. *Introduction to the theory of statistics.* New York: McGraw-Hill, 1950.

Mosteller, F. The mystery of the missing corpus. *Psychometrika*, 1958, **23**, 279–290.

Rao, C. R. *Advanced statistical methods in biometric research.* New York: Wiley, 1952.

Savage, I. R. Nonparametric statistics. *J. Amer. Statist. Ass.*, 1957, **52**, 331–344.

Sawrey, W. L. A distinction between exact and approximate nonparametric methods. *Psychometrika*, 1958, **23**, 171–178.

Senders, V. L. *Measurement and statistics.* New York: Oxford, 1958.

Siegel, S. *Nonparametric statistics.* New York: McGraw-Hill, 1956.

Snedecor, G. W. *Statistical methods.* (5th ed.) Ames: Iowa State Coll. Press, 1956.

Stevens, S. S. Mathematics, measurement, and psychophysics. In S. S. Stevens (Ed.), *Handbook of experimental psychology.* New York: Wiley, 1951.

Tate, M. W., & Clelland, R. C. *Non-parametric and shortcut statistics.* Danville, Ill.: Interstate, 1957.

Wilk, M. B., & Kempthorne, O. Fixed, mixed, and random models. *J. Amer. Statist. Ass.*, 1955, **50**, 1144–1167.

Williams, E. J. The interpretation of interactions in factorial experiments. *Biometrika*, 1952, **39**, 65–81.

2.5 Measurement, Statistics, and the Schemapiric View

S. S. Stevens

A curious antagonism has sometimes infected the relations between measurement and statistics. What ought to proceed as a pact of mutual assistance has seemed to some authors to justify a feud that centers on the degree of independence of the two domains. Thus Humphreys (1) dispenses praise to a textbook because its authors "do not follow the Stevens dictum concerning the precise relationships between scales of measurement and permissible statistical operations." Since that dictum, so-called, lurks as the *bête noire* behind many recurrent complaints, there is need to reexamine its burden and to ask how measurement and statistics shape up in the scientific process—the schemapiric endeavor in which we invent schematic models to map empirical domains.

In those disciplines where measurement is noisy, uncertain, and difficult, it is only natural that statistics should flourish. Of course, if there were no measurement at all, there would be no statistics. At the other extreme, if accurate measurement were achieved in every inquiry, many of the needs for statistics would vanish. Somewhere between the two extremes of no measurement and perfect measurement, perhaps near the psychosocial-behavioral center of gravity, the ratio of statisticizing to measuring reaches its maximum. And that is where we find an acute sensitivity to the suggestion that the type of measurement achieved in an experiment may set bounds on the kinds of statistics that will prove appropriate.

After reviewing the issues Anderson (2) concluded that "the statistical test can hardly be cognizant of the empirical meaning of the numbers with which it deals. Consequently," he continued, "the validity of the statistical inference cannot depend on the type of measuring scale used." This sequitur, if we may call it that, demands scrutiny, for it compresses large issues into a few phrases. Here let me observe merely that, however much we may agree that the statistical test cannot be cognizant of the empirical meaning of the numbers, the same privilege of ignorance can scarcely be extended to experimenters.

Speaking as a statistician, Savage (3) said, "I know of no reason to limit statistical procedures to those involving arithmetic operations consistent with the scale properties of the observed quantities." A statistician, like a computer, may perhaps feign indifference to the origin of the numbers that enter into a statistical computation, but that indifference is not likely to be shared by the scientist. The man in the laboratory may rather suspect that, if something empirically useful is to emerge in the printout, something empirically meaningful must be programed for the input.

Baker, Hardyck, and Petrinovich (4) summed up the distress: "If Stevens' position is correct, it should be emphasized more intensively; if it is incorrect, something should be done to alleviate the lingering feelings of guilt that plague research workers who deliberately use statistics such as *t* on weak measurements." If it is true that guilt must come before repentance, perhaps the age of statistical indifference to the demands of measurement may be drawing to a close. Whatever the outcome, the foregoing samples of opinion suggest that the relation between statistics and measurement is not a settled issue. Nor is it a simple issue, for it exhibits both theoretical and practical aspects. Moreover, peace is not likely to be restored until both the principles and the pragmatics have been resolved.

The Schemapiric Principle

Although measurement began in the empirical mode, with the accent on the counting of moons and paces and warriors, it was destined in modern times to find itself debated in the formal, schematic, syntactical mode, where models can be made to bristle with symbols. Mathematics, which like logic constitutes a formal endeavor, was not always regarded as an arbitrary construction devoid of substantive content, an adventure of postulate and theorem. In early ages mathematics and empirical measurement were as warp and woof, interpenetrating each other so closely that our ancestors thought it proper to prove arithmetic theorems by resort to counting or to some other act of measurement. The divorce took place only in recent times. And mathematics now enjoys full freedom to "play upon symbols," as Gauss phrased it, with no constraints imposed by the demands of empirical measurement.

So also with other formal or schematic systems. The propositions of a formal logic express tautologies that say nothing about the

world of tangible stuff. They are analytic statements, so-called, and they stand apart from the synthetic statements that express facts and relations among empirical objects. There is a useful distinction to be made between the analytic, formal, syntactical propositions of logic and the synthetic, empirical statements of substantive discourse.

Sometimes the line may be hard to draw. Quine (5) the logician denies, in fact, that any sharp demarcation can be certified, and debate on the issue between him and Carnap has reached classic if unresolved proportions. For the scientist, meanwhile, the usefulness of the formal-empirical distinction need not be imperiled by the difficulty of making rigorous decisions in borderline cases. It is useful to distinguish between day and night despite the penumbral passage through twilight. So also is it useful to tune ourselves to distinguish between the formally schematic and the empirically substantive.

Probability exhibits the same double aspect, the same schemapiric nature. Mathematical theories of probability inhabit the formal realm as analytic, tautologous, schematic systems, and they say nothing at all about dice, roulette, or lotteries. On the empirical level, however, we count and tabulate events at the gaming table or in the laboratory and note their relative frequencies. Sometimes the relative frequencies stand in isomorphic relation to some property of a mathematical model of probability; at other times the observed frequencies exhibit scant accord with "expectations."

Those features of statistics that invoke a probabilistic schema provide a further instance of a formal-empirical dichotomy: the distinction between the probability model and the statistical data. E. B. Wilson (6), mathematician and statistician, made the point "that one must distinguish critically between probability as a purely mathematical subject of one sort or another, and statistics which cannot be so regarded." Statistics, of course, is a young discipline—one whose voice changes depending on who speaks for it. Many spokesmen would want to broaden the meaning of statistics to include a formal, mathematical segment.

In another context N. R. Hanson (7) pressed a similar distinction when he said, "Mathematics and physics on this account seem *logically* different disciplines, such that the former can only occasionally solve the latter's problems." Indeed, as Hanson later exlaimed, "Physicists have in unison pronounced, 'Let no man join what nature hath sundered, namely, the *formal creation* of spaces and the physical *description* of bodies.'" Yet it is precisely by way of the proper and judicious joining of the schematic with the empirical that we achieve our beneficial and effective mappings of the universe—the schemapiric mappings known as science. The chronic danger lies in our failure to note the distinction between the map and the terrain, between the simulation and the simulated. The map is an analogue, a schema, a model, a theory. Each of those words has a separate flavor, but they all share a common core of meaning. "Contrary to general belief," wrote Simon and Newell (8), "there is no fundamental, 'in principle,' difference between theories and analogies. All theories are analogies, and all analogies are theories." Indeed, the same can be said for all the other terms that designate the associative binding of schematics to empirics—what I have called the schemapiric bond.

Scales and Invariance

Although it could be otherwise if our choice dictated, most measurement involves the assignment of numbers to aspects of objects or events according to one or another rule or convention. The variety of rules invented thus far for the assignment of numbers has already grown enormous, and novel means of measuring continue to emerge. It has proved possible, however, to formulate an invariance criterion for the classification of scales of measurement (9). The resulting systematization of scale types has found uses in contexts ranging from physics (10) to the social sciences (11), but the conception has not enjoyed immunity from criticism (12).

Let me sketch the theory. It can be done very briefly, because details are given in other places (13). The theory proposes that a scale type is defined by the group of transformations under which the scale form remains invariant, as follows.

A *nominal scale* admits any one-to-one substitution of the assigned numbers. Example of a nominal scale: the numbering of football players.

An *ordinal scale* can be transformed by any increasing monotonic function. Example of an ordinal scale: the hardness scale determined by the ability of one mineral to scratch another.

An *interval scale* can be subjected to a linear transformation. Examples of interval scales: temperature Fahrenheit and Celsius, calendar time, potential energy.

A *ratio scale* admits only multiplication by a constant. Examples of ratio scales: length, weight, density, temperature Kelvin, time intervals, loudness in sones.

The foregoing scales represent the four types in common use. Other types are possible. The permissible transformations defining a scale type are those that keep intact the empirical information depicted by the scale. If the empirical information has been preserved, the scale form is said to remain invariant. The critical isomorphism is maintained. That indeed is the principle of invariance that lies at the heart of the conception. More formal presentations of the foregoing theory have been undertaken by other authors, a recent one, for example, by Lea (14).

Unfortunately, those who demand an abstract tidiness that is completely aseptic may demur at the thought that the decision whether a particular scale enjoys the privilege of a particular transformation group depends on something so ill defined as the preservation of empirical information. For one thing, an empirical operation is always attended by error. Thus Lebesgue (15), who strove so well to perfect the concept of mathematical measure, took explicit note that, in the assignment of number to a physical magnitude, precision can be pushed, as he said, " in actuality only up to a certain error. It never enables us," he continued, "to discriminate between one number and all the numbers that are extremely close to it."

A second disconcerting feature of the invariance criterion lies in the difficulty of specifying the empirical information that is to be preserved. What can it be other than the information that we think we have captured by creating the scale in the first place? We may, for example, perform operations that allow us simply to identify or discriminate a particular property of an object. Sometimes we want to preserve nothing more than that simple outcome, the identification or nominal classification of the items of interest. Or we may go further, provided our empirical operations permit, and determine rank orders, equal intervals, or equal ratios. If we want our number assignments to reflect one or another accrual in information, we are free to transform the scale numbers only in a way that does not lose or distort the desired information. The choice remains ours.

Although some writers have found it possible to read an element of prescription—even proscription—into the invariance principle, as a systematizing device the principle contains no normative force. It can be read more as a description of the obvious than as a directive. It says that, once an isomorphism has been mapped out between aspects of objects or events, on the one hand, and some one or more features of the number system, on the other hand, the isomorphism can be upset by whatever transformations fail to preserve it. Precisely what is preserved or not preserved in a particular circumstance depends upon the empirical operations. Since actual day-to-day measurements range from muddled to meticulous, our ability to classify them in terms of scale type must range from hopelessly uncertain to relatively secure.

The group invariance that defines a scale type serves in turn to delimit the statistical procedures that can be said to be appropriate to a given measurement scale (16). Examples of appropriate statistics are tabulated in Table 1. Under the permissible transformations of a measurement scale, some appropriate statistics remain invariant in value (example: the correlation coefficient r keeps its value under linear transformations). Other statistics change value but refer to the same item or location (example: the median changes its value but continues to refer to mid-distribution under ordinal transformations).

Table 1. Examples of statistical measures appropriate to measurements made on various types of scales. The scale type is defined by the manner in which scale numbers can be transformed without the loss of empirical information. The statistical measures listed are those that remain invariant, as regards either value or reference, under the transformations allowed by the scale type.

Scale Type	Measures of Location	Dispersion	Association or Correlation	Significance Tests
Nominal	Mode	Information H	Information transmitted T	Chi square Fisher's exact test
Ordinal	Median	Percentiles	Rank correlation	Sign test Run test
Interval	Arithmetic mean	Standard deviation Average deviation	Product-moment correlation Correlation ratio	t test F test
Ratio	Geometric mean	Percent variation		
	Harmonic mean	Decilog dispersion		

Reconciliation and New Problems

Two developments may serve to ease the apprehension among those who may have felt threatened by a theory of measurement that seems to place bounds on our freedom to calculate. One is a clearer understanding of the bipartite, schemapiric nature of the scientific enterprise. When the issue concerns only the schema—when, for example, critical ratios are calculated for an assumed binomial distribution—then indeed it is purely a matter of relations within a mathematical model. Natural facts stand silent. Empirical considerations impose no constraints. When, however, the text asserts a relation among such things as measured differences or variabilities, we have a right and an obligation to inquire about the operations that underlie the measurements. Those operations determine, in turn, the type of scale achieved.

The two-part schemapiric view was expressed by Hays (17) in a much-praised book: "If the statistical method involves the procedures of arithmetic used on numerical scores, then the numerical answer is formally correct. . . . The difficulty comes with the interpretation of these numbers back into statements about the real world. If nonsense is put into the mathematical system, nonsense is sure to come out."

At the level of the formal model, then, statistical computations may proceed as freely as in any other syntactical exercise, unimpeded by any material outcome of empirical measurement. Nor does measurement have a presumptive voice in the creation of the statistical models themselves. As Hogben (18) said in his forthright dissection of statistical theory, "It is entirely defensible to formulate an axiomatic approach to the theory of probability as an internally consistent set of propositions, if one is content to leave to those in closer contact with reality the last word on the usefulness of the outcome." Both Hays and Hogben insist that the user of statistics, the man in the laboratory, the maker of measurements, must decide the meaning of the numbers and their capacity to advance empirical inquiry.

The second road to reconciliation winds through a region only partly explored, a region wherein lies the pragmatic problem of appraising the wages of transgression. What is the degree of risk entailed when use is made of statistics that may be inappropriate in the strict sense that they fail the test of invariance under permissible scale transformations? Specifically, let us assume that a set of items can be set in rank order, but, by the operations thus far invented, distances between the items cannot be determined. We have an ordinal but not an interval scale. What happens then if interval-scale statistics are applied to the ordinally scaled items? Therein lies a question of first-rate substance and one that should be amenable to unemotional investigation. It promises well that a few answers have already been forthcoming.

First there is the oft-heeded counsel of common sense. In the averaging of test scores, says Mosteller (19), "It seems sensible to use the statistics appropriate to the type of scale I think I am near. In taking such action we may find the justification vague and fuzzy. One reason for this vagueness is that we have not yet studied enough about classes of scales, classes appropriate to real life measurement, with perhaps real life bias and error variance."

How some of the vagueness of which Mosteller spoke can perhaps be removed is illustrated by the study of Abelson and Tukey (20) who showed how bounds may be determined for the risk involved when an interval-scale statistic is used with an ordinal scale. Specifically, they explored the effect on r^2 of a game against nature in which nature does its best (or worst!) to minimize the value of r^2. In this game of regression analysis, many interesting cases were explored, but, as the authors said, their methods need extension to other cases. They noted that we often know more about ordinal data than mere rank order. We may have reason to believe, they said, "that the scale is no worse than mildly curvilinear, that Nature behaves smoothly in some sense." Indeed the continued use of parametric statistics with ordinal data rests on that belief, a belief sustained in large measure by the pragmatic usefulness of the results achieved.

In a more synthetic study than the foregoing analysis, Baker *et al.* (4) imposed sets of monotonic transformations on an assumed set of data, and calculated the effect on the *t* distribution. The purpose was to compare distributions of *t* for data drawn from an equal-interval scale with distributions of *t* for several types of assumed distortions of the equal intervals. By and large, the effects on the computed *t* distributions were not large, and the authors concluded "that strong statistics such as the *t* test are more than adequate to cope with weak [ordinal] measurements. . . ." It should be noted, however, that the values of *t* were affected by the nonlinear transformations. As the authors said, "The correspondence between values of *t* based on the criterion unit interval scores and values of *t* based on [nonlinear] transformations decreases regularly and dramatically . . . as the departure from linear transformations becomes more extreme."

Whatever the substantive outcome of such investigations may prove to be, they point the way to reconciliation through orderly inquiry. Debate gives way to calculation. The question is thereby made to turn, not on whether the measurement scale determines the choice of a statistical procedure, but on how and to what degree an inappropriate statistic may lead to a deviant conclusion. The solution of such problems may help to refurbish the complexion of measurement theory, which has been accused of proscribing those statistics that do not remain invariant under the transformations appropriate to a given scale. By spelling out the costs, we may convert the issue from a seeming proscription to a calculated risk.

The type of measurement achieved is not, of course, the only consideration affecting the applicability of parametric statistics. Bradley is one of many scholars who have sifted the consequences of violating the assumptions that underlie some of the common parametric tests (21). As one outcome of his studies, Bradley concluded, "The contention that, when its assumptions are violated, a para-

metric test is still to be preferred to a distribution-free test because it is 'more efficient' is therefore a monumental *non sequitur*. The point is not at all academic . . . violations in a test's assumptions may be attended by profound changes in its power." That conclusion is not without relevance to scales of measurement, for when ordinal data are forced into the equal-interval mold, parametric assumptions are apt to be violated. It is then that a so-called distribution-free statistic may prove more efficient than its parametric counterpart.

Although better accommodation among certain of the contending statistical usages may be brought about by computer-aided studies, there remain many statistics that find their use only with specific kinds of scales. A single example may suffice. In a classic textbook, written with a captivating clarity, Peters and Van Voorhis (22) got hung up on a minor point concerning the procedure to be used in comparing variabilities. They noted that Karl Pearson had proposed a measure called the coefficient of variation, which expresses the standard deviation as a percentage of the mean. The authors expressed doubts about its value, however, because it tells "more about the extent to which the scores are padded by a dislocation of the zero point than it does about comparable variabilities." The examples and arguments given by the authors make it plain that the coefficient of variation has little business being used with what I have called interval scales. But since their book antedated my publication in 1946 of the defining invariances for interval and ratio scales, Peters and Van Voorhis did not have a convenient way to state the relationship made explicit in Table 1, namely, that the coefficient of variation, being itself a ratio, calls for a ratio scale.

Complexities and Pitfalls

Concepts like relative variability have the virtue of being uncomplicated and easy for the scientist to grasp. They fit his idiom. But in the current statistics explosion, which showers the investigator with a dense fallout of new statistical models, the scientist is likely to lose the thread on many issues. It is then that the theory of measurement, with an anchor hooked fast in empirical reality, may serve as a sanctuary against the turbulence of specialized abstraction.

"As a mathematical discipline travels far from its empirical source," said von Neumann (23), "there is grave danger that the subject will develop along the line of least resistance, that the stream, so far from its source, will separate into a multitude of insignificant branches, and that the discipline will become a disorganized mass of details and complexities." He went on to say that, "After much 'abstract' inbreeding, a mathematical subject is in danger of degeneration. At the inception the style is usually classical; when it shows signs of becoming baroque, then the danger signal is up."

There is a sense, one suspects, in which statistics needs measurement more than measurement needs statistics. R. A. Fisher alluded to that need in his discourse on the nature of probability (24). "I am quite sure," he said, "it is only personal contact with the business of the improvement of natural knowledge in the natural sciences that is capable to keep straight the thought of mathematically-minded people who have to grope their way through the complex entanglements of error. . . ."

And lest the physical sciences should seem immune to what Schwartz (25) called "the pernicious influence of mathematics," consider his diagnosis: "Thus, in its relations with science, mathematics depends on an intellectual effort outside of mathematics for the crucial specification of the approximation which mathematics is to take literally. Give a mathematician a situation which is the least bit ill-defined—he will first of all make it well defined. Perhaps appropriately, but perhaps also inappropriately. . . . That form of wisdom which is the opposite of single-mindedness, the ability to keep many threads in hand, to draw for an argument from many disparate sources, is quite foreign to mathe-

matics. . . . Quite typically, science leaps ahead and mathematics plods behind."

Progress in statistics often follows a similar road from practice to prescription—from field trials to the formalization of principles. As Kruskal (26) said, "Theoretical study of a statistical procedure often comes after its intuitive proposal and use." Unfortunately for the empirical concerns of the practitioners, however, there is, as Kruskal added, "almost no end to the possible theoretical study of even the simplest procedure." So the discipline wanders far from its empirical source, and form loses sight of substance.

Not only do the forward thrusts of science often precede the mopping-up campaigns of the mathematical schema builders, but measurement itself may often find implementation only after some basic conception has been voiced. Textbooks, those distilled artifices of science, like to picture scientific conceptions as built on measurement, but the working scientist is more apt to devise his measurements to suit his conceptions. As Kuhn (27) said, "The route from theory or law to measurement can almost never be travelled backwards. Numbers gathered without some knowledge of the regularity to be expected almost never speak for themselves. Almost certainly they remain just numbers." Yet who would deny that some ears, more tuned to numbers, may hear them speak in fresh and revealing ways?

The intent here is not, of course, to affront the qualities of a discipline as useful as mathematics. Its virtues and power are too great to need extolling, but in power lies a certain danger. For mathematics, like a computer, obeys commands and asks no questions. It will process any input, however devoid of scientific sense, and it will bedeck in formulas both the meaningful and the absurd. In the behavioral sciences, where the discernment for nonsense is perhaps less sharply honed than in the physical sciences, the vigil must remain especially alert against the intrusion of a defective theory merely because it carries a mathematical visa. An absurdity in full formularized attire may be more seductive than an absurdity undressed.

Distributions and Decisions

The scientist often scales items, counts them, and plots their frequency distributions. He is sometimes interested in the form of such distributions. If his data have been obtained from measurements made on interval or ratio scales, the shape of the distribution stays put (up to a scale factor) under those transformations that are permissible, namely, those that preserve the empirical information contained in the measurements. The principle seems straightforward. But what happens when the state of the art can produce no more than a rank ordering, and hence nothing better than an ordinal scale? The abscissa of the frequency distribution then loses its metric meaning and becomes like a rubber band, capable of all sorts of monotonic stretchings. With each nonlinear transformation of the scale, the form of the distribution changes. Thereupon the distribution loses structure, and we find it futile to ask whether the shape approximates a particular form, whether normal, rectangular, or whatever.

Working on the formal level, the statistician may contrive a schematic model by first assuming a frequency function, or a distribution function, of one kind or another. At the abstract level of mathematical creation, there can, of course, be no quarrel with the statistician's approach to his task. The caution light turns on, however, as soon as the model is asked to mirror an empirical domain. We must then invoke a set of semantic rules—coordinating definitions—in order to identify correspondences between model and reality. What shall we say about the frequency function $f(x)$ when the problem before us allows only an ordinal scale? Shall x be subject to a nonlinear transformation after $f(x)$ has been specified? If so, what does the transformation do to the model and to the predictions it forecasts?

The scientist has reason to feel that a statistical model that specifies the form of a canonical distribution becomes uninterpretable when the empirical domain concerns only ordinal data. Yet many consumers of statistics seem to disregard what to others is a rather obvious and critical problem. Thus Burke (28) proposed to draw "two random samples from populations known to be normal" and then "to test the hypothesis that the two populations have the same mean . . . under the assumption that the scale is ordinal at best." How, we must ask, can normality be known when only order can be certified?

The assumption of normality is repeated so blithely and so often that it becomes a kind of incantation. If enough of us sin, perhaps trangression becomes a virtue. But in the instance before us, where the numbers to be fed into the statistical mill result from operations that allow only a rank ordering, maybe we have gone too far. Consider a permissible transformation. Let us cube all the numbers. The rank order would stand as before. But what do we then say about normality? If we can know nothing about the intervals on the scale of a variable, the postulation that a distribution has a particular form would appear to proclaim a hope, not a circumstance.

The assertion that a variable is normally distributed when the variable is amenable only to ordinal measurement may loom as an acute contradiction, but it qualifies as neither the worst nor the most frequent infraction by some of the practitioners of hypothesis testing. Scientific decision by statistical calculation has become the common mode in many behavioral disciplines. In six psychological journals (29), for example, the proportion of articles that employed one or another kind of inferential statistic rose steadily from 56 percent in 1948 to 91 percent in 1962. In the *Journal of Educational Psychology* the proportion rose from 36 to 100 percent.

What does it mean? Can no one recognize a decisive result without a significance test? How much can the burgeoning of computation be blamed on fad? How often does inferential computation serve as a premature excuse for going to press? Whether the scholar has discovered something or not, he can sometimes subject his data to an analysis of variance, a *t* test, or some other device that will produce a so-called objective measure of "significance." The illusion of objectivity seems to preserve itself despite the admitted necessity for the investigator to make improbable assumptions, and to pluck off the top of his head a figure for the level of probability that he will consider significant. His argument that convention has already chosen the level that he will use does not quite absolve him.

Lubin (30) has a name for those who censure the computational and applaud the experimental in the search for scientific certainty. He calls them stochastophobes. An apt title, if applied to those whose eagerness to lay hold on the natural fact may generate impatience at the gratuitous processing of data. The extreme stochastophobe is likely to ask: What scientific discoveries owe their existence to the techniques of statistical analysis or inference? If exercises in statistical inference have occasioned few instances of a scientific breakthrough, the stochastophobe may want to ask by what magical view the stochastophile perceives glamour in statistics. The charm may stem in part from the prestige that mathematics, however inapposite, confers on those who display the dexterity of calculation. For some stochastophiles the appeal may have no deeper roots than a preference for the prudent posture at a desk as opposed to the harsher, more venturesome stance in the field or the laboratory.

The aspersions voiced by stochastophobes fall mainly on those scientists who seem, by the surfeit of their statistical chants, to turn data treatment into hierurgy. These are not the statisticians themselves, for they see statistics for what it is, a straightforward discipline designed to amplify the power of common sense in the discernment of order amid complexity. By showing how to amend the mismatch in the impedance between question and evidence, the statistician improves the probability that our experiments will speak with greater clarity. And by weighing the entailments of relevant assumptions, he shows us how to milk the most from some of those fortuitous experiments that nature performs once and may never perform again. The stochastophobe should find no quarrel here. Rather he should turn his despair into a hope that the problem of the relevance of this or that statistical model may lead the research man toward thoughtful inquiry, not to a reflex decision based on a burst of computation.

Measurement

If the vehemence of the debate that centers on the nature and conditions of statistical inference has hinted at the vulnerability of the conception, what can be said about the other partner in the enterprise? Is the theory of measurement a settled matter? Apparently not, for it remains a topic of trenchant inquiry, not yet ready to rest its case. And debate continues.

The typical scientist pays little attention to the theory of measurement, and with good reason, for the laboratory procedures for most measurements have been well worked out, and the scientist knows how to read his dials. Most of his variables are measured on well-defined, well-instrumented ratio scales.

Among those whose interests center on variables that are not reducible to meter readings, however, the concern with measurement stays acute. How, for example, shall we measure subjective value (what the economist calls utility), or perceived brightness, or the seriousness of crimes? Those are some of the substantive problems that have forced a revision in our approach to measurement. They have entailed a loosening of the restricted view bequeathed us by the tradition of Helmholtz and Campbell—the view that the axioms of additivity must govern what we call measurement (31). As a related development, new axiomatic systems have appeared, including axioms by Luce and Tukey (32) for a novel "conjoint" approach to fundamental measurement. But the purpose here is not to survey the formal, schematic models that have flowered in the various sciences, for the practice and conception of measurement has as yet been little influenced by them.

As with many syntactical developments, measurement models sometimes drift off into the vacuum of abstraction and become decoupled from their concrete reference. Even those authors who freely admit the empirical features as partners in the formulation of measurement may find themselves seeming to downgrade the empirical in favor of the formal. Thus we find Suppes and Zinnes (33) saying "Some writers . . . appear to define scales in terms of the existence of certain empirical operations. . . . In the present formulation of scale type, no mention is made of the kinds of 'direct' observations or empirical relations that exist. . . . Precisely what empirical operations are involved in the empirical system is of no consequence."

How then do we distinguish different types of scales? How, in particular, do we know whether a given scale belongs among the interval scales? Suppes and Zinnes gave what I think is a proper answer: "We ask if all the admissible numerical assignments are related by a linear transformation." That, however, is not a complete answer. There remains a further question: What is it that makes a class of numerical assignments admissible? A full theory of measurement cannot detach itself from the empirical substrate that gives it meaning. But the theorist grows impatient with the empirical lumps that ruffle the fine laminar flow within his models just as the laboratory fellow may disdain the arid swirls of hieroglyphics that pose as paradigms of his measurements.

Although a congenial conciliation between those two polar temperaments, the modeler and the measurer, may lie beyond reasonable expectations, a tempering détente may prove viable. The two components of schemapirics must both be accredited, each in its own imperative role. To the understanding of the world about us, neither the formal model nor the concrete measure is dispensable.

Matching and Mapping

Instead of starting with origins, many accounts of measurement begin with one or another advanced state of the measuring process, a state in which units and metrics can be taken for granted. At that level, the topic already has the crust of convention upon it, obscuring the deeper problems related to its nature.

If we try to push the problem of measurement back closer to its primordial operations, we find, I think, that the basic operation is always a process of matching. That statement may sound innocent enough, but it contains a useful prescription. It suggests, for example, that if you would understand the essence of a given measuring procedure, you should ask what was matched to what. If the query leads to a pointer reading, do not stop there; ask the same question about the calibration procedure that was applied to the instruments anterior to the pointer: What was matched to what? Diligent pursuit of that question along the chain of measuring operations leads to some of the elemental operations of science.

Or we may start nearer the primordium. The sketchiness of the record forces us to conjecture the earliest history, but quite probably our forefather kept score on the numerosity of his possessions with the aid of piles of pebbles [Latin: *calculi*] or by means of some other tallying device. He paired off items against pebbles by means of a primitive matching operation, and he thereby measured his hoard.

Let us pause at this point to consider the preceding clause. Can the ancestor in question be said to have measured his possessions if he had no number system? Not if we insist on taking literally the definition often given, namely, that measurement is the assignment of numbers to objects or events according to rule. This definition serves a good purpose in many contexts, but it presumes a stage of development beyond the one that we are now seeking to probe. In an elemental sense, the matching or assigning of numbers is a sufficient but not a necessary condition for measurement, for other kinds of matching may give measures.

Numbers presumably arose after our ancestor invented names for the collection of pebbles, or perhaps for the more convenient collections, the fingers. He could then match name to collection, and collection to possessions. That gave him a method of counting, for, by pairing off each item against a finger name in an order decided upon, the name of the collection of items, and hence the numerosity of the items, was specified.

The matching principle leads to the concept of cardinality. Two sets have the same cardinal number if they can be paired off in one-to-one relation to each other. By itself, this cardinal pairing off says nothing about order. (Dictionaries often disagree with the mathematicians on the definition of cardinality, but the mathematical usage recommends itself here.) We find the cardinal principle embodied in the symbols used for the numerals in many forms of writing. Thus the Roman numeral VI pictures a hand V and a finger I.

Let us return again to our central question. In the early cardinal procedure of matching item to item, fingers to items, or names to items, at what point shall we say that measurement began? Perhaps we had best not seek a line of demarcation between measurement and matching. It may be better to go all the way and propose an unstinted definition as follows: Measurement is the matching of an aspect of one domain to an aspect of another.

The operation of matching eventuates, of course, in one domain's being mapped into another, as regards one or more attributes of the two domains. In the larger sense, then, whenever a feature of one domain is mapped isomorphically in some relation with a feature of another domain, measurement is achieved. The relation is potentially symmetrical. Our hypothetical forefather could measure his collection of fish by means of his pile of pebbles, or his pile of pebbles by means of his collection of fish.

Our contemporary concern lies not, of course, with pebbles and fish, but with a principle. We need to break the hull that confines the custom of our thought about these matters. The concern is more than merely academic, however, especially in the field of psychophysics. One justification for the enlarged view of measurement lies in a development in sensory measurement known as cross-modality matching (34). In a suitable laboratory setup, the subject is asked, for example, to adjust the loudness of a sound applied to his ears in order to make it seem equal to the perceived strength of a vibration applied to his finger. The amplitude of the vibration is then changed and the matching process is repeated. An equal sensation function is thereby mapped out ... Loudness has been matched in that manner to ranges of values on some ten other perceptual continua, always with the result that the matching function approximates a power function (35). ...

In the description of a measurement system that rests on cross-modality matching, no mention has been made of numbers. If we are willing to start from scratch in a measurement of this kind, numbers can in principle be dispensed with. They would, to be sure, have practical uses in the conduct of the experiments, but by using other signs or tokens to identify the stimuli we could presumably eliminate numbers completely. It would be a tour de force, no doubt, but an instructive one.

Instead of dispensing with numbers, the practice in many psychophysical studies has been to treat numbers as one of the perceptual continua in the cross-modality matching experiment. Thus in what has come to be known as the method of magnitude estimation, numbers are matched to loudness, say. In the reverse procedure, called magnitude production, the subject adjusts the loudness to match a series of numbers given by the experimenter (36). And as might be expected, despite all the other kinds of cross-modality matches that have been made, it is the number continuum that most authors select as the reference continuum (exponent = 1.0) in terms of which the exponent values for the other perceptual continua are stated. But the point deserves to be stressed: the choice of number as the reference continuum is wholly arbitrary, albeit eminently convenient.

Summary

Back in the days when measurement meant mainly counting, and statistics meant mainly the inventory of the state, the simple descriptive procedures of enumeration and averaging occasioned minimum conflict between measurement and statistics. But as measurement pushed on into novel behavioral domains, and statistics turned to the formalizing of stochastic models, the one-time intimate relation between the two activities dissolved into occasional misunderstanding. Measurement and statistics must live in peace, however, for both must participate in the schemapiric enterprise by which the schematic model is made to map the empirical observation.

Science presents itself as a two-faced, bipartite endeavor looking at once toward the formal, analytic, schematic features of model-building, and toward the concrete, empirical, experiential observations by which we test the usefulness of a particular representation. Schematics and empirics are both essential to science, and full understanding demands that we know which is which.

Measurement provides the numbers that enter the statistical table. But the numbers that issue from measurements have strings attached, for they carry the imprint of the operations by which they were obtained. Some transformations on the numbers will leave intact the information gained by the measurements; other transformations will destroy the desired isomorphism between the measurement scale and the property assessed. Scales of measurement therefore find a useful classification on the basis of a principle of invariance: each of the common scale types (nominal, ordinal, interval, and ratio) is defined by a group of transformations that leaves a particular isomorphism unimpaired.

Since the transformations allowed by a given scale type will alter the numbers that enter into a statistical procedure, the procedure ought properly to be one that can withstand that particular kind of number alteration. Therein lies the primacy of measurement: it sets bounds on the appropriateness of statistical operations. The widespread use on ordinal scales of statistics appropriate only to interval or ratio scales can be said to violate a technical canon, but in many instances the outcome has demonstrable utility. A few workers have begun to assess the degree of risk entailed by the use of statistics that do not remain invariant under the permissible scale transformations.

The view is proposed that measurement can be most liberally construed as the process of matching elements of one domain to those of another domain. In most kinds of measurement we match numbers to objects or events, but other matchings have been found to serve a useful purpose. The cross-modality matching of one sensory continuum to another has shown that sensory intensity increases as the stimulus intensity raised to a power. The generality of that finding supports a psychophysical law expressible as a simple invariance: equal stimulus ratios produce equal sensation ratios.

References and Notes

1. L. Humphreys, *Contemp. Psychol.* **9**, 76 (1964).
2. N. H. Anderson, *Psychol. Bull.* **58**, 305 (1961).
3. I. R. Savage, *J. Amer. Statist. Ass.* **52**, 331 (1957).
4. B. O. Baker, C. D. Hardyck, L. F. Petrinovich, *Educ. Psychol. Meas.* **26**, 291 (1966).
5. W. V. O. Quine, *The Ways of Paradox and Other Essays* (Random House, New York, 1966), pp. 126–134.
6. E. B. Wilson, *Proc. Natl. Acad. Sci. U.S.* **51**, 539 (1964).
7. N. R. Hanson, *Philos. Sci.* **30**, 107 (1963).

8. H. A. Simon and A. Newell, in *The State of the Social Sciences*, L. D. White, Ed. (Univ. of Chicago Press, Chicago, 1956), pp. 66–83.

9. S. S. Stevens, *Science* **103**, 677 (1946).

10. F. B. Silsbee, *J. Wash. Acad. Sci.* **41**, 213 (1951).

11. B. F. Green, in *Handbook of Social Psychology*, G. Lindzey, Ed. (Addison-Wesley, Reading, Mass., 1954), pp. 335–369.

12. Among those who have commented are B. Ellis, *Basic Concepts of Measurement* (University Press, Cambridge, England, 1966); B. Grunstra, "On Distinguishing Types of Measurement," *Boston Studies Phil. Sci.*, vol. 4 (Humanities Press, in press); S. Ross, *Logical Foundations of Psychological Measurement* (Scandinavian University Books, Munksgaard, Copenhagen, 1964); W. W. Rozeboom, *Synthese* **16**, 170–233 (1966); W. S. Torgerson, *Theory and Methods of Scaling* (Wiley, New York, 1958).

13. S. S. Stevens, in *Handbook of Experimental Psychology*, S. S. Stevens, Ed. (Wiley, New York, 1951), pp. 1–49; ———, in *Measurement: Definitions and Theories*, C. W. Churchman and P. Ratoosh, Eds. (Wiley, New York, 1959), pp. 18–64.

14. W. A. Lea, "A Formalization of Measurement Scale Forms" (Technical Memo. KC-T-024, Computer Research Lab., NASA Electronics Res. Ctr., Cambridge, Mass., June 1967).

15. H. Lebesgue, *Measure and the Integral*, K. O. May, Ed. (Holden-Day, San Francisco, 1966).

16. Other summarizing tables are presented by V. Senders, *Measurement and Statistics* (Oxford Univ. Press, New York, 1958). A further analysis of appropriate statistics has been presented by E. W. Adams, R. F. Fagot, R. E. Robinson, *Psychometrika* **30**, 99 (1965).

17. W. L. Hays, *Statistics for Psychologists* (Holt, Rinehart & Winston, New York, 1963).

18. L. Hogben, *Statistical Theory* (Norton, New York, 1958).

19. F. Mosteller, *Psychometrika* **23**, 279 (1958).

20. R. P. Abelson and J. W. Tukey, *Efficient Conversion of Non-Metric Information into Metric Information* (Amer. Statist. Ass., Social Statist. Sec., December 1959), pp. 226–230; see also ———, *Ann. Math. Stat.* **34** 1347 (1963).

21. J. V. Bradley, "Studies in Research Methodology: II. Consequences of Violating Para-metric Assumptions—Facts and Fallacy" (WADC Tech. Rep. 58-574. [II]. Aerospace Med. Lab., Wright-Patterson AFB, Ohio, September 1959).

22. C. C. Peters and W. R. Van Voorhis, *Statistical Procedures and Their Mathematical Bases* (McGraw-Hill, New York, 1940).

23. J. von Neumann, in *The Works of the Mind*, R. B. Heywood, Ed. (Univ. of Chicago Press, Chicago, 1947), pp. 180–196.

24. R. A. Fisher, *Smoking, the Cancer Controversy* (Oliver and Boyd, Edinburgh, 1959).

25. J. Schwartz, in *Logic, Methodology and Philosophy of Science*, E. Nagel *et al.*, Eds., (Stanford Univ. Press, Stanford, Calif., 1962), pp. 356–360.

26. W. R. Kruskal, in *International Encyclopedia of the Social Sciences* (Macmillan and Free Press, New York, 1968), vol. 15, pp. 206–224

27. T. S. Kuhn, in *Quantification*, H. Woolf, Ed. (Bobbs-Merrill, Indianapolis, Ind., 1961), pp. 31–63.

28. C. J. Burke, in *Theories in Contemporary Psychology*, M. H. Marx, Ed. (Macmillan, New York, 1963), pp. 147–159.

29. The journals were tabulated by E. S. Edgington, *Amer. Psychologist* **19**, 202 (1964); also personal communication.

30. A. Lubin, in *Annual Review of Psychology* (Annual Reviews, Palo Alto, Calif., 1962), vol. 13, pp. 345–370.

31. H. v. Helmholtz, "Zählen und Messen," in *Philosophische Aufsätze* (Fues's Verlag, Leipzig, 1887), pp. 17–52; N. R. Campbell, *Physics: the Elements* [1920] (reissued as *The Philosophy of Theory and Experiment* by Dover, New York, 1957); ———, *Symposium: Measurement and Its Importance for Philosophy*. Aristotelian Soc., suppl., vol. 17 (Harrison and Sons, London, 1938).

32. R. D. Luce and J. W. Tukey, *J. Math. Psychol.* **1**, 1 (1964).

33. P. Suppes and J. L. Zinnes, in *Handbook of Mathematical Psychology*, R. D. Luce *et al.*, Eds. (Wiley, New York, 1963), pp. 1–76.

34. S. S. Stevens, *J. Exp. Psychol.* **57**, 201 (1959); *Amer. Sci.* **54**, 385 (1966).

35. ———, *Percept. Psychophys.* **1**, 5 (1966).

36. ——— and H. B. Greenbaum, *ibid.*, p. 439.

37. This article (Laboratory of Psychophysics Rept. PPR-336-118) was prepared with support from NIH grant NB-02974 and NSF grant GB-3211.

2.6 Suggestions for Further Reading*

Stevens, S. S. On the theory of scales of measurement. *Science*, 1946, **103**, 677–680. Scales of measurement are classified and statistical operations appropriate for each scale are listed. This paper began a controversy that was to continue for several decades.

Burke, C. J. Additive scales and statistics. *Psychological Review*, 1953, **60**, 73–75. Defends the position that the use of statistical tests is limited only by the well-known statistical restrictions and that means and standard deviations, for example, can be computed for data whatever the properties of the measurement scale.

Senders, V. L. A comment on Burke's additive scales and statistics. *Psychological Bulletin*, 1953, **60**, 423–424. Senders criticizes the central conclusion presented in the paper by C. J. Burke cited above.

Behan, F. L., and R. A. Behan. Football numbers (continued). *American Psychologist*, 1954, **9**, 262–263. The paper by F. M. Lord included in this chapter is used as a point of departure for examining the operations appropriate for nominal and ordinal scales.

Burke, C. J. Measurement scales and statistical models. In M. H. Marx (Ed.), *Theories*

* Articles are listed chronologically.

in Contemporary Psychology. New York: Macmillan, 1963. Provides a thoughtful examination of the central issues in the "levels of measurement" controversy. The general position stated in his paper "Additive scales and statistics," which was cited previously, is reaffirmed.

Adams, E. W., R. F. Fagot, and R. E. Robinson. A theory of appropriate statistics. *Psychometrika*, 1965, **30**, 99–127. A relatively advanced-level presentation of a theory of appropriateness for statistical operations.

Baker, B. O., C. D. Hardyck, and L. F. Petrinovich. Weak measurement vs. strong statistics: An empirical critique of S. S. Stevens' proscriptions on statistics. *Educational and Psychological Measurement*, 1966, **26**, 291–309. The authors provide an empirical approach to the question of whether or not to restrict the use of statistical procedures to those involving arithmetic operations consistent with the scale properties of sample data.

Labovitz, S. Some observations on measurement and statistics. *Social Forces*, 1967, **46**, 151–160. Arguments are given for using certain interval-type statistics for data measured with ordinal scales. The issue is seen as one requiring the careful weighing of four factors: assumptions, measurement scales, robustness, and power-efficiency.

The following two articles take exception to one or more of Labovitz's comments.

Champion, D. J. "Some observations on measurement and statistics": Comment. *Social Forces*, 1968, **46**, 541.

Morris, R. N. "Some observations on measurement and statistics": Further comment. *Social Forces*, 1968, **46**, 541–542.

Labovitz, S. Reply to Champion and Morris. *Social Forces*, 1968, **46**, 543–544. The author gives a rebuttal of Champion and Morris's criticisms of his article.

3

Probability and Theoretical Distributions

The topics of statistics, probability, and theoretical distributions are inextricably interwoven. The first course in behavioral statistics is often designed to teach just enough of the latter two topics to enable the student to understand the former. The use of a cookbook approach in teaching statistics is common in the behavioral sciences and has been widely accepted for the past three decades. However, during that time the amount of knowledge that must be mastered if a student is to carry out and interpret research has increased exponentially, rendering the time-honored cookbook approach ineffectual. Faced with these realities, many teachers are requiring of their students a more rigorous preparation in mathematics and/or are devoting a greater proportion of their class time to such requisite material as probability and theoretical distributions. The rationale for this emphasis is the belief that such knowledge is necessary if the student is to keep up with advances in statistics and research methodology.

More than likely these are some of the considerations that influenced John C. Flynn as he was writing the following article on probability and theoretical distributions. The article, which constitutes Chapter 3, is divided into nine sections; portions of the following sections are suggested for inclusion in introductory-level courses:

I. Probability and Random Variables

II. Continuous and Discrete Random Variables

III. Probability Density Functions

IV. Some Important Properties of the Probability Density Function

VI. Some Important Probability Density Functions

IX. Some Examples of Statistical Inference

The student will find that this material requires careful study and much contemplation. For the benefit of the student with minimal mathematical preparation, sections or subsections that are mathematically demanding are starred and can be skipped without losing the thread of the development. However, the student who works through the article will be rewarded with a glimpse into some of the mathematical underpinnings of statistics and hopefully will come away with answers to many of his questions that begin with "why."

3.1 Some Basic Concepts of Mathematical Statistics

John C. Flynn

Many if not most users of statistics in the behavioral sciences have learned the use of their quantitative tools in a second- or third-hand fashion. That is to say, statistical textbooks for workers in these areas tend to be distillations of the primary (and sometimes secondary) statistical literature. In this process the heavier mathematical base is left behind while the lighter fare of results and applications is presented in a more or less immediately usable form. This is a time-honored pattern, and not until quite recently have some textbooks (see Hays, 1963) for behavioral scientists made a determined effort to present some of the conceptual underpinnings of statistical methods.

All this is perhaps necessary and not altogether bad. Certainly it would be unreasonable to expect behavioral scientists, or scientists of any variety for that matter, also to be competent mathematical statisticians. On the other hand, a middle ground may be sought wherein the scientist may make more intelligent use of statistical methods if he is acquainted with at least the rudiments of statistical theory. Furthermore, such acquaintance would enable him to consult, when necessary, a wider range of reference works, which might prove beneficial in the solution of some problem. He might also more easily

make use of articles such as are found in the present collection. Such an approach to statistics is in fact more reasonable than it would have been a few years ago. It is the writer's experience, for example, that among undergraduates in the behavioral sciences the level of mathematical sophistication has risen steadily. This seems due in no small part to the change in emphasis on the teaching of mathematics in the public schools. At any rate, these students seem ready and eager to grapple with the conceptual framework of the statistical methods they employ.

The present paper is written in this spirit. It does not pretend to be a statistics textbook. It is, as the title states, a summary of some basic concepts. The summary, however, is of some mathematical substance. A familiarity with the basic algebra of sets is assumed, as is a familiarity with some fundamental notions of the calculus. Most present-day students almost certainly have the former; many will also have had an introductory calculus course. It is hoped that some readers of this paper will be motivated to read in more detail in one of the introductory mathematical statistics texts (Hogg & Craig, 1959; Whitney, 1961).

This paper is a summary. No proofs are given. What is presented is essentially a collection (hopefully in a coherent form) of some important definitions and theorems that will serve as a reference for the present volume and perhaps as a stimulus for further reading. Also, the introductory sections on probability are extremely sketchy and are designed mainly to aid the informed reader in appreciating the motivation behind the section on probability density functions. The reader who is less prepared in this area may, without losing anything that is necessary for subsequent sections, begin his review of statistical concepts with Section III, "Probability Density Functions."

The reader's attention is directed to an organizational device used in the following pages. Some sections and subsections are marked with an asterisk. These "starred" sections contain material that is mathematically more difficult than the material in the unstarred sections. The reader with a more modest background in mathematics may wish to ignore these more difficult sections. This can be done with impunity since the paper has been so arranged that one can obtain a coherent "once over lightly" exposure to basic concepts of mathematical statistics by reading only the unstarred sections. Readers with more mathematics at their command will wish to pursue the starred and unstarred sections alike.

I. Probability and Random Variables

A statistical theory will rest on a theory of probability, and to this we now turn. Actually our concern is not with a theory of probability *per se* but rather with the results of such a theory. While the selection of a definition of probability is an occasion for debate among mathematicians, it will suffice for our purposes to adopt a conventional definition of probability as a number that is related to the long-run relative frequency of occurrence of an event. Thus, to say that the probability of the occurrence of event A is P is to say that in the long run event A occurs on approximately a Pth proportion of those occasions when the event could possibly occur. Thus the statement that on a single roll of a die the probability of a three landing face up is $\frac{1}{6}$ means that over many such rolls of the die a three would land face up on about $\frac{1}{6}$ of the rolls.

As the example of the die indicates, we will in general specify a range of events when we speak of probability. This range will be some sort of specification of the possible events that we could observe. In the case of the die, the possible events of concern to us are the face-up landings of one of the numbers 1, 2, 3, 4, 5, 6. We assume that these are the only possible outcomes of rolling the die. The totality of all possible outcomes is called the *sample space*. A theory of probability assigns to each event or element in a sample space a number that gives the likelihood of its occurrence. We can consider the sample space to be a set of points that has various subsets to which the numbers are assigned. Such numbers are called probabilities; "likelihood" may be interpreted in terms of relative frequency as indicated above.

From the foregoing it may be inferred that probability is a type of measure imposed upon a sample space. In order to clarify some important aspects of this measure we shall use notation from the algebra of sets. We shall also discuss the notion of a *random variable*, one of the basic concepts with which we shall deal.

Suppose that we have an experiment to perform and, as in most experiments, we do not know what the outcome will be. Suppose, however, that we can specify all outcomes that can possibly occur. Thus in the "experiment" of tossing a coin the possible outcomes are heads or tails, neither of which can be predicted with certainty. This type of experiment, one that can be "essentially" replicated as often as necessary, is called a *random experiment* and its outcome is called a *random variable*. Let X be this random variable and let $X = 0$ denote the occurrence of a head and $X = 1$ denote the occurrence of a tail. Using script letters to denote an entire sample space and Roman letters to denote subsets of that space, we now let $\mathscr{A}$ denote the sample space. Then

$$\mathscr{A} = \{x; x = 0, 1\}.$$

This is read, "$\mathscr{A}$ is the set of elements x, such that x is equal to zero or one." In this case $\mathscr{A}$ has the following subsets:

$$A_1 = \mathscr{A} = \{x; x = 0, 1\},$$
$$A_2 = \{x; x = 0\},$$
$$A_3 = \{x; x = 1\},$$
$$A_4 = \varnothing = \text{null set.}$$

The preceding paragraph illustrates a notational convention that we shall observe in discussing random variables. When we speak generally of a random variable—for example, when we discuss a certain subset of the sample space—we shall denote the random variable as x. When our discussion centers specifically on the random variable as an outcome of a random experiment, we shall use the notation X. Thus a reasonable statement would be $P(X \leq x) = K$. This is read, "The probability is K that the outcome X of a random experiment is equal to or less than x."

A probability measure is one that assigns to each of the subsets A_i a number $P(A_i)$, which is the probability that the outcome of the experiment is a member of the ith subset. Obviously in the coin-tossing experiment the numbers are easily assigned. The outcome is always in the sample space, is never in the null set, and is in A_2 or A_3 with probability equal to $\frac{1}{2}$ in both instances. Furthermore $P(A_2) + P(A_3) = 1$.

The Probability Set Function. The example above illustrates some general attributes of any probability measure of a random variable over a sample space. A completely general definition of such a measure, called a *probability set function*, is constructed as follows: given the sample space $\mathscr{A}$ of a random variable X and a function $P(A_i)$ defined for all $A_i \subseteq \mathscr{A}$, then if†

1. $P(\mathscr{A}) = 1$,
2. $P(A_i) \geq 0$, $A_i \subseteq \mathscr{A}$,
3. $P(A_1 \cup A_2 \cup A_3 \cup \cdots)$
   $= P(A_1) + P(A_2) + P(A_3) + \cdots$,

when the A_i are disjoint, $P(A)$ is the probability set function of the random variable. As noted above, $P(A_i)$ is the probability that the

† The meanings of the various set operations used here are probably known to the reader. $A \cup B$ denotes the union of A and B; $A \subseteq B$ means that A is a subset of B. A and B are disjoint sets if they have no elements in common.

outcome of the experiment is a member of the ith subset.

It should be noted that the probabilities in the coin-tossing experiment can be written explicitly as a function of the random variable x. Thus if we let $f(x) = p^x(1-p)^{1-x}$, for all $x \in \mathscr{A} = \{x; x = 0, 1\}$,† we can define the probability set function $P(A_i)$ as $P(A_i) = \sum_{A_i} f(x)$. Then if $p = \frac{1}{2}$, we find the probabilities associated with the above-listed subsets to be

$$P(A_2) = \sum_{A_2} f(x) = p^0(1-p)^{1-0}$$
$$= \left(\frac{1}{2}\right)^0\left(1 - \frac{1}{2}\right) = \frac{1}{2},$$
$$P(A_3) = \sum_{A_3} f(x) = p(1-p)^{1-1}$$
$$= \left(\frac{1}{2}\right)\left(1 - \frac{1}{2}\right)^0 = \frac{1}{2},$$
$$P(A_1) = \sum_{\mathscr{A}} f(x) = p^0(1-p)^{1-0}$$
$$+ p(1-p)^{1-1}$$
$$= \frac{1}{2} + \frac{1}{2} = 1,$$
$$P(A_4) = \sum_{A_4} f(x) = 0,$$

since $\varnothing$ is empty.

If probability is interpreted as relative frequency, the question of whether or not $f(x)$ has been appropriately chosen is an empirical question.

II. Continuous and Discrete Random Variables

It is convenient to distinguish between two broad classes of random variables. We may talk on the one hand of *discrete* random variables and on the other hand of *continuous* random variables. Precise mathematical definitions of these concepts are given in a subsequent starred section. We will supplement those definitions here with a more

† As before, the set notation is probably familiar. Thus, "... for all $x \in \mathscr{A}$" means that all elements of the sample space are under consideration.

intuitive description of the two types of random variables.

Suppose the points of a sample space were plotted along a real line and that the points range over the real line from point a to point d, $a < d$. The sample points would constitute a discrete set if in any finite interval of the line there were only a finite number of sample points that could be placed in correspondence with the points on the line. A discrete random variable would then be defined on this sample space. Since there are points on the line that do not correspond to points in the sample space, these (line) points have zero probability of occurrence. For example, there is some probability $p_1 \geq 0$ that a family will have one child and some probability $p_2 \geq 0$ that a family will have two children. One and one-half children cannot occur, however, and the probability of this event is zero. Number of children is a discrete random variable.

A random variable would be considered continuous, on the other hand, if for every interval on the line, say the interval between k and l, $k < l$, every non-zero interval that is a subset of the points between k and l has a non-zero probability of occurrence. Thus units of time might be a continuous random variable if we suppose that the time measures employed can be expressed to any desired number of decimal points.

The distinction between the two types of random variables has consequences for the mathematics involved. Suppose we wanted to find probabilities, as above, by defining a probability set function in terms of some function $f(x)$ of the random variable X. We can do this provided we are careful in choosing $f(x)$ and in defining the probability set function. It turns out that if $f(x) > 0$ for all points in the sample space and if $\sum_{\mathscr{A}} f(x) = 1$ (that is, if the sum of $f(x)$ over all sample points equals one), then we can define a probability set function for all subsets A_i of the sample space as $P(A_i) = \sum_{A_i} f(x)$.

This type of argument is reasonable enough; it is the approach we took previously in discussing the discrete random variable in the coin-tossing example. But what about random variables that can assume an infinite

number of values? It is difficult to see how one might compute an ordinary summation of the functional values of such a random variable. Actually the problem is easily resolved by making use of some of the techniques of the calculus. For continuous random variables we would choose $f(x) > 0$ and require that the integral over the entire sample space be equal to one—that is, that

$$\int_{\mathscr{A}} f(x)\,dx = 1.$$

We would then define a probability set function for the subset A_i as follows:

$$P(A_i) = \int_{A_i} f(x)\,dx.$$

As far as the mathematics of the present article is concerned, one could view the difference between the two types of random variables in terms of the different mathematical operations required to complete analogous operations on $f(x)$.

** *Continuous and Discrete Random Variables: Definitions.* Suppose that we have a sample space $\mathscr{A}$ of discrete points and a function $f(x)$ which has the following characteristics:

1. $f(x) > 0$ for all $x \in \mathscr{A}$;
2. $\sum_{\mathscr{A}} f(x) = 1$.

If we define a probability set function $P(A_i)$ as $P(A_i) = \sum_{A_i} f(x)$, we would refer to X as a discrete random variable.

Suppose, on the other hand, that we have a sample space $\mathscr{A}$ of a random variable X and a function $f(x)$ of the random variable, and that the following relationships hold:

1. $f(x) > 0$, $x \in \mathscr{A}$,

2. $\int_{\mathscr{A}} f(x)\,dx = 1$,

3. $f(x)$ is continuous except perhaps at a finite number of points in $\mathscr{A}$.

If we define a probability set function $P(A_i)$ as $P(A_i) = \int_{A_i} f(x)\,dx$, we would refer to X as a continuous random variable.

III. Probability Density Functions

In the foregoing section some basic notions were developed concerning random variables. Much of the notation and many of the concepts there developed are cumbersome if employed consistently, but significant notational economies can be effected without loss of meaning. To this we now turn, beginning with a brief statement of what has been accomplished thus far.

A random variable is the outcome of an experiment of a particular type—specifically, one for which the outcome cannot be predicted with certainty but for which all *possible* outcomes can be specified in advance. Different outcomes may be expected on different replications of the experiment. The likelihood of occurrence of any particular outcome is referred to as the probability of observing that outcome. The totality of possible outcomes is called the *sample space*, and a particular outcome of the experiment may be regarded as a subset of the sample space. Since we will always represent the elements or points of the sample space by real numbers, the probability of a given outcome (subset) may be regarded as the probability that the random variable assumes a specified value or range of values.

We said that on a given sample space $\mathscr{A}$ of a random variable x we would impose a measure called the probability set function having the following properties: on the sample space $\mathscr{A}$ a function $P(A_i)$ is defined for $A_i \subseteq \mathscr{A}$ such that

1. $P(\mathscr{A}) = 1$
2. $P(A_i) \geq 0, A_i \subseteq \mathscr{A}$
3. $P(A_1 \cup A_2 \cup A_3 \cdots)$
    $= P(A_1) + P(A_2) + P(A_3) + \cdots$

when the A_i are disjoint. Thus the probability set function was such that to subsets $A_i \subseteq \mathscr{A}$, numbers called probabilities, $P(A_i)$, were made to correspond. These numbers indicated the probability that the random variable X, the outcome of the experiment, took on such a value as to be an element of the ith subset. It was further noted that the probabilities thereby arrived at could be expressed explicitly in terms of a function $f(x)$ of the random variable x.

By way of illustration if we consider the experiment of tossing a coin a single time, it is evident that the sample space consists of two outcomes (elements, points): heads and tails. If we let $X = 0$ denote a head and $X = 1$ denote a tail, then the sample space $\mathscr{A}$ is described in terms of the random variable x as follows:

$$\mathscr{A} = \{x; x = 0, 1\}.$$

If we define a function $f(x) = p^x(1-p)^{1-x}$, $x \in \mathscr{A}$, we can then define a probability set function $P(A_i)$ for this sample space as $P(A_i) = \sum_{A_i} f(x)$.

By this definition of $P(A_i)$, the probabilities associated with the four subsets of the sample space are

for $A_1 = \mathscr{A} = \{x; x = 0, 1\}$,
$$P(A_1) = p + (1 - p),$$
for $A_2 = \{x; x = 0\}$, $P(A_2) = 1 - p$,
for $A_3 = \{x; x = 1\}$, $P(A_3) = p$,
for $A_4 = \varnothing$, $P(A_4) = 0$.

If p is taken to equal $\frac{1}{2}$, the probabilities associated with the four subsets are, respectively, 1, $\frac{1}{2}$, $\frac{1}{2}$, and 0, a result that is certainly intuitively acceptable.

As another example, consider a random variable x on the sample space

$$\mathscr{A} = \{x; 0 < x < \infty\},$$

and a function $f(x) = e^{-x}$, for all $x \in \mathscr{A}$. Now define a probability set function $P(A_i)$, $A_i \subseteq \mathscr{A}$, as

$$P(A_i) = \int_{A_i} f(x)\, dx$$

$$= \int_{A_i} e^{-x}\, dx.$$

Thus, for example, if $A_1 = \{x; 0 < x < 2\}$, then

$$P(A_i) = \int_{A_i} e^{-x}\, dx$$

$$= \int_0^2 e^{-x}\, dx.$$

The first of these illustrations involves a discrete random variable, the second a continuous random variable.

The important thing is that the probability $P(A_i)$ is determined completely by the function $f(x)$, called the *probability density function* (p.d.f.) of the random variable x. We shall see presently how we can deal solely with the p.d.f. of the random variable and obviate any necessity of referring to or specifying the sample space.

As the concept of a probability density function of a random variable is *the* basic notion in this review, a few remarks making it intuitively meaningful are appropriate.

The probability of observing *some* event in the sample space is unity. This probability of one can be considered as being apportioned among the various points or intervals that make up the sample space. A p.d.f. of a random variable is a mathematical description of this apportionment. The p.d.f. details the manner in which the total probability is distributed among the various points (in the case of discrete random variables) or the various intervals between points (in the case of continuous random variables) that constitute the sample space. The necessity for distinguishing between points and intervals of the sample space according to whether the random variable is discrete or continuous will be made clear in the paragraphs that follow. Loosely speaking, then, a p.d.f. is a description of the distribution of the total available probability across the points or intervals of the sample space.

As mentioned above we may simplify our notation and work directly with the p.d.f., avoiding any necessity of discussing the sample space. This simplification is desirable since, as was pointed out, the probability of the occurrence of an event is completely determined by the p.d.f.

Consider a random variable x defined on the sample space $\mathscr{A} = \{x; 0 < x < \infty\}$ which has a p.d.f. $f(x) = e^{-x}$, $x \in \mathscr{A}$. As before the probability of event A_i, $P(A_i)$, where $A_i \subseteq \mathscr{A}$, is $\int_{A_i} f(x)\, dx$.

Suppose in particular that

$$A_i = \{x; 0 < x < 2\}.$$

Then

$$P(A_i) = \int_0^2 f(x)\, dx = \int_0^2 e^{-x}\, dx.$$

This probability remains unchanged if we rewrite the p.d.f. as

$$f(x) = e^{-x}, 0 < x < \infty$$
$$= 0 \text{ elsewhere,}$$

while rewriting $\int_{\mathscr{A}} f(x)\, dx$ as $\int_{-\infty}^{\infty} f(x)\, dx$. That is, if we call

$$f(x) = e^{-x}, 0 < x < \infty$$
$$= 0 \text{ elsewhere}$$

the p.d.f. of the random variable x and agree to consider all real numbers as constituting the sample space, we will not change the distribution of probabilities over the sample space. That is, all $P(A_i)$, $A_i \subseteq \mathscr{A}$, will remain invariant. Note that

$$\int_{-\infty}^{\infty} f(x)\, dx = \int_{-\infty}^{0} f(x)\, dx + \int_{0}^{\infty} f(x)\, dx$$

$$= \int_{-\infty}^{0} 0 \, dx + \int_{0}^{\infty} e^{-x}\, dx$$

$$= \int_{\mathscr{A}} f(x)\, dx$$

by virtue of our revised definition of the probability density function.

The same economy can be accomplished in the discrete case by writing $\sum_x f(x)$ for $\sum_{\mathscr{A}} f(x)$, where the summation on x is over all real values of x and the p.d.f. of x has been extended in a manner analogous to the continuous case discussed above. Specifically, we can replace expressions such as $f(x) = p^x(1-p)^{1-x}$, $x \in \mathscr{A} = \{x; x = 0, 1\}$, by

$$f(x) = p^x(1 - p)^{1-x}, x = 0, 1$$
$$= 0 \text{ elsewhere}$$

while at the same time changing summations as previously indicated. In other words, we can eliminate explicit reference to the sample space by considering the sample space to be the set of all real numbers provided that the p.d.f. of the random variable is suitably defined on this space.

Over and above this economy of notation, there is an important generalization to be obtained from the above discussion. *Any function of a real variable x is a p.d.f. if*

1. $f(x) \geq 0$ for all x

2. $\int_{-\infty}^{\infty} f(x)\, dx = 1$ or $\sum_x f(x) = 1$

in the continuous or discrete case, respectively.

Finally, consider the probability of occurrence of a specific point in the continuous case. Suppose that we ask for the probability of occurrence of the point a where a is a real number—that is, where it is in the sample space. To answer this question we might begin by asking for the probability of observing an event *between a and b, a < b*. If $A_i = \{x; a < x < b\}$, we see from the above that $P(A_i) = \int_{A_i} f(x)\, dx = \int_a^b f(x)\, dx$ when $f(x)$ is the p.d.f. of the random variable x. Further, if $A_i = \{x; x = a\}$, then $P(A_i) = \int_a^a f(x)\, dx = 0$. It is for this reason that we speak of the probabilities of *intervals* in the continuous case.

** *Extension of the p.d.f. to More Than One Random Variable.* The idea of a p.d.f. of a random variable may be easily extended to include the p.d.f. of two or more random variables. Such an extension is often necessary since two or more random variables may be defined on a sample space. As an illustration, consider the experiment of simultaneously tossing a coin and rolling a die. The outcome of such an experiment may be described by stating which face of the coin and which face of the die appeared on a given occasion. If we let the random variable $X = 0$ denote a head and $X = 1$ denote a tail while we let the random variable $Y = 1, 2, 3, 4, 5,$ or 6 denote the appearance of the possible faces of the die, the sample space $\mathscr{A}$ of this experiment may be described as a set of ordered pairs:

$$A = \{(x, y); x = 0, 1; y = 1, 2, 3, 4, 5, 6\}.$$

There are twelve points in this sample space. If a p.d.f. $f(x, y)$ of the two random variables is defined, then by straightforward extension of our previous discussion we can write probabilities for this *two-dimensional* sample space. Thus for a given $A \subseteq \mathscr{A}$ we would define a probability set function by writing

$$P(A) = P[(x, y) \in A] = \sum \sum_A f(x, y).$$

If the two random variables were continuous we would write

$$P(A) = P[(x, y) \in A] = \int_A \int f(x, y)\, dx\, dy.$$

Finally, as might be expected, we can again eliminate reference to the sample space by agreeing to take the entire (x, y) plane as the sample space. Thus we would substitute

$$\sum_y \sum_x f(x, y) \quad \text{for} \quad \sum \sum_A f(x, y)$$

and

$$\int_{-\infty}^{\infty} \int_{-\infty}^{\infty} f(x, y)\, dx\, dy \quad \text{for} \quad \int_A \int f(x, y)\, dx\, dy$$

in the discrete and continuous case, respectively. A similar extension may be made in the case of n random variables having p.d.f. $f(x_1, x_2, \ldots, x_n)$.

IV. Some Important Properties of the Probability Density Function

The determination of the p.d.f. of a random variable is of considerable importance to the statistician, as it allows him to make certain statements concerning probabilities and from these draw useful inferences about the random variable under consideration. Subsequent sections will detail a number of such procedures. First we shall discuss some additional useful concepts related to any p.d.f.

Mathematical Expectation. Given a random variable x having p.d.f. $f(x)$, consider any function $g(x)$ of the random variable. The sum $\sum_x g(x) f(x)$ or the integral $\int_{-\infty}^{\infty} g(x) f(x)\, dx$ over the sample space is termed the *mathematical expectation* of $g(x)$. We will denote the mathematical expectation of $g(x)$ as $E[g(x)]$. Thus by definition

$$E[g(x)] = \int_{-\infty}^{\infty} g(x) f(x)\, dx$$

or

$$= \sum_x g(x)f(x)$$

in the continuous or discrete case, respectively.†

Mathematical expectation is an important concept, and certain specific functions $g(x)$ provide important detailed information about the random variable x. Mathematical expectation has some important properties, independent of any particular p.d.f., which are listed without proof. The proofs are simple and may be constructed by the reader from the definition of mathematical expectation.

1. If $g(x) = c$, a constant, then $E[g(x)] = E[c] = c$.
2. If c is constant and g is a function of x, then $E[cg] = cE[g]$.
3. $E[c_1 g_1 + c_2 g_2 + \cdots + c_n g_n]$
   $= c_1 E[g_1] + c_2 E[g_2] + \cdots + c_n E[g_n]$.

The last fact, that the mathematical expectation of a sum is equal to the sum of the mathematical expectations, proves to be quite useful in a variety of applications.

We shall now consider four specific mathematical expectations—that is, four specific functions $g(x)$ that are of primary importance in statistics. To begin with, let $g(x) = x$. Then

$$E[g(x)] = E[x]$$

$$= \sum_x xf(x) \text{ in the discrete case}$$

$$= \int_{-\infty}^{\infty} xf(x)\,dx \text{ in the continuous case.}$$

This particular expectation is called the *expected value* of the random variable x and is usually denoted μ. That is, $\mu = E[x]$. Readers in the behavioral sciences will recognize μ as the usual symbol for a popula-

tion mean. It is, in fact, the mean of the p.d.f. of the random variable X. Thus the expected value of x is a kind of weighted average. Each value of x in the sample space is weighted by the probability of occurrence (or the probability density) of that value of the random variable.

A second important mathematical expectation is arrived at if we take $g(x) = (x - \mu)^2$. Then

$$E[g(x)] = E[(x - \mu)^2]$$

$$= \int_{-\infty}^{\infty} (x - \mu)^2 f(x)\,dx$$

or

$$= \sum_x (x - \mu)^2 f(x)$$

in the continuous or discrete case, respectively. This expectation can also be viewed as a weighted average. Here the weighting is carried out on the squared deviation of the ith sample point from the mean or expected value of the random variable. This expectation is called the variance and is denoted σ^2. Thus $\sigma^2 = E[(x - \mu)^2]$.

A relationship exists between μ and σ^2 that should be pointed out. First, note that

$$(x - \mu)^2 = x^2 - 2x\mu + \mu^2$$

and

$$\begin{aligned}
\sigma^2 &= E[(x - \mu)^2] \\
&= E[x^2 - 2x\mu + \mu^2] \\
&= E[x^2] - 2\mu E[x] + E[\mu^2] \\
&= E[x^2] - 2\mu^2 + \mu^2
\end{aligned}$$

by the properties of expectation previously outlined. Hence

$$\sigma^2 = E[x^2] - \mu^2$$

or

$$E[x^2] = \sigma^2 + \mu^2.$$

The latter relationship is useful since it is frequently necessary to find the mathematical expectation of the *square* of a random variable. The relationship offers, in addition, an alternative and sometimes easier method of computing σ^2.

---

† Again, this definition may be extended directly to a function $g(x_1, x_2, \ldots, x_n)$ of n random variables. In the continuous case, for example, the expected value of the function $g(x_1, x_2, \ldots, x_n)$ of the random variables $x_1, x_2, \ldots, x_n$ having p.d.f. $f(x_1, x_2, \ldots, x_n)$ is defined as $E[g(x_1, x_2, \ldots, x_n)]$

$$= \int_{-\infty}^{\infty} \cdots \int_{-\infty}^{\infty} \frac{g(x_1, x_2, \ldots, x_n)f(x_1, x_2, \ldots, x_n)}{dx_1, dx_2, \ldots, dx_n.}$$

** *Moment Generating Functions and Characteristic Functions.* A third important mathematical expectation is the *moment generating function* of a distribution. Consider the mathematical expectation of the expression e^{tx}. The expectation of this function of the random variable x is called the moment generating function of x and is denoted $M_X(t)$. That is,

$$M_X(t) = E[e^{tx}] = \int_{-\infty}^{\infty} e^{tx}f(x)dx \text{ or } \sum_x e^{tx}f(x).$$

The moment generating function is important first because of its uniqueness. That is, the distribution of a random variable is completely determined by its moment generating function. In other words, if two random variables have the same moment generating function, they necessarily have the same distribution. The advantage of this relationship becomes apparent when one is unable to determine without difficulty the p.d.f. of a random variable. It may, however, be easy to determine in this circumstance the moment generating function of the random variable. Having done this, one might observe that this moment generating function is the same as the moment generating function of another random variable whose p.d.f. is known. By virtue of the uniqueness cited above, the unknown p.d.f. may be written down immediately.

Another application of the moment generating function is at times of considerable use. To see this we take successive derivatives of $M_X(t)$ as follows:

$$\frac{d}{dt} M_X(t) = M_X'(t) = \frac{d}{dt} \int e^{tx} f(x)\, dx$$

or

$$= \frac{d}{dt} \sum_x e^{tx}f(x)$$

as is appropriate. Now, using the continuous case as an example,

$$M_X'(t) = \frac{d}{dt}\int_{-\infty}^{\infty} e^{tx}f(x)\, dx = \int_{-\infty}^{\infty} xe^{tx}f(x)\, dx$$

and

$$M_X''(t) = \frac{d^2}{dt^2}\int_{-\infty}^{\infty} e^{tx}f(x)\, dx = \int_{-\infty}^{\infty} x^2 e^{tx}f(x)\, dx$$

and so on. If we evaluate expressions like these at $t = 0$ we have

$$M_X'(0) = \int_{-\infty}^{\infty} xf(x)\, dx = E(x)$$

and

$$M_X''(0) = \int_{-\infty}^{\infty} x^2 f(x)\, dx = E(x^2)$$

and so on. In general if $M_X^{(k)}(t)$ denotes the kth derivative with respect to t of $M_X(t)$, then $M_X^{(k)}(0) = E[x^k]$. Expressions like $E[x^k]$, evaluated as integrals or sums as is appropriate, are called raw moments of the distribution. Hence the name "moment generating function" for $M_X(t)$, which does in fact readily yield these moments. Incidentally, $M_X(t)$ often offers a convenient means of calculating σ^2. Thus,

$$\sigma^2 = E[(x - \mu)^2] = E(x^2) - \mu^2$$

as was pointed out previously. But

$$E(x^2) = M_X^{(2)}(0) \quad \text{and} \quad \mu = M_X'(0).$$

Hence

$$\sigma^2 = M_X^{(2)}(0) - [M_X'(0)]^2.$$

The moment generating function is of great importance in statistics. However, the unfortunate fact that it may not always exist is a drawback. That is, $\int_{-\infty}^{\infty} e^{tx}f(x)\, dx$ and $\sum_x e^{tx}f(x)$ may not exist. Another function of the random variable, however, always exists and may be seen to have properties quite analogous to the moment generating function. This is called the *characteristic function* and is defined as

$$\varnothing_X(t) = E[e^{ixt}]$$

$$= \int_{-\infty}^{\infty} e^{ixt}f(x)\, dx$$

$$= \sum_x e^{ixt}f(x)$$

in the continuous or discrete case, respectively. Like the moment generating function, the characteristic function is unique. The technique of generating moments may likewise be generalized to the characteristic function. In the discrete case, for example,

$$\emptyset_X(t) = \sum_x e^{ixt} f(x),$$

$$\emptyset'_X(t) = \sum_x ixe^{ixt} f(x),$$

$$\emptyset''_X(t) = \sum_x (ix)^2 e^{ixt} f(x),$$

$$\vdots$$

etc.

and

$$\emptyset'_X(0) = i \sum_x xf(x) = iE[x],$$

$$\emptyset''_X(0) = i^2 \sum_x x^2 f(x) = -E[x^2],$$

and so on.

From the above it is seen that

$$E[x] = \frac{\emptyset'_X(0)}{i},$$

$$E[x^2] = \frac{\emptyset''_X(0)}{i^2},$$

and in general it is the case that

$$E[x^k] = \frac{\emptyset^{(k)}_X(0)}{i^k}.$$

V. Distribution Functions

We consider finally one additional way of discussing probabilities associated with random variables defined on a sample space. Suppose that we have a random variable X and we wish to represent the probability that the random variable assumes a value equal to or less than some given value x. Our representation would take the form $P(X \le x)$ which, as before, is read, "the probability that X is equal to or less than x." Now suppose we shorten our notation by writing $F(x)$ for $P(X \le x)$. That is, let $F(x) = P(X \le x)$.

Now $F(x)$ is a notation in functional form. Brief reflection will be sufficient to support the conclusion that this is justified, because the probability that X is equal to or less than x is certainly a function of x. To begin with, $F(x) = P(X \le x)$ may be considered for all x in the sample space. Secondly, the points of the sample space may be represented by points on the real line. Finally, the total available probability is distributed among these points (in the discrete case) or intervals (in the continuous case), and the probability that X does not exceed some value x on the line is a function of the value of x chosen.

Moreover, if $F(x)$ is computed as x ranges from its most negative to its most positive value (in some instances $-\infty < x < \infty$), $F(x)$ should not decrease. In other words, as x increases in this fashion, $F(x)$ accumulates the probabilities associated with points in the sample space equal to or less than x.

$F(x)$ is called the *distribution function* of the random variable X. As with the p.d.f. it is possible to describe the general characteristics that any distribution function must have. As noted above, $F(x)$ should be a non-decreasing function of x. In addition, there will be some points on the line more negative than the most negative point in the sample space (even if this occurs at $-\infty$) and some points more positive than the most positive point in the sample space (even if this occurs at $+\infty$). In the former case $F(x) = 0$ and in the latter case $F(x) = 1$. Finally, at any point in the sample space $F(x)$ will be continuous to the right.† Thus any function is a distribution function if $F(x)$ is a non-decreasing function of x, $F(-\infty) = 0$, $F(\infty) = 1$, and $F(x)$ is continuous to the right.

With each experiment whose outcome is a random variable X, we wish to associate a distribution function $F(x)$ having the above properties so that $P(X \le x) = F(x)$.

---

† If the reader cannot attach mathematical meaning to the idea of a function's being continuous at a *point*, it may be of some use to say that in non-mathematical terms a function is continuous on an *interval* if the function is "unbroken" on that interval.

Now if $x_1 < x_2$ it is also the case that

$$P(x_1 < X \le x_2) = F(x_2) - F(x_1).$$

This follows from the definition of $F(x)$ and the additivity of probabilities associated with disjoint subsets of the sample space. That is, the set of points

$$\{x; X \le x_2\} = \{x; X \le x_1\} \cup \{x; x_1 < X \le x_2\}$$

and the sets on the right-hand side are disjoint. Thus

$$P(X \le x_2) = P(X \le x_1) + P(x_1 < X \le x_2)$$

or

$$F(x_2) = F(x_1) + P(x_1 < X \le x_2)$$

or

$$P(x_1 < X \le x_2) = F(x_2) - F(x_1).$$

We have previously seen that if X is continuous and has p.d.f. $f(x)$, then

$$P(x_1 < X < x_2) = \int_{x_1}^{x_2} f(t)\, dt.$$

Since $P(X < x_2) = P(X \le x_2)$, this suggests that there is a relationship between $F(x)$ and $f(x)$. In other words,

$$P(x_1 < X < x_2) = F(x_2) - F(x_1) = \int_{x_1}^{x_2} f(t)\, dt.$$

In completely general terms, in the continuous case

$$F(x) = \int_{-\infty}^{x} f(t)\, dt$$

and

$$\frac{dF(x)}{dx} = f(x).$$

If X is a discrete random variable the relation between $F(x)$ and $f(x)$ is simply that

$$F(x_i) = P(X \le x_i) = \sum_{x \le x_i} f(x).$$

VI. Some Important Probability Density Functions

Workers in the behavioral sciences are accustomed to making and hearing statements like, "Intelligence is normally distributed" or "We assume that the scores on the Ajax Test of Everything are normally distributed." The researcher generally has in mind some *population* of scores from which he has taken or intends to take a random sample. He assumes that the distribution of the measure (Intelligence, Ajax Test) in the population shows certain properties which he subsumes under the heading *normal*. In these cases there is a direct correspondence between his notion of *population* and our concept of a p.d.f. The researcher's statements would be, in our terminology, "Intelligence is a random variable having a normal p.d.f." or "Scores on the Ajax Test of Everything constitute a normal probability density function." The researcher never actually observes the entire population. Indeed, this is why he samples from it. His statements really are *assumptions* regarding the form of the population distribution of scores or, equivalently, regarding the p.d.f. of the random variable under consideration. Statistical inferences regarding the population are made following such assumptions. We will look at examples of such inferences later. We will first discuss a number of different p.d.f.'s of random variables, in particular those of most utility to the behavioral scientist. A mathematical definition of each p.d.f. is given and some important properties of the p.d.f. are listed. We shall, finally, point to relationships that exist among these probability density functions.

Bernoulli Trials and the Binomial Distribution. A *Bernoulli Trial* may be described as a one-shot random experiment in which there are two possible outcomes. One of the outcomes is generally termed "success" and the other is generally termed "failure." Suppose we let the random variable $X = 0$ denote a failure and $X = 1$ denote a success. If now p is the probability of success and $q = 1 - p$ is the probability of failure, these probabilities can be written as a function of the random variable x. Thus, let

$$f(x) = p^x q^{1-x}, \qquad x = 0, 1$$
$$= 0 \text{ elsewhere.}$$

The mean μ and variance σ^2 of the random variable (X is a discrete random variable) are

$$\mu = E[x] = \sum_x xf(x) = p$$

and

$$\sigma^2 = E[(x - \mu)^2] = \sum_x (x - \mu)^2 f(x) = pq.$$

The moment generating function of the random variable is

$$E[e^{tx}] = \sum_x e^{tx}f(x) = pe^t + q.$$

A *binomial* random variable may be regarded as the succession of n independent Bernoulli trials. That is, one may consider the p.d.f. of a random variable x, $x = 0, 1, 2, \ldots, n$, which represents the occurrence of x successes out of n independently conducted Bernoulli trials. The p.d.f. of x is given by

$$f(x) = \binom{n}{x} p^x q^{n-x}, \qquad x = 0, 1, 2, \ldots, n$$

$$= 0 \text{ elsewhere,}$$

where $q = 1 - p$. The name of the p.d.f. derives from the observation that

$$\binom{n}{x} p^x q^{n-x} = \frac{n!}{x!(n-x)!} p^x q^{n-x},$$

$$x = 0, 1, 2, \ldots, n$$

yields for successive values of x the successive terms in the binomial expansion $(q + p)^n$. That $f(x)$ is indeed a p.d.f. is seen by observing that

$$f(x) > 0 \quad \text{and} \quad \sum_x f(x) = 1.$$

The latter equality follows immediately as a consequence of the fact that $p + q = 1$ and from the fact that

$$\sum_x \binom{n}{x} p^x q^{n-x} = (q + p)^n.$$

The mean μ and variance σ^2 of the binomial p.d.f. are

$$\mu = E[x] = np$$
$$\sigma^2 = E[(x - \mu)^2] = npq.$$

The defining characteristics of a binomial experiment may be summarized as follows:

1. There is a finite number of n trials.
2. The probability of success, p, is constant over the n trials.
3. The trials are independent. That is, the outcome of one trial does not influence the outcome of any other trial.

A binomial random variable offers an easy way to illustrate the use of the p.d.f. in computing probabilities. Suppose the binomial experiment under consideration consists of tossing a fair coin n times. Let $X = 0$ denote the occurrence of a tail on a given trial and $X = 1$ denote the occurrence of a head. If there are n independent trials, the number of heads is a random variable X where $x = 0, 1, 2, \ldots, n$. If p is the probability of a head on any one trial, the p.d.f. of X is as given above; that is,

$$f(x) = \binom{n}{x} p^x q^{n-x}, \qquad x = 0, 1, 2, \ldots, n$$

$$= 0 \text{ elsewhere.}$$

Given this information it is a simple matter to compute probabilities such as the following:

$$P(X \leq k) = \sum_{x=0}^{x=k} f(x).$$

The Normal Distribution. Perhaps the most frequently made assumption about the random variables of concern to the behavioral scientist is that they are normally distributed. A normally distributed random variable X is continuous and has p.d.f. $f(x)$ defined as follows:

$$f(x) = \frac{1}{\sigma\sqrt{2\pi}} e^{[-(x-\mu)^2/2\sigma^2]}, \qquad -\infty < x < \infty$$

$$= 0 \text{ elsewhere.}$$

From our earlier discussion it follows that if x is a normally distributed random variable

$$P(a < x < b) = \int_a^b f(x)dx =$$

$$\int_a^b \frac{1}{\sigma\sqrt{2\pi}} e^{[-(x-\mu)^2/2\sigma^2]} \, dx$$

and

$$P(X \leq c) = \int_{-\infty}^{c} f(x)\, dx = F(c)$$

where $F(x)$ is the distribution function of X as previously defined.

It will be noted that there are two parameters, μ and σ^2, in the normal p.d.f. These parameters are called the mean and variance of the random variable. Different values of μ and/or σ^2 will result in p.d.f.'s of the general form given above but differing in their mean values and/or the dispersion of the random variable about the mean. In other words, the normal p.d.f. is actually a *family* of p.d.f.'s, the particular member of the family being determined by a specification of values for μ and σ^2.

The mean, variance, and moment generating function of a normally distributed random variable are defined in the usual way. The results are

$$E[x] = \mu$$

$$E[(x - \mu)^2] = \sigma^2$$

$$M_X(t) = e^{\mu t} + \frac{\sigma^2 t^2}{2}.$$

Now, it happens that integrals of the form

$$\int_{-\infty}^{x} \frac{1}{\sigma\sqrt{2\pi}} e^{[-(t-\mu)^2/2\sigma^2]}\, dt$$

are not easily computed. This would suggest that it would be desirable to compute a table of such values once and for all. Since, however, a different table would be needed for each of the many p.d.f.'s in the family, such tables would not seem to save any effort beyond that necessary to evaluate such integrals whenever they appeared. There is a way out of this apparent impasse. Only *one* such table is necessary: the table that results from letting $\mu = 0$ and $\sigma^2 = 1$. All other normal p.d.f.'s are easily referred to this one table. For emphasis, the p.d.f. of such a random variable is

$$f(x) = \frac{1}{\sqrt{2\pi}} e^{(-x^2/2)}, \quad -\infty < x < \infty$$

$$= 0 \text{ elsewhere.}$$

Suppose we know that a random variable X is normally distributed with mean μ and variance σ^2. For brevity we will denote this fact by writing X is $n(x; \mu, \sigma^2)$. In particular, a random variable which is $n(x; 50, 2.5)$ is a random variable x having mean equal to 50 and variance equal to 2.5.

Now suppose we are given a random variable x which is $n(x; \mu, \sigma^2)$. Consider forming a new random variable z as follows:

$$z = \frac{x - \mu}{\sigma}.$$

It is easily shown that z is also a normally distributed random variable having mean $\mu = 0$ and variance $\sigma^2 = 1$. That is, z is $n(z; 0, 1)$. This random variable z is referred to as a standardized normal random variable. Its distribution function has been tabulated and appears in virtually every statistical textbook in the behavioral sciences. Note that this result—that the p.d.f. of z is $n(z; 0, 1)$— means that probabilities concerning any normally distributed random variable are readily calculated by changing the random variable into a z and referring to the tabled distribution function for z. Thus only one normal curve table is sufficient to calculate probabilities for any of the family of normally distributed random variables.

To illustrate, suppose that X is $n(x; 50, 9)$ and we wish to compute $P(X \leq 60)$. Now $X \leq 60$ occurs when and only when

$$(X - 50)/3 \leq (60 - 50)/3.$$

Thus we have

$$P(X \leq 60) = P\left(\frac{X - 50}{3} \leq \frac{10}{3}\right),$$

but $(X - 50)/3$ is in fact a z random variable, and so $P(X \leq 60) = P(z \leq 3.33)$. The last expression is evaluated by reference to the tabled values of the distribution function of z. Similar calculations are carried out to compute probabilities such as $P(a < X < b)$, where a and b are constants and X is $n(x; \mu, \sigma^2)$. Perhaps it is well to emphasize at this point that the tabled values of z are in fact the functional values of the distribution function of z. That is, the tabled values represent calculations for a number of points, z, of

$$F(z) = \int_{-\infty}^{z} \frac{1}{\sqrt{2\pi}} e^{(-t^2/2)} \, dt.$$

Chi-Square Random Variables. A number of other important distributions of random variables are derived from, or are intimately related to, the normal distribution. A chi-square random variable with one degree of freedom may be defined as the square of a normally distributed random variable having mean $\mu = 0$ and variance $\sigma^2 = 1$. In particular, if z is $n(z; 0, 1)$, then $z^2 = \chi^2_{(1)}$ where the right-hand side of the equation is read as "chi-square with one degree of freedom." More generally, if the random variable X is $n(x; \mu, \sigma^2)$, then $[(x - \mu)/\sigma]^2$ has a chi-square distribution with one degree of freedom. In other words $[(x - \mu)/\sigma]^2 = \chi^2_{(1)}$. What this means is that if X is $n(x; \mu, \sigma^2)$ and if we were able to consider each value of X separately and compute $[(x - \mu)/\sigma]^2$ and then plot the results, the resultant function would be $\chi^2_{(1)}$. Suppose now that we considered random *pairs* (X_1, X_2) of values from $n(x; \mu, \sigma^2)$ and computed $[(x_1 - \mu)/\sigma]^2 + [(x_2 - \mu)/\sigma]^2 = z_1^2 + z_2^2$. If these results are plotted we would obtain a chi-square distribution with two degrees of freedom. That is,

$$z_1^2 + z_2^2 = \chi^2_{(2)}.$$

In general, if one random variable is $\chi^2_{(v_1)}$ and a second *independent* random variable is $\chi^2_{(v_2)}$, then the sum $Y = \chi^2_{(v_1)} + \chi^2_{(v_2)}$ of these two random variables is a chi-square random variable having $v_1 + v_2$ degrees of freedom. Thus Y is $\chi^2_{(v_1 + v_2)}$.

Chi-square random variables therefore constitute a family of random variables, the particular member of the family being determined by the number of degrees of freedom.

We will not write the rather complex p.d.f. of a chi-square random variable, which involves a

Gamma function. Note, however, that if the p.d.f. $f(x)$ of a random variable is $\chi^2_{(v)}$, probabilities are computed in the usual way. Thus $P(X \le x) = \int_0^x f(x) \, dx$ where $f(x)$ is $\chi^2_{(v)}$. The lower limit of integration is zero since chi-square (the ratio of two squared numbers) cannot be negative. Selected values of the above integral are tabled for various degrees of freedom.

The mean, variance, moment generating functions, et cetera are defined in the usual way for random variables having chi-square distributions. We record that for a chi-square random variable having v degrees of freedom,

$$\mu = v$$
$$\sigma^2 = 2v.$$

t and F Distributions. Consider now a random variable X that is $n(x; 0, 1)$ and another independent random variable Y that is chi-square with v degrees of freedom. The random variable

$$W = \frac{X}{\sqrt{Y/v}}$$

has a p.d.f. that is called a t distribution. That is, a random variable has a t distribution if it is the ratio of a normally distributed random variable with mean $\mu = 0$ and variance $\sigma^2 = 1$ to the square root of a chi-square random variable divided by its degrees of freedom. Random variables distributed as t constitute a family of p.d.f.'s, specific members of which are determined by the degrees of freedom, v, associated with the variable having the chi-square distribution. Probabilities such as $P(T \le t)$ are found, as usual, from tabled values of the distribution function $F(t)$. For a t distribution with v degrees of freedom, $\mu = 0$ and $\sigma^2 = v/(v - 2)$.

A final random variable of interest is the ratio of two independent chi-square random variables, each divided by its degrees of freedom. Such a random variable is called an F random variable having v_1 and v_2 degrees of freedom; v_1 and v_2 are the degrees of freedom associated with the numerator and denominator chi-square variables, respectively. Thus if Y is $\chi^2_{(v_1)}$ and U is $\chi^2_{(v_2)}$,

$$F_{(v_1, v_2)} = \frac{Y/v_1}{U/v_2}.$$

Again the distribution function of F is tabled for various combinations of v_1 and v_2, enabling one to compute probabilities such as $P(F \leq f)$.

**VII. Joint and Marginal Probability Density Functions and Statistical Independence

Until now we have concentrated on distributions of a single random variable. We have established the notion of a p.d.f. and of a distribution function and have seen how these may be made to yield probability statements concerning various values that the random variable might assume. We have indicated in passing how the notion of a p.d.f. can be extended to encompass the probability density function of n random variables. This type of extension is necessary in many applications and in establishing a useful definition of an important concept, namely the statistical independence of two or more random variables. We will treat briefly two concepts that grow out of considering the p.d.f. of two or more random variables and then use these concepts to define the statistical independence of such random variables. The two concepts are the *joint probability density function* of two or more random variables and the *marginal probability density function* of a random variable.

Consider two random variables X_1 and X_2 having p.d.f. $f(x_1, x_2)$. Let us agree to call such a p.d.f. of two (or more) random variables the *joint* p.d.f. of these random variables. Now suppose we were concerned with finding the probability that X_1 should assume a value between a and b, $a < b$. That is, we wish to give a value to $P(a < X_1 < b)$. Now $a < X_1 < b$ occurs only when X_2 is some real number—that is, only when $-\infty < X_2 < \infty$. In other words $a < X_1 < b$ only when $a < X_1 < b$ and $-\infty < X_2 < \infty$. Hence,

$$P(a < X_1 < b)$$
$$= P(a < X_1 < b, -\infty < X_2 < \infty).$$

We know how to compute the right-hand side of the above equation. Thus

$$P(a < X_1 < b, -\infty < X_2 < \infty) =$$
$$\sum_{a < x_1 < b} \sum_{x_2} f(x_1, x_2)$$

$$= \int_a^b \int_{-\infty}^{\infty} f(x_1, x_2)\, dx_2\, dx_1$$

in the discrete case and continuous case, respectively. If we carry out the summation or integration with respect to X_2 in the above expressions we are left with a function of x_1 alone. That is,

$$\sum_{x_2} f(x_1, x_2) = f_1(x_1)$$

and

$$\int_{-\infty}^{\infty} f(x_1, x_2)\, dx_2 = f_1(x_1)$$

where f_1 is meant to emphasize the fact that f is a function of x_1 and not of X_1 and X_2. We thus reach the conclusion that

$$P(a < X_1 < b) = \sum_{a < x_1 < b} f_1(x_1)$$

or

$$= \int_a^b f_1(x_1)\, dx_1,$$

depending on whether the variable is discrete or continuous, respectively. In other words, $f_1(x_1)$ is the p.d.f. of X_1 alone. It is called the *marginal* p.d.f. of X_1. By reversing the role of the variables in the above discussion we find the marginal p.d.f. of X_2 to be

$$f_2(x_2) = \sum_{x_1} f(x_1, x_2)$$

$$= \int_{-\infty}^{\infty} f(x_1, x_2)\, dx_1$$

in the discrete and continuous cases, respectively.

It may sometimes be the case that $f(x_1, x_2) = f_1(x_1)f_2(x_2)$; that is, the joint p.d.f. is equal to the product of the marginal p.d.f.'s. This leads to a definition of the statistical independence of random variables. Thus two random variables are statistically *independent* if and only if $f(x_1, x_2) = f_1(x_1)f_2(x_2)$ for all points (x_1, x_2).

This discussion concerning joint and marginal p.d.f.'s and statistical independence is readily generalized to the case of n random variables. In the n variable case, the joint p.d.f. is defined in the natural way and $f_i(x_i)$, the marginal p.d.f. of x_i, is obtained by summing or integrating on the $n - 1$ indices of summation or integration other than the ith index. The n random variables are *mutually* statistically independent if the product of the n marginal p.d.f.'s equals the joint probability density function.

The fact of statistical independence, it can be shown, leads to the following useful results. If X_1 and X_2 are statistically independent,

$$E[g(x_1)h(x_2)] = E[g(x_1)]E[h(x_2)]$$

and

$$P(a < X_1 < b, c < X_2 < d)$$
$$= P(a < X_1 < b)P(c < X_2 < d).$$

Conditional Distributions. We shall examine briefly one additional kind of distribution of a random variable. Suppose that we have two random variables x_1 and x_2 having joint p.d.f. $f(x_1, x_2)$ and marginal p.d.f.'s $f_1(x_1)$ and $f_2(x_2)$. It is reasonable to consider the manner in which probability statements concerning x_1 would be affected if we knew that x_2 had assumed some fixed value. That is, we might want to write a probability statement such as

$$P(a < x_1 < b \,|\, X_2 = x_2) = p.$$

This is read, "The probability is p that x_1 is between a and b given that $X_2 = x_2$." Such a statement is called a *conditional probability*.

We shall now define a *conditional probability density function*. Let

$$f(x_1 \,|\, x_2) = \frac{f(x_1, x_2)}{f_2(x_2)}.$$

Then $f(x_1 \,|\, x_2)$ is called the conditional p.d.f. of x_1 given that $X_2 = x_2$. It is easily seen that $f(x_1 \,|\, x_2)$ is a p.d.f. since $f(x_1 \,|\, x_2) > 0$ and

$$\int_{-\infty}^{\infty} f(x_1 \,|\, x_2)\, dx_1 = \frac{1}{f_2(x_2)} \int_{-\infty}^{\infty} f(x_1, x_2)\, dx_1$$
$$= \frac{f_2(x_2)}{f_2(x_2)}$$
$$= 1.$$

The conditional p.d.f. of x_2, $f(x_2 \,|\, x_1)$, would be defined in an analogous fashion.

Since $f(x_1 \,|\, x_2)$ is a p.d.f., conditional probabilities are defined in the expected fashion. Thus

$$P(a < x_1 < b \,|\, X_2 = x_2) = \int_a^b f(x_1 \,|\, x_2)\, dx_1.$$

Finally, the concept of conditional probability can be made to yield some insight into the meaning of statistical independence of random variables. Suppose that the knowledge of the fixed value of x_2 did not yield any information concerning probabilities associated with x_1. This would imply that conditional probabilities concerning x_1, given $X_2 = x_2$, were equal to the probabilities given by the marginal p.d.f. of x_1 or that $f(x_1 \,|\, x_2) = f_1(x_1)$. However since $f(x_1 \,|\, x_2) = \dfrac{f(x_1, x_2)}{f_2(x_2)}$, the above reasoning leads us to conclude that $f(x_1, x_2) = f_1(x_1)f_2(x_2)$.

To say that a knowledge of the value of x_2 does not change probability statements concerning x_1 is to say that x_1 and x_2 are "unrelated." Thus, we have outlined the reason behind our previous definition of statistical independence.

VIII. Linear Functions of Statistically Independent Random Variables

We enumerate here a number of very useful results concerning statistically independent random variables. If $X_1, X_2, \ldots,$ X_n are mutually statistically independent random variables, we can define a new random variable

$$Y = c_1 X_1 + c_2 X_2 + \cdots + c_n X_n,$$

c_i constant. Although the p.d.f. of Y may not be readily determined, the following may be shown to be true:

$$\mu_y = \sum_{i=1}^{n} c_i \mu_i$$

$$\sigma_y^2 = \sum_{i=1}^{n} c_i^2 \sigma_i^2$$

where μ_i and σ_i^2 represent the mean and variance of X_i, respectively.

In particular, if the X_i have the *same* p.d.f., $\mu_i = \mu$ and $\sigma_i^2 = \sigma^2$. Therefore

$$\mu_y = \mu \sum_{i=1}^{n} c_i$$

and

$$\sigma_y^2 = \sigma^2 \sum_{i=1}^{n} c_i^2.$$

IX. *Some Examples of Statistical Inference*

In this section we look briefly at some of the ways in which our statistical theory is put to work in the process of making *inferences*. The two major types of statistical inference are estimation of parameters and hypothesis testing. We shall take a brief look at both. First, however, what is meant by statistical inference? There are many ways of answering this question, some of which are already familiar to the reader, but we shall attempt to give a mathematical answer that is consistent with our development up to this point.

Consider the distribution function $F(x)$ of a random variable X. To say that we want to make an *inference* about X implies that we either cannot or do not wish to examine empirically all possible outcomes of a random experiment. Indeed, if we examined all cases there would be no need for an inferential process. Rather we content ourselves with a random sample of the variable X and attempt to draw certain conclusions (inferences) concerning the entire range of values, including those (the bulk of the cases) that we have not examined. Another way of stating this is that we take a sample of, say, n cases of the random variable X. We can observe, *in this sample*, $P(X \leq x) = F_n(X)$, where $F_n(X)$ may be regarded as the *proportion* of sample values satisfying the equality. It seems natural to call $F_n(X)$ the *sample distribution function*. We would like to use the sample values, in particular $F_n(X)$, to conclude that an *hypothesized* distribution function $F(x)$ of the random variable X is or is not a "good" model of the distribution of all instances of the random variable X. The particular way in which such an inference is made will vary with the problem under consideration. However, in general, a large deviation between $F_n(X)$ and $F(x)$ will result in the rejection of $F(x)$ as a "reasonable" model of the distribution of the values of X. In other words, although specific inferences may differ in form, the major inferential problem resolves itself into the test of making reasonable conclusions concerning $F(x)$ on the basis of an observed $F_n(X)$.

We may use still another way to describe the inferential task. On the observations in the sample of the random variable we may compute various numerical quantities. Such quantities—that is, numbers computed from a finite sample of observations of a random variable—are called *statistics*. Some familiar statistics are the sample mean $\bar{X}$, the sample variance S^2, and so on. One type of inferential task lies in using these known *statistics* to make probabilistic statements concerning the value of related *parameters* of the p.d.f. of the random variable. These parameters are in practice or in principle unobservable and one is faced with the necessity of making probabilistic judgments concerning their value or range of values on the basis of the finite information contained in the sample. By way of illustration, consider a random variable X that is assumed to be $n(x; \mu, \sigma^2)$. We might find it desirable to be able to say that we are 95% confident that an unknown mean μ lies between two constants a and b, $a < b$. Or we might wish to test the reasonableness of the hypothesis that $\mu \leq k$. Specific illustrations of such problems are given subsequently.

The above remarks illustrate only a sample of the range of inferential problems encountered. More involved problems are frequently met (e.g., the problem of deciding whether the means μ_1 and μ_2 of two random variables differ or whether $\sigma_1^2 = \sigma_2^2$ for these random variables), but in each case the basic statistical reasoning is essentially the same as that outlined above. We consider only a few very simple cases in this discussion, which we hope will suffice to indicate the ways in which the abstract structure of mathematical statistics can be put to work in the solution of empirical problems or in the making of empirical decisions.

Confidence Intervals. Suppose we take a random sample of N independent observations of a random variable x which is $n(x; \mu, 9)$. We thus have a situation (not a very realistic one) where $\sigma^2 = 9$ is a known quantity. Suppose further that we compute the sample mean $\overline{X}$. We will now demonstrate that we can find two numbers c_1 and c_2 such that we are able to say we are 95% sure that the *unknown* value of μ lies between $\overline{X} + c_1$ and $\overline{X} - c_2$. Put another way, we will be able to make a statistical inference that, prior to drawing the random sample,

$$P(\overline{X} - c_2 < \mu < \overline{X} + c_1) = .95$$

(or any other probability we choose). If we compute $\overline{X} = (1/N)(X_1 + X_2 + \cdots + X_n)$ it is evident that $\overline{X}$ is a random variable. Hence, it is meaningful to discuss the p.d.f. of $\overline{X}$. Now from the previous discussion of linear combinations of independent random variables it follows that whatever the p.d.f. of $\overline{X}$ is, $\overline{X}$ has mean $\mu_{\overline{x}} = \mu$ and $\sigma_{\overline{x}}^2 = \sigma^2/N = 9/N$. Although we have not done so, it can be shown that if the X_i have normal p.d.f.'s, as is the case in our example, then so does $\overline{X}$. Thus

$\overline{X}$ is a random variable that is $n(\overline{x}; \mu, \sigma^2/N)$, or in our example $\overline{X}$ is $n(\overline{x}; \mu, 9/N)$.

From our discussion of z we observe immediately that

$$z = \frac{\overline{X} - \mu}{\dfrac{\sigma}{\sqrt{N}}}$$

or specifically that $(\overline{X} - \mu)/(3/\sqrt{N})$ is $n(z; 0, 1)$. Referring to the tabled values of z it is possible to find two numbers a and b such that

$$P\left(a < \frac{\overline{X} - \mu}{\dfrac{3}{\sqrt{N}}} < b\right) = 0.95.$$

Actually we would have to find only one number $b > 0$ in the table because, in order for z to be in the interval between a and b with probability equal to .95, z must exceed these limits with probability equal to .05. If we allocate this probability of .05 equally to the areas beyond both ends of the interval, then $P(z > b) = .025$ and $P(z < a) = .025$, and due to the symmetry of the normal curve $a = -b$. Therefore the event

$$a < \frac{\overline{X} - \mu}{\dfrac{3}{\sqrt{N}}} < b$$

occurs only when the event

$$\overline{X} - \frac{3a}{\sqrt{N}} > \mu > \overline{X} - \frac{3b}{\sqrt{N}}$$

occurs; thus the events have the same probability, $p = .95$. Since $a = -b$ we can write

$$P\left(\overline{X} - \frac{3b}{\sqrt{N}} < \mu < \overline{X} + \frac{3b}{\sqrt{N}}\right) = .95.$$

Thus we have found the two numbers $c_1 = 3b/\sqrt{N}$ and $c_2 = -c_1$ that we started out to find. The interval that brackets μ is called a confidence interval (in our case 95%) for μ.

In most cases we will not know the value for σ^2. It is still possible to find the required confidence interval for μ when σ^2 is unknown.

We shall indicate briefly how this is accomplished and in the process state a basic result of mathematical statistics. We begin by considering a random sample of N independent random variables X from a p.d.f. that is $n(x; \mu, \sigma^2)$, μ and σ^2 unknown. We refer to the following quantities:

$$\overline{X} = \frac{1}{N} \sum_{i=1}^{N} X_i,$$

the sample mean, and

$$S^2 = \frac{1}{N} \sum_{i=1}^{N} (X_i - \overline{X})^2,$$

the sample variance. We know that

$$\frac{\overline{X} - \mu}{\sigma/\sqrt{N}}$$

is normal with mean equal to zero and variance equal to one. We state without proof that NS^2/σ^2 is $\chi^2_{(N-1)}$ and is independent of $(\overline{X} - \mu)/(\sigma/\sqrt{N})$. Given the last item of information, we conclude from our discussion of the t distribution that

$$t = \frac{(\overline{X} - \mu)/(\sigma/\sqrt{N})}{\sqrt{NS^2/\sigma^2(N-1)}} = \frac{\overline{X} - \mu}{S/\sqrt{N-1}}.$$

From the tabled values of t for $n-1$ degrees of freedom it is a simple matter to find two numbers c_1 and c_2 such that

$$P\left(c_1 < \frac{\overline{X} - \mu}{S/\sqrt{N-1}} < c_2\right) = .95$$

or any other desired probability. The inequalities in parentheses are manipulated as before to find a 95% confidence interval for μ. (The reader can determine to his own satisfaction that NS^2/σ^2 can be used to find a confidence interval for σ^2.)

Hypothesis Testing. We will illustrate in this section a simple example of the second major type of statistical inference—*hypothesis testing*. Our example will be quite elementary. We will touch only briefly upon some of the interesting and subtle issues involved in the test of a statistical hypothesis. Nevertheless, it is hoped that the basic logic involved in the test of a statistical hypothesis will be made clear.

Let X be $n(x; \mu, 9)$. Suppose we have some notion that the unknown mean $\mu \leq 52$, and we wish to assess the tenability of this notion. We can state our notion in the form of a hypothesis and then make a statistical inference concerning this hypothesis. Such a hypothesis is called a *null hypothesis* and is denoted $H_0 : \mu \leq 52$.

Let us now consider a random sample of size N from $f(x)$ and compute

$$\overline{X} = \frac{\sum_{i=1}^{N} X_i}{N}.$$

From our discussion of linear combinations of independent random variables we know that $\mu_{\overline{X}} = \mu$ and that

$$\sigma_{\overline{X}}^2 = \frac{\sigma^2}{N} = \frac{9}{N}$$

or that $\sigma_{\overline{X}} = 3/\sqrt{N}$. We know also that

$$z = \frac{\overline{X} - \mu}{3/\sqrt{N}}$$

is $n(z; 0, 1)$.

A *statistical test* of the null hypothesis $H_0 : \mu \leq 52$ is constructed by beginning with the assumption that H_0 is true. To assume the truth of the null hypothesis is to assume that X is $n(x; 52, 9)$. (See remarks at the end of this section.) A corollary of this assumption is that

$$z = \frac{\overline{X} - 52}{3/\sqrt{N}}$$

is $n(z; 0, 1)$.

We now reason that if H_0 is true, the statistic $\overline{X}$ should not be very much greater than $\mu = 52$. If $\overline{X}$ is very much greater than its expected value, we might wish to conclude that our assumed value of $\mu \leq 52$ under the null hypothesis is not a reasonable assumption. We would in this case reject the null hypothesis. In other words, since $\overline{X}$ is a random variable we would expect the probability to be low that $\bar{X}$ is very much greater than $\mu = 52$.

In the above line of reasoning, it is necessary to define "very much." To do this we choose a value for $\bar{X}$ which will be exceeded by the computed $\bar{X}$ only with low probability (under a true H_0) and which, if exceeded, will lead us to decide to reject the null hypothesis. The specification of a range of values of $\bar{X}$ which leads to a rejection of H_0 if the computed $\bar{X}$ falls within this range is called a *statistical test* of the null hypothesis. The range itself is referred to as the *rejection region*. Such a range of values is easily found for the statistic $\bar{X}$. Suppose we will reject H_0 if the probability of the observed $\bar{X}$ equals 0.01. That is, we will reject H_0 if, given the specified null hypothesis, $P(\bar{X} \geq \bar{x}) = .01$. Now

$$z = \frac{\bar{X} - 52}{3/\sqrt{N}}$$

is $n(z; 0, 1)$. From the tabulated distribution function z it is easy to determine a value z_0 such that $P(z \geq z_0) = .01$. This is, of course, a value z_0 such that $P(z \leq z_0) = .99$. Thus since to any $\bar{X}$ we can associate a z, let

$$z_0 = \frac{\bar{X} - 52}{3/\sqrt{N}}$$

or

$$\bar{X} = \frac{3z_0}{\sqrt{N}} + 52$$

and $P(\bar{X} \leq (3z_0/\sqrt{N}) + 52) = .99$. Thus any computed values of $\bar{X} \geq (3z_0/\sqrt{N}) + 52$ fall in the rejection region and lead to a rejection of the null hypothesis.

The above situation is artificial in that we will not usually know the value of σ^2. In such a situation we can make use of the fact that the random variable

$$\frac{\bar{X} - \mu}{S/\sqrt{N-1}}$$

is distributed as t with $N - 1$ degrees of freedom. From this point on the reasoning underlying the test of H_0 is analogous to that above; the p.d.f. with respect to which a rejection region is determined is t rather than z.

A complete discussion of hypothesis testing would involve many subtle issues, most of which revolve about the problem of choosing a rejection region in the "optimal" fashion. Obviously a definition of "optimal" is crucial here; the reader is referred to any standard text for details. The alert reader may have noted that in the above example of hypothesis testing a null hypothesis $\mu \leq 52$ was stated, while in finding a value z_0 we acted as if the null hypothesis were actually $\mu = 52$. The statistician would call $H_0 : \mu \leq 52$ an *inexact* hypothesis. We constructed our statistical test on the basis of an *exact* hypothesis $H_0 : \mu = 52$. While the problems here are not formidable, the detailed issues involved in this situation are beyond the scope of the present article.

Point Estimation. The previous sections demonstrating that certain statistics can be used to make inferences or test hypotheses concerning parameters suggest that there may exist certain relationships between statistics and parameters that may be further explored. We shall discuss this topic very briefly.

From a very general point of view, the statistical task always revolves about making estimates of a parametric state of affairs. For example, the construction of a confidence interval can be viewed as an estimation procedure. In some cases, however, we would like to make more precise estimates of parameters —always, of course, basing such estimates on statistics. We may, for example, want to make an estimate of the *specific* value of a parameter rather than constructing a confidence *interval*. This is a problem in *point estimation*; we will now discuss some basic considerations involved in such estimates.

Suppose we have a sample of a random variable X of size N and have computed $\bar{X}$. Consider $E[\bar{X}]$. We know from previous considerations that $E[\bar{X}] = \mu$, the mean of X.

When the expected value of a statistic is equal to a parameter, the statistic is said to be an *unbiased estimate* of the parameter. Thus $\overline{X}$ is an unbiased estimate of μ. In general, a statistic Q is unbiased for a parameter θ if $E[Q] = \theta$.

If a statistic is to be used as an estimator, it should be unbiased. If more than one statistic is unbiased for a given parameter, the choice among them will depend upon other speci-fications for a "good" estimator. Since this question is too involved to pursue in detail we shall indicate in a general way the proper-ties of a "good" estimator. Consider a random variable X having p.d.f. $f(x)$ con-taining an unknown parameter θ which we wish to estimate. For example, θ might be equal to μ in a normal p.d.f., to p in a bino-mial p.d.f., and so on. If a statistic Q is a "good" estimate of θ, then $E[Q] = \theta$; that is, it is unbiased. R may be another statistic that is unbiased for θ; that is, $E[R] = \theta$. The problem is choosing the best estimate of θ.

If X is a random variable having mean μ and variance σ^2, then $\overline{X}$ is unbiased for μ; that is, $E[\overline{X}] = \mu$. Note that X is *also* unbiased for μ; that is, $E[X] = \mu$. We might intuitively feel that $\overline{X}$, being based on more observations, is a better estimate of μ than is X. If we follow our intuition we find that

$$\sigma_{\overline{X}}^2 < \sigma_X^2, \quad \text{or} \quad \frac{\sigma^2}{N} < \sigma^2.$$

We thus arrive at an essential defining characteristic of the "best" estimate of a parameter. The basic requirements of a good estimator of a parameter θ are: (1) the statistic Q should be unbiased for θ, that is, $E[Q] = \theta$; (2) the variance of Q, σ_Q^2, should be less than that of other unbiased statistics R_i, that is, $\sigma_Q^2 < \sigma_{R_i}^2$ for all i where $E[R_i] = \theta$. In essence we are looking for a statistic that has a p.d.f. whose mean is equal to θ and whose variance about θ is less than other statistics having the same expected value. In the example of $\overline{X}$ and X, both of which are unbiased for μ, the choice was easy; however, we glossed over the fact that in addition to $\overline{X}$ and X there might exist many such unbiased estimates of μ. The mathematical considerations necessary for choosing a "best" estimate under these conditions are too involved to consider here. The essential point remains that we are looking for a statistic having the described characteristics. For more precisely defined mathematical properties, the interested reader should consult any standard text.

References

1. Hays, W. L. *Statistics for Psychologists.* New York: Holt, Rinehart & Winston, 1963.
2. Hogg, R. V. & Craig, A. T. *Introduction to Mathematical Statistics.* New York: Mac-millan, 1959.
3. Whitney, D. R. *Elements of Mathematical Statistics.* New York: Holt, Rinehart & Winston, 1961.

3.2 Suggestions for Further Reading

Dwass, M. *First Steps in Probability*. New York: McGraw-Hill, 1967.

Hays, W. L. *Statistics for Psychologists*. New York: Holt, Rinehart, and Winston, 1963.

Mosteller, F., R. E. K. Rourke, and G. B. Thomas, Jr. *Probability with Statistical Applications*. Reading, Mass.: Addison-Wesley, 1961.

Stilson, D. W. *Probability and Statistics*. San Francisco: Holden-Day, 1966.

4

Hypothesis Testing
and
Confidence Intervals

Basic to the scientific method is the process of formulating and testing hypotheses. It is important to distinguish between two kinds of hypotheses: scientific hypotheses and statistical hypotheses. The former correspond to scientists' hunches about phenomena in the universe and are normally stated in general terms, at least in the initial stages of an inquiry. Hypotheses in this form are not amenable to evaluation by the powerful tools and techniques of statistical theory. In order to utilize these tools and techniques it is necessary for the scientist through deductive logic to translate his scientific hypothesis into a statistical hypothesis. Statistical hypotheses are statements about one or more character-istics of population distributions, and as such they refer to situations that might be true. Although the scientist's hunches may be originally stated in a form that is identical to statistical hypotheses, this is unlikely.

In a very real sense this chapter constitutes the heart of this book, for it is concerned with procedures whereby (1) scientific hypotheses are transformed into statistical hypotheses, (2) decisions regarding the statistical hypoth-eses are made, and (3) the scientist draws conclusions concerning the probable truth or falsity of his scientific hypotheses. In addition, such related topics as confidence-interval procedures and the use of non-random samples are examined.

The student will discover as he reads the chapter that in spite of the importance of hypothesis-testing methodology in the scien-tific enterprise, many theoretical issues concerning hypothesis testing remain con-troversial. This is understandable considering that basic hypothesis-testing concepts as developed by R. A. Fisher, Jerzy Neyman, and Egon Pearson are less than fifty years old.

Editor's comments: The statistical concepts of significance and confidence have often been confused in the psychological literature. Probably the most common example of this confusion is the use of the phrase, "The results were significant at the .05 level of *confidence*." Robert E. Chandler in the following article discusses the meaning of the concepts of significance and confidence and makes a clear distinction between them.

4.1 The Statistical Concepts of Confidence and Significance

Robert E. Chandler

Recently there have been at least three different book reviewers who have commented on the confusion that currently exists in the psychological literature regarding the statistical concepts of confidence and significance (2, 7, 9). Although this confusion can be partially explained as a semantic problem, it behooves the psychologist to examine these two concepts rather closely and to adopt pristine terminology for the benefit of beginning students and individuals of other disciplines that draw rather heavily upon the psychological literature.

Confidence and Confidence Coefficients

Confidence, a concept customarily reserved for discussions of interval estimation, is the faith which one is willing to place in a statement that an interval established by a sampling process actually contains or bounds a para-

meter of interest. One generally expresses this faith statistically by affixing to each interval a confidence coefficient, or confidence probability, which can be written as $1 - \varepsilon$, where $\varepsilon = p/100$ for $0 \le p \le 100$, and p is usually taken to be a very small number (1, 3, 6, 8). For example, if $p = 5$ the confidence coefficient would be .95, and one would refer to the interval with which this coefficient is associated as the 95% confidence interval.

The confidence coefficient is frequently interpreted in the following manner: If one were to draw samples of size K from a population of N elements (K naturally being $<N$) and from each sample establish a 95% confidence interval on some specified parameter of the population, then in the long run about 95% of the totality of these intervals would actually contain the parameter of interest, and approximately 100 ε%, or 5%, of them would not (3). This interpretation is correct, but of course assumes $\binom{N}{K}$ to be a rather large number.[1]

Reprinted from *Psychological Bulletin*, 1957, **54**, 429–430, by permission of the publisher and author. Copyright 1957 by the American Psychological Association.

---

[1] The notation $\binom{N}{K}$ is used here as a combinational symbol to indicate the number of ways that K objects can be selected from N.

Significance and Significance Levels

Significance, as contrasted to confidence, is given to the testing of hypotheses. Here one makes a statement, i.e., states an hypothesis, which will hereafter in this discussion be represented as H, that may be either true or false and then takes action on this H by accepting or rejecting it. Clearly, any one of the following actions is a likely outcome as a result of testing an H: (*a*) rejection of a false H; (*b*) acceptance of a true H; (*c*) rejection of a true H; or (*d*) acceptance of a false H. It is quite evident that actions (*a*) and (*b*) are desirable, while (*c*) and (*d*) carry the connotation of committing an error—*c* being the familiar Type I error or an error of the first kind, while *d* is called a Type II error or an error of the second kind (4, 8).

When one tests an H, the probability that he will take action *c* is defined as the significance level, which we will represent as α (8). Although α is generally of the same order of magnitude as ε, α and ε differ in the amount of information which they convey, for while ε completely tells all there is to know about "being wrong" in interval estimation, α only gives information about a very particular type of error, i.e. the action described by *c*. To emphasize the contrast made here between α and ε, one merely needs to examine the other type of error that can be made in the test of an H.

For this purpose, let β represent the probability that action *d* is taken, i.e. a Type II error is committed; then, by definition, $1 - \beta$ is known as the power of the statistical test or the probability that action *a* will occur (4, 8). Although texts in psychological statistics do not seem to place a great deal of emphasis upon the power of a test, power is the basic concept responsible for one's employing statistical tests as a basis for taking action on an H. If this were not so, to test an H at the 5% level of significance, one could simply draw from a box of 100 beads—95 white and 5 red—a bead at random and adopt the convention that he would reject the H whenever a red bead appeared. With such a test, one can readily see that not only α but also $1 - \beta$ always equals .05, or $\beta = .95$. It is this large value of β that precludes one's employing the bead-box test. For an excellent discussion of β and its relation to the alternative H against which one might be testing, the reader is referred to Dixon and Massey (4, pp. 244–261).

Summary and Discussion

The admixing of the concepts of confidence and significance has become so prevalent in the psychological literature that one typically reads statements, in the reports of psychological research, indicating that certain experimental results were significant, at say, the 5% "level of confidence."

It may be that this confusion arises from the fact that one can utilize a confidence interval as a significance test (e.g., see 5, p. 241), and in doing so may hastily, but incorrectly, conclude that there is no difference between the two concepts.

Inasmuch as explicit terminology is needed to convey the probabilities of committing statistical errors in the respective areas of interval estimation and testing of hypotheses, the concept of confidence should never be associated with the statistical test of an H regardless of the nature of the test being employed.

References

1. Anderson, R. L., & Bancroft, T. A. *Statistical theory in research.* New York: McGraw-Hill, 1952.
2. Chandler, R. E. A review of Guilford's *Fundamental statistics in psychology and education.* (3rd ed.) *Personnel Psychol.,* 1957, **10**, 272–273.
3. Cramér, H. *Mathematical methods of statistics.* Princeton: University Press, 1951.
4. Dixon, W. J., & Massey, F. J., Jr. *Introduction to statistical analysis.* (2nd ed.) New York: McGraw-Hill, 1957.

5. Edwards, A. L. *Statistical methods for the behavioral sciences.* New York: Rinehart, 1954.
6. Hoel, P. G. *Introduction to mathematical statistics.* (2nd ed.) New York: Wiley, 1954.
7. Milton, T. E. A review of Edwards' *Statistical methods for the behavioral sciences. J. Amer. statist. Ass.*, 1956, **51**, 382.
8. Mood, A. M. *Introduction to the theory of statistics.* New York: McGraw-Hill, 1950.
9. Walker, H. M. A review of Adams' *Basic statistical concepts. Educ. psychol. Measmt.*, 1956, **16**, 554–557.

4.2 Statistical Significance—What ?

Thomas C. O'Brien and Bernard J. Shapiro

The purpose of many research studies in education is to determine which of two (or more) educational policies is more effective. Excellent articles on this topic by Johnson and Brownell have appeared in the recent past.[1] The purpose of this article is to bring to light an important though relatively neglected aspect of such research.

In most statistical comparisons of educational policies—and in many other statistical comparisons, for that matter—the researcher makes use of the concept of statistical significance. Differences among means or other statistics are statistically significant if the likelihood that they happened by chance is

Reprinted from *Mathematics Teacher*, 1968, **61**, 673–676, by permission of the publisher and authors. Copyright 1968 by the National Council of Teachers of Mathematics.

[1] Donovan A. Johnson, "A Pattern for Research in the Mathematics Classroom," *The Mathematics Teacher*, LIX (May 1966), 418–25; and W. A. Brownell, "The Evaluation of Learning Under Dissimilar Systems of Instruction," *The Arithmetic Teacher*, XIII (April 1966), 267–74.

less than some experimenter-established limit. For example, if differences are significant at the .05 level, there are 5 or less chances in 100 that they could have occurred by chance. If the experimenter had previously adopted this .05 limit and finds that his outcomes had .05 or less chance to occur if only chance effects were involved, he then rejects the hypothesis that only chance effects were involved, says that some systematic influence was at work (the effect of a particular educational policy, for example), and takes the risk that 5 times out of 100 he may be wrong.

In the professional literature, it would appear that a great many studies report statistically significant results. This in itself may not be cause for concern. We suggest, however, that this situation may reflect a tendency on the part of both editors and researchers to identify statistical significance with success and importance and at the same time to reject out-of-hand any statistically nonsignificant results. The burden of this article is that statistical significance has all too

Table 1. Penny-tossing experiments in which the obtained proportion of heads was .52.

Experiment	Number of tosses	Expected proportion of heads	Obtained proportion of heads	p	Reject H_0
1a	25	.50	.52	.84	No
1b	250	.50	.52	.52	No
1c	2,500	.50	.52	.04	Yes
1d	25,000	.50	.52	<.0001	Yes

Table 2. Penny-tossing experiments in which the obtained proportion of heads was .70.

Experiment	Number of tosses	Expected proportion of heads	Obtained proportion of heads	p	Reject H_0
2a	10	.50	.70	.20	No
2b	20	.50	.70	.07	No
2c	40	.50	.70	.01	Yes
2d	80	.50	.70	<.001	Yes

often become an end in itself rather than a basis for further inquiry.

For the purpose of studying statistical significance in terms of concrete examples, let us imagine two series of four penny-tossing experiments in each of which the obtained proportion of heads was constant: .52 in one case and .70 in the other. Admittedly, such perfectly constant ratios are extremely unlikely to occur in practice, but they will add a great deal to the discussion that follows.

A test in all cases was the trial hypothesis (H_0) that the expected ratio of heads to the total number of tosses is .5 or, in other words, that the penny used in each experimental series was unbiased. The 5 percent significance level was chosen. That is, if under H_0 there were 5 or less chances in 100 of obtaining deviations in either direction from .5 which were as large as or larger than the observed deviations, H_0 was rejected and it was concluded that the coin was biased. The probability (p) of obtaining deviations from .5 as large as or larger than that of each of the experimental outcomes was determined by the normal approximation to the binomial as given by Edwards.[2] The experimental data are presented in Tables 1 and 2.

In these tables, there are basically four types of findings:

1. Small differences exist, and they are not statistically significant (experiments 1a and 1b).
2. Small differences exist, but they are statistically significant (experiments 1c and 1d).
3. Large differences exist, but they are not statistically significant (experiments 2a and 2b).
4. Large differences exist, and they are statistically significant (experiments 2c and 2d).

For the moment, let us consider only those findings (2 and 4) in which the observed differences were statistically significant, that is, where the null hypothesis was rejected. The data indicate that, other things being equal, statistical significance can be demonstrated (at least for an infinite population) for both small (Table 1) and large (Table 2) differences provided only that the number of observations (in this case, tosses) is large enough.[3]

[2] A. L. Edwards, *Experimental Design in Psychological Research* (rev. ed.; New York: Holt, Rinehart & Winston, 1964), p. 49.

[3] That is as it should be. In tests of statistical significance we are concerned with inferences from a sample to a parent population. As more and more members of the parent population become known, the role of chance and error is diminished.

However, the data also indicate that not all differences which are statistically significant are of the same magnitude, for while one of the pennies would be regarded as only slightly biased, the other would be seen as rather definitely loaded. That is, the detection of statistical significance merely indicates that an unlikely event has taken place. Nothing is indicated about the magnitude or importance of the event. It is this distinction between statistical significance and practical importance that seems often to be overlooked by many researchers.

There are several ways in which the magnitude of observed differences can be assessed. The experimenter can speak in terms of (1) standard deviations (e.g., group A's mean exceeded group B's mean by x standard deviations), (2) ω^2, the percent of the variance accounted for by a particular experimental factor,[4] or (3) one or more of the various indices reported by Cohen.[5] At least, the experimenter can always describe the differences in terms of their practical significance. For example, differences can be assessed in terms of grade scores for normed tests (e.g., group A's mean is three months higher than Group B's) or even in terms of the number of test items the difference represents. Which one of these or other approaches is the most appropriate will depend on the particular problem at hand.

Now let us consider the experiments in which statistically nonsignificant results occurred, i.e., where H_0 was not rejected. One's first impulse is to either reformulate or abandon the alternative hypotheses which gave birth to the research in the first place, for without statistical significance the researcher cannot generalize to a parent population on the basis of statistical inference.[6]

One must not, however, move too quickly. In addition to statistical significance, the experimenter must also be concerned with the concept of power, which is the probability of correctly rejecting the trial hypothesis, i.e., rejecting H_0 when it is in fact false.

Imagine, for the moment, a biologist using a microscope in a search for microbes. The more powerful his microscope, the greater his chance of seeing the microbes. The more powerful the microscope, the better he can see the smaller microbes.

In the case of nonsignificant differences, the experimenter had better be aware of the type of microscope he was using, i.e., the power of his test, before he makes any hasty decisions. If he determines that his microscope (i.e., his power) was only 20 percent capable of detecting the microbes (the difference) in question, he might well consider replicating his experiment and increasing his power by using a larger N. Far from having found nothing of importance, as seems to be believed by many researchers, he may have been on the verge of uncovering something very important. It is just that his microscope was not strong enough. It would seem that the least this researcher can do is to report his findings—including the fact that his microscope was weak—and let others take up the cudgel where he left off. Five parameters—sample size, N; the significance level; the probability that a given difference will be detected (power); the magnitude of the differences; and σ, the population standard deviation—are interrelated in such a way that the establishment of any four of them determines the fifth. Having worked with an actual N, having obtained a given difference and estimate of σ, and having preset a significance level, the experimenter can estimate the power of his test by mere computation. Computation methods for obtaining estimates of power for various statistical tests

---

[4] W. L. Hays, *Statistics for Psychologists* (New York: Holt, Rinehart & Winston, 1963), ch. x, xii, xiii.

[5] J. Cohen, "The Statistical Power of Abnormal-Social Psychological Research," *Journal of Abnormal and Social Psychology*, LXV (July 1962), 145–53.

[6] Although no inferences can be made about the parent population without statistical significance, the

researcher can always assess the magnitude of the difference judgmentally in terms of the situation as it exists. An IQ difference of 50 between husband and wife cannot ever be tested for statistical significance, but it certainly is likely to be of practical importance.

are discussed in Walker and Lev,[7] Guilford,[8] Hays,[9] Dixon and Massey,[10] Feldt,[11] and Edwards.[12]

Thus far we have only considered the results of experiments already completed. It must be noted, however, that rather than rely only on *post hoc* interpretations, it is both possible and preferable for the experi-

menter to plan in advance for possible outcomes. In the pre-experimental planning he can preset a significance level (in most current research usually the only preset value), a desired power value, a difference he is interested in detecting; and he can often at least estimate, from previous experience, the value of the population standard deviation. Then it is again merely a matter of computation to determine the sample size (N) required to realize these conditions.

At best, the N should be determined by preselection of significance level, power, and magnitude of the differences, and the estimation of the population standard deviation. At worst, the experimenter should at least make use of the *post hoc* approaches suggested above. In either case, considerable care should be given to the description of the magnitude of the obtained differences.

[7] H. Walker and J. Lev, *Statistical Inference* (New York: Holt, Rinehart & Winston, 1953), ch. iii, iv.

[8] J. P. Guilford, *Fundamental Statistics in Psychology and Education* (4th ed.; New York: McGraw-Hill, 1965), pp. 210–14.

[9] Hays, *op. cit.*, pp. 269–80.

[10] W. Dixon and F. Massey, *Introduction to Statistical Analysis* (New York: McGraw-Hill, 1951), pp. 207–20.

[11] L. S. Feldt and N. W. Mahmoud, "Power Function Charts for Specification of Sample Size in Analysis of Variance," *Psychometrika*, XXIII (1958), 203–10.

[12] Edwards, *op. cit.*, pp. 94–100.

Editor's comments: An examination of literature in the behavioral sciences reveals that tests of significance are commonplace whereas confidence interval procedures are rarely ever used. This is somewhat surprising in view of the preference generally expressed by mathematical statisticians for confidence-interval estimation. Mary G. Natrella discusses the relative merits of the two procedures and points out some of the reasons why confidence-interval procedures are recommended by statisticians.

4.3   The Relation between Confidence Intervals and Tests of Significance

Mary G. Natrella

1. Introduction

The advertising sheet for a recent revision of a classical text book on statistical methods states "The author has diverted emphasis from tests of significance to point and interval estimates." This author is not alone. Many statistical consultants, analyzing an experiment for the purpose of testing a statistical hypothesis, e.g., in comparing means of normal populations, find that they prefer to present results in terms of the appropriate confidence interval.

It must be noted of course that not every statistical test can be put in the form of a confidence interval. Kendall, [5], for example, speaks of two broad classes of statistical tests, "those which give a direct test of a given value of a parent parameter, and those which do not." Berkson [2] also distinguishes these

The author is indebted to Dr. Churchill Eisenhart and Dr. Norman C. Severo who encouraged the writing of this paper.

Reprinted from *American Statistician*, 1960, **14**, 20–22, 33, by permission of the publisher and author.

two classes of tests in discussing tests of normality and says "I suggest tentatively that the two classes I have in mind can be differentiated as (1) those which in principle can be alternatively stated in terms of an estimate and its confidence interval and (2) those which cannot be so stated." It is this first class of tests which will be discussed in this paper. Tests such as tests of normality, tests of goodness-of-fit, and tests of randomness fall into the second class.

When the results of a statistical test can alternatively be stated in terms of a confidence interval for a parameter, is there any reason to prefer the confidence interval statement? An early indication of dissatisfaction with the logic of tests of significance as experimental evidence is given in another paper by Berkson [3]. He stresses the point that experimenters are not typically engaged in disproving things, but are looking for evidence for affirmative conclusions, and that after rejecting the null hypothesis, they will then look for a reasonable hypothesis to accept. The relation between confidence intervals and tests of significance is mentioned only briefly by most textbooks, and ordinarily no insight is given as to which conclusion might be more appropriate. (A notable exception is Wallis and Roberts [7].)

In the present note, we draw attention to how these two approaches are related and how they differ. One reason for preferring to present a confidence interval statement (where possible) is that the confidence interval, by its width, tells more about the reliance that can be placed on the results of the experiment than does a YES-NO test of significance. Of course, a test of significance, when accompanied by its appropriate Operating Characteristic curve, provides much the same kind of information as does a confidence interval. In practice, however, the associated *OC* curve is often ignored by and may be unknown to the experimenter. We feel that the experimenter himself finds the confidence interval more natural and more appealing, but generally has little notion of how the two concepts are related.

2. An Example

Let us review both procedures with reference to a numerical example.

For a certain type of shell, specifications state that amount of powder should average 0.735 lb. In order to determine whether the average for the present stock meets the specification, twenty shells are taken at random and the weight of powder is determined. The sample average ($\overline{X}$) is 0.710 lb. The estimated standard deviation (s) is 0.0504 lb. The question is whether or not the average of present stock differs from the specification value. In order to do a two-sided test of significance at the $(1 - \alpha)$ probability level, we compute a critical value, to be called for example, C. Let $C = t^* s / \sqrt{n}$ where t^* is the positive number exceeded by $100\,(\alpha/2)\%$ of the t-distribution with $n - 1$ degrees of freedom.

In the example above with $\alpha = .05$, $t^* = 2.09$, $C = 0.0236$ lb. The test of significance says that if $|\overline{X} - 0.735| > C$, we decide

that the average for present stock differs from the specified average. Since $|0.710 - 0.735| > 0.0236$, we decide that there *is* a difference.

We can also compute from the data a 95% confidence interval for the average of present stock. This confidence interval is $\overline{X} \pm C = 0.710 \pm 0.0236$ or 0.686 to 0.734 lb. The confidence interval can be used for a test of significance; since it *does not include* the standard value 0.735, we conclude that the average for the present stock *does differ* from the standard.

Comparisons of two materials (both means unknown and equal variances) may be made similarly. In computing a test of significance we compare the observed difference $|\overline{X}_A - \overline{X}_B|$ with a C' (a computed critical quantity similar to C above). If $|\overline{X}_A - \overline{X}_B|$ is larger than C' we declare that the means differ significantly at the chosen level. We also note that the interval $(\overline{X}_A - \overline{X}_B) \pm C'$ is a confidence interval for the difference between the true means $(\mu_A - \mu_B)$. If then this interval does not include zero, we conclude from the experiment that the two materials differ in mean value.

3. Do the Two Approaches Differ?

Here then are two ways to get the same answer to the original question. We may present the result of a test of significance, or we may present a confidence interval. Are there any differences between the two? The significance test is a "go no-go" decision. We compute a critical value C, and we compare it with an observed difference. If the difference exceeds C, we announce a "difference"; if it does not, we announce no "difference." If we had no *OC* curve for the test, our decision would be a yes-no proposition with no shadowland of indifference. The test may say NO, but only the *OC* curve can qualify this by saying that this particular experiment had only a ghost of a chance of saying "Yes" to this particular question. For example, see Fig. 2. If the true value $d = |(\mu_1 - \mu_0)/\sigma|$ is equal to 0.5, a sample of 10 is not likely to detect a difference, but a sample of 100 is almost certain to.

Using a rejection criterion alone is not the proper way to think of a significance test. One should always think of the associated *OC* curve as part and parcel of the test. Unfortunately this has not always been the case, and the significance test without its *OC* curve has distorted the thinking in some experimental problems. As a matter of fact many experimenters who use significance tests are using them as though there were no such thing as an *OC* curve. For this reason, it may be preferable for the experimenter to approach the problem of testing hypotheses by using confidence intervals.

4. Why Prefer the Confidence Interval?

A confidence interval procedure contains information similar to the appropriate *OC* curve, and at the same time is intuitively more appealing than the combination of a test of significance and its *OC* curve. If the standard value is contained in the confidence interval, one can announce "no difference." The *width* of the confidence interval gives a good idea of how firm is the Yes or No answer.[1]

Suppose that the standard value for some property is known to be 0.735, and that a $100(1 - \alpha)\%$ confidence interval for the same property of a possibly different material is determined to be 0.600 to 0.800. It is true that the standard value is in the interval, and that we would say that there is no difference. All that we really know about the new product, however, is that its mean probably is between 0.6 and 0.8. If a much more extensive experiment gave a $100(1 - \alpha)\%$ confidence interval for the new mean of $0.60 - 0.70$, our previous decision of no difference would be reversed.

On the other hand, if the computed confidence interval, for the same confidence coefficient, had been $.710 - .750$, our answer would still have been "no difference," but we

would have said "No" more loudly and firmly. The confidence interval not only gives a Yes or No answer, but also, by its width, gives an indication of whether the answer should be whispered or shouted.

This is certainly true when the width of the interval, for a given confidence coefficient, is a function only of n and the appropriate dispersion parameter (e.g., known σ). When the width itself is a random variable (e.g., is a fixed multiple of s, the estimate of σ from the sample), one can occasionally be misled by unusually short or long intervals. But the *average width* of the entire *family* of intervals associated with a given confidence-interval procedure is a definite function of the appropriate dispersion parameter, so that *on the average* the random widths do give similar information. See [1] for a graphical illustration of confidence intervals computed from 100 random samples of $n = 4$ (actually random normal deviates).[2] Figure 14 in reference [6] shows a similar illustration of 100 intervals for $n = 4$, and in addition shows 40 intervals for $n = 100$, and 4 intervals for $n = 1000$. The fluctuation in size and position is of course very much reduced in the latter cases.

The significance test gives the same answer, and a study of the *OC* curve of the test indicates how firm is the answer. If the test is dependent on the value of σ, the *OC* curve has to be given in terms of the unknown σ. In such a situation, one has to make use of an upper bound for σ in order to interpret the *OC* curve, and again one may be misled by a poor choice of this upper bound. On the other hand the *width* of the confidence interval is part and parcel of the information provided by that method. No *a priori* estimates need be made of σ as would be necessary to interpret the *OC* curve. Furthermore, a great advantage of

---

[1] There is a caution in this regard as explained a little further on.

[2] This picture is an excellent teaching aid in itself. Despite the fluctuation in size and position of the individual intervals, a proportion of the intervals remarkably close to the specified proportion do include the known population average. If σ were known rather than estimated from the individual sample, the intervals would fluctuate only in position, of course.

confidence intervals is that the width of the interval is in the same units as the parameter itself. The experimenter finds this information easy to grasp, and to compare with previous information he may have had.

5. What Does the Confidence Interval Show?

The most striking illustration of information provided by confidence intervals is shown in the charts of confidence limits for a binomial parameter. In this case the limits depend only on n and the parameter itself, and one cannot be misled in an individual sample. Figure 1 shows the "central" 95% confidence limits for proportions. These "central" limits are the well-known Clopper-Pearson limits, such that each tail probability is not greater than .025. The central limits correspond to an equal-tail significance test at the $(1 - \alpha)$ probability level, and to each of the two "central" limits there corresponds a single-tail significance test at the $(1 - \alpha/2)$ probability level. In constructing a system of confidence limits there is no unique method of subdividing between the two tails. Limits which are not "central" may have other optimum properties—e.g., the recently developed system of E. L. Crow [4] gives

limits which are shorter than the "central" limits.

Suppose that a new item is being tested for comparison with a standard. In a sample of 10 we observed two defectives and therefore estimate the proportion defective for the new item as 0.20. The central 95% confidence interval corresponding to an observed proportion of 0.20 ($n = 10$) is $0.02 - 0.56$. Assume that the known proportion defective for the standard (P_0) is 0.10. Our experiment with 10 gives a confidence interval which includes P_0, and therefore we announce "no difference" between the new item and the standard in this regard. Intuitively, however, we feel that the interval $0.02 - 0.56$ is so wide that our experiment was not very indicative. Suppose then we test 100 new items and observe 20 defectives. The observed proportion defective is again 0.20. The confidence interval now is $0.13 - 0.29$, and does *not* include $P_0 = 0.10$. This time we are forced to announce that the new item "is different" from the standard, and the narrower width of the confidence interval ($0.13 - 0.29$) gives us some confidence in doing so.

6. What Does the Operating Characteristic Curve Show?

The foregoing has shown that it is possible to get some notion of the discriminatory power of the test from the size of confidence intervals. Is it also possible in reverse, to deduce from the *OC* curve what kind of confidence interval we would get for the new mean? Although we cannot deduce the exact width of the confidence interval, we can infer the order of magnitude. Suppose that we have measured 100 items, have performed a two-sided t-test (does the average μ_1 differ from μ_0?), and have obtained a significant result. Look at the curve for $n = 100$ in Figure 2, which plots the probability of accepting H_0 (the null hypothesis) against $d = |(\mu_1 - \mu_0)/\sigma|$. From the curve we see that, when d is larger than 0.4, the probability of accepting the null hypothesis is practically zero. Since our significance test

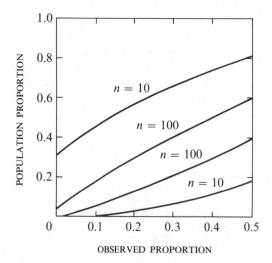

Fig. 1. 95% Confidence limits for population proportion.

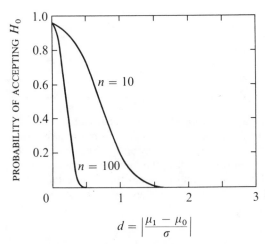

Fig. 2. Operating characteristics of the two-sided t test ($\alpha = 0.05$).

did reject the null hypothesis, we may reasonably assume that our $d = |(\mu_1 - \mu_0)/\sigma|$ is larger than 0.4, and thus perhaps infer a bound for the true value of $|\mu_1 - \mu_0|$, in other words, some "confidence interval" for μ_1.

On the other hand, suppose that only 10 items were tested and a significant result was obtained. If we look at the curve for $n = 10$ in Fig. 2, we see that the value of d which is practically certain to be picked up on a significance test is now $d = 1.5$ or larger. A significant result from an experiment which tested only 10 items thus, as expected, corresponds to a wider confidence interval for μ_1 than that inferred from the test of 100 items. A rough comparison of the relative widths may be made. More quantitative comparisons could be made, but the purpose here is to show a broad general relationship.

7. Relation to the Problem of Determining Sample Size

The problem of finding the required sample size to detect differences between means can also be approached in two ways. We can specify tolerable risks of making either kind of "wrong" decision (errors of the first and second kind)—thereby fixing two points on the OC curve of the required test. Matching these two points with computed curves for various n enables one to pick the proper sample size for the experiment.

Alternatively, we can specify the magnitude of difference between means which is of importance. We then compute the sample size required to give a confidence interval of fixed length equal to the specified difference.

8. Conclusion

Presentation of results in terms of confidence intervals is often more meaningful than is the presentation of the usual tests of significance (if the test result is not considered in connection with its OC curve). Things are rarely black or white, and decisions are rarely made on one-shot tests, but usually in conjunction with other information. Confidence intervals give a feeling of the uncertainty of experimental evidence, and (very important) give it in the same units, metric or otherwise, as the original observations.

References

1. *ASTM Manual on Quality Control of Materials*, (1951), p. 45. Available from the American Society for Testing Materials, 1916 Race Street, Philadelphia, Pennsylvania.
2. Joseph Berkson, "Comments on Dr. Madow's 'Note on Tests of Departure from Normality' with Some Remarks Concerning Tests of Significance." *J. Amer. Statist. Assoc.* **36**, No. 216, pp. 539–541 (1941).
3. Joseph Berkson, "Tests of Significance Considered as Evidence," *J. Amer. Statist. Assoc.* **37**, No. 219, pp. 325–335 (1942).
4. Edwin L. Crow, "Confidence Intervals for a Proportion," *Biometrika*, **43**, pp. 423–435 (1956).
5. Maurice G. Kendall, *The Advanced Theory of Statistics*, Vol. II, pp. 134–136. Hafner Publishing Co., New York (1951).
6. Walter A. Shewhart, edited by W. Edwards Deming, *Statistical Method from the Viewpoint of Quality Control.* The Graduate School, Department of Agriculture, Washington, D.C. (1939).
7. W. Allen Wallis and Harry V. Roberts, *Statistics, A New Approach*, pp. 461–463. The Free Press, Glencoe, Ill. (1956).

Editor's comments: Statistical hypothesis-testing procedures play an important role in the evaluation of scientific theories. Some of the issues that arise in the use of such procedures in accepting or rejecting scientific theories are discussed in the following article. The author, Arnold Binder, places the issue in historical perspective and provides the persevering reader with valuable insights into the logic of hypothesis testing.

4.4   Further Considerations on Testing the Null Hypothesis and the Strategy and Tactics of Investigating Theoretical Models

Arnold Binder

Abstract. David A. Grant has argued that it is inappropriate to design experiments such that support for a theory comes from acceptance of the null hypothesis. The present article points out that while this position could be defended in Fisher's approach to testing statistical hypotheses, it could not in the Neyman-Pearson approach or on more general scientific grounds. It is emphasized that one optimally designs experiments with enough sensitivity for rejecting poor theories and accepting useful theories, whether acceptance or rejection of the null hypothesis leads to empirical support. The argument that, in the procedure to which Grant objects, an insensitive experiment is more likely to lead to support for a theory is shown to be only a special case of the argument against bad experimentation.

The arguments in a recent article by Grant (1962) are directed against experimental designs oriented toward acceptance of the null hypothesis, that is, where support for an empirical hypothesis depends upon acceptance of the null hypothesis. Atkinson and Suppes (1958) provide an excellent example of the type of experimental logic to which Grant objects. These investigators postulated a one-stage Markov model for a zero-sum,

Reprinted from *Psychological Review*, 1963, **70**, 107–115, by permission of the publisher and author. Copyright 1963 by the American Psychological Association.

two-person game. On the basis of the model they predicted, first, the mean proportion of various responses over asymptotic trials and, second, that the probability of State k given States i and j on the two previous trials is equal to the probability of State k given only State j on the immediately preceding trial (i.e., that a one-stage Markov model accounts for the data). The predictions were then compared with the obtained results by means of a series of t tests, in the former case, and a χ^2 test, in the latter. One of the t tests, for example, involved a comparison of the predicted proportion of .600 against the observed mean proportion of .605, while another a comparison of a predicted value of .667 and an observed value of .670. Support for the one-stage Markov model was then inferred by the failure of the t tests and the χ^2 to reach the .05 level of significance. That is, support for the empirical model came from acceptance of the null hypotheses. Other examples may be found in Binder and Feldman, 1960; Bower, 1962; Brody, 1958; Bush and Mosteller, 1955; Grant and Norris, 1946; Harrow and Friedman, 1958; Weinstock, 1958; and Witte, 1959.

To facilitate future discussion it is convenient to refer to the procedure where acceptance of the null hypothesis leads to support for an empirical hypothesis as acceptance-support (a-s), and to the procedure where empirical support comes from rejection of the null hypothesis as rejection-support (r-s).

In addition to the objections to a-s, Grant argues that the method of testing statistical hypotheses may not be a very good idea in any case. He thus argues it is wise to shift away from the current emphasis in psychological research on hypothesis testing in the direction of statistical estimation.

Statistical Logic

There have been two principal schools of thought in regard to the logical and procedural ramifications of statistical inference. The older of these stems from the writings of Yule, Karl Pearson, and Fisher, while the other comes from the early work of Neyman and Pearson and the more recent developments of Wald. The respective influences of each of these schools on experimental statistics is abundantly evident, but a difficulty in separating these influences is that the actual recommendations for tests and interval estimates in a field like psychology are similar for both.

In the Fisher school one starts the testing process with a hypothesis, called the "null hypothesis," which states that the sample at issue comes from a hypothetical population with a sampling distribution in a certain known class. Using this distribution one rejects the null hypothesis whenever the discrepancy between the statistic and the relevant parameter of the distribution of interest is so large that the probability of obtaining that discrepancy or a larger one is less than the quantity designated α (the significance level). No clear statement is provided for the manner in which the null hypothesis is chosen, but the tests with which Fisher (1949) has been associated are in the form where the null hypothesis is equated with the statement " the phenomenon to be demonstrated is in fact absent " (p. 13).

The concept " rejection of the null hypothesis " is therefore unambiguous in the context of Fisher's viewpoint, but what about " acceptance of the null hypothesis ? " Fisher (1949) provides the following statement " the null hypothesis is never proved or established, but is possibly disproved, in the course of experimentation. Every experiment may be said to exist only in order to give the facts a chance of disproving the null hypothesis " (p. 16). This is not very edifying since one does not expect to prove any hypothesis by the methods of probabilistic inference. Hogben (1957) has interpreted these and similar statements of the Yule-Fisher group to mean that a test of significance can lead to one of two decisions: the null hypothesis is rejected at the α level or judgment is reserved in the absence of sufficient basis for rejecting the null hypothesis.

Papers by Neyman and Pearson (1928a, 1928b) pointed out that the choice of a statistical test must involve consideration of alternative hypotheses as well as the hypothesis of central concern. They introduced the distinction between the error of falsely rejecting the null hypothesis and the error of falsely accepting it (rejecting its alternative). Neyman and Pearson's (1933) general theory of hypothesis testing, based on the concepts Type I error, Type II error, power, and critical region, was presented later.

The possible parameters for the distribution of the random variable or variables in a given investigation are conceptually represented by a set of points in what is called a parameter space. This space is considered to be divided into two or more subsets, but we shall restrict our present discussion to the classical case in which there are exactly two subsets of points.

The statistical hypothesis specifies that the parameter point lies in a particular one of these two subsets while the alternative hypothesis specifies the other subset for the point. A statistical test is a procedure for deciding, on the basis of a set of observations, whether to accept or reject the hypothesis. Acceptance of the hypothesis is precisely the same as deciding that the parameter point lies in the set encompassed by the hypothesis, while rejection of the hypothesis is deciding that the point lies in the other subset. A typical test procedure assigns to each possible value of the random variable (statistic) one of the two possible decisions.

Sets of distributions (or their associated parameters), in this mathematical model, may be considered to correspond to the explanations in the empirical world which may account for the possible outcomes of a given experiment. Empirical hypotheses, which specify values or relationships in the scientific world, are translatable on this basis into statistical hypotheses. But the distinction between empirical and statistical hypotheses is quite important: the former refer to scientific results and relationships, the latter to subsets of points in a parameter space; they are related by a set of correspondences between scientific events and parameter sets.

The term "null hypothesis" does not occur in the writings of many of the advocates of the Neyman-Pearson view. Except for one pejorative footnote I was unable to find the term used by Neyman (1942), for example, in any of an extensive array of his publications. In general, these people prefer the term "statistical hypothesis" or simply "hypothesis" in designating the subset of central concern and alternative hypothesis for the other subset. However, null hypothesis has

taken on meaning over the years in the context of the Neyman-Pearson tradition among many writers of statistics, particularly those with expository proclivities. In the *Dictionary of Statistical Terms* (Kendall & Buckland, 1957) we find the following definition for null hypothesis: "In general, this term relates to a particular hypothesis under test, as distinct from the alternative hypotheses which are under consideration. It is therefore the hypothesis which determines the Type I Error" (p. 202).

An Evaluation[1]

Grant's position in regard to *a-s* is certainly not new or novel since it has been implicit in the writings of Fisher for the past 25 years. Moreover, it has been part of the folklore of statistical advising in psychology at least as

---

[1] There is a third viewpoint, represented in the psychological literature by Rozeboom's (1960) recent article, from which Grant's position could be evaluated. This viewpoint emphasizes the importance of the *a posteriori* probabilities of alternative explanations, in the Bayes sense, rather than the decision aspects of experimentation. However, the philosophical and practical problems of this approach remain enormous as is evident in the debates on this and related topics over the years. See, for example, Jeffreys (1957), Neyman (1952), Hogben (1957), Savage (1954), Chernoff and Moses (1959), von Mises (1942, 1957), and particularly Parzen (1960) who discusses the dangers of using Bayesian inverse probability in applied problems. It is typically not the case in basic research that one can assume that an unknown parameter is a random variable with some specified a priori distribution, and in such cases this approach does not presently provide any adequate answers to the problems of hypothesis evaluation.

While of a markedly different philosophical persuasion than the present writer, Rozeboom (1960) is equally unsympathetic with the inferential bias represented by Grant. He cuts into an essential component of this bias in the following succinct and effective manner:

> Although many persons would like to conceive NHD [the null hypothesis decision procedure] testing to authorize only rejection of the hypothesis, not, in addition, its acceptance when the test statistic fails to fall in the rejection region, if failure to reject were not taken as grounds for acceptance, then NHD procedure would involve no Type II error, and no justification would be given for taking the rejection region at the extremes of the distribution, rather than in its middle (p. 419).

far back as my initial exposure to psychological statistics (see Footnote 2). And, in fact, if Grant wishes to argue that his position holds only in the very narrowest interpretation of the Yule-Karl Pearson-Fisher structure, I see no grounds for contesting it. If there are only two possible decisions—reject the null hypothesis or reserve judgment—one would surely not wish to equate the null hypothesis with the empirical hypothesis designating a specific value. Using this logic an investigator could just as well discard as retain a theory when it has led to perfect predictions over a wide range.

In this context I would like to point out that there are many logical difficulties connected with the Fisher formulations which have been brought out dramatically in years of debate (Fisher, 1935, 1950, 1955, 1959, 1960; Neyman, 1942, 1952, 1956, 1961). Moreover there are some people who, while generally sympathetic with the Fisher viewpoint, are quite willing to accept the null hypothesis and conclude that this provides support for an empirical hypothesis (Mather, 1943; Snedecor, 1956).

In the pursuit of evaluating Grant's position from the Neyman-Pearson theory we must remember that the null hypothesis is a statistical hypothesis which designates a particular subset of parameter points. Moreover, the null hypothesis and the alternative hypothesis (the other subset) are mutually exhaustive so that rejection of the one implies acceptance of the other; acceptance of a hypothesis being the belief, at a certain probability level, that the subset specified by the hypothesis includes the parameter point. There can be no question about the legitimacy or acceptability of acceptance of the null hypothesis within this purely mathematical scheme since acceptance and rejection are perfectly complementary.

Consequently any interpretive difficulties which result from accepting the null hypothesis must be in the rules for or manner of relating empirical and statistical (null) hypotheses. The null hypothesis is of course that hypothesis for which the probability of erroneous rejection is fixed as α (or set at a maximum of α); the test (critical region) is chosen so as to maximize power for the given α and the alternative hypothesis. Since therein lies the only feature of the process that differentiates the null hypothesis from the other subset, the relating of empirical and statistical hypotheses must be based upon it.

While there are no firm rules for deciding with which of the two subsets a given empirical hypothesis should be associated, there have been certain practices or conventions used by different writers. Neyman (1942), for example, suggested a most reasonable convention for relating empirical and statistical hypotheses which is to equate with the null hypothesis that empirical hypothesis for which the error of erroneous rejection is more serious than the error of erroneous acceptance so that the more important error is under the direct control of the experimenter. There are a few other conventions based upon the derivational advantages of fixing α for a simple (rather than a composite) hypothesis, but it is quite clear that Grant has not merely restated any of these. In fact, Grant's (1962) strong statement that "using these predictions as the values in H_0 [the null hypothesis] is tactically inappropriate, frustrating, and self-defeating," (p. 61) indicates that his position is much more than a convention of convenience.

The position which I will develop over the remainder of this paper is not that *a-s* is preferable to *r-s*, but that there are no sound foundations for damning *a-s*. In this process let me initially point out that one can be led astray unless he recognizes that when one tests a point prediction he usually knows before the first sample element is drawn that his empirical hypothesis is not precisely true. Consider testing the hypothesis that two groups differ in means by some specified amount. We might test the hypothesis that the difference in means is 0, or perhaps 12, or perhaps even 122.5. But in each case we are certain that the difference is not precisely 0.0000 ... ad inf., or 12.0000 ..., or 122.50000 ... ad inf.

Recognition of this state of affairs leads to thinking in terms of differences or deviations that are or are not of importance for a given stage of theory construction or of application. Some express this in terms of differences which do and do not have practical importance, but I prefer the term zone of indifference which is used with important implications in sequential analysis. That is, if, for example, the difference in mean performance between two groups is less than, say, ε the two means may be considered equivalent for the given stage of theoretical development. In the case of a prediction of one-third for the proportion of right turns of rats in a maze, one would expect the same courses of action to be followed if the figure were actually .334 or .335. Thus, although we may specify a point null hypothesis for the purpose of our statistical test, we do recognize a more or less broad indifference zone about the null hypothesis consisting of values which are essentially equivalent to the null hypothesis for our present theory or practice.

While the formal procedures for testing statistical hypotheses are based upon the assumption that the sample size (n) is fixed prior to consideration of alternative test procedures, the user of statistical techniques is faced with the problem of choosing n and does so with regard for the magnitude of the discriminations which are or are not important for his particular application or level of theory development. In the typical case we choose the conditions of experimentation, including sample size, such that we will reject the null hypothesis with a given probability when the parameter difference is a certain magnitude. This is frequently done very formally in fields like agriculture, although rather informally in psychology. For example, in Cochran and Cox (1957) there is an extended discussion of the procedures for choosing the number of replications for an experiment on the basis of the practical importance of true differences. Thus, in one of their examples, a difference of 20% of the mean of two values is considered sufficiently important to warrant a sensitive enough

experiment to have an .80 probability of detecting it; that is, if the difference is 20% a large enough n is desired to insure that the power of the test is .80. Although it may happen that the required sample size is a function of an unknown distribution and not determinable in advance, it can usually be approximated with the tests used most frequently by psychologists.

The choice of sample size is but one feature in the overall planning to obtain an experiment of the desired precision with due consideration for the level of theory development (including alternate theories), the zones of indifference, and the related consequences of decision. However, such other features as the standard error per unit observation and the design efficiency do not have the flexibility of sample size, and, moreover, are usually chosen to maximize precision for reasons of economy. The choice of optimum sample size applies to all experimental strategies, including the non-objectionable (to Grant) and more usual *r-s*. It is surely apparent that anyone who wants to obtain a significant difference badly enough can obtain one—if his only consideration is obtaining that significant difference. Accepting that the means, for example, of two groups are never perfectly equal, the difference between them is some value ε. It is obviously an easy matter to choose a sample size large enough for the ε, such that we will reject the null hypothesis with a given probability. But the difference may be so slight as to have no practical or theoretical consequences for the given stage of measurement and theory construction. As McNemar (1950) has recently pointed out, in his objections to the use of extreme groups, significant differences may be obtained even when the underlying correlation is as low as .10 which implies a proportion of predicted variance equal to .01.

After arguing against *a-s* on the basis of the dangers of tests that tend toward leniency, Grant points out that the procedure may be equally objectionable when the test is too stringent. He illustrates the latter by an example of a theory which is useful though far from perfect in its predictions. This particular point is perfectly in accord with my arguments since it demonstrates the parallelism of *a-s* and *r-s*. First, one does not usually want an experiment that is too stringent: in *a-s* because it may not be desirable to reject a useful, though inaccurate theory; in *r-s* because one may accept an extremely poor and practically useless theory. Second, one does not want an experiment that is too lenient or insensitive; in *a-s* because one may accept an extremely poor and practically useless theory; in *r-s* because it may not be desirable to reject a useful, though inaccurate theory (that is, to accept the null hypothesis which implies rejection of its alternative). The identical terms were chosen in the preceding sentences to dramatize the parallel implications for *a-s* and *r-s* of the general desirability of a test that is neither too stringent nor too insensitive. Whether or not the experiment is precise enough is, then, a function of theoretical and practical consequences, and not of whether acceptance or rejection of the null hypothesis leads to support for an empirical theory.

But, one may argue, while there is logical equivalence as stated above, there is not motivational equivalence. That is, while it is agreed that ideally investigators design their experiments (including their choice of sample sizes) in order to be reasonably certain of detecting only differences which are of practical or theoretical importance, in actual practice they are neither so wise nor so pure as to be influenced by these factors to the exclusion of social motivations. And it is indeed much easier to do insensitive rather than precise experimentation. This phenomenon is of course what Grant (1962) referred to in his statement,

The tactics of accepting H_0 as proof and rejecting H_0 as disproof of a theory lead to the anomalous results that a small-scale, insensitive experiment will most often be interpreted as favoring a theory, whereas a large-scale sensitive experiment will usually yield results opposed to the theory! (p. 56).

Perhaps that reflects the essential point of Grant's presentation—merely to caution imprudent experimenters that the combination of personal desire to establish one's hypothesis and the ease of performing insensitive experimentation produce a particularly troublesome interaction.

Before proceeding it should be remembered that scientific considerations may be made secondary to personal desires to establish a theory whether the procedure be *a-s* or *r-s* in a perfectly analogous fashion. The only difference involves such practical considerations as the fact that it is usually easier to run 5 or 10 subjects than 100 or 500.

If Grant (1962) merely intended his article to convey this obvious warning, I cannot understand the discussions which involve such statements as the following:

Unfortunately most of the procedures used to date in testing the adequacy of such theoretical predictions [from mathematical models] set rather bad examples. Probably the least adequate of these procedures has been that in which an H_0 of exact correspondence between theoretical and empirical points is tested against H_1 covering any discrepancy between predictions and experimental results (p. 55).

If one is pointing out the dangers of using insensitive *a-s* tests (rather than condemning *a-s* on logical grounds), one would be expected to object to a particular or general use of *a-s* only if the use involved insensitive tests. Thus, it might be argued that experimenter WKE obtained support for his quantitative prediction by the use of *a-s* with a test so insensitive that it could not reasonably detect important discrepancies between predictions and observations. Or, as another fictional example, it might be stated that RRB always used *a-s* and always found support for his linear models, but his *n* was uniformly less than 5. But, unless there were almost uniform use of insensitive tests with *a-s*, this cautionary position could not reasonably lead to a condemnation of *a-s*.

As I see it, moreover, the argument against insensitive *a-s* tests is nothing but a particular form of the more general argument against bad experimentation. It is unquestionably the case that an *a-s* experiment that is too small and insensitive is poor, but the poorness is a property of the insensitivity and not of the *a-s* procedure. An *r-s* experiment that is too small and insensitive is equally poor. Due to the interaction between personal achievement desires and the ease of sloppy experimentation, as referred to previously, it may be necessary to be particularly alert to the usual scientific safeguards when using *a-s*, but that is a trivial matter and hardly worthy of an article.

In summary, it would be perfectly justifiable to argue that *n* is too small (or even too large) for a particular degree of sensitivity required at a given level of scientific development, but that is far from a proscription of designs where acceptance of the null hypothesis is in some way to the experimenter's social or personal advantage.

Grant's Position from the Viewpoint of Scientific Development

In the process of concluding this discussion I would like to emphasize and expand on certain factors which seem most critical in the process of evaluating scientific theories, as well as to indicate that my objections to Grant's apparent position are justifiable beyond the confines of the Neyman-Pearson theory.

It is surely clear that at various phases in the development of a scientific field one is faced with the problem of deciding about the suitability of different theories. When a discipline is at an early stage of development, knowledge of empirical relationships is crude so that broad isolation of explanatory constructs may be the most that is obtainable. At this stage one might consider as a significant accomplishment the ruling out of the hypothesis that observed differences are chance phenomena. The empirical hypothesis of central concern would be that there is some relationship of unknown magnitude, while its

alternative would be the chance or noise explanation.

With increasing sophistication in the discipline the alternative hypotheses may represent different, but more or less equally well-developed theories. One does not choose between theory and chance, but between theory and theory or between theory and theories. Another aspect of increased sophistication is frequently the greater precision in the prediction of empirical results for the various theories.

The decision as to which of the theories is admissible on the basis of the available data may be accomplished directly within the Neyman-Pearson framework, but that is not necessarily the case. Sometimes the choice among theories depends upon a succession of tests of hypotheses or possibly even upon quite informal considerations; as an example of the latter, one theory may lead to a prediction which is perfectly in accord (within rounding errors) with the observations while the other theory is off by quite a margin—a statistical test would be considered foolish indeed. In disciplines that have markedly smaller observational variability than psychology the most common procedure consists of a subjective comparison between predictions and observations. Moreover, the point that one chooses among alternative hypotheses at various stages of scientific development (whether by statistical methods or otherwise) most certainly does not imply that his efforts stop once he has accepted or rejected a given hypothesis as Grant implies; if the accepted theory, for example, is of any interest he proceeds to make finer analyses and comparisons which may range from orthogonal sub-comparisons in the analysis of variance to intuitive rumination. This provides a basis for objecting to Grant's arguments to the effect that hypothesis testing should be replaced (not supplemented) by estimation. The point is that both are usable, but at different phases of investigation.

I will again refer to the Atkinson and Suppes (1958) experiment to illustrate the relative roles of hypothesis testing and subsequent analysis in scientific advancement. Their first strategy was to decide which of two theories—game theory or the Markov model—was most adequate in the given experimental context. This clearly was a problem of testing hypotheses; a choice had to be made and the procedures of estimation could at best provide a substage on the way to the decision. The Markov model was accepted and game theory rejected, as noted above, but this certainly did not lead to a cessation of activity. Instead the investigators initially compared theoretical and observed transition matrices (and found them distinctly different), they then tested the more specific hypothesis of a one-stage Markov model against the alternative of a two-stage model, and finally they investigated the stationarity of the Markov process.

During its early phases, Einstein's general theory of relativity was equivalent to Newtonian theory in the success of explaining various common phenomena and a choice between them could not be made. But the Einstein theory led to certain predictions differing from Newtonian and these in turn led to a series of "crucial" tests. Among these were the exact predictions as to the magnitude of the bending of a light ray from a star by the gravitational field of the sun and the shift of wavelength of light emitted from atoms at the surface of stars. The general theory of relativity, thus, led to predictions which differed from the predictions of the alternative theory (Newton's), and the ultimate correspondence between these predictions and empirical observations (acceptance of no difference between predicted and obtained results) let to support for general relativity. While agreements between theory and observational results have been close they certainly have not been perfect—even physicists have problems of measurement precision and intricacy of mathematical derivation. But to the best judgment of the scientists the closeness of the fit between predictions and observations warrants the conclusion that the data provide support for the theory. Surely, however, despite its tremendous

power, physicists do not claim that Einstein's general theory has been proved nor are they convinced that it will not be ultimately replaced by a better theory.

It does not seem reasonable to argue that this method of scientific procedure is not suitable for psychology—just because our measurement precision happens to be lower than in physics and we use statistical tests rather than purely observational comparison.

References

Atkinson, R. C., & Suppes, P. An analysis of two-person game situations in terms of statistical learning theory. *J. exp. Psychol.*, 1958, **55**, 369–378.

Binder, A., & Feldman, S. E. The effects of experimentally controlled experience upon recognition responses. *Psychol. Monogr.*, 1960, **74** (9, Whole No. 496).

Bower, G. H. An association model for response and training variables in paired-associate learning. *Psychol. Rev.*, 1962, **69**, 34–53.

Brody, A. L. Independence in the learning of two consecutive responses per trial. *J. exp. Psychol.*, 1958, **56**, 16–20.

Bush, R. R., & Mosteller, F. *Stochastic models for learning*. New York: Wiley, 1955.

Chernoff, H., & Moses, L. E. *Elementary decision theory*. New York: Wiley, 1959.

Cochran, W. G., & Cox, Gertrude M. *Experimental designs*. (2nd ed.) New York: Wiley, 1957.

Fisher, R. A. The fiducial argument in statistical inference. *Ann. Eugen.*, 1935, **6**, 391–398.

Fisher, R. A. *Statistical methods for research workers*. (10th ed.) Edinburgh: Oliver & Boyd, 1948.

Fisher, R. A. *The design of experiments*. (5th ed.) Edinburgh: Oliver & Boyd, 1949.

Fisher, R. A. The comparison of samples with possibly unequal variances. In *Contributions to mathematical statistics*. New York: Wiley, 1950.

Fisher, R. A. Statistical methods and scientific induction. *J. Roy. Statist. Soc., Ser. B.*, 1955, **17**, 69–78.

Fisher, R. A. *Statistical methods and scientific inference*. (2nd ed.) Edinburgh: Oliver & Boyd, 1959.

Fisher, R. A. Scientific thought and the refinement of human reasoning. *J. Operat. Res. Soc., Japan*, 1960, **3**, 1–10.

Grant, D. A. Testing the null hypothesis and the strategy and tactics of investigating theoretical models. *Psychol. Rev.*, 1962, **69**, 54–61.

Grant, D. A., & Norris, Eugenia B. Dark adaptation as a factor in the sensitization of the beta response of the eyelid to light. *J. exp. Psychol.*, 1946, **36**, 390–397.

Harrow, M., & Friedman, G. B. Comparing reversal and nonreversal shifts in concept formation with partial reinforcement controlled. *J. exp. Psychol.*, 1958, **55**, 592–598.

Hogben, L. *Statistical theory*. London: Allen & Unwin, 1957.

Jeffreys, H. *Scientific inference*. (2nd ed.) Cambridge: Cambridge Univer. Press, 1957.

Kendall, M. G., & Buckland, W. R. *A dictionary of statistical terms*. Edinburgh: Oliver & Boyd, 1957.

Mather, K. *Statistical analysis in biology*. New York: Interscience, 1943.

McNemar, Q. At random: Sense and nonsense. *Amer. Psychologist*, 1960, **15**, 295–300.

Neyman, J. Basic ideas and theory of testing statistical hypothesis. *J. Roy. Statist. Soc.*, 1942, **105**, 292–327.

Neyman, J. *Lectures and conferences on mathematical statistics and probability*. (2nd ed.) Washington: United States Department of Agriculture, Graduate School, 1952.

Neyman, J. Note on article by Sir Ronald Fisher. *J. Roy. Statist. Soc.*, 1956, **18**, 288–294.

Neyman, J. Silver jubilee of my dispute with Fisher. *J. Operat. Res. Soc., Japan*, 1961, **3**, 145–154.

Neyman, J., & Pearson, E. S. On the use and interpretation of certain test criteria for purposes of statistical inference. Part I. *Biometrika*, 1928, **20A**, 175–240.   (a)

Neyman, J., & Pearson, E. S. On the use and interpretation of certain test criteria for purposes of statistical inference. Part II. *Biometrika*, 1928, **20A**, 263–294.   (b)

Neyman, J., & Pearson, E. S. On the problem of the most efficient tests of statistical hypotheses. *Phil. Trans. Roy. Soc., Ser. A*, 1933, **231**, 289–337.

Parzen, E. *Modern probability theory and its applications*. New York: Wiley, 1960.

Rozeboom, W. W. The fallacy of the null-hypothesis significance test. *Psychol. Bull.*, 1960, **57**, 416–428.

Savage, L. J. *The foundations of statistics*. New York: Wiley, 1954.

Snedecor, G. W. *Statistical methods*. (5th ed.) Ames: Iowa State Coll. Press, 1956.

von Mises, R. On the correct use of Bayes' formula. *Ann. math. Statist.*, 1942, **13**, 156–165.

von Mises, R. *Probability, statistics and truth*. (2nd ed.) London: Allen & Unwin, 1957.

Weinstock, S. Acquisition and extinction of a partially reinforced running response at a 24-hour intertrial interval. *J. exp. Psychol.*, 1958, **56**, 151–158.

Witte, R. S. A stimulus-trace hypothesis for statistical learning theory. *J. exp. Psychol.*, 1959, **57**, 273–283.

Editor's comments: What is the appropriate relationship between a scientific hypothesis and the statistical hypotheses of classical hypothesis-testing procedures? This question is not an easy one to answer. In the previous article Arnold Binder set forth the basic issues in this controversy. Further clarification of the issues and alternative points of view are provided in the following paper by Ward Edwards and a subsequent paper by Warner Wilson, Howard L. Miller, and Jerold S. Lower.

According to Edwards, classical hypothesis-testing procedures, or any other statistical procedures for that matter, can never satisfactorily test the goodness of fit of a single model to data. As an alternative he recommends that likelihood-ratio procedures be used to compare the fit of several models to the same data. On the other hand Wilson, Miller, and Lower recommend continued use of classical procedures combined with the identification of a scientific theory with the alternative hypothesis. While the main threads of the various arguments are examined in these papers, they also contain references to other papers that provide further insight concerning this controversy.

4.5 Tactical Note on the Relation between Scientific and Statistical Hypotheses

Ward Edwards

Abstract. Grant, Binder, and others have debated what should be the appropriate relationship between the scientific hypotheses that a scientist is interested in and the customary procedures of classical statistical inference. Classical significance tests are violently biased against the null hypothesis. A conservative theorist will therefore associate his theory with the null hypothesis, while an enthusiast will not—and they may often reach conflicting conclusions, whether or not the theory is correct. No procedure can satisfactorily test the goodness of fit of a single model to data. The remedy is to compare the fit of several models to the same data. Such procedures do not compare null with alternative hypotheses, and so are in this respect unbiased.

Grant (1962), Binder (1963), and Wilson and Miller (1964) have been debating the question of what should be the appropriate relationship between the scientific hypotheses or theories that a scientist is interested in and the statistical hypotheses, null and alternative,

Reprinted from *Psychological Bulletin*, 1965, **63**, 400–402, by permission of the publisher and author. Copyright 1965 by the American Psychological Association.

This research was supported by the United States Air Force under Contract AF 19 (628)-2823 monitored by the Electronics Systems Division, Air Force Systems Command. The author is grateful to L. J. Savage, D. A. Grant, and W. R. Wilson for helpful criticisms of an earlier draft.

that classical statistics invites him to use in significance tests. Grant rightly notes that using the value predicted by a theory as a null hypothesis puts a premium on sloppy experimentation, since small numbers of observations and large variances favor acceptance of the null hypothesis and " confirmation " of the theory, while sufficiently precise experimentation is likely to reject any null hypothesis and so the theory associated with it, even when that theory is very nearly true. Grant's major recommendation for coping with the problem is to use confidence intervals around observed values; if the theoretical values do not lie within these limits, the theory is suspect. With this technique also, sloppy experimentation will favor acceptance of the theory—but at least the width of the intervals will display sloppiness. Grant also suggests testing the hypothesis that the correlation between predicted and observed values is zero (in cases in which a function rather than a point is being predicted), but notes that an experiment of reasonable precision will nearly always reject this hypothesis for theories of even very modest resemblance to the truth. Binder, defending the more classical view, argues that the inference from outcome of a statistical procedure to a scientific conclusion must be a matter of judgment, and should certainly take the precision of the experiment into account, but that there is no reason why the null hypothesis should not, given an experiment of reasonable precision, be identified with the scientific hypothesis of interest. Wilson and Miller point out that the argument concerns not only statistical procedures but also choice of theoretical prediction to be tested, since some predictions are of differences and some of no difference. Their point seems to apply primarily to loosely formulated theories, since precise theories will make specific numerical predictions of the sizes of differences and it would be natural to treat these as null hypothesis values.

Edwards, Lindman, and Savage (1963), in an expository paper on Bayesian statistical inference, have pointed out that from a Bayesian point of view, classical procedures for statistical inference are always violently biased against the null hypothesis, so much so that evidence that is actually in favor of the null hypothesis may lead to its rejection by a properly applied classical test. This fact implies that, other things being equal, a theory is likely to look better in the light of experimental data if its prediction is associated with the alternative hypothesis than if it is associated with the null hypothesis.

For a detailed mathematical exposition of the bias of classical significance tests, see Edwards, Lindman, and Savage (1963) and Lindley (1957). Lindley has proven a theorem frequently illustrated in Edwards, Lindman, and Savage (1963) that amounts to the following. An appropriate measure of the impact of evidence on one hypothesis as against another is a statistical quantity called the likelihood ratio. Name any likelihood ratio in favor of the null hypothesis, no matter how large, and any significance level, no matter how small. Data can always be invented that will simultaneously favor the null hypothesis by at least that likelihood ratio and lead to rejection of that hypothesis at at least that significance level. In other words, data can always be invented that highly favor the null hypothesis, but lead to its rejection by an appropriate classical test at any specified significance level. That theorem establishes the generality and ubiquity of the bias. Edwards, Lindman, and Savage (1963) show that data like those found in psychological experiments leading to .05 or .01 level rejections of null hypotheses are seldom if ever strong evidence against null hypotheses, and often actually favor them.

The following example gives the flavor of the argument, though it is extremely crude and makes no use of such tools as likelihood ratios. The boiling point of statistic acid is known to be exactly 150° C. You, an organic chemist, have attempted to synthesize statistic acid; in front of you is a beaker full of foul-smelling glop, and you would like to know whether or not it is indeed statistic acid. If it is not, it may be any of a large number of related compounds with boiling points diffusely (for the example, that means uniformly) distributed over the region from 130° C to 170° C. By one of those happy accidents so common in statistical examples, your thermometer is known to be unbiased and to produce normally distributed errors with a standard deviation of 1°. So you measure the boiling point of the glop, once.

The example, of course, justifies the use of the classical critical ratio test with a standard deviation of 1°. Suppose that the glop really is statistic acid. What is the probability that the reading will be 151.96° or higher? Since 1.96 is the .05 level on a two-tailed critical ratio test, but we are here considering only the upper tail, that probability is .025. Similarly, the probability that the reading will be 152.58° or greater is .005. So the probability that the reading will fall between 151.96° and 152.58°, if the glop is really statistic acid, is .025–.005 = .02.

What is the probability that the reading will fall in that interval if the glop is not statistic acid? The size of the interval is .62°. If the glop is not statistic acid, the boiling points of the other compounds that it might be instead are uniformly distributed over a 40° region. So the probability of any interval within that region is simply the width of the interval divided by the width of the region, .62/40 = .0155. So if the compound is statistic acid, the probability of a reading between 151.96° and 152.58° is .02, while if it is not statistic acid that probability is only .0155. Clearly the occurrence of a reading in the region, especially a reading near its lower end, would favor the null hypothesis, since a reading in that region is more likely if the null hypothesis is true than if it is false. And yet, any such reading would lead to a rejection of the null hypothesis at the .05 level by the critical ratio test.

Obviously the assumption made about the alternative hypothesis was crucial to the calculation. (Such special features as normality, the literal uniformity of the distribution under the alternative hypothesis, and the particular regions and significance levels chosen are not at all important; they affect only the numerical details, not the basic phenomenon.) The narrower the distribution under the alternative hypothesis, the less striking is the paradox; the wider that distribution, the more striking. That distribution is narrowest if it is a single point, and favors the alternative hypothesis most if that point happens to coincide with the datum. And yet Edwards, Lindman, and Savage (1963) show that even a single-point alternative hypothesis located exactly where the data fall cannot bias the likelihood ratio against the null hypothesis as severely as classical significance tests are biased.

This violent bias of classical procedures is not an unmitigated disaster. Many null hypotheses tested by classical procedures are scientifically preposterous, not worthy of a moment's credence even as approximations. If a hypothesis is preposterous to start with, no amount of bias against it can be too great. On the other hand, if it is preposterous to start with, why test it?

The implication of this bias of classical procedures against null hypotheses seems clear. If classical procedures are to be used, a theory identified with a null hypothesis will have several strikes against it just because of that identification, whether or not the theory is true. And the more thorough the experiment, the larger that bias becomes. The scientific conservative, eager to make sure that error is scotched at any cost, will therefore prefer to test his theories as null hypotheses— to their detriment. The scientific enthusiast, eager to make sure that his good new ideas do not die premature or unnecessary deaths, will if possible test his theories as alternative hypotheses—to their advantage. Often, these men of different temperament will reach different conclusions.

The subjectivity of this conclusion is distressing, though realistic. There should be a better, less subjective approach—and there is. The trouble is that in classical statistics the alternative hypothesis is essentially undefined, and so provides no standard by means of which to judge the congruence between datum and null hypothesis; hence the arbitrariness of the .05, .01, and .001 levels, and their lack of agreement with less arbitrary measures of congruence. A man from Mars, asked whether or not your suit fits you, would have trouble answering. He could notice the discrepancies between its measurements and yours, and might answer no; he could notice that you did not trip over it, and might answer yes, But give him two suits and ask him which fits you better, and his task starts to make sense, though it still has its difficulties. I believe that the argument between Grant and Binder is essentially unresolvable; no procedure can test the goodness of fit of a single model to data in any satisfactory way. But procedures for comparing the goodness of fit of two or more models to the same data are easy to come by, entirely appropriate, and free of the difficulties Binder and Grant have been arguing about. (They do have difficulties. Most important, either these models must specify to some extent the error characteristics of the data-generating process, or else a special model of the data-generating process, such as the normality assumption concerning the thermometer in the statistic acid example, must also be supplied. But of course, this difficulty is common to all of statistics, and is fully as much a difficulty for the approaches I am rejecting as for those I am espousing.) The likelihood-ratio procedures I advocate do not make any use of classical null-hypothesis testing, and so the question of which model to associate with the null hypothesis does not arise. While there is nothing essentially Bayesian about such procedures, I naturally prefer their Bayesian to their non-Bayesian versions, and so refer you to Savage (1962), Raiffa and Schlaifer (1961), Schlaifer (1959, 1961), and Edwards, Lindman, and Savage (1963) as appropriate introductions to them. Unfortunately, I cannot refer you to literature telling you how to invent not just one but several plausible models that might account for your data.

References

Binder, A. Further considerations on testing the null hypothesis and the strategy and tactics of investigating theoretical models. *Psychological Review*, 1963, **70**, 107–115.

Edwards, W., Lindman, H., & Savage, L. J. Bayesian statistical inference for psychological research. *Psychological Review*, 1963, **70**, 193–242.

Grant, D. A. Testing the null hypothesis and the strategy and tactics of investigating theoretical models. *Psychological Review*, 1962, **69**, 54–61.

Lindley, D. V. A statistical paradox. *Biometrika*, 1957, **44**, 187–192.

Raiffa, H., & Schlaifer, R. *Applied statistical decision theory*. Boston: Harvard University, Graduate School of Business Administration, Division of Research, 1961.

Savage, L. J., et al. *The foundations of statistical inference: A discussion*. New York: Wiley, 1962.

Schlaifer, R. *Probability and statistics for business decisions*. New York: McGraw-Hill, 1959.

Schlaifer, R. *Introduction to statistics for business decisions*. New York: McGraw-Hill, 1961.

Wilson, W. R., & Miller, H. A note on the inconclusiveness of accepting the null hypothesis. *Psychological Review*, 1964, **71**, 238–242.

4.6 Much Ado about the Null Hypothesis

Warner Wilson, Howard L. Miller,
and Jerold S. Lower

Abstract. Edwards has charged that classical statistics, in contrast to Bayesian statistics, is always violently biased against the null hypothesis. Edwards has advised the conservative classical investigator that he should, therefore, always identify his theory with the null hypothesis, so as to minimize specious claims for theoretical support. This paper reinterprets the so-called bias in terms of differential assumptions about the nature of the alternatives which must be considered; its main purpose, however, is to point out that insensitive experiments, in contrast to sensitive ones, are always biased for, rather than against, the null hypothesis. It is this 2nd bias (which exists independently of the 1st) that prompts the conservative investigator not to identify his theory with the null hypothesis.

Grant (1962) and Binder (1963) have clarified the fact that two strategies can be used in theory testing. First, one can identify the theory under test with the null hypothesis and claim support for the theory if the null hypothesis is accepted. The second presumably more orthodox and traditional approach is to identify the theory under test with the alternative hypothesis and claim support for the theory if the null hypothesis is rejected. Binder has referred to these two approaches as acceptance support and rejection support strategies. In this paper, however, the authors will follow Edwards (1965) and speak of

The authors wish to express their thanks to Ward Edwards for his personal communications with them and for the patient forbearance he has demonstrated in attempting to help clarify the ideas that are discussed in this paper. The order of authorship should not be interpreted as implying a greater contribution on the part of the senior author.

Reprinted from *Psychological Bulletin*, 1967, **67**, 188–196, by permission of the publisher and authors. Copyright 1967 by the American Psychological Association.

"identifying one's theory with the null hypothesis" and basing support on the acceptance of the null versus "identifying with the alternative hypothesis" and basing support on the rejection of the null and, of course, the subsequent acceptance of the alternative.

Binder (1963) has ably pointed out that either strategy may be used effectively under some circumstances. Wilson and Miller (1964a, 1964b) have joined with Grant, however, in arguing that it is generally better to identify with the alternative and, hence, base support for a theory on the rejection of some null hypothesis. These writers pointed out that while the probability of *rejecting* the null hypothesis wrongly is held constant, for example, at the .05 level, the probability of *accepting* the null hypothesis wrongly varies with the precision of the experiment. To the extent that error is large, the study in question is biased for the null hypothesis and for any theories identified with it. According to this view, the conservative, cautious approach is to identify one's theory with the alternative hypothesis.

This essentially orthodox (Fisherian) view of classical statistical procedures has seemed so reasonable to the present authors that they were surprised to find Edwards, who identifies himself with Bayesian statistics, taking a dramatically opposed view. Edwards' article is only one of several considering the relative virtues of classical versus Bayesian statistics (see Binder, 1964, for an excellent review). Edwards' article seems especially important, however, due to the fact that it strongly urges changes in the tactics of the orthodox classical statistician—changes which might prove to be ill-advised in some cases and impossible in others. Edwards' (1965) paper seems to make or imply the following points: (*a*) Classical procedures, in fact, "are always violently biased against the null hypothesis [p. 400]." (*b*) The cautious, conservative approach, therefore, is to identify one's theory with the null hypothesis and, hence, base support for one's theory on the acceptance of the null hypothesis. (*c*) The ideal solution, however, is to compare the goodness of fit of several models to the same data, thus avoiding the whole problem of null hypothesis testing.

Edwards apparently believes that Bayesian analysis is always feasible or that if it is not, the experiment in question is not worth doing. The present writers do not agree with this point. They do, however, find much that is admirable in Edwards' position and certainly agree that Bayesian procedures are to be preferred—when they can be used. The purpose of the present paper, therefore, is not to disagree with Edwards so much as to suggest clarification and qualification.

In connection with Point *a*, it is conceded that Edwards is commenting on differences between classical and Bayesian statistics that really exist. It is suggested, however, that the term "bias" is perhaps not the best way to sum up these differences. A Bayesian analysis typically assumes that the datum comes from a null distribution or from some other distribution. Classical statistics assumes that the datum comes from a null distribution or from some one of all other possible distributions. The assumption that the datum may come from *any* distribution does, indeed, always increase the apparent probability that it comes from some distribution other than the null. The difference lies in the nature of the alternatives which are to be taken into account. Classical procedures happen to assume that all possible alternatives must be taken into account. Granted this assumption, any bias for or against the null is then expressed by the probability level, which, even in the case of the .05 level, clearly favors the null.

In relation to Point *b*, it is suggested that the bias for theories identified with the null hypothesis in imprecise experiments, which Grant talked about, exists independently of and is logically distinct from the bias against the null hypothesis which Edwards talked about. Even when Edwards' bias *against* the null hypothesis exists, it does not imply the absence of Grant's bias *for* the null hypothesis. These considerations make the choice of tactics more complex than Edwards' article indicated. Edwards' (1965) tactical advice was that, "If classical procedures are to be used, a theory identified with a null hypothesis will have several strikes against it. . . . And the more thorough the experiment, the larger that bias becomes [p. 402]." An attempt will be made to show that this advice is valid only in experiments of extreme precision, a type presumably rare in psychology. As experiments become imprecise, just the opposite tactical advice becomes appropriate.

In relation to Point *c*, it is suggested that no matter how many models we may have, many people will still find null hypotheses which will seem to them to need testing.

Is Classical Statistics Always Biased against the Null Hypothesis?

Edwards' first point was that classical statistics is always violently biased against the acceptance of the null hypothesis. He argued persuasively for this point in several different ways. He presented, for one thing, the following example, which supposedly illustrates the bias. This example and other points made by Edwards will be considered here in hopes that the discussion will help clarify the circumstances under which classical procedures can and cannot meaningfully be said to be biased, and also give some indication of how often inappropriate rejections of the null hypothesis may, in fact, occur.

The following example gives the flavor of the argument, though it is extremely crude and makes no use of such tools as likelihood ratios. The boiling point of statistic acid is known to be exactly 150° C. You, an organic chemist, have attempted to synthesize statistic acid; in front of you is a beaker full of foul-smelling glop, and you would like to know whether or not it is indeed statistic acid. If it is not, it may be any of a large number of related compounds with boiling points diffusely (for the example, that means uniformly) distributed over the region from 130° C. to 170° C. By one of those happy accidents so common in statistical examples, your thermometer is known to be unbiased and to produce normally distributed errors with a standard deviation of 1°. So you measure the boiling point of the glop, once.

The example, of course, justifies the use of the classical critical ratio test with a standard deviation of 1°. Suppose that the glop really is statistic acid. What is the probability that the reading will be 151.96° or higher? Since 1.96 is the .05 level on a two-tailed critical ratio test, but we are here considering only the upper tail, that probability is .025. Similarly, the probability that the reading will be 152.58° or greater is .005. So the probability that the reading will fall between 151.96° and 152.58°, if the glop is really statistical acid, is .025 − .005 = .02.

What is the probability that the reading will fall in that interval if the glop is not statistic acid? The size of the interval is .62°. If the glop is not statistic acid, the boiling points of the other compounds that it might be instead are uniformly distributed over a 40° region. So the probability of any interval within that region is simply the width of the interval divided by the width of the region, .62/40 = .0155. So if the compound is statistic acid, the probability of a reading between

151.96° and 152.58° is .02, while if it is not statistic acid that probability is only .0155. Clearly the occurrence of a reading in that region, especially a reading near its lower end, would favor the null hypothesis, since a reading in that region is more likely if the null hypothesis is true than if it is false. And yet, any such reading would lead to a rejection of the null hypothesis at the .05 level by the critical ratio test [Edwards, 1965, p. 401].

If we follow the mode of analysis Edwards used, that of comparing an area of the null distribution to an area of the alternative distribution, we can note first that the probability of actually rejecting the null when the data in fact favor it is quite low, even in this contrived example. It would seem that we would reject the null when the data actually favor it when the observation falls in the interval 151.96–152.40. When the null is true, the data will fall in this interval, or in the corresponding lower interval, only about 3 % of the time.

In addition, the type of analysis Edwards used has two aspects that seem intuitively unappealing: (*a*) *All* outcomes have a low probability under the alternative hypotheses, and (*b*) this probability is equally low no matter where an observation occurs. For example, if a datum occurs at 150° the implied inference is that the probability of a hit in this segment of .62 is .62/40 or .0155; likewise, if a datum occurs at 169° the implied inference is that the probability of a hit in this segment of .62 is still .62/40 or .0155.

Another way of looking at this example avoids both of the aspects noted above. Suppose, for simplification, that 1.96 is rounded off to 2.00. The null hypothesis then implies that the observations should fall in a segment of 4° between 148° and 152° 95 % of the time. Suppose the actual reading is 152°. The probability of a hit as far as 2° away from 150°, if the null is true, is only .05; therefore, the null is rejected. It would seem that in order to test the alternative hypothesis, one could ask "What is the probability of a hit as close as 2° to 150° if the acid is not statistic?" The answer would seem to be 4/40 or .10. It would seem, then, that a reading of 151.96° or 152° is more probable if the null hypothesis is false, and rejection of the null would seem appropriate after all. Perhaps, then, the bias is not as prevalent as Edwards would lead us to believe. Indeed, if a person working at the .05 level of significance is to run any danger of rejecting the null when the data actually support it, it would seem that the width of the distribution under the alternative hypothesis must be more than 80 standard deviations. The present authors will leave it up to Edwards to show that situations of this sort occur frequently enough to justify concern.

The reader may wish to note, however, that the apparent bias in classical procedures can be manipulated at will. If either a broad alternative distribution or a small error term is assumed, the bias will be increased.

Lindley (1957), another Bayesian, presumably had just such considerations in mind when he stated that data can always be invented that highly favor the null hypothesis yet lead to its rejection by a properly applied classical test. Although such data can be invented, it is still meaningful to ask whether such data are inevitable or even likely in reality. By assuming a wide enough alternative, a bias can be created; however, in reality the alternative distribution is supposedly based on some theoretical or empirical consideration, and its width cannot be set arbitrarily. Error terms, on the other hand, can be reduced, even in reality, by the expedient of collecting more cases. However, in order for a reduced error term to lead inevitably to a bias, it is necessary to assume that as the error term is reduced, the absolute deviation from chance becomes less, so that the probability of the deviation's occurrence remains constant. This convenient constancy of probability in the face of a shrinking error term cannot be expected in reality. Consider again the statistic acid example, assuming a deviation of 2° and an error term of 1°. As indicated above, the probability of a hit at 152° is .05 under the null and .10 under the alternative. In real life, any observation is

more likely than not to be near the true value, so if data are collected until the error term becomes .1°, the mean value of all the observations may still be near 152°. Now the probability under the alternative would still be .10, but the probability under the null would have become far, far less, since the critical ratio would now be 20. Thus, although it is necessary to admit that data can always be invented which will make a classical test look biased, it is also possible to point out that such data are not necessarily obtainable in reality and that certainly it is also easy to invent data that make classical tests look unbiased.

The considerations up to this point seem to support the assertion that classical statistics if blindly applied will sometimes be biased against the null hypothesis. Support for the assertion that classical statistics is always biased seems lacking. Instead, it appears that the extent of such bias depends in great part on the width of the alternative distribution relative to the width of the null distribution.

In further exploring Edwards' position, it may be noted that Edwards conceded that the bias becomes less as the relative width of the alternative distribution decreases. He firmly insisted, however, that no matter how narrow the alternative distribution, the bias will persist. Edwards seems to have come to this conclusion through an inappropriate comparison of probability levels with likelihood ratios.

The likelihood ratio is the ratio of one probability density (at the point of the data) to another. In a case such as the acid example, the Bayesian would apparently rather base his conclusions on the likelihood ratio than on the probability level, and the present authors would have no objection so far as such cases are concerned. In commenting on the relation of likelihood ratios to classical significance tests, Edwards (1965) said:

Even a single-point alternative hypothesis located exactly where the data fall [the form of alternative distribution that most violently biases the likelihood ratio against the null hypothesis] cannot bias the likelihood ratio against the null hypothesis as severely as classical significance tests are biased [p. 401].

Whether we formally use likelihood ratios (ratios of ordinates) or simply other ratios of conditional probabilities (areas), what Edwards was saying seems to have a certain surface validity. If we consider only the probability (or probability density at a point) under one distribution, the null hypothesis, and do not take into consideration how likely this event might be under some specified alternative distribution, we are biasing ourselves against our null hypothesis by comparing it essentially to all possible alternatives instead of comparing it to just one. It seems to us that this is exactly what the classical statistician *intends* to do. If we have a clearly defined alternative, and we can say that " reality " must be one or the other, then we can justify a likelihood ratio or some similar procedure. If we do not have such alternative models, we cannot invent them to avoid a theoretical bias.

Edwards, Lindman, and Savage (1963) implied that in any inference based on statistics, the decision involved must be a joint function of a prior probability estimate (what you thought before about the likelihood of your hypothesis), the likelihood ratio, and the payoff matrix (the relative rewards and costs of being right and wrong about your hypothesis). Edwards seemed to imply that "classical statisticians" use no such considerations. It is doubtful that this is so. For one thing, a choice of significance level can, under some circumstances, be construed as a prior probability estimate—not the subjective one of the individual scientist but an admittedly arbitrary attempt to standardize a bias against alternative hypotheses (not altogether different from a bias for the null). It appears to be a deliberate attempt to offer a standardized, public method for objectifying an individual scientist's willingness to make an inference. An undisputed goal of science is objectivity—public reproducibility.

To introduce, into inferential statistics, a methodology dependent on partially subjective estimates of probability would seem to undermine this goal.

It is apparently impossible to say whether the choice of low probability levels implies a low prior probability or a payoff matrix biased for retention of the null hypothesis, but one or both seem implicit. The natural tendency is for investigators to believe that their hypotheses are correct and that the world can ill afford to ignore them. If such subjective inclinations were allowed full sway, experimentation could become superfluous, and science might well degenerate into controversy. Classical statistics wisely resolves prior probability and payoff considerations in a conservative and standard rather than in a subjective and variable manner.

In answer to the obvious criticism concerning the subjectivity of invented alternative hypotheses, Edwards[1] stated essentially that a scientist always has some information regarding alternatives. Bayesian statistics considers this information; classical statistics does not. It seems that the bias that Edwards was discussing is a function of the failure to use (assume?) this additional information. In point of fact, the choice of one particular alternative distribution with which to compare the null excludes other possible alternatives. It is precisely this exclusion (which may be justifiable under some circumstances) which increases the probability that you will accept the null.

In this section we have attempted to discuss such questions as "Are classical procedures biased against the null hypothesis?" and "How often might such a bias result in rejections of the null when the data actually favor it or discredit it only weakly?" We see no reason to view classical procedures as biased against the null in any absolute sense —the opposite seems to be the case. We see no reason to view classical procedures as biased against the null hypothesis relative to

[1] W. Edwards, personal communication, October 1966.

Bayesian statistics since this comparison is at best unsatisfactory due to the difference in the procedures which are used and in the information which is assumed to be available. We can concede only that under special conditions, including presumably a specifiable alternative distribution, blind use of a classical analysis might result in a rejection of the null when a defensible Bayesian analysis, considering only the specifiable alternative, might show that the data actually support the null. We know of not one real-life instance in which the above has been demonstrated. It would seem to us that circumstances making such errors likely are not frequent, and it is suggested that the burden of proof is on those who think that such errors occur frequently—in the literature, as opposed to in the examples used in articles written by Bayesians.

Should the Scientific Conservative Always Identify His Theory with the Null Hypothesis?

Edwards' second main point was that if one does use classical statistics, the more conservative strategy is to identify one's theory *not* with the alternative hypothesis—as Wilson, Miller, and Grant advocated—but with the null hypothesis. This point, too, is apparently incorrect. What is more, even if the first point, about classical statistics being always biased, were correct, the second point would not necessarily follow. In order to see the actual independence of the two points, it is helpful to realize that if something is biased, it is biased relative to something else. Edwards presumably meant that classical statistics is always biased relative to Bayesian statistics.

On the other hand, when Grant, Wilson, and Miller said that an insensitive experiment is biased for the acceptance of the null hypothesis and any theory identified with it, they meant that an insensitive experiment is biased relative to a sensitive one. A classical analysis can be biased against the null in comparison to a Bayesian analysis, and an insensitive experiment can still be biased for the null in

comparison to a sensitive experiment. The two biases exist independently. What is more, the bias for the null hypothesis in insensitive experiments is just as true with a Bayesian analysis as with a classical analysis.

The following example (see Table 1) is offered as an illustration of the bias in insensitive versus sensitive experiments—a bias which, prior to the Edwards article, the present writers would have considered not in need of illustration. The example views experiments as perfectly sensitive or as completely insensitive; null hypotheses as true or false; and theories as identified with the alternative and supported if the null is rejected, or as identified with the null and supported if the null is not rejected.

Edwards said that if you are a conservative, that is, if you wish to minimize undeserved successes, you should always identify your theory with the null. Wilson, Miller, and Grant said that in the case of insensitive experiments, such as are common in psychology, just the opposite tactic is to be recommended to the conservative. Table 1 indicates the number of deserved and undeserved successes achieved by theorists depending on whether they do identify with the alternative or with the null. Overall, identification with the null clearly promotes success for one's theory, a total of 285 (out of 400) successes versus 115 for those who identify with the alternative. The same is true in the case of total undeserved successes. Those who identify with the null get a total of 95 undeserved successes (out of 200) versus 10 for those who identify with the alternative. The most dramatic difference occurs, of course, in the case of insensitive experiments. The score is 95 illegitimate successes for those who test their theories as null hypotheses and only 5 for those who test their theories as alternative hypotheses. Hopefully, no one will wish to attempt to reconcile these outcomes with Edwards' (1965) assertion: "The scientific conservative, eager to make sure that error is scotched at any cost, will therefore [always] prefer to test his theories as null hypotheses—to their detriment [p. 402]."

The authors would like to concede, however, that in recommending continued use of classical statistics combined with identification of theories with alternative hypotheses, they are operating on the basis of several beliefs which are not mathematically demonstrable.

Perhaps the most critical belief in this context is that most experiments in psychology are insensitive. It may be noted that a conservative would not favor identification with the alternative on the basis of Table 1

Table 1. Limits of deserved and undeserved successes assuming identification with the null versus the alternative in perfectly sensitive and insensitive experiments.

	Identification of theory with the null		Identification of theory with the alternative	
	Percentage deserved successes	Percentage undeserved successes	Percentage deserved successes	Percentage undeserved successes
Sensitive experiments				
True null	95	0	0	5
False null	0	0	100	0
Insensitive experiments				
True null	95	0	0	5
False null	0	95	5	0

unless he held this belief. In a perfectly sensitive experiment, only identification with the alternative leads to accepting one's theory when it is false. This would occur because no matter how small the error, t's of 1.96 and greater will occur 5% of the time. These writers would point out, however, that in very sensitive experiments, specious deviations from chance, though technically significant, would be so small that they could hardly mislead anyone. It might also be argued that belief in ESP, for example, survives partly on just such deviations. Such a consideration might justify more concern about such errors, greater use of the .01 or .001 significance level, and greater interest in likelihood ratios when meaningful ones can be computed.

Although the good showing of "acceptance support" in sensitive experiments is in its favor, it should still be noted that, seemingly, grounds are seldom available for deciding if experiments are going to be sensitive. In the face of this inevitable equivocality, investigators are encouraged to identify their theories with the alternative and so put an upper limit on error of 5% rather than 95%. One is not justified, after all, in assuming that favorable circumstances—true theories and sensitive experiments—will occur generally. A sensible strategy must assume the possibility of unfavorable circumstances—false theories and insensitive experiments—and must provide protection against these unfavorable circumstances.

Intuitively, it might seem impossible to base any inference, or express any bias, on the basis of a completely insensitive experiment. This consideration, however, is just the point of the objection to identification with the null. Identification with the null allows one to base positive claims for theoretical confirmation on the acceptance of the null hypothesis, which is, of course, virtually assured ($p = .95$) in the completely insensitive experiment.

Yet another consideration relates to the fact that it makes relatively little difference which approach you use in precise experiments. With great precision, you cannot go too far wrong. As experiments become imprecise, however, the tactical choice makes an increasingly large difference. All the more reason, therefore, for the conservative to choose the tactic that protects him even when experiments are imprecise.

The second belief is that statistics can sometimes reveal facts worth knowing even if they are not apparent to the naked eye. The relationships between smoking and lung cancer and between obesity and heart disease are examples. It is true that statistics may lead one to be overly optimistic about the importance of effects that are more significant than important, and certainly this tendency is to be deplored. On the other hand, although the human observer, unbeguiled by statistics, may indeed discount many trivial effects, he may also infer strong effects when only a trivial effect or no effect at all is present. Many people are, in the opinion of these authors, overly optimistic about the existence of ESP and the efficacy of psychotherapy. This overoptimism exists, however, in spite of statistics, not because of them.

Those who favor the so-called interocular test should realize that situations in which effects are large relative to error will yield their secrets quickly. This consideration suggests that investigators will inevitably spend most of their time on ambiguous situations in which the effects of interest are small relative to the precision of measurement so far achieved. Statistics will be needed in such situations.

The third belief is that false positives are more damaging than false negatives. In the present context, this statement means that it is worse to view a false theory as already proved than to view a true theory as not yet proved. This belief is widespread (see, e.g., Campbell, 1959) and will not be belabored here. Granted the belief that false positives cause more trouble, identification of one's theory with the alternative is only natural. The traditional .05 significance level then limits the investigator to 5% error no matter how insensitive his experiments and no matter how false his hypotheses. When one identifies his theory with the null, however, the natural conservatism inherent in the traditional use of small probability levels works for, rather than against, hasty claims for support. Indeed, if an investigator identifies his theories with the null and if his theories are false, his number of false positives approaches not 5%, but 95%, as his experiments become increasingly imprecise.

The fourth belief is that many null hypotheses are worth testing. Edwards questioned this belief by suggesting that most null hypotheses are obviously false and that the testing of them is, therefore, meaningless. One of several possible replies is that the real question is frequently not whether the data deviate from the null, but whether the devia-

tion is positive or negative. In such a case the testing of the null is obviously meaningful.

If the conservative investigator believes that false positives are the greater threat and that his experiment will be insensitive, he will surely choose to identify his theory with the null hypothesis. On the other hand, none of the beliefs discussed so far necessarily justifies a preference for probability levels over likelihood ratios, and likelihood ratios are, in fact, strongly recommended whenever it is possible to compute them. The last belief, however, is that in most psychological experiments it is not possible to use a Bayesian approach based on likelihood ratios. The calculation of a ratio requires at least one specifiable alternative distribution. In most psychology experiments, no grounds are available for arriving at such a distribution. In such cases, classical statistics appears to be the only alternative, and, under this condition, the possibility of classical statistics being biased relative to Bayesian statistics is a meaningless issue.

Naturally, classical statisticians as well as Bayesian statisticians should consider alternative distributions. The point is that such considerations may or may not yield enough information to clearly justify a Bayesian analysis. So long as such information is often, if not usually, lacking, Edwards' rejection of classical approaches seems premature.

Edwards implied, to be sure, that investigators using classical tests have often rejected true null hypotheses without real evidence even when Bayesian statistics were potentially available. Bayesians could perhaps make a definite contribution by analyzing a number of published experiments and pointing out instances in which classical statistics has led to null hypothesis rejection when a potentially available Bayesian analysis would not have. It seems entirely possible, however, that such examples might be hard to find. At least they seem to be conspicuously absent from Bayesian critiques. It is also suggested that it would be most intructive if Bayesians would supply a precise alternative distribution to accompany any of the following null hypotheses, which are presented as typical psychological problems: (*a*) partially reinforced subjects extinguish at the same rate as continuously reinforced subjects; (*b*) patients show no change on tests of adjustment as a function of counseling; (*c*) punishment does not influence response rate; and (*d*) students working with teaching machines learn no faster and no better than students reading ordinary textbooks. Furthermore, the likelihood ratio, even if available, is not likely to disagree with the classical test unless the alternative distribution is very broad. If the meaningful alternative is not uniform, as it was in the statistic acid example, but instead has a mode or modes somewhere in the neighborhood of the null value, the likelihood ratio and significance test are even more likely to point in the same direction.

Another tactical point that merits attention is the question of whether a null hypothesis rejection resulting from a sensitive experiment should always be viewed as a legitimate success. As Table 1 indicates, if a null hypothesis is false, a sufficiently sensitive experiment will always reject it, and theories identified with the alternative will always be supported. Table 1 views such successes as legitimate. The possibility always exists, however, that the difference, though real, may be so slight that recording it in the literature is a complete waste of effort. The present writers think that the indiscriminate cataloguing of trivial effects is, in fact, a major problem in psychology today, and they would certainly regret it if their position were in any way interpreted as encouraging this unfortunate practice. On the other hand, as Wilson and Miller have pointed out, one must weigh one problem against another. If investigators identify their theories with the alternative hypothesis, they may be tempted to run *many* subjects and report theories to be true, even though the theories have no predictive utility. On the other hand, if investigators identify their theories with the null hypothesis, they may be tempted to run *few*

subjects and report theories to be true, even though they are completely false. The present writers find no difficulty in deciding that they would prefer to confront investigators with the first temptation rather than the second.

A consideration of the problem of accepting trivial effects does not, therefore, greatly modify this paper's inclination towards identification with the alternative hypothesis in combination with the use of classical tests. This strategy may not be the best imaginable, but it is still often the best available, and it is recommended to the scientific conservative.

In summing up this section, we wish to remind the reader that Edwards apparently recommended that investigators switch to Bayesian techniques altogether. He went on to say, however, that if one *does* use classical statistics, the more conservative tactic, that is, the tactic that minimizes erroneous claims for theoretical support, is identification with the null. Our perspective on this advice can be summed up very briefly: If a conservative is a person who wishes to abandon a tactic that puts a ceiling on error of 1 in 20 and adopt instead a tactic that puts a ceiling on error of 19 in 20, then this is excellent advice.

Does the Development of Multiple Models Avoid the Need for Null Hypothesis Testing?

Edwards' last point was that the better tactic is to compare your data to several plausible models, and hence avoid null hypothesis testing altogether. Multiple models seem thoroughly desirable, and it seems worthwhile to note that in a traditional two-tailed test, classical statistics always implies three families of models: one predicting no difference, one predicting a positive difference, and one predicting a negative difference. On the other hand, it is hard to see how multiple models avoid the need for null hypothesis testing. Again an example from Edwards' (1965) paper may be helpful.

A man from Mars, asked whether or not your suit fits you, would have trouble answering. He could notice the discrepancies between its measurements and yours, and might answer no; he could notice that you did not trip over it, and might answer yes. But give him two suits and ask him which fits you better, and his task starts to make sense [p. 402].

But ask him if he's sure of his decision or if he might reverse it if he saw you model the suits again, and you have a null hypothesis to test. In other words, although on one occasion one model (suit) was judged to fit the data (you) better, one must still ask if the difference in fit is significant relative to the potential sources of error. However stated and however tested, this question still seems to constitute a null hypothesis. Bayesian statistics may offer a meaningful alternative to null hypothesis testing, but it will take more than this example to convince the present authors.

Are Undefined Alternative Hypotheses the Fault of Classical Statistics?

From the current vantage point, the classical bias against the null hypothesis does not appear as obvious as Edwards' bias against classical statistics. It is instructive to note one aspect of the rationale behind Edwards' (1965) bias: "The trouble is that in classical statistics the alternative hypothesis is essentially undefined, and so provides no standard by means of which to judge the congruence between datum and null hypothesis [p. 402]." Edwards seemed to see the absence of a specifiable alternative as a shortcoming of classical procedure. It seems appropriate to point out that, in fact, the absence of a well-defined alternative is not a vice of classical statistics. The absence of a well-defined alternative is a problem, a problem which is unavoidable in many cases, a problem which Bayesian statistics presumably cannot handle, and also a problem which classical statistics is especially designed to avoid.

Summary

Some of the main points of the position of the present writers may be summed up as follows: (a) Identification with the alternative limits erroneous claim for theoretical support to 5%; identification with the null limits such errors to 95%. The conservative is therefore, presumably better advised to identify with the alternative. (b) Classical procedures assume that only the null distribution can be specified, and ask if the data are from this distribution or some other. Bayesian procedures assume that two distributions can be specified, and ask if the data are from the null distribution or the alternative. Granted the difference in procedure and in the information assumed, discussion of a bias in one procedure relative to the other seems of dubious meaningfulness. (c) Granted the information assumed by classical procedures, they show an absolute bias in favor of the null hypothesis.

References

Binder, A. Further considerations on testing the null hypothesis and the strategy and tactics of investigating theoretical models. *Psychological Review*, 1963, **70**, 107–115.

Binder, A. Statistical theory. *Annual Review of Psychology*, 1964, **15**, 277–310.

Campbell, D. T. Methodological suggestions from a comparative psychology of knowledge processes. *Inquiry*, 1959, **2**, 152–182.

Edwards, W. Tactical note on the relation between scientific and statistical hypotheses. *Psychological Bulletin*, 1965, **63**, 400–402.

Edwards, W., Lindman, H., & Savage, L. J. Bayesian statistical inference for psychological research. *Psychological Review*, 1963, **70**, 193–242.

Grant, D. A. Testing the null hypothesis and the strategy and tactics of investigating theoretical models. *Psychological Review*, 1962, **69**, 54–61.

Lindley, D. V. A statistical paradox. *Biometrika*, 1957, **44**, 187–192.

Wilson, W., & Miller, H. The negative outlook. *Psychological Reports*, 1964, **15**, 977–978. (a)

Wilson, W. R., & Miller, H. A note on the inconclusiveness of accepting the null hypothesis. *Psychological Review*, 1964, **71**, 238–242. (b)

Editor's comments: Students in statistics courses usually have little difficulty understanding that the level of significance adopted in an experiment determines the probability of rejecting the null hypothesis when in fact it is true. However, students frequently raise the question, "Why the .05 level of significance and not the .06 or .10 level?" They are puzzled by the apparent arbitrariness of setting the level of significance equal to, say, .05. James K. Skipper, Jr., Anthony L. Guenther, and Gilbert Nass in the following paper examine some of the factors that should be considered in selecting a level of significance. Their paper should help to dispel the mystical aura that surrounds the .05 level of significance.

4.7 The Sacredness of .05: A Note Concerning the Uses of Statistical Levels of Significance in Social Science

James K. Skipper, Jr., Anthony L. Guenther, and Gilbert Nass

Decisions regarding the uses of statistical levels of significance have typically been rendered by social scientists with specialized training and considerable expertise. However, a strong case can be made for more involvement in such issues by researchers whose activities include tests of significance in an applied context. The issue of whether inference should be made at all has been rather thoroughly debated,[1] but there are indications that the use of present "recommended" levels of significance is due for reassessment. This note attempts to state the problem, explore the issues involved, and suggest an alternative to current policy.

The choice of a statistical level of significance, that is, establishing the probability of rejecting the null hypothesis when in fact it is true, apparently demands little psychic energy on the part of researchers. Casual examination of the literature discloses that the common, arbitrary and virtually sacred levels of .05, .01, and .001 are almost universally

Reprinted from *The American Sociologist*, 1967, **2**, 16–18, by permission of the publisher and authors.

[1] See for example, the classic statement by Hanan Selvin, "A Critique of Tests of Significance in Survey Research," *American Sociological Review*, **22**, 1957, pp. 519–527, followed by David Gold, "Comment on 'A Critique of Tests of Significance,'" *American Sociological Review*, **23**, 1958, pp. 85–86, and Leslie Kish, "Some Statistical Problems in Research Design," *American Sociological Review*, **24**, 1959, pp. 328–338. For a discussion of the more general question of the current usefulness of mathematics and statistics in social science, the recent comments by Parsons, Sibly, Selvin, and Etzioni are of special interest. See: Talcott Parsons, "The Sibley Report on Training in Sociology," *American Sociological Review*, **29**, 1964, pp. 747–748; Elbridge Sibley, "Parsons On the Sibley Report," *American Sociological Review*, **30**, 1965, p. 110; Hanan Selvin, "Mathematics and Sociology," *American Sociological Review*, **30**, 1965, pp. 264–265; Amitai Etzioni, "Mathematics for Sociologists?" *American Sociological Review*, **30**, 1965, pp. 943–945.

selected regardless of the nature and type of problem. Of these three, .05 is perhaps most sacred.

Although statistically-inclined methodologists do not often explicitly recommend use of these arbitrary levels, their positions frequently suggest that these levels are *conventional*. Prominent textbooks bear this out:

But the agony of making a fresh selection of the level of significance in each instance would be a painful business. Hence, the statistical worker finds welcome relief in the .05 *convention*, which prescribes that the null hypothesis is automatically to be rejected whenever the probability of being wrong in that decision is 5 percent or less. [2]

By locating the observed findings in the theoretical distribution of all possible findings, the investigator determines the probability of obtaining the finding by chance if the null hypothesis is actually correct. If this probability turns out to be small enough (less than a predetermined level such as .05 of .01), he then decides to reject the null hypothesis. [3]

In the social sciences, it is more or less *conventional* to reject the null hypothesis when the statistical analysis indicates that the observed difference would not occur more than 5 times out of 100 by chance alone. [4]

A *convention* frequently followed is to state the result *significant* if the hypothesis is rejected with $\propto = .05$ and *highly significant* if it is rejected with $\propto = .01$. [5]

The current obsession with .05, it would seem, has the consequence of differentiating significant research findings and those best forgotten, published studies from unpublished ones, and renewal of grants from

termination. It would not be difficult to document the joy experienced by a social scientist when his F ratio or t value yields significance at the .05, nor his horror when the table reads " only " .10 or .06. One comes to internalize the difference between .05 and .06 as " right " *vs.* " wrong," " creditable " *vs.* " embarrassing," " success " *vs.* " failure." Tradition notwithstanding, there seems to be little justifiable reason for such a state of affairs. [6] We find it hard to believe that social scientists simply wish to avoid the inconvenience of selecting significance level as one of the parameters of the problem under investigation.

Most textbooks in research methods and statistics recommend that the levels of significance associated with tests on the data be set in advance of actual data collection. In part, this decision is one of minimizing the probability of committing Type I or Type II error. (Type I error occurs when one rejects the null hypothesis which in fact is true. Type II error is committed when one fails to reject the null hypothesis which in fact is false). Since the two types of errors are inversely related to each other, it is impossible to minimize both of them at the same time without increasing the sample size. Therefore, it is the *nature of the problem under study which ought to dictate which type of error is to be minimized*.

Some social scientists feel that too much emphasis has been placed upon the level of significance of a test at the expense of *power* considerations. Representative of these is the managing editor of *Psychometrika* who comments:

The frequent use of the .05 and .01 levels of significance is a matter of convention having little

---

[2] John Mueller and Karl Schuessler, *Statistical Reasoning in Sociology*, Boston: Houghton Mifflin, 1961, p. 395. Emphasis ours.

[3] Matilda White Riley, *Sociological Research*, New York: Harcourt, Brace and World, 1963, p. 638. Emphasis ours.

[4] Claire Selltiz, Marie Jahoda, Morton Deutsch and Stuart W. Cook, *Research Methods in Social Relations*, New York: Holt, Rinehart and Winston, 1961, rev. one volume ed., p. 418. Emphasis ours.

[5] Wilfred Dixon and Frank Massey, *Introduction to Statistical Analysis*, New York: McGraw-Hill, 1957, p. 91. Emphasis ours.

[6] The use of the arbitrary .05 level of significance seems to have originated with the statistician R. A. Fisher who used this level with experimental situations in agriculture and biology. Subsequently it was adopted by social scientists. See R. A. Fisher, *Statistical Methods for Research Workers*, Edinburgh: Oliver and Boyd, 1935; and the discussion by Quinn McNemar, *Psychological Statistics*, New York: Wiley, 1955, p. 64.

scientific or logical basis. When the power of tests is likely to be low under these levels of significance, and when type 1 and type 2 errors are of approximately equal importance, the .30 and .20 levels of significance may be more appropriate than the .05 and .01 levels.[7]

It is appropriate to consider at this point an example illustrating the dependence of optimal alpha level upon the problem investigated. A social scientist may face the problem of whether to recommend a group of college seniors having high I.Q.'s but poor academic indexes, for graduate school. Controlling other relevant variables, he sets up H_0: There will be no difference in rate of achievement in graduate school between individuals with (a) high I.Q.'s but low grades, and (b) high I.Q.'s and outstanding grades. A decision must be made regarding the circumstances under which H_0 may be rejected. If a low level of alpha, say .001, is established, the probability of unfairly discriminating in favor of the good students is minimized. On the other hand, $\propto = .001$ implies a much greater probability of sending students to graduate school who could not do the work than would be the case with, e.g., $\propto = .25$.

In our opinion, there is no "right" or "wrong" level here—the decision must be made in full consideration of parameters inherent in the problem itself. It is doubtful that setting *a priori* levels of .05, .01 or what have you, settles the matter.

The issue becomes clearer when we compare the above problems with another. A decision may be needed, for example, whether to recommend that military personnel in combat zones are more effective when fighting alone or when accompanied by a "buddy." Again, controlling other relevant variables, H_0 states: There will be no difference in the combat effectiveness of military personnel fighting alone, as opposed to fighting with a "buddy." What alpha level is optimal for this hypothesis? In contrast with the first problem, a higher level (say, .20) would be

justified, that is, we feel it is more desirable to risk rejection of the null hypothesis when it is true. In terms of the example, it is more desirable to place combat personnel with buddies when in fact they are no more effective fighters this way.

In the first example (college students with poor grades), it seems best to minimize Type I error, while in the second case (combat effectiveness) it is more important to minimize Type II error. Regardless of whether one agrees with the values we have employed in interpreting these examples or not is of little importance. The point is that blind adherence to the .05 level denies any consideration of alternative strategies, and it is a serious impediment to the interpretation of data.

There are those, of course, who may feel that the cases we have cited are essentially moot problems, since they believe social scientists *should* deal only in " pure " and not " applied " research. This view contends that one should design, execute and report upon his research, but not serve to implement his findings. Only in his role as "moral man" should the scientist-as-layman take responsibility for practical decisions affecting the community.[8] Where one stands on the "pure or applied " issue is also of little consequence for this report. Either view implies a responsible consideration of the various factors underlying decisions of significance levels.

Even if one accepts the pure research orientation there are still important reasons for choosing optimal alpha levels. Blalock, for example, points out that usually the only real decision facing a social scientist is whether to publish or suppress his findings.[9] In this circumstance he suggests a rule of thumb: "The researcher should lean over backwards to prove himself wrong or to

---

[7] B. J. Winer, *Statistical Principles in Experimental Design*, New York: McGraw-Hill, 1962, p. 13.

[8] A view most elaborately put forth by George Lundberg, *Can Science Save Us*, New York: Longmans, Green, 1947. An antithetical position, of course, was set forth by Robert S. Lynd, *Knowledge for What*, Princeton, N.J.: Princeton University Press, 1939.

[9] Hubert Blalock, *Social Statistics*, New York: McGraw-Hill, 1960, p. 135. Pragmatically, whether findings are acceptable may depend upon achieving significance at .05 or better.

obtain results that he usually does not want to obtain."[10] He warns, however, that this too may involve the decision to minimize chances of making Type I or Type II error depending on the type of problem.[11] It is inadvisable, therefore, in any type of problem, to state *a priori* the " most appropriate " level of significance. The risk of so doing may be evinced by the paucity of criteria or guidelines for establishing levels of significance for different classes of problems.

Blalock's remarks are most appropriate for research which is designed to test hypotheses. There is, of course, much social science research which seeks to *develop* hypotheses rather than attempt verification. The rubrics pilot study and exploratory research are well known, and frequently provide for the collection and analysis of data not "explained " in advance by an explicit theoretical framework. Past experience has proven this to be an effective strategy and often such serendipitous findings outweigh the original purpose of the research.[12] In cases where the goal is generation rather than testing hypotheses, it would seem advisable to tailor level of significance to the open-ended character of research design. Again, an arbitrary .05 level may not be optimal.

With the increasing involvement of social scientists in the very structures and processes they study, e.g., complex organizations, ethnic relations and political activity, it is expected that published research findings will be disseminated to audiences of laymen untrained in the interpretation of data. When the researcher places an asterisk (or maybe even *two* asterisks) after his obtained *F*, Chi-square or *t* value, and thereby decrees *significance*, this may constitute the basis for

social action by statistically unsophisticated decision makers. On the other hand, failure to achieve significance at the prearranged level may lead to immediate dismissal of the hypothesized relationship.[13]

As social scientists increasingly publish research findings in scholarly journals having a wider circulation than the scientific community of their own particular discipline, this problem becomes of increasing importance. Journals of practicing professions and occupations often reach an audience who is relatively unsophisticated in scientific methodology and techniques of statistical analysis. For example, in the area of medical sociology a study concerning interaction and communication in a hospital, say between nurses and patients, published in a sociological journal might reach anywhere from 2,000–12,000 subscribers, mostly sociologists and other social scientists. The same findings published in the *American Journal of Nursing* would come to the attention of more than 160,000 registered nurses, a portion of whom might try to make practical use of findings which are termed "significant." It is not at all improbable that this may be characteristic of many other areas of specialization.

Finally, we would like to suggest that a more suitable procedure for setting and reporting levels of significance might be: (1) For the social scientist to reflect upon the *arbitrary* nature of .05, .01 and .001 levels now in common use, and to recognize that selection of one of these is not a panacea for the interpretation of evidence. Different classes of research problems (or components thereof) may require different levels of alpha. (2) For the social scientist to recall that even well educated lay groups are inexpert in the interpretation of statistical significance. The

[10] *Ibid.*, p. 125.

[11] McNemar suggests: "If the findings of a study are to be used as the basis either for theory and further hypotheses or for social action, it does not seem unreasonable to require a higher level of significance than the .05 level." McNemar, *op. cit.*, p. 65.

[12] See Robert K. Merton, *Social Theory and Social Structure*, Glencoe, Ill.: The Free Press, 1957, pp. 103–108.

[13] Yates points out that this may be equally true of the scientists themselves: ". . . scientific workers have often regarded the execution of a test of significance on an experiment as the ultimate objective. Results are significant or not significant and this is the end of it." See Frank Yates, "The Influence of *Statistical Methods for Research Workers* on the Development of the Science of Statistics." *Journal of the American Statistical Association*, **46**, March, 1951, p. 33.

demand by such groups upon professional researchers must be met by a response that is ethical and communicative. (3) For social scientists reporting research findings (especially in journals indigenous to their discipline) to do away with arbitrary levels of significance and the calling of one test result "significant" and another "not significant."[14] We recommend a procedure whereby the actual level of significance associated with each research finding be stated.[15] In other words, if a finding yields significance at the .30 level, or the .09, or whatever, report it *at that level* together with other essential information (e.g., degrees of freedom) and let the reader determine whether for his purposes it has any practical significance. A statement of opinion regarding support or non-support of the data for the relevant hypothesis may be made by the researcher.

We believe that this type of research reporting would have several advantages over present procedure: (1) Non-social scientists using the professional literature would be spared the judgments of "significant" and "non-significant" as the basis for deciding whether findings have important implications in their applied situations. In a sense this would force laymen to acquire some expertise to offset an absence of "spoon-fed" decisions regarding data analysis. (2) Much the same function would be performed for individual social scientists whose time for critical thinking about verification of research propositions is frequently negligible. (3) It would tend to intensify the search for appropriate significance levels, given attributes of the research problem. If, in contrast with present policy, it were conventional that editorial readers for professional journals routinely ask "What justification is there for this level of significance?" authors might be less likely to indiscriminately select an alpha level from the field of popular eligibles.

In summary, there is a need for social scientists to choose levels of significance with full awareness of the implications of Type I and Type II error for the problem under 'investigation. The current use of arbitrary levels of alpha, while appropriate for some designs, detracts from interpretive power in others. Moreover, the tendency to dichotomy resulting from judging some results "significant" and others "non-significant" can be misleading both to professional and lay audiences. It is suggested that a more rational approach might be to report the actual level of significance, placing the burden of interpretive skill upon the reader. Such a policy would also encourage social scientists to give higher priority to selecting appropriate levels of significance for a given problem.

---

[14] Kish has recommended that: "Significance should stand for meaning and refer to substantive matter," in Leslie Kish, *op. cit.*, p. 336. His *Survey Sampling*, New York: Wiley, 1965, treats the importance of significance under a variety of sampling designs.

[15] Sidney Siegel, *Nonparametric Statistics for the Behavioral Sciences*, New York: McGraw-Hill, 1956, p. 9.

Editor's comments: In conducting an experiment, two kinds of randomization procedures for subjects can be used either singularly or in combination. One can *randomly sample* subjects from some population about which inferences are to be made, or one can *randomly assign* subjects to treatment conditions. Eugene S. Edgington points out in the following article that subjects in psychological experiments are rarely selected on the basis of random sampling, although random sampling is one of the assumptions of inferential statistics. His article provides a justification for using inferential statistics with non-randomly selected subjects. Of course random sampling is a necessary condition for making statistical inferences about a population of subjects. Edgington shows, however, that even though random sampling is not used, random assignment of subjects to the treatment conditions is a sufficient condition for drawing statistical inferences about the effects of treatment conditions. Such inferences apply only to the subjects actually used in the experiment.

Thus if an experimenter wants to use inferential statistics there must be some randomization procedure in the experiment. Which randomization procedure he uses will depend upon the objectives of his experiment.

4.8   Statistical Inference and Nonrandom Samples

Eugene S. Edgington

Abstract. Psychological experimentation rarely meets the assumption of random sampling that conventional statistical hypothesis-testing procedures are generally believed to require. A rationale is presented to justify the current procedure of testing statistical hypotheses with nonrandom samples. Random assignment of Ss to treatments allows statistical analysis with nonrandomly selected Ss, although the statistical inferences apply only to the Ss actually used in the experiment. Randomization tests can be used as exact tests for this purpose and parametric tests can be used as approximations to randomization tests. Nonstatistical inferences about persons not used in an experiment can be made on the basis of logical considerations.

Psychologists use statistical hypothesis testing procedures that, according to statisticians,

Reprinted from *Psychological Bulletin*, 1966, **66**, 485–487, by permission of the publisher and author. Copyright 1966 by the American Psychological Association.

require the assumption of random sampling of the population or populations about which inferences are to be made. Since experimental psychologists seldom sample randomly, it is important to see whether there is any justification for them to use such hypothesis-testing procedures.

Failure to Obtain Random Samples

A pollster can enumerate the population to be sampled and select a random sample by means of a procedure that gives every person in the population an equal chance of being selected, but an experimental psychologist has neither the time and money nor the knowledge to take a random sample of the world population in order to draw statistical inferences about people in general. Few psychological experiments use randomly selected subjects, and in those cases the population sampled often is so specific as to be of little interest. Whenever human subjects for psychology experiments are randomly selected, they are frequently drawn from a population of persons attending a certain college, enrolled in a particular class, willing to serve as subjects, etc. Animal psychologists do not even pretend to take random samples, although they use the same hypothesis testing procedures and reject null hypotheses about populations. These well-known facts are mentioned here as a reminder of the specificity and nonrandomness of samples in psychology. Statistical inferences cannot be made concerning populations that have not been randomly sampled; therefore, few experiments would be published if it were necessary to show that the experiment permitted statistical inference concerning an important population, a population of general interest to the readers.

Testing Statistical Hypotheses with Nonrandom Samples

In much psychological experimentation, the principal interest in statistical analysis is to distinguish between treatment effects and effects arising from individual differences. The notion of population comes into the statistical analysis not because the experimenter has in mind a particular population to which he wishes to generalize, but because the only way he has been taught to interpret statistical results is in terms of inferences about popula-

tions. This section will discuss an alternative interpretation of statistical results, one that is applicable to nonrandom samples or to samples that have been systematically rather than randomly selected. This alternative view will not permit statistical inferences about persons not used in the experiment, but it will permit statistical inferences about treatment effects for the experimental subjects.

Although the following procedure will not allow drawing *statistical* inferences about persons not used in the experiment, *nonstatistical* inferences can, of course, be drawn on the basis of logical considerations. For example, if the effects of a particular experimental treatment depend mainly on physiological functions that are almost unaffected by the social or physical environment, we might draw inferences about persons in other cultures than that from which the subjects came. On the other hand, if the experimental effects were easily modified by social conditions, we would be more cautious in generalizing to other cultures.

A rational justification for carrying out statistical tests in the usual way, in which the nonrandomness of the samples is ignored, follows: Take as an example an experiment with two treatments, A and B, where we have predicted that subjects perform better under A than under B. We use the first 10 available subjects or 10 specially selected subjects, making no attempt at random sampling. Names are drawn from a hat to randomly assign five subjects to Treatment A and five to B. Suppose all five A subjects performed better than any of the B subjects. The null hypothesis is that the performances are independent of the treatments, that is, that the performance of a subject is the same under one treatment as it would have been under the other. Under this hypothesis, any difference between the performance under the two treatments is due solely to subject differences in performance. Since the subjects were randomly assigned to the two treatments, the null hypothesis attributes large differences between treatments to "randomization error," discrepancies resulting from the chance assignment of subjects. To test the null hypothesis, then, we consider the distribution of differences between the treatments under every equally likely partition of the 10 performance measurements into two sets of five measurements each. If we did this, we would find that only .004 of the partitions assign the five high performing subjects to Treatment A and the five lowest performing subjects to Treatment B. It is not necessary to actually make all of the partitions of the data; the numerical value .004 can be determined from the Mann-Whitney U probability table (Siegel, 1956, p. 271), which is based on the complete distribution of partitions. We can thus reject the null hypothesis at the .01 level and accept the alternative hypothesis that, for the experimental subjects, performance was affected by the treatment. This conclusion is justified by the random assignment of subjects to treatments, no matter how biased our procedure of selecting subjects. For example, our conclusion would be justified even if we had selected the subjects on the basis of pretests with Treatments A and B.

Although correlation coefficients are seldom computed in psychological experiments, it will be instructive to consider how to determine the significance of a correlation when there is nonrandom sampling. In this hypothetical example, we measure the intelligence of prospective foster parents who have applied at an orphanage to adopt a baby. The babies for adoption are assigned by a random procedure to the applicants. We want to find out whether a baby adopted by an intelligent person will become a more intelligent child than if he had been adopted by a less intelligent person. After a number of years, we test the intelligence of children and correlate their IQs with the IQs of their foster parents, as measured at the time of adoption. This experiment is analogous to the previous one, the "subjects" in this case being the babies and the "treatments" being the intelligence of the foster parents. If the intelligence of the foster parents is independent of the intelligence of the children, any correla-

tion will be the result of "randomization error," the result of random assignment of children to foster parents. A randomization type correlation test such as a rank correlation technique could be used to test this hypothesis, since such techniques are based on the random pairing of one set of scores with another. If the correlation is significant, we reject the null hypothesis and conclude that, within our sample, the intelligence of a child is not independent of the intelligence of the person who adopted him.

To recognize the importance of the random assignment of subjects, consider a variation of the above study. We are investigating the same problem, but instead of randomly assigning babies to foster parents, we take a group of foster parents and the children they had adopted years before and correlate their present IQs. If we did not randomly select our sample, there is no statistical inference we can make, even about the subjects within the sample. A systematically selected sample may favor the selection of foster parents who are willing to have their own intelligence and their adopted child's intelligence measured. To illustrate how bias could complicate the statistical interpretation in this case, assume that the only foster parents who would participate were those who thought their child's IQ was about the same as their own IQ. This could result from the desire of some parents to conceal what they consider to be the dullness of the child or from their desire not to appear duller than their child. With such a sample, we might indeed get a high correlation. However, this correlation would indicate nothing of interest to us. Since we do not have a random sample of a population, it cannot indicate anything about the correlation within a population. A high correlation would indicate a computed correlation coefficient which is significantly greater than would be expected through random pairing of the children's IQs with the IQs of the foster parents, but the statistical significance is not relevant since no such randomization was carried out. This study would not justify concluding that any of the children would have had different intelligence if he had been adopted by a parent whose intelligence was different from that of his actual foster parent. Thus, randomization of subjects is essential if we are to draw statistical inferences from nonrandomly selected subjects.

If a person wants to use more information from the experimental measurements than simply the relative rank, which is used in the rank-order statistical techniques, he can find exact randomization test counterparts (Edgington, 1964; Siegel, 1956, pp. 88–92, 152–156), or approximate randomization tests (Edgington, 1964).

The testing procedures given above for nonrandom samples have been described in terms of randomization tests, but this does not imply that parametric tests cannot be used. We can use parametric tests as approximations to randomization tests. The closeness of the approximation under certain conditions has been shown theoretically (Silvey, 1954; Wald & Wolfowitz, 1944) and by numerical examples (Eden & Yates, 1933; Fisher, 1935, Section 21; Kempthorne, 1952, p. 152; Pitman, 1937; Welch, 1937).

In order to see what it means for a parametric test to be an approximation to a randomization test, let us consider the *t* test for a difference of means. We take a nonrandom sample of 40 subjects and randomly assign 25 to Treatment A and 15 to Treatment B. We use an exact randomization test in which we partition the 40 scores in every possible way into two groups, with 25 subjects in Treatment A and 15 in B. For each

partition we calculate a value of *t*. We can say that the *t* test is a close approximation to the randomization test if the distribution of *t* on which the probabilities in the table of *t* are based is approximately the same as the distribution of *t* obtained by the randomization test.

References

Eden, T., & Yates, F. On the validity of Fisher's z-test when applied to an actual sample of nonnormal data. *Journal of Agricultural Science,* 1933, **23**, 6–16.

Edgington, E. S. Randomization tests. *Journal of Psychology,* 1964, **57**, 445–449.

Fisher, R. A. *The design of experiments.* Edinburgh: Oliver & Boyd, 1935.

Kempthorne, O. *The design and analysis of experiments.* New York: Wiley, 1952.

Pitman, E. J. G. Significance tests which may be applied to samples from any populations: III. The analysis of variance test. *Biometrika,* 1937, **29**, 322–335.

Siegel, S. *Nonparametric statistics for the behavioral sciences.* New York: McGraw-Hill, 1956.

Silvey, S. D. The asymptotic distributions of statistics arising in certain nonparametric tests. *Proceedings of the Glasgow Mathematics Association,* 1954, **2**, 47–51.

Wald, A., & Wolfowitz, J. Statistical tests based on permutations of the observations. *Annals of Mathematical Statistics,* 1944, **15**, 358–372.

Welch, B. L. On the z-test in randomized blocks and Latin squares. *Biometrika,* 1937, **29**, 21–52.

4.9 Statistical Significance in Psychological Research

David T. Lykken

Abstract. Most theories in the areas of personality, clinical, and social psychology predict no more than the direction of a correlation, group difference, or treatment effect. Since the null hypothesis is never strictly true, such predictions have about a 50–50 chance of being confirmed by experiment when the theory in question is false, since the statistical significance of the result is a function of the sample size. Confirmation of a single directional prediction should usually add little to one's confidence in the theory being tested. Most theories should be tested by multiple corroboration and most empirical generalizations by constructive replication. Statistical significance is perhaps the least important attribute of a good experiment; it is never a sufficient condition for claiming that a theory has been usefully corroborated, that a meaningful empirical fact has been established, or that an experimental report ought to be published.

In a recent journal article Sapolsky (1964) developed the following substantive theory: Some psychiatric patients entertain an unconscious belief in the "cloacal theory of birth" which involves the notions of oral impregnation and anal parturition. Such patients should

Reprinted from *Psychological Bulletin*, 1968, **70**, 151–159, by permission of the publisher and author. Copyright 1968 by the American Psychological Association.

be inclined to manifest eating disorders: compulsive eating in the case of those who wish to get pregnant and anorexia in those who do not. Such patients should also be inclined to see cloacal animals, such as frogs, on the Rorschach. This reasoning led Sapolsky to predict that Rorschach frog responders show a higher incidence of eating disorders than patients not giving frog responses. A test of this hypothesis in a psychiatric hospital showed that 19 of 31 frog responders had eating disorders indicated in their charts, compared to only 5 of the 31 control patients. A highly significant chi-square was obtained.

It will be an expository convenience to analyze Sapolsky's article in considerable detail for purposes of illustrating the methodological issues which are the real subject of this paper. My intent is not to criticize a particular author but rather to examine a kind of epistemic confusion which seems to be endemic in psychology, especially, but by no means exclusively, in its " softer " precincts. One would like to demonstrate this generality with multiple examples. Having just combed the latest issues of four well-known journals in the clinical and personality areas, I could undertake to identify several papers in each issue wherein, because they were able to reject a directional null hypothesis at some high level of significance, the authors claimed to have usefully corroborated some rather general theory or to have demonstrated some important empirical relationship. To substantiate that these claims are overstated and that much of this research has not yet earned the right to the reader's overburdened attentions would require a lengthy analysis of each paper. Such profligacy of space would ill become an essay one aim of which is to restrain the swelling volume of the psychological literature. Therefore, with apologies to Sapolsky for subjecting this one paper to such heavy handed scrutiny, let us proceed with the analysis.

Since I regarded the prior probability of Sapolsky's theory (that frog responders unconsciously believe in impregnation per os) to be nugatory and its likelihood unenhanced by the experimental findings, I undertook to check my own reaction against that of 20 colleagues, most of them clinicians, by means of a formal questionnaire. The 20 estimates of the prior probability of Sapolsky's theory, which these psychologists made before being informed of his experimental results, ranged from 10^{-6} to 0.13 with a median value of 0.01, which can be interpreted to mean, roughly, " I don't believe it." Since the prior probability of many important scientific the-

ories is considered to be vanishingly small when they are first propounded, this result provides no basis for alarm. However, after being given a fair summary of Sapolsky's experimental findings, which " corroborate " the theory by confirming the operational hypothesis derived from it with high statistical significance, these same psychologists attached posterior probabilities to the theory which ranged from 10^{-5} to 0.14, with the median unchanged at 0.01. I interpret this consensus to mean, roughly, " I still don't believe it." This finding, I submit, *is* alarming because it signifies a sharp difference of opinion between, for example, the consulting editors of the journal and a substantial segment of its readership, a difference on the very fundamental question of what constitutes good (i.e., publishable) clinical research.

The thesis of the present paper is that Sapolsky and the editors were in fact following, with reasonable consistency, our traditional rules for evaluating psychological research, but that, as the Sapolsky paper exemplifies, at least two of these rules should be reconsidered. One of the rules examined here asserts roughly the following: " When a prediction or hypothesis derived from a theory is confirmed by experiment, a nontrivial increment in one's confidence in that theory should result, especially when one's prior confidence is low." Clearly, my 20 colleagues were violating this rule here since their confidence in the frog responder-cloacal birth theory was not, on the average, increased by the contemplation of Sapolsky's highly significant chi-square. From their comments it seems that they found it too hard to accept that a belief in oral impregnation could lead to frog responding merely because the frog has a cloacus. (One must, after all, admit that few patients know what a cloacus is or that a frog has one and that those few who do know probably will also know that the frog's eggs are both fertilized and hatched externally so neither oral impregnation nor anal birth are in any way involved. Hence, *neither* the average patient *nor* the biologically sophisticated patient should logically be expected to employ the frog as a symbol for an unconscious belief in oral conception.) My colleagues, on the contrary, found it relatively easy to believe that the observed association between frog responding and eating problems might be due to some other cause entirely (e.g., both symptoms are immature or regressive in character; the frog, with its disproportionately large mouth and voice may well constitute a common orality totem and hence be associated with problems in the oral sphere; "squeamish" people might tend both to see frogs and to have eating problems; and so on.)

Assuming that this first rule *is* wrong in this instance, perhaps it could be amended to allow one to make exceptions in cases resembling this illustration. For example, one could add the codicil: "This rule may be ignored whenever one considers the theory in question to be overly improbable or whenever one can think of alternative explanations for the experimental results." But surely such an amendment would not do. ESP, for example, could never become scientifically respectable if the first exception were allowed, and one consequence of the second would be that the importance attached to one's findings would always be inversely related to the ingenuity of one's readers. The burden of the present argument is that this rule is wrong not only in a few exceptional instances *but as it is routinely applied to the majority of experimental reports in the psychological literature.*

Corroborating Theories by Experimental Confirmation of Theoretical Predictions[1]

Most psychological experiments are of three kinds: (*a*) studies of the effect of some treatment on some output variables, which can be regarded as a special case of (*b*) studies of the difference between two or more groups of individuals with respect to some variable,

---

[1] Much of the argument in this section is based upon ideas developed in certain unpublished memoranda by P. E. Meehl (personal communication, 1963) and in a recent article (Meehl, 1967).

which in turn are a special case of (*c*) the study of the relationship or correlation between two or more variables within some specified population. Using the bivariate correlation design as paradigmatic, then, one notes first that the strict null hypothesis must always be assumed to be false (this idea is not new and has recently been illuminated by Baken, 1966). Unless one of the variables is wholly unreliable so that the values obtained are strictly random, it would be foolish to suppose that the correlation between any two variables is identically equal to 0.0000 ... (or that the effect of some treatment or the difference between two groups is exactly *zero*). The molar dependent variables employed in psychological research are extremely complicated in the sense that the measured value of such a variable tends to be affected by the interaction of a vast number of factors, both in the present situation and in the history of the subject organism. It is exceedingly unlikely that any two such variables will not share at least some of these factors and equally unlikely that their effects will exactly cancel one another out.

It might be argued that the more complex the variables the smaller their average correlation ought to be since a larger pool of common factors allows more chance for mutual cancellation of effects in obedience to the Law of Large Numbers. However, one knows of a number of unusually potent and pervasive factors which operate to unbalance such convenient symmetries and to produce correlations large enough to rival the effect of whatever causal factors the experimenter may have had in mind. Thus, we know that (*a*) "good" psychological and physical variables tend to be positively correlated; (*b*) experimenters, without deliberate intention, can somehow subtly bias their findings in the expected direction (Rosenthal, 1963); (*c*) the effects of common method are often as strong as or stronger than those produced by the actual variables of interest (e.g., in a large and careful study of the factorial structure of adjustment to stress among officer candidates, Holtzman & Bitterman, 1956, found that their 101 original variables contained five main common factors representing, respectively, their rating scales, their perceptual-motor tests, the McKinney Reporting Test, their GSR variables, and the MMPI); (*d*) transitory state variables such as the subject's anxiety level, fatigue, or his desire to please, may broadly affect all measures obtained in a single experimental session.

This average shared variance of "unrelated" variables can be thought of as a kind of ambient noise level characteristic of the domain. It would be interesting to obtain empirical estimates of this quantity in our field to serve as a kind of Plimsoll mark against which to compare obtained relationships predicted by some theory under test. If, as I think, it is not unreasonable to suppose that "unrelated" molar psychological variables share on the average about 4% to 5% of common variance, then the expected correlation between any such variables would be about .20 in absolute value and the expected difference between any two groups on some such variable would be nearly 0.5 standard deviation units. (Note that these estimates assume zero measurement error. One can better explain the near-zero correlations often observed in psychological research in terms of unreliability of measures than in terms of the assumption that the true scores are in fact unrelated.)

Suppose now that an investigator predicts that two variables are positively correlated. Since we expect the null hypothesis to be false, we expect his prediction to be confirmed by experiment with a probability of very nearly 0.5; by using a large enough sample, moreover, he can achieve any desired level of statistical significance for this result. If the ambient noise level for his domain is represented by correlations averaging, say, .20 in absolute value, then his chances of finding a statistically significant confirmation of his prediction with a reasonable sample size will be quite high (e.g., about 1 in 4 for $N = 100$) even if there is no truth whatever to the theory on which the prediction was based. Since most theoretical predictions in psychol-

ogy, especially in the areas of clinical and personality research, specify no more than the direction of a correlation, difference or treatment effect, we must accept the harsh conclusion that a single experimental finding of this usual kind (confirming a directional prediction), no matter how great its statistical significance, will seldom represent a large enough increment of corroboration for the theory from which it was derived to merit very serious scientific attention. (In the natural sciences, this problem is far less severe for two reasons: (*a*) theories are powerful enough to generate point predictions or at least predictions of some narrow range within which the dependent variable is expected to lie; and (*b*) in these sciences, the degree of experimental control and the relative simplicity of the variables studied are such that the ambient noise level represented by unexplained and unexpected correlations, differences, and treatment effects is often vanishingly small.)

The Significance of Large Correlations

It might be argued that, even where only a weak directional prediction is made, the obtaining of a result which is not only statistically significant but large in absolute value should constitute a stronger corroboration of the theory. For example, although Sapolsky predicted only that frog responding and eating disorders would be positively related, the fourfold point correlation (phi coefficient) between these variables in his sample was about .46, surely much larger than the average relationship expected between random pairs of molar variables on the premise that "everything is related to everything else." Does not such a large effect therefore provide stronger corroboration for the theory in question ?

One difficulty with this reasonable sounding doctrine is that, in the complex sort of research considered here, *really large* effects, differences, or relationships are not usually to be expected and, when found, may ever argue *against* the theory being tested. To illustrate this, let us take Sapolsky's theory seriously and, by making reasonable guesses concerning the unknown base rates involved, attempt to estimate the actual size of the relationship between frog responding and eating disorders which the theory should lead us to expect. Sapolsky found that 16% of his control sample showed eating disorders; let us take this value as the base rate for this symptom among patients who do not hold the cloacal theory of birth. Perhaps we can assume that all patients who do hold this theory will give frog responses but surely not all of these will show eating disorders (any more than will all patients who believe in vaginal conception be inclined to show coital or urinary disturbances); it seems a reasonable assumption that no more than 50% of the believers in oral conception will therefore manifest eating problems. Similarly, we can hardly suppose that the frog response *always* implies an unconscious belief in the cloacal theory; surely this response can come to be emitted now and then for other reasons. Even with the greatest sympathy for Sapolsky's point of view, we could hardly expect more than, say, 50% of frog responders to believe in oral impregnation. Therefore, we might reasonably predict that 16 of 100 nonresponders would show eating disorders in a test of this theory, 50 of 100 frog responders would hold the cloacal theory and half of these show eating disorders, while 16% or 8 of the remaining 50 frog responders will show eating problems too, giving a total of 33 eating disorders among the 100 frog responders. Such a finding would produce a significant chi-square but the actual degree of relationship as indexed by the phi coefficient would be only about .20. In other words, if one considers the supplementary assumptions which would be required to make a theory compatible with the actual results obtained, it becomes apparent that the finding of a really strong association may actually embarrass the theory rather than support it (e.g., Sapolsky's finding of 61% eating disorders among his frog responders is *significantly larger* ($p < .01$)

than the 33% generously estimated by the reasoning above).

Multiple Corroboration

In the social, clinical, and personality areas especially, we must expect that the size of the correlations, differences, or effects which might reasonably be predicted from our theories will typically not be very large relative to the ambient noise level of correlations and effects due solely to the "all-of-a-pieceness of things." The conclusion seems inescapable that the only really satisfactory solution to the problem of corroborating such theories is that of *multiple corroboration*, the derivation and testing of a number of separate, quasi-independent predictions. Since the prior probability of such a multiple corroboration may be on the order of $(0.5)^n$, where n is the number of independent[2] predictions experimentally confirmed, a theory of any useful degree of predictive richness should in principle allow for sufficient empirical confirmation through multiple corroboration to compel the respect of the most critical reader or editor.

The Relation of Experimental Findings to Empirical Facts

We turn now to the examination of a second popular rule for the evaluation of psychological research, which states roughly that "When no obvious errors of sampling or experimental method are apparent, one's confidence in the general proposition being tested (e.g., Variables A and B are positively correlated in Population C) should be proportional to the degree of statistical significance obtained." We are following this rule when we say, "Theory aside, Sapolsky has at least demonstrated an empirical fact, namely, that frog responders have more eating disturbances

---

[2] Tests of predictions from the same theory are seldom strictly independent since they often share some of the same supplementary assumptions, are made at the same time on the same sample, and so on

than patients in general." This conclusion means, of course, that in the light of Sapolsky's highly significant findings we should be willing to give very generous odds that any other competent investigator (at another hospital, administering the Rorschach in his own way, and determining the presence of eating problems in whatever manner seems reasonable and convenient for him) will also find a substantial positive relationship between these two variables.

Let us be more specific. Given Sapolsky's fourfold table showing 19 of 31 frog responders to have eating disorders (61%), it can be shown by chi-square that we should have 99% confidence that the true population value lies between $13\!/\!31$ and $25\!/\!31$ (between 42% and 81%). With 99% confidence that the population value is at least 13 in 31, we should have $.99(99) = 98\%$ confidence that a new sample from that population should produce at least 6 eating disorders among each 31 frog responders, assuming that 5 of each 31 nonresponders show eating problems also as Sapolsky reported. That is, we should be willing to bet $98 against only $2 that a replication of this experiment will show *at least as many* eating disorders among frog responders as among nonresponders. The reader may decide for himself whether his faith in the " empirical fact " demonstrated by this experiment can meet the test of this gambler's challenge.

Three Kinds of Replication

If, as suggested above, "demonstrating an empirical fact" must involve a claim of confidence in the replicability of one's findings, then to clearly understand the relation of statistical significance to the probability of a " successful " replication it will be helpful to distinguish between three rather different methods of replicating or cross-validating an experiment. *Literal replication*, of course, would involve exact duplication of the first investigator's sampling procedure, experimental conditions, measuring techniques, and methods

of analysis; asking the original investigator to simply run more subjects would perhaps be about as close as we could come to attaining literal replication and even this, in psychological research, might often not be close enough. In the case of *operational replication*, on the other hand, one strives to duplicate exactly just the sampling and experimental procedures given in the first author's report of his research. The purpose of operational replication is to test whether the investigator's " experimental recipe "—the conditions and procedures he considered salient enough to be listed in the " Methods " section of his report —will in other hands produce the results that he obtained. For example, replication of the " Clever Hans " experiment revealed that the apparent ability of that remarkable horse to add numbers had been due to an uncontrolled and unsuspected factor (the presence of the horse's trainer within his field of view). This factor, not being specified in the "methods recipe" for the result, was omitted in the replication which for that reason failed. Operational replication would be facilitated if investigators would accept more responsibility for specifying what they believe to be the minimum essential conditions and controls for producing their results. Psychologists tend to be inconsistently prolix in describing their experimental methods; thus, Sapolsky tabulates the age, sex, and diagnosis for each of his 62 subjects. Does he mean to imply that the experiment will not work if these details are changed ?—surely not, but then why describe them ?

In the quite different process of *constructive replication*, one deliberately avoids imitation of the first author's methods. To obtain an ideal constructive replication, one would provide a competent investigator with *nothing more than* a clear statement of the empirical "fact" which the first author would claim to have established—for example, "psychiatric patients who give frog responses on the Rorschach have a greater tendency toward eating disorders than do patients in general" —and then let the replicator formulate his own methods of sampling, measurement, and data analysis. One must keep in mind that the data, the specific results of a particular experiment, are only seldom of any real interest in themselves. The "empirical facts" which we value so highly consist usually of confirmed conceptual or constructive (not operational) hypotheses of the form "Construct A is positively related to Construct B in Population C." We are interested in the *construct* "tendency toward eating disorders," not in the *datum* "has reference made to overeating in the nurse's notes for May 15th." An operational replication tests whether we can duplicate our findings using the same methods of measurement and sampling; a constructive replication goes further in the sense of testing the validity of these methods.

Thus, if I cannot confirm Sapolsky's results for patients from my hospital, assessing eating disorders by means of informant interviews, say, or actual measurements of food intake, then clearly Sapolsky has *not* demonstrated any "fact" about eating disorders among psychiatric patients in general. I could then revert to an operational replication, assessing eating problems from the psychiatric notes as Sapolsky did and selecting my sample to conform with the age, sex, and diagnostic properties of his, although I might not regard this endeavor to be worth the effort since, under these circumstances, even a successful operational replication could not establish an empirical conclusion of any great generality or interest. Just as a reliable but invalid test can be said to measure something, but not what it claimed to measure, so an experiment which replicates operationally but not constructively could be said to have demonstrated something, but not the relation between meaningful constructs, generalizable to some broad reference population, which the author originally claimed to have established.[3]

Relation of the Significance Test to the Probability of a "Successful" Replication

The probability values resulting from significance testing can be directly used to measure one's confidence in expecting a "successful" literal replication only. Thus, we can be 98% confident of finding at least 6 of 31 frog responders to have eating problems only if we reproduce all of the conditions of Sapolsky's experiment with absolute fidelity, something that he himself could not undertake to do at this point. Whether we are entitled to anything approaching such high confidence that we could obtain such a result from an operational replication depends entirely upon whether Sapolsky has accurately specified all of the conditions which were in fact determinative of his results. That he did not in this instance is suggested by the fact that, investigating the feasibility of replicating his experiment at the University of Minnesota Hospitals, I found that I should have to review several thousand case records in order to turn up a sample of 31 frog responders like his. Although he does not indicate how many records he examined, one strongly suspects that the base rate of Rorschach frog responding must have been higher at Sapolsky's hospital, either because of some difference in the patient population or, more probably, because an investigator's being interested in some class of responses will tend to subtly elicit such responses at a higher rate unless the testing procedure is very rigorously controlled. If the base rates for frog responding are so different at the two hospitals, it seems doubtful that the response can have the same correlates or meaning in the two populations and therefore one would be reckless indeed to offer high odds on the outcome of even the most careful operational replication. The

---

[3] This distinction between operational and constructive replication seems to have much in common with that made by Sidman (1960) between what he calls "direct" and "systematic" replication. However, in the operant research context to which Sidman directs his attention, "replication" means to run another animal or the same animal again; thus, direct

replication involves maintaining the same experimental conditions in detail whereas in systematic replication one allows all supposedly irrelevant factors to vary from one subject to the next in the hope of demonstrating that one has correctly identified the variables which are really in control of the behavior being studied.

likelihood of a successful constructive replication is, of course, still smaller since it depends on the additional assumptions that Sapolsky's samples were truly representative of psychiatric patients in general and that his method of assessing eating problems was truly valid, that is, would correlate highly with a different, equally reasonable appearing method.

Another Example

It is not my purpose, of course, to criticize statistical theory or method but rather to suggest ways in which these tools are sometimes misused or misinterpreted by writers or readers of the psychological literature. Nor do I mean to abuse a particular investigator whose research report happened to serve as a convenient illustration of the components of the argument. An abundance of articles can be found in the journals which exemplify these points quite as well as Sapolsky's but space limitations forbid multiple examples. As a compromise, therefore, I offer just one further illustration, showing how the application of these same critical principles might have increased a reader's—and perhaps even an editor's—skepticism concerning some research of my own.

The purpose of the experiment in question (Lykken, 1957) was to test the hypothesis that the "primary" psychopath has reduced ability to condition anxiety or fear. To segregate a subgroup in which such primary psychopaths might be concentrated, I asked prison psychologists to separate inmates already diagnosed as psychopathic personalities into one group that met 14 rather specific clinical criteria specified by Cleckley (1950, pp. 355–392) and to identify another group which clearly did not fit some of these criteria. The normal control subjects were comparable to the psychopathic groups in age, IQ, and sex. Fear conditioning was assessed using the GSR as the dependent variable and a rather painful electric shock as the unconditioned stimulus (UCS). On the index used to measure rate of conditioning, the primary psychopathic group scored significantly lower than did the controls. By the usual reasoning, therefore, one might conclude that this result demonstrates that primary psychopaths are abnormally slow to condition the GSR, at least with an aversive UCS, and this empirical fact in turn provides significant support for the theory that primary psychopaths have defective fear-learning ability (i.e., a low "anxiety IQ").

But to anyone who has actually participated in research of this kind, this seemingly straightforward reasoning must appear appallingly oversimplified. It is quite impossible to obtain anything resembling a truly random sample of psychopaths (or of nonpsychopathic normals either, for that matter) and it is a matter of unquantifiable conjecture how a sample obtained by a different investigator using equally defensible methods might perform on the tests which I employed. Even with the identical sample, no two investigators are likely to measure the GSR in the same way, use the same conditioned stimulus (CS) and UCS or the same pattern of reinforced and CS-only trials. Given even the same set of protocols, there is no standard formula for obtaining an index of degree or rate of conditioning; the index I used was essentially arbitrary and whether it was a good one is a matter of opinion. My own evaluation of the methods used, together with a complex set of supplementary assumptions difficult to explicate, leads me to believe that these results increase the likelihood that primary psychopaths have slower GSR conditioning with an aversive UCS; I might now give odds of two to one that this empirical generalization is true and odds of three to two that another investigator would be able to confirm it by means of a constructive replication. But this already biased claim is far more modest than the one which is implicit in the significance testing operation, namely, "such a mean difference would only be expected 5 times in 100 if the [generalization] is not true."

This empirical generalization, about GSR conditioning, is derivable from the hypothesis of interest, that psychopaths have a low anxiety IQ, by a chain of reasoning so complex and elliptical and so burdened with accessory assumptions as to be quite impossible to spell out in the detail required for rigorous logical analysis. Psychologists knowledgeable in the area can evaluate whether it is a reasonable derivation but their opinions will not necessarily agree. Moreover, even if the derivation could pass the scrutiny of some "Certified Public Logician," confirmation of the prediction about GSR conditioning should add only very slightly to our confidence in the hypothesis about fear conditioning. Even if this confirmation were made relatively more firm by, for example, constructive replication of the generalization, "aversive GSR conditioning is retarded in primary psychopaths," the hypothesis that these individuals have a low anxiety IQ could still be said to have passed only the weakest kind of test. This is so because such simple directional predictions about group differences have nearly a 50–50 chance of being true a priori even if our particular hypothesis is false. There are doubtless many possible explanations for low GSR conditioning scores in psychopaths other than the possibility of defective fear conditioning. Indeed, some of my subjects whose conditioning scores were nearly as low as those of the most extreme primary psychopaths seemed to me to be clearly neurotic with considerable anxiety and I attempted to account for their GSR performance with an ad hoc conjecture involving a kind of repression phenomenon, that is, a denial that a low GSR index implied poor fear conditioning in their cases.

A redeeming feature of this study was that two other related but distinguishable predictions from the same hypothesis were tested at the same time, namely, that primary psychopaths should do as well as normals on a learning task involving positive reward but less well on an avoidance learning problem, and that they should be more willing than normals to choose embarrassing or frightening situations in preference to alternatives involving tedium, frustration, physical discomfort, and the like. Tests of these predictions gave affirmative results also, thus providing some of the multiple corroboration

necessary for the hypothesis to claim the attention of other experimenters.

Obviously, I do not mean to criticize the editor's decision to publish my (1957) paper. The tendency to evaluate research in terms of mechanical rules based on the results of the significance tests should not be replaced by equally rigid requirements concerning replication or corroboration. This study, like Sapolsky's or most others in this field, can be properly evaluated only by a qualified reader who can substitute his own informed judgment and scientific intuition for the rigorous reasoning and experimental control that is usually not achievable in clinical and personality research. As it happens, subsequent work has provided some encouraging support for my 1957 findings. The two additional predictions mentioned above have received operational replication (i.e., the same test methods used in a different context) by Schachter and Latené (1964). The prediction that psychopaths show slower GSR conditioning with an aversive UCS has been constructively replicated (i.e., independently tested with no attempt to copy my procedures) by Hare (1965a). Finally, two additional predictions from the theory that the primary psychopath has a low anxiety IQ have been tested with affirmative results (Hare, 1965b; 1966). All told, then, this hypothesis can now boast of having led to at least five quasi-independent predictions which have been experimentally confirmed and three of which have been replicated. The hypothesis is therefore entitled to serious consideration although one would be rash still to regard it as proven. At least one alternative hypothesis, that the psychopath has an unusually efficient mechanism for inhibiting emotional arousal, can account equally well for the existing findings so that, as is usually the case, further research is called for.

Conclusions

The moral of this story is that the finding of statistical significance is perhaps the least important attribute of a good experiment;

it is *never* a sufficient condition for concluding that a theory has been corroborated, that a useful empirical fact has been established with reasonable confidence—or that an experimental report ought to be published. The value of any research can be determined, not from the statistical results, but only by skilled, subjective evaluation of the coherence and reasonableness of the theory, the degree of experimental control employed, the sophistication of the measuring techniques, the scientific or practical importance of the phenomena studied, and so on. Ideally, all experiments would be replicated before publication but this goal is impractical. "Good" experiments will tend to replicate better than poor ones (and, when they do not, the failures will tend to be informative in themselves, which is not true for poor experiments) and should be published so that they may stimulate replication and extension by others. Editors must be bold enough to take responsbility for deciding which studies are good and which are not, without resorting to letting the *p* value of the significance tests determine this decision. There is little real danger that anything of value will be lost through this approach since the unpublished investigator can always resort to constructive replication to induce editorial acceptance of his empirical conclusions or to multiple corroboration to compel editorial respect for his theory. Since operational replication must really be done by an independent second investigator and since constructive replication has greater generality, its success strongly implying that an operational replication would have succeeded also, one should usually replicate one's own work constructively, using different sampling and measurement procedures within the purview of the same constructive hypothesis. If only unusually well done, provocative, and important research were published without such prior authentica-tion, operational replication of such research by others would become correspondingly more valuable and entitled to the respect now accorded capable replication in the other experimental sciences.

References

Baken, D. The test of significance in psychological research. *Psychological Bulletin*, 1966, **66**, 423–437.

Cleckley, H. *The mask of sanity.* Saint Louis: C. V. Mosby, 1950.

Hare, R. D. Acquisition and generalization of a conditioned fear response in psychopathic and nonpsychopathic criminals. *Journal of Psychology*, 1965, **59**, 367–370. (a)

Hare, R. D. Temporal gradient of fear arousal in psychopaths. *Journal of Abnormal Psychology*, 1965, **70**, 442–445. (b)

Hare, R. D. Psychopathy and choice of immediate versus delayed punishment. *Journal of Abnormal Psychology*, 1966, **71**, 25–29.

Holtzman, W. H., & Bitterman, M. E. A factorial study of adjustment to stress. *Journal of Abnormal and Social Psychology*, 1956, **52**, 179–185.

Lykken, D. T. A study of anxiety in the sociopathic personality. *Journal of Abnormal and Social Psychology*, 1957, **55**, 6–10.

Meehl, P. E. Theory-testing in psychology and physics: A methodological paradox. *Philosophy of Science*, 1967, **34**, 103–115.

Rosenthal, R. On the social psychology of the psychological experiment: The experimentor's hypothesis as unintended determinant of experimental results. *American Scientist*, 1963, **51**, 268–283.

Sapolsky, A. An effort at studying Rorschach content symbolism: The frog response. *Journal of Consulting Psychology*, 1964, **28**, 469–472.

Schachter, S., & Latené, B. Crime, cognition and the automatic nervous system. *Nebraska Symposium on motivation*, 1964, **12**, 221–273.

Sidman, M. *Tactics of scientific research.* New York: Basic Books, 1960.

4.10 Suggestions for Further Reading

I. Hypothesis Testing: Classical Approaches versus Alternative Approaches*

Nunnally, J. The place of statistics in psychology. *Educational and Psychological Measurement*, 1960, **20**, 641–650. Gives a critical examination of the null-hypothesis model and recommends the use of measures of strength of association.

Rozeboom, W. W. The fallacy of the null-hypothesis significance test. *Psychological Bulletin*, 1960, **57**, 416-428. Gives a critical examination of the null-hypothesis model and recommends the use of Bayesian procedures.

Grant, D. A. Testing the null hypothesis and the strategy and tactics of investigating theoretical models. *Psychological Review*, 1962, **69**, 54–61. Gives a critical examination of the null-hypothesis model and recommends the use of interval estimation procedures.

Bolles, R. C. The difference between statistical hypotheses and scientific hypotheses. *Psychological Reports*, 1962, **11**, 639–645. A lucid comparison of the hypotheses of the statistician and those of the scientist.

Wilson, W. R., and H. Miller. A note on the inconclusiveness of accepting the null hypothesis. *Psychological Review*, 1964, **71**,

238–242. Alternative strategies for investigating theoretical models are examined.

Bakan, D. The test of significance in psychological research. *Psychological Bulletin*, 1966, **66**, 423–437. Gives a critical examination of the null-hypothesis model and recommends the use of Bayesian procedures.

LaForge, R. Confidence intervals or tests of significance in scientific research. *Psychological Bulletin*, 1967, **68**, 446–447. Outlines the advantages of the classical Neyman-Pearson approach relative to the Bayesian approach and recommends the use of interval estimation procedures.

II. How Experimenters Interpret Levels of Significance*

The following three articles examine the degree of confidence of psychology faculty and graduate students in making judgments about research findings on the basis of probability information.

Rosenthal, R., and J. Gaito. The interpretation of levels of significance by psychological researchers. *The Journal of Psychology*, 1963, **55**, 33–38.

Beauchamp, K., and R. B. May. Replication report: Interpretation of levels of significance by psychological researchers. *Psychological Reports*, 1964, **14**, 272.

---

* Articles listed chronologically.

Rosenthal, R., and J. Gaito. Further evidence for the cliff effect in the interpretation of levels of significance. *Psychological Reports*, 1964, **15**, 570.

III. Multiple Tests of Significance and Probability Pyramiding

Brožek, J., and K. Tiede. Reliable and questionable significance in a series of statistical tests. *Psychological Bulletin*, 1952, **49**, 339–341. Authors illustrate the use of the normal curve approximation to the binomial distribution for determining the probability of obtaining n or more significant statistics out of $N > 100$ calculated statistics.

Jones, L. V., and D. W. Fiske. Models for testing the significance of combined results. *Psychological Bulletin*, 1953, **50**, 375–382. Two models are presented for jointly evaluating the statistical significance of a collection of experiments.

Sakoda, J. M., B. H. Cohen, and G. Beall. Test of significance for a series of statistical tests. *Psychological Bulletin*, 1954, **51**, 172–175. Provides tables for determining the probability of obtaining n or more significant statistics out of $N < 100$ calculated statistics.

Wilkinson, B. A statistical consideration in psychological research. *Psychological Bulletin*, 1951, **48**, 156–158. Provides tables for determining the probability of obtaining n or more significant statistics out of $N < 25$ calculated statistics.

Journal editors rarely publish papers unless one or more of the tests of significance reach at least the .05 level. The implications of this practice in terms of probability pyramiding are examined in the following two articles.

Neher, A. Probability pyramiding, research error and the need for independent replication. *The Psychological Record*, 1967, **17**, 257–262.

Sterling, T. D. Publication decisions and their possible effects on inferences drawn from tests of significance or vice versa. *Journal of the American Statistical Association*, 1959, **54**, 30–34.

5

Correlation

Correlational methods provide a means for measuring the degree of relationship between variables. The place of assumptions in the interpretation of a Pearsonian correlation coefficient, one of the many kinds of correlational methods, is examined by Arnold Binder in this chapter. He points out that no assumptions are necessary for computing a Pearsonian correlation coefficient; however, the appropriateness of a particular interpretation of the coefficient depends upon the degree to which certain assumptions have been fulfilled.

5.1 Considerations of the Place of Assumptions in Correlational Analysis

Arnold Binder

In a recent article, Nefzger and Drasgow (1957) questioned whether it is necessary to assume that two variables are normally distributed in order to compute a Pearson product-moment correlation coefficient. They then assumed a linear relationship between continuous variables expressed in standard form and proceeded to derive the least squares estimate of the slope of a relationship line. Thereupon they pointed out that this estimate was of the form of the product-moment correlation coefficient and argued that this derivation showed that it was not necessary to assume normality to compute Pearson's *r*.

The arguments of the article are not well-supported at many points and are inaccurate at others. Some of these faults have been pointed out in the comments of Milholland (1958), Furfey (1958), and LaForge (1958). While the comments have been accurate and helpful, they have been far from complete and they have necessarily lacked logical development and organization because of the method of expression.

For present purposes, the weaknesses of the Nefzger and Drasgow presentation may be

The author is grateful to J. R. Blum, C. J. Burke, D. G. Ellson, and S. E. Feldman for their comments on the manuscript.

conveniently considered from an over-all viewpoint as well as from an evaluation of specific features. Before a detailed discussion of the various issues raised by the article, let us consider a summary of, first, the over-all criticisms and, second, the specific criticisms:

1. Over-all: (*a*) The role of models and derivations in mathematical analysis and the methods by which the empirical scientist uses these models are generally ignored by the writers. This deficiency is nowhere more evident than in their discussions relative to the status of assumptions and interpretations in the analysis of data. Thus, they ignore the intimate relationship between assumptions and interpretations and give the impression that assumptions are nuisances to be avoided wherever possible rather than potentially invaluable assets in terms of deductive power. Moreover, the central question which they raise and for which they claim an answer—Is it necessary "... to assume normality for the *computation* of *r*?"—reflects some confusion, since assumptions are never needed for computations. (*b*) The writers have neglected the more important and widely used models for correlation analysis in favor of one of limited utility.

2. Specific: (*a*) The model for correlation presented by Nefzger and Drasgow is neither developed nor presented in a precise way. In addition, some of their statistical statements and arguments dealing with the model are in error. (*b*) The writers have misinterpreted some of the literature on correlation.

The Role and Uses of Mathematical Models

Statistics, as a mathematical discipline, involves the setting up of models or miniature deductive systems on the basis of primitive terms, definitions, axioms, and the theorems of preceding disciplines. Using these undefined terms, definitions, axioms, and so forth, the mathematical statistician derives or deduces certain consequences which are called theorems. The process of establishing a theorem is called a proof.

The empirical scientist using statistical models attempts to find an interpretation of the purely formal system within his discipline of interest. In this process, the scientist substitutes specific expressions and constants for the free terms of the formal model. If the axioms are valid with these substitutions, he has an interpretation (or better, realization) of the formal system. The scientist is then in a very advantageous position since all of the theorems which were deduced on the basis of the mathematical model are valid (with proper substitution of terms) for his realization of the model.

The assumptions of which we speak in statistical analysis are nothing but the axioms of certain mathematical models. Fulfilling assumptions, thus, implies the condition that our realization satisfies the axioms of our model. We can perform all the transformations indicated by our model and be sure that the theorems are applicable. In statistical analysis, the transformations are computations; and the theorems are statements of population estimates with certain properties, of confidence limits, of the significance of differences, of conditional probabilities, of the probable characteristics of future samples, and so forth.

Thus, it is most desirable to satisfy the assumptions of a mathematical deductive system in order to make use of its deductive power for interpretations following data manipulation. While it is true that a mathematician may frequently be able to generate all the theorems of a given system with fewer axioms (and consequently fewer assumptions), the empirical scientist must work with models which are available. And to do so he must of course meet their assumptions. In correlational analysis, as we shall see, the most interesting and important interpretations depend upon a few models which do involve restrictive axioms.

It must be emphasized that this position does not imply that the axioms of statistical models must be met for the results of computations to be interpretable. In fact, an example below will show many interpretations of the correlation coefficient that may be made without any assumptions (other than those involved in the algorithms of ordinary algebra). But introducing assumptions usually leads to vastly more interesting and important interpretations.

One last point should be made before illustrating the above comments by means of the mathematical models suitable for correlational analysis. The assumptions made prior to computation and the potential interpretations of analyzed data are directly related, but there is never need for assumptions in order to perform any computations whatsoever on a set or sets of numbers. One can compute a mean, a standard deviation, a correlation coefficient, standard scores, or even take the logarithm of the square root of every odd number without the necessity for assumptions of any sort.

When a research worker has a set of numbers at hand, he may perform computations upon it for one of two purposes: description of the data or generalization from the data. While descriptions or descriptive interpretations are limited to statements about the characteristics of the actual data at hand, generalizations or inferential interpretations refer to a larger population from which the given data were obtained. The following are examples of descriptive interpretations: the standard deviation can never be negative; the greater the dispersion of points about the mean, the larger the standard deviation; and, when a set of sample points is symmetrical, its mean is equal to its median. This type of interpretation is frequently quite useful in many statistical applications, but inferential interpretations are usually much more interesting and important for the empirical scientist. Inferential interpretations are involved in statistical estimation, setting confidence limits, testing statistical hypotheses, and prediction.

To be able to interpret the results of computations descriptively one must assume that the axioms of ordinary algebra[1] are applicable, but these are quite different from the assumptions (like those of normality and linearity) to which Nefzger and Drasgow refer. Statistical models and their accompanying assumptions are important boons, however, in broadening the scope and power of possible interpretations after data manipulation.

Correlational Models

Three well-developed and widely used mathematical models appropriate for interpretations involving the correlation coefficient are: the bivariate normal distribution, the linear regression model, and the randomization model.

In the *bivariate normal* distribution, the two variables have the following joint probability density function:

$$f(X, Y) = \frac{1}{2\pi\sigma_x\sigma_y\sqrt{1 - \rho^2}}$$

$$\times \exp\left\{-\frac{1}{2(1 - \rho^2)}\left[\frac{(X - \mu_x)^2}{\sigma_x^2}\right.\right.$$

$$\left.\left. - 2\rho\frac{(X - \mu_x)(Y - \mu_y)}{\sigma_x\sigma_y} + \frac{(Y - \mu_y)^2}{\sigma_y^2}\right]\right\}$$

where:

σ_x = standard deviation of marginal X distribution

σ_y = standard deviation of marginal Y distribution

μ_x = mean of marginal X distribution

μ_y = mean of marginal Y distribution

ρ = population correlation coefficient

---

[1] That is, one assumes a complete, ordered field for the real numbers.

This distribution has many interesting features, including the properties that (*a*) any plane perpendicular to the X, Y plane will slice the surface in such a way that the curve produced by the intersection will be normal, and (*b*) both marginal distributions are normal. But this must be said with all due emphasis: The condition of marginal normality in a bivariate distribution does not imply bivariate normality. That is, two variables may have a joint distribution function other than $f(X, Y)$ above and still have normal marginal distributions. Such a counter-example is most easy to construct.

The axioms (or required assumptions) of a *linear regression* model stated in summary form are, essentially:

1. $Y = \alpha + \beta X + \varepsilon$.
2. A total of nX's are selected and the corresponding values of Y determined.
3. The nX's are considered fixed—that is, they are chosen prior to making experimental observations and do not vary from sample to sample.
4. The ε's are normally and independently distributed with means equal to zero and equal variances.

In other words, there is assumed a normal distribution of Y's for each value of the fixed X's, and the means of these distributions (which have equal variances) lie on a straight line. In this model, the marginal distribution of Y may or may not be normal, while the fixed X's can obviously never be normal since only finitely many values are involved.

The *randomization* model is scarcely worthy of the name because of the paucity of its axioms. The only assumption necessary is that an obtained correlation coefficient between two variables is one of a larger array consisting of those which could be obtained from every possible pairing of the observed values of the variables. The possible population of values is thus based on all possible permutations of the obtained pairs, $n!$ in number. In this model there are no assumptions of normality, homoscedasticity, continuity, or even of linearity.

To demonstrate the desirability of satisfying assumptions (or meeting the axioms of a formal system), let us consider the increase in our interpretive possibilities resulting from the deductive power of each of these models. Without any statistical assumptions whatsoever, one may make the following descriptive interpretations of the correlation coefficient purely as a result of working with real numbers according to the procedures of ordinary algebra in connection with the least squares method (these were shown very early by Yule, 1897, who will be discussed later): (a) $-1 \leq r \leq 1$; (b) if $|r| = 1$, all plotted points will lie on a straight line; (c) if two variates are independent, $r = 0$; and (d) the closer the absolute value of r to 1, the less the scatter of points about the least squares regression line.

In this connection one should note carefully the referral by Nefzger and Drasgow to the assumption of linearity as a "crucial assumption in the derivation of the product-moment correlation coefficient" (p. 623). As stated earlier, no assumptions whatsoever are needed to compute r. And beyond that, we have just seen that the correlation coefficient as computed in the usual way has many interpretive properties without any (other than algebraic) assumptions. These interpretations hold for all bivariate number sets, and certainly are independent of any linearity assumption.

But if we do satisfy the assumptions of the bivariate normal distribution, many more interesting interpretations are open to us because of the theorems of the model. Thus our theorems tell us that r is a maximum likelihood estimator of ρ, and they tell us various properties of ρ as an indicator of the relationship between the two variables in the population. We know further that independence and lack of correlation are synonymous. We know a good deal about the sampling behavior of r. We know a great deal about the marginal and conditional distributions. We know the maximum likelihood estimators of means, variances, and covariance, and the joint sampling distributions of these estimators. In psychology, the bivariate normal model is probably most appropriate for measurement theory and test analysis.

In the case of the linear regression model, which is appropriate for much experimental work in psychology, the theorems deduced from our model tell us perhaps even more interesting things. We know the maximum likelihood estimators for the means and variances of the Y distributions. We know the sampling behavior of these estimators and can set confidence limits and test hypotheses. We can set confidence limits and test hypotheses about the parameters α and β. We understand the close relationship between the given method of analysis and the analysis of variance. We can test for linearity of regression and can set confidence limits for the Y-value of a new individual when his X-value is known.

To dramatize the potentially great value of assumptions we can contrast the above with the increase in interpretive possibilities (above those possible purely on the basis of algebraic procedures) when the randomization model is used. This model of course involves the very minimum in the way of assumptions. Our interpretation is restricted to a statement to the effect that, of all correlations that could have been obtained by all pairings of the observed values of the variables, ours was among the M most extreme ones.

The assumption which is of primary concern to Nefzger and Drasgow is that of marginal normality. Their preoccupation with this assumption is hard to understand from a formal viewpoint. There is no model in general use that has as an axiom the condition of marginal normality for the variables of interest in correlational problems. Marginal normality is a theorem in the case of the model involving the bivariate normal distribution, to be sure, but certainly not an assumption.

It is a most interesting incidental fact that the three correlational models make use of the very same statistic to test the hypothesis that an obtained r is significant. That is the t test with $n - 2$ degrees of freedom where:

$$t = \frac{r}{\sqrt{1 - r^2}} \sqrt{n - 2}$$

The specific hypotheses to be accepted or rejected are different for the three models, as are the interpretations resulting from a statistical decision. But the form of the test is precisely as above in each case.

The Model of Nefzger and Drasgow

There are of course a vast array of other possible models in addition to the bivariate normal, linear regression, and randomization models. But these must be systematized and developed before they are of much use for the empirical scientist. Nefzger and Drasgow present a model involving only the assumption of a linear relationship between two continuous variables. The model is not stated precisely and contains such faults as the failure to specify clearly the population and to differentiate the sample from it, the lack of a distinction between parameters and their estimators, and general confusion in dealing with measurement error and sampling fluctuation.

The writers state their basic axiom in the following way: "... in the population the true relation between the two variates can be represented by an equation of the form $Y = AX + B$" (p. 623). For simplicity of derivation these are then converted to standard scores with zero means and unit variances, and the resulting linearity assumption takes the form $z_y = az_x + b$. (Incidentally, their statement that conversion from raw to z scores does not alter the distributions is inaccurate.) Although the transition from population to sample is not made clear, they end up with a sample size of N pairs. Moreover, somehow between population and sample, discrepancies or errors enter into the picture. In the writers' terms such discrepancies lie "between the observed values in z_y and the corresponding values computed from $z_y' = az_x + b$" (p. 623), which is an extremely strange statement since no difference between z_y and z_y' is specified.

When an axiom of linearity is ordinarily introduced, it is put in some such form as:

$$Y = \beta X + \alpha + \varepsilon$$

where ε is the source of the discrepancies. An alternate form which is frequently used is:

$$\mu_{y/x} = \beta X + \alpha$$

where $\mu_{y/x}$ is the mean of the conditional distribution of Y's for a given X.[2] Each Y distribution, it is assumed, contains a set of values deviating from the distribution mean because of errors. Since data are always subject to errors of measurement, such errors must be described in the axioms of the model. As Furfey (1958) has pointed out, the defining equation of the Nefzger and Drasgow model ($z_y = az_x + b$) cannot be accepted at face value since it implies $|Y| = 1$.

After a statement of their axiom of linearity Nefzger and Drasgow state:

If it is now required that the linear constants a and b be so selected that the sum of the squared residuals in z_y is minimal, the required values of the constants can be obtained through the simultaneous solution of two partial differential equations (p. 623).

And further on in the article they argue that by this process they are

... merely requiring that the straight line selected to represent the relation between the two variables satisfy the least squares criterion of a best fit line (p. 624).

But in their model a and b seem to be parameters and not subject to sampling fluctuation. The values obtained for slope and intercept from a sample would surely not be expected to be equal to the population values. If a and b are considered sample values

---

[2] For the bivariate normal distribution:

$$\beta = \frac{\rho \sigma_y}{\sigma_x}$$

$$\alpha = \mu_y - \frac{\rho \sigma_y}{\sigma_x} \mu_x$$

(which of course fluctuate from sample to sample), it is inaccurate to state as an axiom that "the *z*-scores in *Y* are *linearly* related to the *z*-scores in *X*, i.e., $z_y = az_x + b$" (p. 623).

Models like that used by Nefzger and Drasgow involving very few assumptions have appeared and continue to appear in many textbooks (as Nefzger and Drasgow rightly pointed out). They have generally been presented, however, in a more explicit and accurate manner. The usage of such models has been restricted to descriptive statistics, and they are for the most part found in the exposition of correlational material for elementary course students with little mathematical sophistication. Such simple models have not been developed for uses involving statistical inference or the testing of hypotheses, and so the possible interpretations for the empirical scientist are generally meager. Some of the more restricted uses of *r* for prediction and as a measure of relationship, where few assumptions are involved, have been indicated by Milholland (1958), Furfey (1958), and LaForge (1958).

Nefzger and Drasgow then argue that two further interpretations are possible with the additional assumptions of "normal and homogeneous dispersions in at least one of the variables over all levels of the other" (p. 624). Since these interpretations involve prediction and the linear regression model is most appropriate for purposes of prediction and, in addition, has axioms of normality and homogeneity for all conditional distributions, this departure seems to bring them in line with the linear regression model. The words "at least" are not easy to account for, however. The linear regression model does not contain an axiom of marginal normality, but it is nevertheless quite different from the one which the authors originally set forth. Moreover, it can easily be shown that one of the two interpretations claimed to be permissible with this more restricted model is not accurate and the other is of doubtful utility. First, contrary to their contention, the standard error of estimate as defined in the

usual way is generally a poor indicator of the accuracy of prediction. While the assumption of homoscedasticity is applicable to the population of values in the linear regression model, the sampling fluctuation of the regression line is such that the confidence bands for prediction are hyperbolic in nature. Estimation of the means of the various *Y* distributions (one for each fixed *X*) when accomplished by means of the regression line has a variance of:

$$\left[\frac{1}{n} + \frac{(X_i - \overline{X})^2}{\sum_{i=1}^{n} (X_i - \overline{X})^2} \right] \sigma^2$$

And prediction of the *Y*-value for a *new* individual for a given *X* when similarly accomplished has variance:

$$\left[1 + \frac{1}{n} + \frac{(X_i - \overline{X})^2}{\sum_{i=1}^{n} (X_i - \overline{X})^2} \right] \sigma^2$$

The values of these variances become larger and larger as the particular X_i deviates more and more from the mean of all *X*'s, and so the accuracy of prediction diminishes correspondingly. The standard error of estimate does provide a measure of column variability for the individuals included in the sample from which it is computed, but this is of little interest indeed for prediction purposes. While the sample standard error of estimate does provide an estimate of the population standard deviation for the various *Y* distributions, it must be emphasized that the regression line (and hence each predicted value) varies from sample to sample.

The other claimed interpretation of Nefzger and Drasgow for the model is that r^2 is a measure of the proportion of the variance of one variable which is predictable from the other. No one can question this usage if the purpose is purely one of description for the data on which *r* was computed, but the inferential potential is not at all clear. The restricted use as a descriptive statistic, however, makes the stated assumptions unnecessary. And there is not a parameter for which r is an estimate in the axioms of the linear regression model.

The Literature on Correlation

Nefzger and Drasgow claim that Dixon and Massey (1951) and Mood (1950) indicate that "it is necessary for each of the variables to be normally distributed for proper application of the Pearson technique" (p. 623). The arguments in the other portions of the article make it relatively clear that the authors use "Pearson technique" and "product-moment correlation technique" synonymously. Both Dixon and Massey, and Mood discuss the bivariate normal and the linear regression models, pointing out their areas of applicability and limitations. But, as noted above, in the linear regression model X is certainly not normally distributed and the marginal Y distribution may not be (and usually is not) normal. It is true that Dixon and Massey entitle their discussion of the linear regression model "Regression" and their discussion of the bivariate normal model "Correlation Problems." But a casual glance at the chapter containing these discussions will convince one that the topics covered under "Regression" are closer to the analyses of Nefzger and Drasgow than the topics under "Correlation Problems." Thus, the standard error of estimate, least squares, linear prediction equations, and the test for linearity of regression are discussed under "Regression," while the emphasis under "Correlation Problems" is upon r as a measure of relationship.

Mood presents the following definition: "The *correlation* between two variates, say X and Z, is denoted by ρ_{xz} and is defined by

$$\rho_{xz} = \frac{\sigma_{xz}}{\sigma_x \sigma_z}$$

where σ_x and σ_z are the standard deviations of X and Z" (p. 103). And, as Mood stated previously, X and Z are variates from any multivariate distribution whatever. Moreover, Mood, in the following statement, provides

further evidence that he is a long way from the position attributed to him:

In recent years it has come to be realized that most (though not all) correlation problems which arise in practice can be handled more appropriately by regression methods. The latter require only the assumption that deviations from the regression function be normal, whereas the correlation analysis requires that the variate and what we have called the observable parameters all be *jointly* [italics added] normally distributed (p. 312).

On the preceding page, Mood described "correlation analysis" as a special class of problems involving the estimation of multiple regression coefficients in the particular case where "one considers the conditional distribution of, say, X_1 in a k-variate normal distribution" (p. 311).

Nefzger and Drasgow seem to imply that Pearson restricted his analysis to a bivariate normal model because of oversight, while Yule emphasized that important characteristics of correlation could be deduced from less restrictive assumptions. In their words:

An objection to Pearson's choice of a poor example and his failure to generalize beyond the restrictive limits of normal data were pointed out at the time by Yule. But evidently Yule's comments passed without sufficient attention to prevent the development of the present confusion (p. 623).

Late in the nineteenth century, Yule (1897) did publish a method for correlational analysis that involved no assumptions of normality or linearity. He started out with a scatter of points representing correlated pairs and obtained the slope and intercept of the least squares regression line. His derivation was contingent only upon the adequacy of the least squares method of curve fitting and the assumptions involved in ordinary algebraic computation with real numbers. By his procedure, Yule showed that the following descriptive interpretations could be made of the product-moment correlation coefficient without any statistical assumptions: (*a*) when standard scores (mean = 0, standard deviation = 1) are used, the regression of X on Y is equal to the regression of Y on X and both are equal to r; (*b*) the value of r always lies between -1 and $+1$; (*c*) when $|r|$ is equal to 1, the values of the two variates show a simple linear relationship; (*d*) the greater the absolute value of r, the less the deviations from the regression line; (*e*) the condition $r = 0$ is a necessary but not a sufficient condition for independence; and (*f*) if the true regression is not linear, the absolute value of r must be less than 1.

Yule's procedure was hardly exciting, but it did serve to point out certain descriptive interpretations which were possible with a very minimum of assumptions. It could not be used for setting parameter confidence limits, or for testing hypotheses, or for prediction, or even for testing r for significance.

Pearson (1920) questioned the ultimate utility of Yule's method, but not because he was blind to the need for a model more general than the bivariate normal distribution. He argued against Yule's use of the least squares method and pointed out the restriction of possible interpretations of the results without normal assumptions. Moreover, while Pearson's correlational work did rely most heavily upon the bivariate normal distribution, he very early indicated his interest in a more general approach. Thus, he published (1911) a method for multiple correlation which was entirely free of normal assumptions, and later (1920) he wrote:

This seems to me the desideratum of the theory of correlation at the present time: the discovery of an appropriate system of surfaces, which will give bivariate skew frequency. We want to free ourselves from the limitations of the normal surface, as we have from the normal curve of errors (p. 44).

To say that Yule's earlier work "passed without sufficient attention" is not accurate.

His original approach has been reproduced, in various forms, innumerable times and is the ultimate basis for the descriptive expositions in at least a dozen books at the present time. Indeed the very model of Nefzger and Drasgow is an offspring (though illegitimate) of the Yule approach. Whatever "present confusion" exists reflects a deficiency in understanding correlational models, and certainly no lack of attention to Yule. The preference for the bivariate normal model to a more general model stems from its great deductive power and its usefulness in many empirical situations (even when the assumptions are not fully met, as Fisher, 1946, has pointed out) and not from ignorance of the development and relevance of other models.

References

Dixon, W. J., & Massey, F. J., Jr. *Introduction to statistical analysis.* New York: McGraw-Hill, 1951.

Fisher, R. A. *Statistical methods for research workers.* (10th ed.) Edinburgh: Oliver and Boyd, 1946.

Furfey, P. H. Comment on "The needless assumption of normality in Pearson's *r.*" *Amer. Psychologist*, 1958, **13**, 545–546.

LaForge, R. Comment on "The needless assumption of normality in Pearson's *r.*" *Amer. Psychologist*, 1958, **13**, 546.

Milholland, J. E. Comment on "The needless assumption of normality in Pearson's *r.*" *Amer. Psychologist*, 1958, **13**, 544–545.

Mood, A. M. *Introduction to the theory of statistics.* New York: McGraw-Hill, 1950.

Nefzger, M. D. & Drasgow, J. The needless assumption of normality in Pearson's *r. Amer. Psychologist*, 1957, **12**, 623–625.

Pearson, K. On the general theory of the influence of selection on correlation and variation. *Biometrika*, 1911–12, **8**, 437–443.

Pearson, K. Notes on the history of correlation. *Biometrika*, 1920–21, **13**, 25–45.

Yule, G. U. On the theory of correlation. *J. Roy. Statist. Soc.*, 1897, **60**, 812–854.

5.2 Suggestions for Further Reading*

Nefzger, M. D., and J. Drasgow. The needless assumption of normality in Pearson's *r*. *The American Psychologist*, 1957, **12**, 623–625. This article focuses attention on the place of assumptions in correlational analysis.

Many of the points as well as the emphasis of Nefzger and Drasgow's article are challenged in the following three articles, as well as in Binder's article in this chapter.

Milholland, J. E. Comment on "The needless assumption of normality in Pearson's *r*." *The American Psychologist*, 1958, **13**, 544–545.

Furfey, P. H. Comment on "The needless assumption of normality in Pearson's *r*." *The American Psychologist*, 1958, **13**, 545–546.

LaForge, R. Comment on "The needless assumption of normality in Pearson's *r*." *The American Psychologist*, 1958, **13**, 546.

Carrol, J. B. The nature of data, or how to choose a correlation coefficient. *Psycho-metrika*, 1961, **26**, 347–372. A scholarly discussion of issues involved in the descriptive use of the correlation coefficient.

Darlington, R. B. Multiple regression in psychological research and practice. *Psychological Bulletin*, 1968, **69**, 161–182. A number of common practices and beliefs concerning the multiple regression method are examined. The author criticizes many of these beliefs and practices and offers alternative conceptions.

Gordon, R. A. Issues in multiple regression. *The American Journal of Sociology*, 1968, **73**, 592–616. Issues of particular interest to sociologists concerning the misuse of multiple regression are examined.

Linn, R. L., and C. E. Werts. Assumptions in making causal inferences from part correlations, partial correlations, and partial regression coefficients. *Psychological Bulletin*, 1969, **72**, 307–310. This article describes four techniques for making causal inferences in, say, naturalistic studies where it is necessary to statistically partial out the effects of antecedent variables. It also discusses the assumptions underlying the techniques.

* Articles are listed chronologically.

6

The Design of Experiments

The term *experimental design* refers to five interrelated activities that are involved in the investigation of scientific hypotheses. These activities, listed in the order that they are performed, are:

1. Formulate statistical hypotheses and make plans for the collection and analysis of data to test the hypotheses. (The reader will recall from Chapter 4 that a scientist's hunches or scientific hypotheses are usually stated in general terms, at least in the initial stages of an inquiry. In this form they are not amenable to evaluation through the use of inferential statistics. Statistical hypotheses, on the other hand, represent testable formulations of scientific hypotheses.)
2. State decision rules to be followed in testing the statistical hypotheses.
3. Collect data according to plan.
4. Analyze data according to plan.
5. Make decisions concerning the statistical hypotheses based on the decision rules, and make inductive inferences concerning the

probable truth or falsity of the scientific hypotheses.

In a more restricted sense the term *experimental design* is used to designate a particular type of plan for assigning subjects to experimental conditions and the statistical analysis associated with the plan.[1]

An experimental design can be compared to a road map; it shows the route between the experimenter's initial scientific hunch (the starting point) and his inductive inference about the probable truth or falsity of the hunch (the destination). Although some useful end might be accomplished by randomly collecting data without following such a systematic course of action, research that is guided by a carefully conceived and executed design is more likely to accomplish the experimenter's purpose. In practice, research is rarely conducted without some thought

---

[1] R. E. Kirk. *Experimental Design: Procedures for the Behavioral Sciences.* Belmont, California: Brooks/Cole Publishing Co., 1968, p. 1.

given to the experimental design; that is, most experimenters know what they want to investigate and have at least a general notion of how to proceed. However, experimenters often fail to work out all the details of their design before collecting data, or they do not consider the relative merits of the alternative designs that are usually available. The various designs are rarely equally efficient in terms of cost of data collection, time required for data analysis, facilities, equipment required, et cetera.

The articles in this chapter focus on some of the factors that an experimenter should consider as he designs an experiment. The first selection, "Hiawatha Designs an Experiment," is a satire showing the amusing consequences that follow when an experimenter is preoccupied with research techniques to the exclusion of research objectives.

6.1 Hiawatha Designs an Experiment

Maurice G. Kendall

1. Hiawatha, mighty hunter
   He could shoot ten arrows upwards
   Shoot them with such strength and swiftness
   That the last had left the bowstring
   Ere the first to earth descended.
   This was commonly regarded
   As a feat of skill and cunning.

2. One or two sarcastic spirits
   Pointed out to him, however,
   That it might be much more useful
   If he sometimes hit the target.
   Why not shoot a little straighter
   And employ a smaller sample?

3. Hiawatha, who at college
   Majored in applied statistics
   Consequently felt entitled
   To instruct his fellow men on
   Any subject whatsoever,
   Waxed exceedingly indignant
   Talked about the law of error,
   Talked about truncated normals,
   Talked of loss of information,
   Talked about his lack of bias
   Pointed out that in the long run
   Independent observations
   Even though they missed the target
   Had an average point of impact
   Very near the spot he aimed at

(With the possible exception
Of a set of measure zero.)

4. This, they said, was rather doubtful.
   Anyway, it didn't matter
   What resulted in the long run;
   Either he must hit the target
   Much more often than at present
   Or himself would have to pay for
   All the arrows that he wasted.

5. Hiawatha, in a temper
   Quoted parts of R. A. Fisher
   Quoted Yates and quoted Finney
   Quoted yards of Oscar Kempthorne
   Quoted reams of Cox and Cochran
   Quoted Anderson and Bancroft
   Practically in extenso
   Trying to impress upon them
   That what actually mattered
   Was to estimate the error.

6. One or two of them admitted
   Such a thing might have its uses
   Still, they said, he might do better
   If he shot a little straighter.

7. Hiawatha, to convince them
   Organized a shooting contest
   Laid out in the proper manner
   Of designs experimental
   Recommended in the textbooks
   (Mainly used for tasting tea, but

Reprinted from *American Statistician*, 1959, **13**, 23–24, by permission of the publisher.

Sometimes used in other cases)
Randomized his shooting order
In factorial arrangements
Used in the theory of Galois
Fields of ideal polynomials
Got a nicely balanced layout
And successfully confounded
Second-order interactions.

8. All the other tribal marksmen
Ignorant, benighted creatures,
Of experimental set-ups
Spent their time of preparation
Putting in a lot of practice
Merely shooting at a target.

9. Thus it happened in the contest
That their scores were most impressive
With one solitary exception
This (I hate to have to say it)
Was the score of Hiawatha,
Who, as usual, shot his arrows
Shot them with great strength and swiftness
Managing to be unbiased
Not, however, with his salvo
Managing to hit the target.

10. There, they said to Hiawatha,
That is what we all expected.

11. Hiawatha, nothing daunted,
Called for pen and called for paper
Did analyses of variance
Finally produced the figures
Showing beyond peradventure
Everybody else was biased

And the variance components
Did not differ from each other
Or from Hiawatha's
(This last point, one should acknowledge
Might have been much more convincing
If he hadn't been compelled to
Estimate his own component
From experimental plots in
Which the values all were missing.
Still, they didn't understand it
So they couldn't raise objections
This is what so often happens
With analyses of variance).

12. All the same, his fellow tribesmen
Ignorant, benighted heathens,
Took away his bow and arrows,
Said that though my Hiawatha
Was a brilliant statistician
He was useless as a bowman,
As for variance components
Several of the more outspoken
Made primeval observations
Hurtful to the finer feelings
Even of a statistician.

13. In a corner of the forest
Dwells alone my Hiawatha
Permanently cogitating
On the normal law of error.
Wondering in idle moments
Whether an increased precision
Might perhaps be rather better
Even at the risk of bias
If thereby one, now and then, could
Register upon the target.

Editor's comments: The following article constitutes a primer for designing experiments. The author, Julian C. Stanley, defines most of the terms that the reader will encounter in the experimental-design literature—terms such as replicate, crossed factors, within-group variation, additive effects, and fixed-effects model. A lucid discussion of basic design principles and concepts makes this article a source of valuable information for the introductory student.

6.2 Elementary Experimental Design—an Expository Treatment

Julian C. Stanley

What is a Controlled, Variable-Manipulating Experiment?

The word "experimentation" has come to have many meanings for educationists. The most common one of these might be termed "experi*en*tation," trying new approaches and subjectively evaluating their effectiveness. My concern here is with a more structured type of inquiry, akin to that carried out in many of the sciences. It involves control by the experimenter of at least one variable, such as method of teaching arithmetic, that he can *manipulate*. Thus, experimentation of this kind differs from observation of naturally occurring events in that the stage has been set by the experimenter so that the possibly

Based on a paper read at the Eighteenth Annual State Conference on Educational Research in San Francisco on 18 November 1966. The author thanks Jason Millman for helpful comments about an earlier version.

Reprinted from *Psychology in the Schools*, 1967, **4**, 195–203, by permission of the publisher and author.

differential effects of at least two "treatments" can be observed in a situation where assignment of *experimental units* (often, pupils or classes) to the several treatments has been made without bias. Nature almost always makes biased assignments to its treatments; even before a natural experiment, the experimental units to be subjected to one treatment are usually not comparable to those to be subjected to another treatment.

In brief, nature rarely assigns experimental units to treatments randomly, whereas the careful experimenter almost always does. One can define a controlled, variable-manipulating, comparative experiment as a study in which the available experimental units are assigned at random (either simply or restrictively) to the various treatments. More generally, if we consider each *factor* to be manipulated—e.g., several ways to teach arithmetic, overt versus covert response in programmed instruction, or 100% reinforcement versus 50% reinforcement—as having two or more *levels* or categories (e.g., SMSG versus two other ways to teach mathematics would constitute three levels of the teaching-of-mathematics factor), we can then talk about *factor-level combinations*, such as the six generated by three ways to teach mathematics *crossed* with two levels of reinforcement. This is a 3 × 2 factorial design—one factor at three levels crossed with a second factor at two levels.

For concreteness, let us consider an experiment involving four styles of printing type crossed with three sizes of printing type, a 4 × 3 factorial design yielding 12 factor-level combinations: each style is tried with every size, and each size is tried with every style. This is a *complete* design. If we take 12 pupils and assign one at random to each of the 12 combinations, we produce one *replicate*. Usually, as a minimum we shall assign 24 pupils, two at random to each of the 12 combinations, and thereby create two replicates. In order to have a powerful enough experiment, we may need to assign more than two pupils to each combination. Some number of replicates, *n*, will yield the power we require. Methods for determining *n* exist.

How do we conduct this *factorial-design* experiment involving four styles crossed with three sizes? Of course, we must decide which styles and sizes to use. Probably we have firmly in mind four styles that are candidates for use in the textbook or test we plan to prepare. We also know which three sizes to try for our purposes—perhaps 8-point, 12-point, and 16-point type. Those four styles and three sizes are the only ones of interest to us in this experiment. In the jargon of experimental design, we have two *fixed-effects factors* and therefore employ a *fixed-effects model*, because we have "drawn" the four styles from a target population of just four styles and the three sizes from a target population of just three sizes.

Where do we get the experimental units with which to do the experiment? We might secure a "grab-group" consisting of the first 12*n* individuals who happen our way, or we might define a population of individuals,

such as all fourth graders in a large school system, and draw 12*n* individuals at random from that population. Using a grab-group limits statistical generalization from the outcome of the experiment to just those 12*n* persons, whereas drawing the experimental units randomly from a population permits statistical generalization to that population, thereby increasing *external validity*—i.e., generalizability. (We may, however, be able to generalize non-statistically from the grab-groupers to other persons "like them," if we know enough about the adventitiously chosen individuals to be reasonably sure that none of their characteristics determining the outcome of the experiment differs enough from those of the target population to change the results there. This is difficult to ascertain, and in any event we have no *probabilistic* warrant for generalizing from the grab-groupers to anyone else whatsoever.) See Millman (1966).

After we secure the 12*n* experimental units by one of the above methods, we assign at random *n* of them to read each of the 12 sets of materials, which have common content but different combinations of style and size. We wish to vary only style and size. All other variables should be held constant (as, for example, by using just males in the experiment, thereby keeping the sex factor at one level) or randomized over all 12 factor-level combinations. This is where experimental control becomes crucial. It might be practicable, for instance, to seat all 12*n* pupils randomly in the room. (That could be done by passing out random seat assignments at the door as the pupils arrived.) If a systematic nonrandom room arrangement were used, this would have to be considered as part of the experimental design, making it more complex than a 4 × 3 factorial. Control of extraneous variables calls for great care and ingenuity so that the experiment will be *internally valid*—i.e., free from bias. See Campbell and Stanley (1966).

After the 12*n* pupils have read the same passage with the same time limit under conditions randomized except for style and size of printing type, they will be given a

Table 1. Outline of the design of the size-style experiment.

Size	Style			
	1	2	3	4
1	X_{111} X_{211} . . . X_{n11}	X_{112} X_{212} . . . X_{n12}	X_{113} X_{213} . . . X_{n13}	X_{114} X_{214} . . . X_{n14}
2	X_{121} X_{221} . . . X_{n21}	X_{123} X_{222} . . . X_{n22}	$X_{12\bar{3}}$ X_{223} . . . X_{n23}	X_{124} X_{224} . . . X_{n24}
3	X_{131} X_{231} . . . X_{n31}	X_{132} X_{232} . . . X_{n32}	X_{133} X_{233} . . . X_{n33}	X_{134} X_{234} . . . X_{n34}

common test to determine how much each learned from the particular combination of style and size of type to which he was exposed. The total score of each pupil on this outcome test will constitute the values of the *dependent variable* to be analyzed. These scores are represented schematically in Table 1, where X_{ips} represents the outcome score (X) of the *i*th individual ($i = 1, 2, \ldots, n$) on the *p*th size or points ($p = 1, 2, 3$) and the *s*th style ($s = 1, 2, 3, 4$). Because this paper is concerned with experimental design rather than statistical analysis, I shall state merely that the $12n$ scores are analyzed for four sources of variation: among each set of *n* individuals who all had the same factor level combination in the experiment (this is called *within-group* variation); between the mean of the $4n$ individuals who had Size 1, the mean of the $4n$ who had Size 2, and the mean of the $4n$ who had Size 3 (called between-size variation); between the mean of the $3n$ individuals who had Style 1, the mean of the $3n$ who had Style 2, the mean of the $3n$ who had Style 3, and the mean of the $3n$ who had Style 4 (called between-style variation); and whatever remains, called *interaction* of size with style.

Characteristics of the Factorial Design

Complete, *balanced* factorial designs (i.e., where each of the possible factor-level combinations occurs and has $n \geq 1$ experimental units assigned to it) permit testing more than one hypothesis about *main effects* (e.g., the influence of type size or the influence of style) efficiently in the same experiment. They also make it possible, where two or more factors are used, to study how the factors interact. Perhaps the least effective of the three sizes when combined with the least effective of the four styles does not produce the least effective of the twelve factor-level combinations. If the effects are *additive*, so that knowing the effectiveness of a certain size factor level and a certain style factor level one can predict the effectiveness of that factor-level combination as well as "chance" permits, then we say that the two factors do not interact. One cannot study interaction statistically in one-factor studies, and yet appreciable interaction may greatly limit generalization of the results of the experiment.

The factorial design is relatively simple. It does not require any pre-measurements or "matching" of the experimental units, because bias is avoided (in the probabilistic sense) by the random assignment of the experimental units to the factor-level combinations. On the other hand, it permits the within-factor-level-combination variability to be as great as true-score variation among individuals treated alike plus errors of measurement dictate. Thus the *signals* (the genuine effects) may be drowned out by the *noise* (the within-combination variability) if the signal-to-noise ratio is low. This noise (or *error*, as it is usually called) lowers the *power* of the significance tests used and increases the width of the confidence intervals computed. True-score error can be lessened by various techniques such as *blocking*, *stratifying*, *leveling*, and *covarying*, all of which depend on classification or pre-measurement of the experimental units and are used to reduce the within-combination variability of true scores; measurement error can be reduced somewhat by using a more reliable outcome measure.

Blocking on a Nominal-Scale Variable

A familiar example of blocking is the classification of each pupil as being either male or female and the introduction of this two-level factor explicitly into the experimental design. Physiological sex is not a manipulated variable, but in the experimental design it is treated like the manipulated factors. With two levels of sex, three levels of type size, and four levels of type style one has $2 \times 3 \times 4 = 24$ factor-level combinations and needs $24n$ experimental units, where n is at least 2. Main effects of sex, size, and style can be estimated. Also, one can study the interaction of sex with size, sex with style, size with style, and sex with size and style. If the main effect of sex is significant, then by having a sex factor one has reduced the error variance significantly. If sex interacts with either or both of the manipulated factors, then by having sex as an explicit factor one has learned how to limit his

generalizations appropriately. For example, it *might* be discovered that women find Style 3 easiest to read, whereas men find Style 1 easiest. When pursued further, this might have practical consequences for the design of textual materials.

Another example of blocking is the use of identical twins in an experiment where one factor is manipulated at two levels. Twin A of each pair is assigned at random to a level of the manipulated factor, and Twin B of that same pair then receives the other level. With P pairs of twins, one has $P \times 2 = 2P$ experimental units and generates $2P$ observations. The three sources of variation are between twin pairs, between the two treatments, and interaction of pairs with treatments. Note that this is just one replicate, so no direct statistical test of the interaction is afforded. If the variation among the twin pairs is significant, one has reduced the error term significantly, but because one sacrifices half his degrees of freedom for error in so doing, the power of the statistical test of the two-level treatment effect may not be improved. Incidentally, this is a social- or biological-science version of the *randomized-block design*, widely used in agriculture for more than 40 years (Fisher, 1925, pp. 226–229). The twin pairs are unordered, having been "measured" on a nominal scale.

Stratifying on Ordinal-Scale Variables

If we have an ordinal-scale variable, such as socioeconomic status of each experimental unit, as an explicit factor in the experimental design I call it a stratifying variable. There might be five levels, such as high, upper-middle, middle-middle, lower-middle, and low socio-economic status, creating a five-level *classificatory* (i.e., not manipulated) factor.

Leveling on Interval- or Ratio-Scale Variable

An interval or nearly interval scale or a ratio scale can be used to yield what I call a leveling variable. To reduce within-factor-level-combination variability one may group the experimental units before the experiment begins on something such as measured reading comprehension or height that is expected to correlate well within treatments with the outcome measure of the experiment. If there are T levels of a treatment factor and LT experimental units, one would arrange the experimental units from highest to lowest on the pre-measured factor into L levels. Within each such level, one experimental unit would be assigned at random to each of the T treatments (i.e., the T levels of the manipulated variable), creating one replicate of an $L \times T$ design. If the measures of the leveling factor do correlate significantly greater than zero with the outcome measures, then (as in the twin design outlined above) the within-treatment variability will be reduced significantly.

For example, one might have $LT = 50$ experimental units. If $T = 5$, there would be $L = 50/5 = 10$ levels, each level containing 5 experimental units randomly assigned, one to each treatment.

Alternatively, one might choose to use $N = nLT$ experimental units, where n is greater than 1. (In the above paragraph, $n = 1$.) Then he would group the N experimental units, from highest to lowest, into $L = N/nT$ sets, and would assign at random n experimental units to each treatment within each level. This would permit testing the interaction of levels with treatments, which the $n = 1$ design does not allow directly.

For example, when $N = 50$, $n = 2$, and $T = 5$, $L = 50/(2)(5) = 5$. Thus there are 10 experimental units within each of the 5 levels. Two of these are assigned at random to each of the 5 treatments within the level.

Error Reduction by Analysis of Covariance

Even though an experimenter has pre-measures of one or more relevant variables, he may nevertheless assign all the experimental units at random to the factor-level combinations, without "leveling" on the basis of the pre-measures, and then in the analysis use the pre-measures statistically to improve the power of his statistical tests. The technique for doing this is known as the *analysis of covariance*, a simple extension of the well-known analysis of variance. Various combinations of blocking, stratifying, leveling, and covarying may be used.

These are some of the basic principles of experimental design underlying the complete, balanced designs. There are a number of incomplete designs, the best-known to educationists and psychologists being Latin squares and Greco-Latin squares. Fractional-factorial designs may become more common in education during the next decade. (See McLean, 1966.) They do, however, have certain limitations that should be studied carefully by the prospective user before he rushes into the experiment with less than a complete design.

Ordered Levels of Factors

If one has three equally spaced sizes of printing type, such as 8, 12, and 16 points, he has three equally spaced levels of an *ordered* factor. A significant *trend* for size of type might be linear, representing an equal increment (or decrement) as one goes from 8-point type to 12-point type and from 12-point type to 16-point type. The trend might be quadratic (i.e., second-order), as when 8 and 16 are equally effective but 12 is much better, or it might combine both linear and quadratic components. Generally, the order of the highest possible trend for an L-level factor is $L - 1$; for size of printing type in our example this is $3 - 1 = 2$.

Style, you note, is a nominal-scale factor, not ordered. It is quite possible to have two or more ordered factors in the same study, however, as for example if one introduced five equally spaced brightnesses of print into the size-style experiment; the various trends that result therefrom can be evaluated.

Random Selection of Factor Levels

Earlier in this paper it was hinted that the four styles of printing type might have been drawn at random from a larger target population of printing styles to which one wished to generalize. If that population contained, say, 40 styles, then the four drawn would be 10% of the entire population, a small but hardly negligible percentage.[1] If, on the other hand, we drew four schools out of a target population of 4000 schools, the one-tenth of 1% that the schools in the experiment constitute of the population would be tiny, so we might choose to consider that essentially the four schools had been drawn from an infinite population of schools, in which case we would (using the jargon of the field) say that the schools are a *random-effects factor*. (Recall the fixed-effects factor defined earlier.)

If one has both fixed-effects and random-effects factors in an experiment, we say that he should use a *mixed-model* analysis of his results. Genuine random-effects factors seem rare in educational research, but often we choose to act as if the levels of a factor such as teachers, schools, classes, ratees, or twin pairs had been drawn randomly from a virtually infinite population of such levels. We do this by arguing that the particular levels used are plausibly a random sample from a hypothetical population to which we wish to generalize: teachers "like these," schools "like these," etc. This raising-one-self-by-one's-own-bootstraps logic has the endorsement of several top-level mathematical statisticians who have debated it rather hotly with each other. If we are going to generalize to other teachers, other schools, and the like, anyway, then we should use the analytical model that fits such generalization, rather than using the model that applies just to the particular levels in the experiment itself.

Natural and Controlled Experiments

Sampling factor levels from a population of factor levels larger than the number to be

[1] From these 4 one could estimate the variation among the population means of the 40 styles.

used in the experiment brings controlled experimentation closer to the methodology of sample surveys than it was until the early 1950's. A difference that persists is the manipulation by the experimenter of the levels of one or more factors, along with the random assignment of the experimental units to the factor-level combinations. Sampling of occupations might occur in a survey of salaries, where respondents are classified by occupation, sex, and marital status, as when the occupations to be studied are drawn at random from a large list of occupations. The analysis of the results of such a study might proceed formally in much the same way as that for an experiment involving styles of printing type, sex, and marital status, but in the status study nothing was manipulated, whereas in the experiment style of type was. Of course, one can *conceive* of an experiment in which experimental units representing the crossing of the two sexes and three marital statuses are assigned at random to occupations, thereby distributing ability, interests, education, age, and the like randomly across the occupations and in this way removing the confounding of occupation with those personal characteristics.

Generally, interpretation of the results of a controlled experiment will be easier than that of the analogous "natural experiment." A more familiar example than the occupational one is investigation of the effects on general English vocabulary of studying Latin in high school. If students elect (i.e., volunteer) to take Latin or not to take it, the inputs for the two conditions will almost always be substantially different, the students taking Latin being better *initially* on English vocabulary, IQ, and a host of other cognitive and affective variables. If, however, half the prospective enrollees in Latin could be assigned at random to Latin in the ninth and tenth grades and the other half in, say, the eleventh and twelfth grades, nonreactively so that disappointment and frustration did not upset the operation of the school, it should be possible to compare both groups unbiasedly at the end of the tenth grade, after half had completed two years of Latin and the other half had taken none. Because of the random assignment, there would be no systematic confounding of any antecedent variables with the experimental variable (i.e., took Latin versus did not take it). Results should be much more readily interpretable than in the natural experiment.

One has only to recall the great difficulties encountered by statisticians when analyzing the results of the vast natural-smoking experiment that has been going on for many years. Is cigarette smoking one of the potent "causes" of lung cancer? Of other ills? After much comparison of human subgroups and animal experimentation in order to discredit plausible alternative hypotheses, most researchers in this area have concluded that smoking cigarettes does increase the probability that a person will develop lung cancer and have certain other ailments, but because the work with humans was not controlled experimentation, no proof overwhelmingly convincing to all intelligent persons has even yet been provided. Indeed, associational analyses cannot eliminate all plausible alternative hypotheses, whereas controlled experiments *if conducted impeccably* can rule out all systematic ones, leaving only chance fluctuations (usually of quite low probability) as the alternative explanation. This is not to say that a single experiment can be definitive or perfect. Often it will raise more new questions than it answers old ones, but at least the process of randomized assignment of experimental units to factor-level combinations removes the chief source of systematic bias that afflicts natural experiments, so with respect to internal validity the controlled experiment starts out well ahead of most natural experiments.

Control may exact a high price in terms of lowered external validity (i.e., generalizability), however. For example, one probably can't assign persons at random to unidentified-flying-object (UFO) versus non-UFO clubs and preserve the sense of the distinction as it occurs naturally. One might try paying some persons to smoke and others to refrain from smoking, but very likely this would not simulate well enough the natural situation

where persons of certain temperaments and backgrounds cannot resist smoking several packages of cigarettes daily, whereas other types of individuals aren't tempted. One can't very well assign occupations randomly in a meaningful way. Even assigning Latin versus no Latin to eligible high-school students by deferring this subject for half of them has never been done, so far as I am aware. There do not even seem to be any controlled experiments involving cursive handwriting versus manuscript handwriting in the primary grades, despite the fact that all school children throughout the country are taught to write by some combination of such methods.

I have dealt only with attempts to make causal inferences from comparative data. Questions such as "What is the distribution of the ages of female first-grade teachers in California?" and "Are there more male than female elementary-school principals in the United States?" are examples of perfectly reputable inquiries not directly concerned with causal influences. Without a great store of such information, one is hardly ready to design controlled experiments. We get answers to important questions by the use of questionnaires, interviews, and various documents. Survey researchers have developed powerful analytical techniques for inferring causation from status data, too. For excellent recent examples, see Sieber and Lazarsfeld (1966), Trow (1966), and Webb, Campbell, Schwartz, and Sechrest (1966).

Quasi-Experimentation

Although randomization is the method of choice in experimentation,

There are many natural social settings in which the research person can introduce something like experimental design into his scheduling of data collection procedures (e.g., the *when* and *to whom* of measurement), even though he lacks the full control over the scheduling of experimental stimuli (the *when* and *to whom*) of exposure and the ability to randomize exposures which makes a true experiment possible. Collectively, such situations can be regarded as quasi-experimental designs. One purpose of this [monograph] is to encourage the utilization of such quasi-experiments and to increase the awareness of the kinds of settings in which opportunities to employ them occur. But just because full experimental control *is* lacking, it becomes imperative that the researcher be thoroughly aware of which specific variables his particular design fails to control. It is for this need in evaluating quasi-experiments, more than for understanding true experiments, that the check lists of sources of invalidity [in this monograph] were developed. (Campbell & Stanley, 1966, p. 34.)

Quasi-experimentation, as devised by Campbell (1957) and developed further by Campbell and Stanley (1966), seems to offer a middleground between the controlled experiment of the laboratory and the uncontrolled experiment of nature. If not used carelessly and faddishly, as " action research " often was, it may helpfully augment the armamentarium of the educational experimenter.

Conclusion

Controlled and quasi-experimentation are only two of *many* promising ways to tackle difficult and important research problems. As Bridgman (1945, p. 45) said, " The scientific method, as far as it is a method, is nothing more than doing one's damnedest with one's mind, no holds barred." In that mighty struggle at the frontier of knowledge we cannot afford to relegate associational analyses or any other approach to the intellectual junkyard. Experimentation seems especially well suited for attacking certain problems. Perhaps it is used too little in educational research (see Stanley, 1966), but just because I have devoted this article to experimentation, you must not conclude that other procedures are inferior. For balance and supplementation, read Trow (1967), Cronbach (1957), and Platt (1964).

References

Bridgman, P. W. The prospect for intelligence. *Yale Review*, 1945, **34**, 444–461.

Campbell, D. T. Factors relevant to the validity of experiments in social settings. *Psychological Bulletin*, 1957, **54**, 297–312.

Campbell, D. T., & Stanley, J. C. *Experimental and quasi-experimental designs for research*. Chicago: Rand McNally, 1966. Supplemented version of Chapter 5 (pp. 171–246) in N. L. Gage (Ed.), *Handbook of research on teaching*. Chicago: Rand McNally, 1963.

Cronbach, L. J. The two disciplines of scientific psychology. *American Psychologist*, 1957, **12**, 671–684.

Fisher, R. A. *Statistical methods for research workers*. Edinburgh: Oliver & Boyd, 1925.

McLean, L. D. Some important principles for the use of incomplete designs in behavioral research. In J. C. Stanley (Ed.), *Improving experimental design and statistical analysis*. Chicago: Rand McNally, 1966. Also see L. D. McLean, Phantom classrooms. *School Review*, 1966, **74**, 139–149.

Millman, J. In the service of generalization. *Psychology in the Schools*, 1966, **3**, 333–339.

Platt, J. R. Strong inference. *Science*, 1964, **146**, 347–352.

Sieber, S. D., with the collaboration of P. F. Lazarsfeld. *The organization of educational research*. New York: Bureau of Applied Social Research, Columbia University, 1966.

Stanley, J. C. On improving certain aspects of educational experimentation. In J. C. Stanley (Ed.), *Improving experimental design and statistical analysis*. Chicago: Rand McNally, 1966.

An earlier version, without the discussion, appeared with the same title in the *Delta Journal*, Stanford University, March 1966, pp. 17–34.

Trow, M. Education and survey research. In C. Y. Glock (Ed.), *Survey research in the social sciences*. Russell Sage Foundation, 1967, in press.

Webb, E. J., Campbell, D. T., Schwartz, R. D., & Sechrest, L. *Unobtrusive measures: nonreactive research in the social sciences*. Chicago: Rand McNally, 1966.

Supplementary References

Cochran, W. G., & Cox, Gertrude M. *Experimental designs*. (2nd ed.) New York: Wiley, 1957.

Cox, D. R. *Planning of experiments*. New York: Wiley, 1958.

Lerner, D. (Ed.) *Cause and effect.* New York: Free Press, 1965.

Stanley, J. C. Controlled experimentation in the classroom. *Journal of Experimental Education,* 1957, **25**, 195–201. Reprinted on pp. 209–218 of J. M. Seidman (Ed.), *Educating for mental health.* New York: Crowell, 1963.

Stanley, J. C. Educational experimentation. In R. A. Price (Ed.), Needed research in the teaching of the social studies. *Research Bulletin No. 1, National Council for the Social Studies.* Washington, D.C. 20036: The Council, 1964. Pp. 119–124.

Stanley, J. C. Quasi-experimentation. *School Review,* 1965, **73**, 197–205.

Stanley, J. C. A common class of pseudo-experiments. *American Educational Research Journal,* 1966, **3**, 79–87.

Stanley, J. C. The influence of Fisher's *The design of experiments* on educational research thirty years later. *American Educational Research Journal,* 1966, **3**, 223–229.

Stanley, J. C. Quasi-experimentation in educational settings. *School Review,* in press.

Editor's comments: Experiments always include an independent variable, a dependent variable, and one or more extraneous or nuisance variables. Procedures for manipulating independent variables and for measuring dependent variables are familiar to most students in statistics. Less familiar are procedures for controlling extraneous variables, those undesired sources of variation that can confound the interpretation of the experiment. Before extraneous variables can be controlled they must be identified, and this is often a difficult task. To aid the experimenter Donald T. Campbell has classified the most common extraneous variables into seven categories. Campbell's classification and suggested experimental designs for controlling different kinds of extraneous variables are contained in the following article.

6.3   Factors Relevant to the Validity of Experiments in Social Settings

Donald T. Campbell

What do we seek to control in experimental designs? What extraneous variables which would otherwise confound our interpretation of the experiment do we wish to rule out? The present paper attempts a

A dittoed version of this paper was privately distributed in 1953 under the title "Designs for Social Science Experiments." The author has had the opportunity to benefit from the careful reading and suggestions of L. S. Burwen, J. W. Cotton, C. P. Duncan, D. W. Fiske, C. I. Hovland, L. V. Jones, E. S. Marks, D. C. Pelz, and B. J. Underwood, among others, and wishes to express his appreciation. They have not had the opportunity of seeing the paper in its present form, and bear no responsibility for it. The author also wishes to thank S. A. Stouffer (33) and B. J. Underwood (36) for their public encouragement.

specification of the major categories of such extraneous variables and employs these categories in evaluating the validity of standard designs for experimentation in the social sciences.

Validity will be evaluated in terms of two major criteria. First, and as a basic minimum, is what can be called *internal validity*: did in fact the experimental stimulus make some significant difference in this specific instance? The second criterion is that of *external validity, representativeness* or *generalizability*: to what populations, settings, and variables can this effect be generalized? Both criteria are obviously important although it turns out that they are to some extent incompatible, in that the controls required for internal validity often tend to jeopardize representativeness.

The extraneous variables affecting internal validity will be introduced in the process of analyzing three pre-experimental designs. In the subsequent evaluation of the applicability of three true experimental designs, factors leading to external invalidity will be introduced. The effects of these extraneous variables will be considered at two levels: as simple or main effects, they occur independently of or in addition to the effects of the experimental variable; as interactions, the effects appear in conjunction with the experimental variable. The main effects typically turn out to be relevant to internal validity, the interaction effects to external validity or representativeness.

The following designation for experimental designs will be used: X will represent the exposure of a group to the experimental variable or event, the effects of which are to be measured; O will refer to the process of observation or measurement, which can include watching what people do, listening, recording, interviewing, administering tests, counting lever depressions, etc. The Xs and Os in a given row are applied to the same specific persons. The left to right dimension indicates temporal order. Parallel rows represent equivalent samples of persons unless otherwise specified. The designs will be numbered and named for cross-reference purposes.

Three Pre-experimental Designs and Their Confounded Extraneous Variables

The One-Shot Case Study. As Stouffer (32) has pointed out, much social science research still uses Design 1, in which a single individual or group is studied in detail only once, and in which the observations are attributed to exposure to some prior situation.

X O 1. One-Shot Case Study

This design does not merit the title of experiment, and is introduced only to provide a reference point. The very minimum of useful

scientific information involves at least one formal comparison and therefore at least two careful observations (2).

The One-Group Pretest-Posttest Design. This design does provide for one formal comparison of two observations, and is still widely used.

O_1 X O_2 2. One-Group Pretest-Posttest Design

However, in it there are four or five categories of extraneous variables left uncontrolled which thus become rival explanations of any difference between O_1 and O_2, confounded with the possible effect of X.

The first of these is the main effect of *history*. During the time span between O_1 and O_2 many events have occurred in addition to X, and the results might be attributed to these. Thus in Collier's (8) experiment, while his respondents[1] were reading Nazi propaganda materials, France fell, and the obtained attitude changes seemed more likely a result of this event than of the propaganda.[2] By history is meant the specific event series other than X, i.e., the extra-experimental uncontrolled stimuli. Relevant to this variable is the concept of experimental isolation, the employment of experimental settings in which all extraneous stimuli are eliminated. The approximation of such control in much physical and biological research has permitted the satisfactory employment of Design 2. But in social psychology and the other social sciences, if history is confounded with X the results are generally uninterpretable.

The second class of variables confounded with X in Design 2 is here designated as *maturation*. This covers those effects which are systematic with the passage of time, and not, like history, a function of the specific events

[1] In line with the central focus on social psychology and the social sciences, the term *respondent* is employed in place of the terms *subject*, *patient*, or *client*.

[2] Collier actually used a more adequate design than this, an approximation to Design 4.

involved. Thus between O_1 and O_2 the respondents may have grown older, hungrier, tireder, etc., and these may have produced the difference between O_1 and O_2, independently of X. While in the typical brief experiment in the psychology laboratory, maturation is unlikely to be a source of change, it has been a problem in research in child development and can be so in extended experiments in social psychology and education. In the form of "spontaneous remission" and the general processes of healing it becomes an important variable to control in medical research, psychotherapy, and social remediation.

There is a third source of variance that could explain the difference between O_1 and O_2 without a recourse to the effect of X. This is the effect of *testing* itself. It is often true that persons taking a test for the second time make scores systematically different from those taking the test for the first time. This is indeed the case for intelligence tests, where a second mean may be expected to run as much as five IQ points higher than the first one. This possibility makes important a distinction between *reactive* measures and *nonreactive* measures. A reactive measure is one which modifies the phenomenon under study, which changes the very thing that one is trying to measure. In general, any measurement procedure which makes the subject selfconscious or aware of the fact of the experiment can be suspected of being a reactive measurement. Whenever the measurement process is *not* a part of the normal environment it is probably reactive. Whenever measurement exercises the process under study, it is almost certainly reactive. Measurement of person's height is relatively nonreactive. However, measurement of weight, introduced into an experimental design involving adult American women, would turn out to be reactive in that the process of measuring would stimulate weight reduction. A photograph of a crowd taken in secret from a second story window would be nonreactive, but a news photograph of the same scene might very well be reactive, in that the presence of the photographer would modify the behavior of people seeing themselves being photographed. In a factory, production records introduced for the purpose of an experiment would be reactive, but if such records were a regular part of the operating environment they would be nonreactive. An English anthropologist may be nonreactive as a participant-observer at an English wedding, but might be a highly reactive measuring instrument at a Dobu nuptials. Some measures are so extremely reactive that their use in a pretest-posttest design is not usually considered. In this class would be tests involving surprise, deception, rapid adaptation, or stress. Evidence is amply present that tests of learning and memory are highly reactive (35, 36). In the field of opinion and attitude research our well-developed interview and attitude test techniques must be rated as reactive, as shown, for example, by Crespi's (9) evidence.

Even within the personality and attitude test domain, it may be found that tests differ in the degree to which they are reactive. For some purposes, tests involving voluntary self-description may turn out to be more reactive (especially at the interaction level to be discussed below) than are devices which focus the respondent upon describing the external world, or give him less latitude in describing himself (e.g., 5). It seems likely that, apart from considerations of validity, the Rorschach test is less reactive that the TAT or MMPI. Where the reactive nature of the testing process results from the focusing of attention on the experimental variable, it may be reduced by imbedding the relevant content in a comprehensive array of topics, as has regularly been done in Hovland's attitude change studies (14). It seems likely that with attention to the problem, observational and measurement techniques can be developed which are much less reactive than those now in use.

Instrument decay provides a fourth uncontrolled source of variance which could produce an O_1—O_2 difference that might be mistaken for the effect of X. This variable can be exemplified by the fatiguing of a spring scales, or the condensation of water vapor in a cloud chamber. For psychology and the social sciences it becomes a particularly acute problem when human beings are used as a part of the measuring apparatus, as judges, observers, raters, coders, etc. Thus O_1 and O_2 may differ because the raters have become more experienced, more fatigued, have acquired a different adaptation level, or have learned about the purpose of the experiment, etc. However infelicitously, this term will be used to typify those problems introduced when shifts in measurement conditions are confounded with the effect of X, including such crudities as having a different observer at O_1 and O_2, or using a different interviewer or coder. Where the use of different interviewers, observers, or experimenters is unavoidable, but where they are used in large numbers, a sampling equivalence of interviewers is required, with the relevant N being the N of interviewers, not interviewees, except as refined through cluster sampling considerations (18).

A possible fifth extraneous factor deserves mention. This is statistical *regression*. When, in Design 2, the group under investigation has been selected for its extremity on O_1, O_1-O_2 shifts toward the mean will occur which are due to random imperfections of the measuring instrument or random instability within the population, as reflected in the test-retest reliability. In general, regression operates like maturation in that the effects increase systematically with the O_1-O_2 time interval. McNemar (22) has demonstrated the profound mistakes in interpretation which failure to control this factor can introduce in remedial research.

The Static Group Comparison. The third pre-experimental design is the Static Group Comparison.

X O_1
– – – – –   3. The Static Group Comparison
O_2

In this design, there is a comparison of a group which has experienced X with a group which has not, for the purpose of establishing the effect of X. In contrast with Design 6, there is in this design no means of certifying that the groups were equivalent at some prior time. (The absence of sampling equivalence of groups is symbolized by the row of dashes.) This design has its most typical occurrence in the social sciences, and both its prevalence and its weakness have been well indicated by Stouffer (32). It will be recognized as one form of the correlational study. It is introduced here to complete the list of confounding factors. If the Os differ, this difference could have come through biased *selection* or recruitment of the persons making up the groups; i.e., they might have differed anyway without the effect of X. Frequently, exposure to X (e.g., some mass communication) has been voluntary and the two groups have an inevitable systematic difference on the factors determining the choice involved, a difference which no amount of matching can remove.

A second variable confounded with the effect of X in this design can be called experimental *mortality*. Even if the groups were equivalent at some prior time, O_1 and O_2 may differ now not because individual members have changed, but because a biased subset of members have dropped out. This is a typical problem in making inferences from comparisons of the attitudes of college freshmen and college seniors, for example.

True Experimental Designs

The Pretest-Posttest Control Group Design. One or another of the above considerations led psychologists between 1900 and 1925 (2, 30) to expand Design 2 by the addition of a control group, resulting in Design 4.

O_1 X O_2 4. Pretest-Posttest Control Group
O_3 O_4 Design

Because this design so neatly controls for the main effects of history, maturation, testing, instrument decay, regression, selection, and mortality, these separate sources of variance are not usually made explicit. It seems well to state briefly the relationship of the design to each of these confounding factors, with particular attention to the application of the design in social settings.

If the differences between O_1 and O_2 were due to intervening historical events, then they should also show up in the O_3–O_4 comparison. Note, however, several complications in achieving this control. If respondents are run in groups, and if there is only one experimental session and one control session, then there is no control over the unique internal histories of the groups. The O_1–O_2 difference, even if not appearing in O_3–O_4, may be due to a chance distracting factor appearing in one or the other group. Such a design, while controlling for the shared history or event series, still confounds X with the unique session history. Second, the design implies a simultaneity of O_1 with O_3 and O_2 with O_4 which is usually impossible. If one were to try to achieve simultaneity by using two experimenters, one working with the experimental respondents, the other with the controls, this would confound experimenter differences with X (introducing one type of instrument decay). These considerations make it usually imperative that, for a true experiment, the experimental and control groups be tested and exposed individually or in small subgroups, and that sessions of both types be temporally and spatially intermixed.

As to the other factors: if maturation or testing contributed an O_1–O_2 difference, this should appear equally in the O_3–O_4 comparison, and these variables are thus controlled for their main effects. To make sure the design controls for instrument decay, the same individual or small-session approximation to simultaneity needed for history is required. The occasional practice of running the experimental group and control group at different times is thus ruled out on this ground as well as that of history. Otherwise the observers may have become more experienced, more hurried, more careless, the maze more redolent with irrelevant cues, the lever-tension and friction diminished, etc.

Only when groups are effectively simultaneous do these factors affect experimental and control groups alike. Where more than one experimenter or observer is used, counterbalancing experimenter, time, and group is recommended. The balanced Latin square is frequently useful for this purpose (4).

While regression is controlled in the design as a whole, frequently secondary analyses of effects are made for extreme pretest scorers in the experimental group. To provide a control for effects of regression, a parallel analysis of extremes should also be made for the control group.

Selection is of course handled by the sampling equivalence ensured through the randomization employed in assigning persons to groups, perhaps supplemented by, but not supplanted by, matching procedures. Where the experimental and control groups do not have this sort of equivalence, one has a compromise design rather than a true experiment. Furthermore, the O_1–O_3 comparison provides a check on possible sampling differences.

The design also makes possible the examination of experimental mortality, which becomes a real problem for experiments extended over weeks or months. If the experimental and control groups do not differ in the number of lost cases nor in their pretest scores, the experiment can be judged internally valid on this point, although mortality reduces the generalizability of effects to the original population from which the groups were selected.

For these reasons, the Pretest-Posttest Control Group Design has been the ideal in the social sciences for some thirty years. Recently, however, a serious and avoidable imperfection in it has been noted, perhaps first by Schanck and Goodman (29). Solomon (30) has expressed the point as an *interaction* effect of testing. In the terminology of analysis of variance, the effects of history, maturation and testing, as described so far, are all *main* effects, manifesting themselves in mean differences independently of the presence of other variables. They are effects that could be added on to other effects, including the effect of the experimental variable. In contrast, interaction effects represent a joint effect, specific to the concomitance of two or more conditions, and may occur even when no main effects are present. Applied to the testing variable, the interaction effect might involve not a shift due solely or directly to the measurement process, but rather a sensitization of respondents to the experimental variable so that when X was preceded by O there would be a change, whereas both X and O would be without effect if occurring alone. In terms of the two types of validity, Design 4 is internally valid, offering an adequate basis for generalization to other sampling-equivalent *pretested* groups. But it has a serious and systematic weakness in representativeness in that it offers, strictly speaking, no basis for generalization to the *unpretested* population. And it is usually the *unpretested* larger universe from which these samples were taken to which one wants to generalize.

A concrete example will help make this clearer. In the NORC study of a United Nations information campaign (31), two equivalent samples, of a thousand each, were drawn from the city's population. One of these samples was interviewed, following which the city of Cincinnati was subjected to an intensive publicity campaign using all the mass media of communication. This included special features in the newspapers and on the radio, bus cards, public lectures, etc. At the end of two months, the second sample of 1,000 was interviewed and the results compared with the first 1,000. There were no differences between the two groups except that the second group was somewhat more pessimistic about the likelihood of Russia's cooperating for world peace, a result which was attributed to history rather than to the publicity campaign. The second sample was no better informed about the United Nations nor had it noticed in particular the publicity campaign which had been going on. In connection with a program of research on panels and the reinterview problem, Paul Lazarsfeld and the Bureau of Applied Social Research

arranged to have the initial sample reinterviewed at the same time as the second sample was interviewed, after the publicity campaign. This reinterviewed group showed significant attitude changes, a high degree of awareness of the campaign and important increases in information. The inference in this case is unmistakably that the initial interview had sensitized the persons interviewed to the topic of the United Nations, had raised in them a focus of awareness which made the subsequent publicity campaign effective for them but for them only. This study and other studies clearly document the possibility of interaction effects which seriously limit our capacity to generalize from the pretested experimental group to the unpretested general population. Hovland (15) reports a general finding which is of the opposite nature but is, nonetheless, an indication of an interactive effect. In his Army studies the initial pretest served to reduce the effects of the experimental variable, presumably by creating a commitment to a given position. Crespi's (9) findings support this expectation. Solomon (30) reports two studies with school children in which a spelling pretest reduced the effects of a training period. But whatever the direction of the effect, this flaw in the Pretest-Posttest Control Group Design is serious for the purposes of the social scientist.

The Solomon Four-Group Design. It is Solomon's (30) suggestion to control this problem by adding to the traditional two-group experiment two unpretested groups as indicated in Design 5.

$$
\begin{array}{ccl}
O_1 & X & O_2 \\
O_3 & & O_4 \qquad \text{5. Solomon Four-Group Design} \\
& X & O_5 \\
& & O_6
\end{array}
$$

This Solomon Four-Group Design enables one both to control and measure both the main and interaction effects of testing and the main effects of a composite of maturation and history. It has become the new ideal design for social scientists. A word needs to be said about the appropriate statistical analysis.

In Design 4, an efficient single test embodying the four measurements is achieved through computing for each individual a pretest-posttest difference score which is then used for comparing by t test the experimental and control groups. Extension of this mode of analysis to the Solomon Four-Group Design introduces an inelegant awkwardness to the otherwise elegant procedure. It involves assuming as a pretest score for the unpretested groups the mean value of the pretest from the first two groups. This restricts the effective degrees of freedom, violates assumptions of independence, and leaves one without a legitimate base for testing the significance of main effects and interaction. An alternative analysis is available which avoids the assumed pretest scores. Note that the four posttests form a simple two-by-two analysis of variance design:

	No X	X
Pretested	O_4	O_2
Unpretested	O_6	O_5

The column means represent the main effect of X, the row means the main effect of pretesting, and the interaction term the interaction of pretesting and X. (By use of a t test the combined main effects of maturation and history can be tested through comparing O_6 with O_1 and O_3.)

The Posttest-Only Control Group Design. While the statistical procedures of analysis of variance introduced by Fisher (10) are dominant in psychology and the other social sciences today, it is little noted in our discussions of experimental arrangements that Fisher's typical agricultural experiment involves no pretest: equivalent plots of ground receive different experimental treatments and the subsequent yields are measured.[3] Applied

to a social experiment as in testing the influence of a motion picture upon attitudes, two randomly assigned audiences would be selected, one exposed to the movie, and the attitudes of each measured subsequently for the first time.

A	X	O_1	6. Posttest-Only Control Group
A		O_2	Design

In this design the symbol A had been added, to indicate that at a specific time prior to X the groups were made equivalent by a random sampling *assignment*. A is the point of selection, the point of allocation of individuals to groups. It is the existence of this process that distinguishes Design 6 from Design 3, the Static Group Comparison. Design 6 is not a static cross-sectional comparison, but instead truly involves control and observation extended in time. The sampling procedures employed assure us that at time A the groups were equal, even if not measured. A provides a point of prior equality just as does the pretest. A point A is, of course, involved in all true experiments, and should perhaps be indicated in Designs 4 and 5. It is essential that A be regarded as a specific point in time, for groups change as a function of time since A, through experimental mortality. Thus in a public opinion survey situation employing probability sampling from lists of residents, the longer the time since A, the more the sample underrepresents the transient segments of society, the newer dwelling units, etc. When experimental groups are being drawn from a self-selected extreme population, such as applicants for psychotherapy, time since A introduces maturation (spontaneous remission) and regression factors. In Design 6 these effects would be confounded with the effect of X if the As as well as the Os were not contemporaneous for experimental and control groups.

Like Design 4, this design controls for the effects of maturation and history through the practical simultaneity of both the As and the Os. In superiority over Design 4, no main or interaction effects of pretesting are involved. It is this feature that recommends it in

---

[3] This is not to imply that the pretest is totally absent from Fisher's designs. He suggests the use of previous year's yields, etc., in covariance analysis. He notes, however, "with annual agricultural crops, knowledge of yields of the experimental area in a previous year under uniform treatment has not been found sufficiently to increase the precision to warrant the adoption of such uniformity trials as a preliminary to projected experiments" (10, p. 176).

particular. While it controls for the main and interaction effects of pretesting as well as does Design 5, the Solomon Four-Group Design, it does not measure these effects, nor the main effect of history-maturation. It can be noted that Design 6 can be considered as the two unpretested "control" groups from the Solomon Design, and that Solomon's two traditional pretested groups have in this sense the sole purpose of measuring the effects of pretesting and history-maturation, a purpose irrelevant to the main aim of studying the effect of X (25). However, under normal conditions of not quite perfect sampling control, the four-group design provides in addition greater assurance against mistakenly attributing to X effects which are not due it, inasmuch as the effect of X is documented in three different fashions (O_1 vs. O_2, O_2 vs. O_4, and O_5 vs. O_6). But, short of the four-group design, Design 6 is often to be preferred to Design 4, and is a fully valid experimental design.

Design 6 has indeed been used in the social sciences, perhaps first of all in the classic experiment by Gosnell, *Getting Out the Vote* (11). Schanck and Goodman (29), Hovland (15) and others (1, 12, 23, 24, 27) have also employed it. But, in spite of its manifest advantages of simplicity and control, it is far from being a popular design in social research and indeed is usually relegated to an inferior position in discussions of experimental designs if mentioned at all (e.g., 15, 16, 32). Why is this the case?

In the first place, it is often confused with Design 3. Even where Ss have been carefully assigned to experimental and control groups, one is apt to have an uneasiness about the design because one "doesn't know what the subjects were like before." This objection must be rejected, as our standard tests of significance are designed precisely to evaluate the likelihood of differences occurring by chance in such sample selection. It is true, however, that this design is particularly vulnerable to selection bias and where random assignment is not possible it remains suspect. Where naturally aggregated units, such as classes, are employed intact, these should be

used in large numbers and assigned at random to the experimental and control conditions; cluster sampling statistics (18) should be used to determine the error term. If but one or two intact classrooms are available for each experimental treatment, Design 4 should certainly be used in preference.

A second objection to Design 6, in comparison with Design 4, is that it often has less precision. The difference scores of Design 4 are less variable than the posttest scores of Design 6 if there is a pretest-posttest correlation above .50 (15, p. 323), and hence for test-retest correlations above that level a smaller mean difference would be statistically significant for Design 4 than for Design 6, for a constant number of cases. This advantage to Design 4 may often be more than dissipated by the costs and loss in experimental efficiency resulting from the requirement of two testing sessions, over and above the considerations of representativeness.

Design 4 has a particular advantage over Design 6 if experimental mortality is high. In Design 4, one can examine the pretest scores of lost cases in both experimental and control groups and check on their comparability. In the absence of this in Design 6, the possibility is opened for a mean difference resulting from differential mortality rather than from individual change, if there is a substantial loss of cases.

A final objection comes from those who wish to study the relationship of pretest attitudes to kind and amount of change. This is a valid objection, and where this is the interest, Design 4 or 5 should be used, with parallel analysis of experimental and control groups. Another common type of individual difference study involves classifying persons in terms of amount of change and finding associated characteristics such as sex, age, education, etc. While unavailable in this form in Design 6, essentially the same correlational information can be obtained by subdividing both experimental and control groups in terms of the associated characteristics, and examining the experimental-control difference for such subtypes.

For Design 6, the Posttest-Only Control

Group Design, there is a class of social settings in which it is optimally feasible, settings which should be more used than they now are. Whenever the social contact represented by X is made to single individuals or to small groups, and where the response to that stimulus can be identified in terms of individuals or type of X, Design 6 can be applied. Direct mail and door-to-door contacts represent such settings. The alternation of several appeals from door-to-door in a fund-raising campaign can be organized as a true experiment without increasing the cost of the solicitation. Experimental variation of persuasive materials in a direct-mail sales campaign can provide a better experimental laboratory for the study of mass communication and persuasion than is available in any university. The well-established, if little-used, split-run technique in comparing alternative magazine ads is a true experiment of this type, usually limited to coupon returns rather than sales because of the problem of identifying response with stimulus type (20). The split-ballot technique (7) long used in public opinion polls to compare different question wordings or question sequences provides an excellent example which can obviously be extended to other topics (e.g., 12). By and large these laboratories have not yet been used to study social science theories, but they are directly relevant to hypotheses about social persuasion.

Multiple X designs. In presenting the above designs, X has been opposed to No-X, as is traditional in discussions of experimental design in psychology. But while this may be a legitimate description of the stimulus-isolated physical science laboratory, it can only be a convenient shorthand in the social sciences, for any No-X period will not be empty of potentially change-inducing stimuli. The experience of the control group might better be categorized as another type of X, a control experience, an X_C instead of No-X. It is also typical of advance in science that we are soon no longer interested in the qualitative fact of effect or no-effect, but want to specify degree of effect for varying degrees of X. These considerations lead into designs in which multiple groups are used, each with a different X_1, X_2, X_3, X_n, or in multiple factorial design, as X_{1a}, X_{1b}, X_{2a}, X_{2b}, etc. Applied to Designs 4 and 6, this introduces one additional group for each additional X. Applied to 5, The Solomon Four-Group Design, two additional groups (one pretested, one not, both receiving X_n) would be added for each variant on X.

In many experiments, X_1, X_2, X_3, and X_n are all given to the same group, differing groups receiving the Xs in different orders. Where the problem under study centers around the effects of order or combination, such counterbalanced multiple X arrangements are, of course, essential. Studies of transfer in learning are a case in point (34). But where one wishes to generalize to the effect of each X as occurring in isolation, such designs are not recommended because of the sizable interactions among Xs, as repeatedly demonstrated in learning studies under such labels as proactive inhibition and learning sets. The use of counterbalanced sets of multiple Xs to achieve experimental equation, where natural groups not randomly assembled have to be used, will be discussed in a subsequent paper on compromise designs.

Testing for effects extended in time. The researches of Hovland and his associates (14, 15) have indicated repeatedly that the longer range effects of persuasive Xs may be qualitatively as well as quantitatively different from immediate effects. These results emphasize the importance of designing experiments to measure the effect of X at extended periods of time. As the misleading early research on reminiscence and on the consolidation of the memory trace indicate (36), repeated measurement of the same persons cannot be trusted to do this if a reactive measurement process is involved. Thus, for Designs 4 and 6, two separate groups must be added for each posttest period. The additional control group cannot be omitted, or the effects of intervening history, maturation, instrument decay, regression, and mortality are confounded with the delayed effects of X. To follow fully the logic of Design 5, four additional groups are required for each posttest period.

True experiments in which O *is not under* E's *control.* It seems well to call the attention of the social scientist to one class of true experiments which are possible without the full experimental control over both the "when" and "to whom" of both X and O. As far as this analysis has been able to go, no such true experiments are possible without the ability to control X, to withhold it from carefully randomly selected respondents while presenting it to others. But control over O does not seem so indispensable. Consider the following design.

A X O_1 6. Posttest Only Design, where O
A O_2 cannot be withheld from any
   (O) respondent
   (O)
   (O)

The parenthetical Os are inserted to indicate that the studied groups, experimental and control, have been selected from a larger universe all of which will get O anyway. An election provides such an O, and using "whether voted" rather than "how voted," this was Gosnell's design (11). Equated groups were selected at time A, and the experimental group subjected to persuasive materials designed to get out the vote. Using precincts rather than persons as the basic sampling unit, similar studies can be made on the content of the voting (6). Essential to this design is the ability to create specified randomly equated groups, the ability to expose one of these groups to X while withholding it (or providing X_2) from the other group, and the ability to identify the performance of each individual or unit in the subsequent O. Since such measures are natural parts of the environment to which one wishes to generalize, they are not reactive, and Design 4, the Pretest-Posttest Control Group Design, is feasible if O has a predictable periodicity to it. With the precinct as a unit, this was the

design of Hartmann's classic study of emotional vs. rational appeals in a public election (13). Note that 5, the Solomon Four-Group Design, is not available, as it requires the ability to withhold O experimentally, as well as X.

Further Problems of Representativeness

The interaction effect of testing, affecting the external validity or representativeness of the experiment, was treated extensively in the previous section, inasmuch as it was involved in the comparison of alternative designs. The present section deals with the effects upon representativeness of other variables which, while equally serious, can apply to any of the experimental designs.

The interaction effects of selection. Even though the true experiments control selection and mortality for internal validity purposes, these factors have, in addition, an important bearing on representativeness. There is always the possibility that the obtained effects are specific to the experimental population and do not hold true for the populations to which one wants to generalize. Defining the universe of reference in advance and selecting the experimental and control groups from this at random would guarantee representativeness if it were ever achieved in practice. But inevitably not all those so designated are actually eligible for selection by any contact procedure. Our best survey sampling techniques, for example, can designate for potential contact only those available through residences. And, even of those so designated, up to 19 per cent are not contactable for an interview in their own homes even with five callbacks (37). It seems legitimate to assume that the more effort and time required of the respondent, the larger the loss through nonavailability and noncooperation. If one were to try to assemble experimental groups away from their own homes it seems reasonable to estimate a 50 per cent selection loss. If, still trying to extrapolate to the general public, one further limits oneself to docile preassembled groups, as in schools, military units, studio audiences, etc., the proportion of the universe systematically excluded through the sampling process must approach 90 per cent or more. Many of the selection factors involved are indubitably highly systematic. Under these extreme selection losses, it seems reasonable to suspect that the experimental groups might show reactions not characteristic of the general population. This point seems worth stressing lest we unwarrantedly assume that the selection loss for experiments is comparable to that found for survey interviews in the home at the respondent's convenience. Furthermore, it seems plausible that the greater the cooperation required, the more the respondent has to deviate from the normal course of daily events, the greater will be the possibility of nonrepresentative reactions. By and large, Design 6 might be expected to require less cooperation than Design 4 or 5, especially in the natural individual contact setting. The interactive effects of experimental mortality are of similar nature. Note that, on these grounds, the longer the experiment is extended in time the more respondents are lost and the less representative are the groups of the original universe.

Reactive arrangements. In any of the experimental designs, the respondents can become aware that they are participating in an experiment, and this awareness can have an interactive effect, in creating reactions to X which would not occur had X been encountered without this "I'm a guinea pig" attitude. Lazarsfeld (19), Kerr (17), and Rosenthal and Frank (28), all have provided valuable discussions of this problem. Such effects limit generalizations to respondents having this awareness, and preclude generalization to the population encountering X with nonexperimental attitudes. The direction of the effect may be one of negativism, such as an unwillingness to admit to any persuasion or change. This would be comparable to the absence of any immediate effect from dis-

credited communicators, as found by Hovland (14). The result is probably more often a cooperative responsiveness, in which the respondent accepts the experimenter's expectations and provides pseudoconfirmation. Particularly is this positive response likely when the respondents are self-selected seekers after the cure that X may offer. The Hawthorne studies (21) illustrate such sympathetic changes due to awareness of experimentation rather than to the specific nature of X. In some settings it is possible to disguise the experimental purpose by providing plausible façades in which X appears as an incidental part of the background (e.g., 26, 27, 29). We can also make more extensive use of experiments taking place in the intact social situation, in which the respondent is not aware of the experimentation at all.

The discussion of the effects of selection on representativeness has argued against employing intact natural preassembled groups, but the issue of conspicuousness of arrangements argues for such use. The machinery of breaking up natural groups such as departments, squads, and classrooms into randomly assigned experimental and control groups is a source of reaction which can often be avoided by the use of preassembled groups, particularly in educational settings. Of course, as has been indicated, this requires the use of large numbers of such groups under both experimental and control conditions.

The problem of reactive arrangements is distributed over all features of the experiment which can draw the attention of the respondent to the fact of experimentation and its purposes. The conspicuous or reactive pretest is particularly vulnerable, inasmuch as it signals the topics and purposes of the experimenter. For communications of obviously persuasive aim, the experimenter's topical intent is signaled by the X itself, if the communication does not seem a part of the natural environment. Even for the posttest-only groups, the occurrence of the posttest may create a reactive effect. The respondent may say to himself, "Aha, now I see why we got that movie." This consideration justifies the practice of disguising the connection between O and X even for Design 6, as through having different experimental personnel involved, using different façades, separating the settings and times, and embedding the X-relevant content of O among a disguising variety of other topics.[4]

Generalizing to other Xs. After the internal validity of an experiment has been established, after a dependable effect of X upon O has been found, the next step is to establish the limits and relevant dimensions of generalization not only in terms of populations and settings but also in terms of categories and aspects of X. The actual X in any one experiment is a specific combination of stimuli, all confounded for interpretative purposes, and only some relevant to the experimenter's intent and theory. Subsequent experimentation should be designed to purify X, to discover that aspect of the original conglomerate X which is responsible for the effect. As Brunswik (3) has emphasized, the representative sampling of Xs is as relevant a problem in linking experiment to theory as is the sampling of respondents. To define a category of Xs along some dimension, and then to sample Xs for experimental purposes from the full range of stimuli meeting the specification while other aspects of each specific stimulus complex are varied, serves to untie or unconfound the defined dimension from specific others, lending assurance of theoretical relevance.

In a sense, the placebo problem can be understood in these terms. The experiment without the placebo has clearly demonstrated that some aspect of the total X stimulus complex has had an effect; the placebo experiment serves to break up the complex X

into the suggestive connotation of pill-taking and the specific pharmacological properties of the drug—separating two aspects of the X previously confounded. Subsequent studies may discover with similar logic which chemical fragment of the complex natural herb is most essential. Still more clearly, the sham operation illustrates the process of X purification, ruling out general effects of surgical shock so that the specific effects of loss of glandular or neural tissue may be isolated. As these parallels suggest, once recurrent unwanted aspects of complex Xs have been discovered for a given field, control groups especially designed to eliminate these effects can be regularly employed.

Generalizing to other Os. In parallel form, the scientist in practice uses a complex measurement procedure which needs to be refined in subsequent experimentation. Again, this is best done by employing multiple Os all having in common the theoretically relevant attribute but varying widely in their irrelevant specificities. For Os this process can be introduced into the initial experiment by employing multiple measures. A major practical reason for not doing so is that it is so frequently a frustrating experience, lending hesitancy, indecision, and a feeling of failure to studies that would have been interpreted with confidence had but a single response measure been employed.

Transition experiments. The two previous paragraphs have argued against the *exact* replication of experimental apparatus and measurement procedures on the grounds that this continues the confounding of theory-relevant aspects of X and O with specific artifacts of unknown influence. On the other hand, the confusion in our literature generated by the heterogeneity of results from studies all on what is nominally the "same" problem but varying in implementation, is leading some to call for exact replication of initial procedures in subsequent research on a topic. Certainly no science can emerge without dependably repeatable experiments. A suggested resolution is the *transition experiment*,

---

[4] For purposes of completeness, the interaction of X with history and maturation should be mentioned. Both affect the generalizability of results. The interaction effect of history represents the possible specificity of results to a given historical moment, a possibility which increases as problems are more societal, less biological. The interaction of maturation and X would be represented in the specificity of effects to certain maturational levels, fatigue states, etc.

in which the need for varying the theory-independent aspects of X and O is met in the form of a multiple X, multiple O design, one segment of which is an " exact " replication of the original experiment, exact at least in those major features which are normally reported in experimental writings.

Internal vs. external validity. If one is in a situation where either internal validity or representativeness must be sacrificed, which should it be? The answer is clear. Internal validity is the prior and indispensable consideration. The optimal design is, of course, one having both internal and external validity. Insofar as such settings are available, they should be exploited, without embarrassment from the apparent opportunistic warping of the content of studies by the availability of laboratory techniques. In this sense, a science is as opportunistic as a bacteria culture and grows only where growth is possible. One basic necessity for such growth is the machinery for selecting among alternative hypotheses, no matter how limited those hypotheses may have to be.

Summary

In analyzing the extraneous variables which experimental designs for social settings seek to control, seven categories have been distinguished: history, maturation, testing, instrument decay, regression, selection, and mortality. In general, the simple or main effects of these variables jeopardize the internal validity of the experiment and are adequately controlled in standard experimental designs. The interactive effects of these variables and of experimental arrangements affect the external validity or generalizability of experimental results. Standard experimental designs vary in their susceptibility to these interactive effects. Stress is also placed upon the differences among measuring instruments and arrangements in the extent to which they create unwanted interactions. The value for social science purposes of the Posttest-Only Control Group Design is emphasized.

References

1. Annis, A. D., & Meier, N. C. The induction of opinion through suggestion by means of planted content. *J. soc. Psychol.*, 1934, **5**, 65–81.
2. Boring, E. G. The nature and history of experimental control. *Amer. J. Psychol.*, 1954, **67**, 573–589.
3. Brunswik, E. *Perception and the representative design of psychological experiments.* Berkeley: Univer. of California Press, 1956.
4. Bugelski, B. R. A note on Grant's discussion of the Latin square principle in the design and analysis of psychological experiments. *Psychol Bull.*, 1949, **46**, 49–50.
5. Campbell, D. T. The indirect assessment of social attitudes. *Psychol. Bull.*, 1950, **47**, 15–38.
6. Campbell, D. T. On the possibility of experimenting with the " Bandwagon " effect. *Int. J. Opin. Attitude Res.*, 1951, **5**, 251–260.
7. Cantril, H. *Gauging public opinion.* Princeton: Princeton Univer. Press, 1944.
8. Collier, R. M. The effect of propaganda upon attitude following a critical examination of the propaganda itself. *J. soc. Psychol.*, 1944, **20**, 3–17.
9. Crespi, L. P. The interview effect in polling. *Publ. Opin. Quart.*, 1948, **12**, 99–111.
10. Fisher, R. A. *The design of experiments.* Edinburgh: Oliver & Boyd, 1935.
11. Gosnell, H. F. *Getting out the vote: an experiment in the stimulation of voting.* Chicago: Univer. of Chicago Press, 1927.
12. Greenberg, A. Matched samples. *J. Marketing*, 1953–54, **18**, 241–245.
13. Hartmann, G. W. A field experiment on the comparative effectiveness of "emotional" and "rational" political leaflets in determining election results. *J. abnorm. soc. Psychol.*, 1936, **31**, 99–114.
14. Hovland, C. E., Janis, I. L., & Kelley, H. H. *Communication and persuasion.* New Haven: Yale Univer. Press, 1953.
15. Hovland, C. I., Lumsdaine, A. A., & Sheffield, F. D. *Experiments on mass communication.* Princeton: Princeton Univer. Press, 1949.

16. Jahoda, M., Deutsch, M., & Cook, S. W. *Research methods in social relations.* New York: Dryden Press, 1951.

17. Kerr, W. A. Experiments on the effect of music on factory production. *Appl. Psychol. Monogr.*, 1945, No. 5.

18. Kish, L. Selection of the sample. In L. Festinger and D. Katz (Eds.), *Research methods in the behavioral sciences.* New York: Dryden Press, 1953, 175–239.

19. Lazarsfeld, P. F. Training guide on the controlled experiment in social research. Dittoed. Columbia Univer., Bureau of Applied Social Research, 1948.

20. Lucas, D. B., & Britt, S. H. *Advertising psychology and research.* New York: McGraw-Hill, 1950.

21. Mayo, E. *The human problems of an industrial civilization.* New York: Macmillan, 1933.

22. McNemar, Q. A critical examination of the University of Iowa studies of environmental influences upon the IQ. *Psychol. Bull.*, 1940, **37**, 63–92.

23. Menefee, S. C. An experimental study of strike propaganda. *Soc. Forces*, 1938, **16**, 574–582.

24. Parrish, J. A., & Campbell, D. T. Measuring propaganda effects with direct and indirect attitude tests. *J. abnorm. soc. Psychol.*, 1953, **48**, 3–9.

25. Payne, S. L. The ideal model for controlled experiments. *Publ. Opin. Quart.*, 1951, **15**, 557–562.

26. Postman, L., & Bruner, J. S. Perception under stress. *Psychol. Rev.*, 1948, **55**, 314–322.

27. Rankin, R. E., & Campbell, D. T. Galvanic skin response to Negro and white experimenters. *J. abnorm. soc. Psychol.*, 1955, **51**, 30–33.

28. Rosenthal, D., & Frank, J. O. Psychotherapy and the placebo effect. *Psychol. Bull.*, 1956, **53**, 294–302.

29. Schanck, R. L., & Goodman, C. Reactions to propaganda on both sides of a controversial issue. *Publ. Opin. Quart.*, 1939, **3**, 107–112.

30. Solomon, R. W. An extension of control group design. *Psychol. Bull.*, 1949, **46**, 137–150.

31. Star, S. A., & Hughes, H. M. Report on an educational campaign: The Cincinnati plan for the United Nations. *Amer. J. Sociol.*, 1949–50, **55**, 389.

32. Stouffer, S. A. Some observations on study design. *Amer. J. Sociol.*, 1949–50, **55**, 355–361.

33. Stouffer, S. A. Measurement in sociology. *Amer. sociol. Rev.*, 1953, **18**, 591–597.

34. Underwood, B. J. *Experimental psychology.* New York: Appleton-Century-Crofts, 1949.

35. Underwood, B. J. Interference and forgetting. *Psychol. Rev.*, 1957, **64**, 49–60.

36. Underwood, B. J. *Psychological research.* New York: Appleton-Century-Crofts, 1957.

37. Williams, R. Probability sampling in the field: A case history. *Publ. Opin. Quart.*, 1950, **14**, 316–330.

Editor's comments: Leslie Kish distinguishes four classes of variables: explanatory, controlled, confounded, and randomized. Within this classification he examines many of the issues involved in the use of statistical tests and the determination of causation; he concludes with a discussion of three common misuses of statistical tests. This article should be of particular benefit to those students who someday will do survey research.

6.4   Some Statistical Problems In Research Design

Leslie Kish

Abstract. Several statistical problems in the design of research are discussed: (1) The use of statistical tests and the search for causation in survey research are examined; for this we suggest separating four classes of variables: explanatory, controlled, confounded, and randomized. (2) The relative advantages of experiments, surveys, and other investigations are shown to derive respectively from better control, representation, and measurement. (3) Finally, three common misuses of statistical tests are examined: "hunting with a shot-gun for significant differences," confusing statistical significance with substantive importance, and overemphasis on the primitive level of merely finding differences.

Statistical inference is an important aspect of scientific inference. The statistical consultant spends much of his time in the borderland between statistics and the other aspects, philosophical and substantive, of the scientific search for explanation. This marginal life is rich both in direct experience and in discussions of fundamentals; these have stimulated my concern with the problems treated here.

I intend to touch on several problems dealing with the interplay of statistics with the more general problems of scientific inference. We can spare elaborate introductions because these problems are well known. Why then discuss them here at all? We do so because, first, they are problems about which there is a great deal of misunderstanding, evident in current research; and, second, they are *statistical* problems on which there is broad agreement among research statisticians—and on which these statisticians generally disagree with much in the current practice of research scientists.[1]

This research has been supported by a grant from the Ford Foundation for Development of the Behavioral Sciences. It has benefited from the suggestions and encouragement of John W. Tukey and others. But the author alone is responsible for any controversial opinions.

Reprinted from *American Sociological Review*, 1959, **24**, 328–338, by permission of the publisher and author.

[1] *Cf.* R. A. Fisher, *The Design of Experiments*, London: Oliver and Boyd, 6th edition, 1953, pp. 1–2: "The statistician cannot evade the responsibility for understanding the processes he applies or recommends. My immediate point is that the questions involved can be disassociated from all that is strictly technical in the statistician's craft, and *when so detached*, are questions only of the right use of human reasoning powers, with which all intelligent people,

Several problems will be considered briefly, hence incompletely. The aim of this paper is not a profound analysis, but a clear elementary treatment of several related problems. The footnotes contain references to more thorough treatments. Moreover, these are not *all* the problems in this area, nor even necessarily the most important ones; the reader may find that his favorite, his most annoying problem, has been omitted. The problems selected are a group with a common core, they arise frequently, yet they are widely misunderstood.

Statistical Tests of Survey Data

That correlation does not prove causation is hardly news. Perhaps the wittiest statements on this point are in George Bernard Shaw's preface to *The Doctor's Dilemma*, in the sections on "Statistical Illusions," "The Surprises of Attention and Neglect," "Stealing Credit from Civilization," and "Biometrika." (These attack, alas, the practice of vaccination.) The excellent introductory textbook by Yule and Kendall[2] deals in three separate chapters with the problems of advancing from correlation to causation. Searching for causal factors among survey data is an old, useful sport; and the attempts to separate true explanatory variables from extraneous and "spurious" correlations have taxed scientists since antiquity and will undoubtedly continue to do so. Neyman and Simon[3] show that beyond

common sense, there are some technical skills involved in tracking down spurious correlations. Econometricians and geneticists have developed great interest and skill in the problems of separating the explanatory variables.[4]

The researcher designates the explanatory variables on the basis of substantive scientific theories. He recognizes the evidence of other *sources of variation* and he needs to separate these from the explanatory variables. Sorting all sources of variation into four classes seems to me a useful simplification. Furthermore, no confusion need result from talking about sorting and treating "variables," instead of "sources of variation."

I. The *explanatory* variables, sometimes called the "experimental" variables, are the objects of the research. They are the variables among which the researcher wishes to find and to measure some specified relationships. They include both the "dependent" and the "independent" variables, that is, the "predictand" and "predictor" variables.[5] With respect to the aims of the research all

who hope to be intelligible, are equally concerned, and on which the statistician, as such, speaks with no special authority. The statistician cannot excuse himself from the duty of getting his head clear on the principles of scientific inference, but equally no other thinking man can avoid a like obligation."

[2] G. Undy Yule and M. G. Kendall, *An Introduction to the Theory of Statistics*, London: Griffin, 11th edition, 1937, Chapters 4, 15, and 16.

[3] Jerzy Neyman, *Lectures and Conferences on Mathematical Statistics and Probability*, Washington, D.C.: Graduate School of Department of Agriculture, 1952, pp. 143–154. Herbert A. Simon, "Spurious Correlation: A Causal Interpretation," *Journal of the American Statistical Association*, **49** (September, 1954), pp. 467–479; also in his *Models of Man*, New York: Wiley, 1956.

[4] See the excellent and readable article, Herman Wold, "Causal Inference from Observational Data," *Journal of the Royal Statistical Society* (A), **119** (Part 1, January, 1956), pp. 28–61. Also the two-part technical article, M. G. Kendall, "Regression, Structure and Functional Relationship," *Biometrika*, **38** (June, 1951), pp. 12–25; and **39** (June, 1952), pp. 96–108. The interesting methods of "path coefficients" in genetics have been developed by Wright for inferring causal factors from regression coefficients. See, in Oscar Kempthorne *et al.*, *Statistics and Mathematics in Biology*, Ames, Iowa: The Iowa State College Press, 1954; Sewall Wright, "The Interpretation of Multi-Variate Systems," Chapter 2; and John W. Tukey, "Causation, Regression and Path Analysis," Chapter 3. Also C. C. Li, "The Concept of Path Coefficient and Its Impact on Population Genetics," *Biometrics*, **12** (June, 1956), pp. 190–209. I do not know whether these methods can be of wide service in current social science research in the presence of numerous factors, of large unexplained variances, and of doubtful directions of causation.

[5] Kendall points out that these latter terms are preferable. See his paper cited in footnote 4, and M. G. Kendall and W. R. Buckland, *A Dictionary of Statistical Terms*, prepared for the International Statistical Institute with assistance of UNESCO, London: Oliver and Boyd, 1957. I have also tried to follow in IV below his distinction of "variate" for random variables from "variables" for the usual (nonrandom) variable.

other variables, of which there are three classes, are extraneous.

II. There are extraneous variables which are *controlled*. The control may be exercised in either or both the selection and the estimation procedures.

III. There may exist extraneous uncontrolled variables which are *confounded* with the Class I variables.

IV. There are extraneous uncontrolled variables which are treated as *randomized* errors. In "ideal" experiments (discussed below) they are actually randomized; in surveys and investigations they are only assumed to be randomized. Randomization may be regarded as a substitute for experimental control or as a form of control.

The aim of efficient design both in experiments and in surveys is to place as many of the extraneous variables as is feasible into the second class. The aim of randomization in experiments is to place all of the third class into the fourth class; in the "ideal" experiment there are no variables in the third class. And it is the aim of controls of various kinds in surveys to separate variables of the third class from those of the first class; these controls may involve the use of repeated cross-tabulations, regression, standardization, matching of units, and so on.

The function of statistical "tests of significance" is to test the effects found among the Class I variables against the effects of the variables of Class IV. An "ideal" experiment here denotes an experiment for which this can be done through randomization without any possible confusion with Class III variables. (The difficulties of reaching this "ideal" are discussed below.) In survey results, Class III variables are confounded with those of Class I; the statistical tests actually contrast the effects of the random variables of Class IV against the explanatory variables of Class I confounded with unknown effects of Class III variables. In both the ideal experiment and in surveys the statistical tests serve to separate the effects of the random errors of Class IV from the effects of other variables. These, in surveys, are a mixture of explanatory and confounded variables;

their separation poses severe problems for logic and for scientific methods; statistics is only one of the tools in this endeavor. The scientist must make many decisions as to which variables are extraneous to his objectives, which should and can be controlled, and what methods of control he should use. He must decide where and how to introduce statistical tests of hypotheses into the analysis.

As a simple example, suppose that from a probability sample survey of adults of the United States we find that the level of political interest is higher in urban than in rural areas. A test of significance will show whether or not the difference in the "levels" is large enough, compared with the sampling error of the difference, to be considered "significant." Better still, the confidence interval of the difference will disclose the limits within which we can expect the "true" population value of the difference to lie.[6] If families had been sent to urban and rural areas respectively, after the randomization of a true experiment, then the sampling error would measure the effects of Class IV variables against the effects of urban *versus* rural residence on political interest; the difference in levels beyond sampling errors could be ascribed (with specified probability) to the effects of urban *versus* rural residence.

Actually, however, residences are not assigned at random. Hence, in survey results, Class III variables may account for some of the difference. If the test of significance rejects the null hypothesis of no difference, *several* hypotheses remain in addition to that of a simple relationship between urban *versus* rural residence and political interest. Could differences in income, in occupation, or in family life cycle account for the difference in the levels? The analyst may try to remove (for example, through cross-tabulation, re-

---

[6] The sampling error measures the chance fluctuation in the difference of levels due to the sampling operations. The computation of the sampling error must take proper account of the actual sample design, and not blindly follow the standard simple random formulas. See Leslie Kish, "Confidence Intervals for Complex Samples," *American Sociological Review*, **22** (April, 1957), pp. 154–165.

gression, standardization) the effects due to such variables, which are extraneous to his expressed interest; then he computes the difference, between the urban and rural residents, of the levels of interest now free of several confounding variables. This can be followed by a proper test of significance—or, preferably, by some other form of statistical inference, such as a statement of confidence intervals.

Of course, other variables of Class III may remain to confound the measured relationship between residence and political interest. The separation of Class I from Class III variables should be determined in accord with the nature of the hypothesis with which the researcher is concerned; finding and measuring the effects of confounding variables of Class III tax the ingenuity of research scientists. But this separation is beyond the functions and capacities of the statistical tests, the tests of null hypotheses. Their function is not explanation; they cannot point to causation. Their function is to ask: "Is there anything in the data that *needs* explaining?"—and to answer this question with a certain probability.

Agreement on these ideas can eliminate certain confusion, exemplified by Selvin in a recent article:

Statistical tests are unsatisfactory in nonexperimental research for two fundamental reasons: it is almost impossible to design studies that meet the conditions for using the tests, and the situations in which the tests are employed make it difficult to draw correct inferences. The basic difficulty in design is that sociologists are unable to randomize their uncontrolled variables, so that the difference between "experimental" and "control" groups (or their analogs in nonexperimental situations) are a mixture of the effects of the variable being studied and the uncontrolled variables or correlated biases. Since there is no way of knowing, in general, the sizes of these correlated biases and their directions, there is no point in asking for the probability that the observed differences could have been produced by random errors. The place for significance tests is after all relevant correlated biases have been controlled.... In design and in interpretation, in principle and in practice, tests of statistical significance are inapplicable in nonexperimental research.[7]

Now it is true that in survey results the explanatory variables of Class I are confounded with variables of Class III; but it does not follow that tests of significance should not be used to separate the random variables of Class IV. Insofar as the effects found "are a mixture of the effects of the variable being studied and the uncontrolled variables"; insofar as "there is no way of knowing, in general, the sizes" and directions of these uncontrolled variables, Selvin's logic and advice should lead not only to the rejection of statistical tests; it should lead one to refrain altogether from using survey results for the purposes of finding explanatory variables. *In this sense*, not only tests of significance but any comparisons, any scientific inquiry based on surveys, any scientific inquiry other than an "ideal" experiment, is "inapplicable." That advice is most unrealistic. In the (unlikely) event of its being followed, it would sterilize social research—and other nonexperimental research as well.

Actually, much research—in the social, biological, and physical sciences—must be based on nonexperimental methods. In such cases the rejection of the null hypothesis leads to several alternate hypotheses that may explain the discovered relationships. It is the duty of scientists to search, with painstaking effort and with ingenuity, for bases on which to decide among these hypotheses.

As for Selvin's advice to refrain from making tests of significance until "after all relevant" uncontrolled variables have been controlled—this seems rather farfetched to scientists engaged in empirical work who consider themselves lucky if they can explain

[7] Hanan C. Selvin, "A Critique of Tests of Significance in Survey Research," *American Sociological Review*, **22** (October, 1957), p. 527. In a criticism of this article, McGinnis shows that the separation of explanatory from extraneous variables depends on the type of hypothesis at which the research is aimed. Robert McGinnis, "Randomization and Inference in Sociological Research," *American Sociological Review*, **23** (August, 1958), pp. 408–414.

25 or 50 per cent of the total variance. The control of all relevant variables is a goal seldom even approached in practice. To postpone to that distant goal all statistical tests illustrates that often "the perfect is the enemy of the good."[8]

Experiments, Surveys, and Other Investigations

Until now, the theory of sample surveys has been developed chiefly to provide descriptive statistics—especially estimates of means, proportions, and totals. On the other hand, experimental designs have been used primarily to find explanatory variables in the analytical search of data. In many fields, however, including the social sciences, survey data must be used frequently as the analytical tools in the search for explanatory variables.

---

[8] Selvin performs a service in pointing to several common mistakes: (a) The mechanical use of "significance tests" can lead to false conclusions. (b) Statistical "significance" should not be confused with substantive importance. (c) The probability levels of the common statistical tests are not appropriate to the practice of "hunting" for a few differences among a mass of results. However, Selvin gives poor advice on what to do about these mistakes, particularly when, in his central thesis, he reiterates that "tests of significance are inapplicable in nonexperimental research," and that "the tests are applicable only when all relevant variables have been controlled." I hope that the benefits of his warnings outweigh the damages of his confusion.

I noticed three misleading references in the article. (a) In the paper which Selvin appears to use as supporting him, Wold (*op. cit.*, p. 39) specifically disagrees with Selvin's central thesis, stating that "The need for testing the statistical inference is no less than when dealing with experimental data, but with observational data other approaches come to the foreground." (b) In discussing problems caused by complex sample designs, Selvin writes that "Such errors are easy enough to discover and remedy" (p. 520), referring to Kish (*op. cit.*). On the contrary, my article pointed out the seriousness of the problem and the difficulties in dealing with it. (c) "Correlated biases" is a poor term for the confounded uncontrolled variables and it is not true that the term is so used in literature. Specifically, the reference to Cochran is misleading, since he is dealing there only with errors of measurement which may be correlated with the "true" value. See William C. Cochran, *Sampling Techniques*, New York: Wiley, 1953, p. 305.

Furthermore, in some research situations, neither experiments nor sample surveys are practical, and other investigations are utilized.

By "experiments" I mean here "ideal" experiments in which all the extraneous variables have been randomized. By "surveys" (or "sample surveys"), I mean probability samples in which all members of a defined population have a known positive probability of selection into the sample. By "investigations" (or "other investigations"), I mean the collection of data—perhaps with care, and even with considerable control—without either the randomization of experiments or the probability sampling of surveys. The differences among experiments, surveys, and investigations are not the consequences of statistical techniques; they result from different methods for introducing the variables and for selecting the population elements (subjects). These problems are ably treated in recent articles by Wold and Campbell.[9]

In considering the larger ends of any scientific research, only part of the total means required for inference can be brought under objective and firm control; another part must be left to more or less vague and subjective—however skillful—judgment. The scientist seeks to maximize the first part, and thus to minimize the second. In assessing the ends, the costs, and the feasible means, he makes a strategic choice of methods. He is faced with the three basic problems of scientific research: measurement, representation, and control. We ignore here the important but vast problems of measurement and deal with representation and control.

Experiments are strong on control through randomization; but they are weak on representation (and sometimes on the "naturalism" of measurement). Surveys are strong on representation, but they are often weak on control. Investigations are weak on control and often on representation; their use is due frequently to convenience or low cost

---

[9] Wold, *op. cit.*; Donald T. Campbell, "Factors Relevant to the Validity of Experiments in Social Settings," *Psychological Bulletin*, **54** (July, 1957), pp. 297–312.

and sometimes to the need for measurements in "natural settings."

Experiments have three chief advantages: (1) Through randomization of extraneous variables the confounding variables (Class III) are eliminated. (2) Control over the introduction and variation of the "predictor" variables clarifies the *direction* of causation from "predictor" to "predictand" variables. In contrast, in the correlations of many surveys this direction is not clear—for example, between some behaviors and correlated attitudes. (3) The modern design of experiments allows for great flexibility, efficiency, and powerful statistical manipulation, whereas the analytical use of survey data presents special statistical problems.[10]

The advantages of the experimental method are so well known that we need not dwell on them here. It is the scientific method *par excellence*—when feasible. In many situations experiments are not feasible and this is often the case in the social sciences; but it is a mistake to use this situation to separate the social from the physical and biological sciences. Such situations also occur frequently in the physical sciences (in meteorology, astronomy, geology), the biological sciences, medicine, and elsewhere.

The experimental method also has some shortcomings. First, it is often difficult to choose the "control" variables so as to exclude *all* the confounding extraneous variables; that is, it may be difficult or impossible to design an "ideal" experiment. Consider the following examples: The problem of finding a proper control for testing the effects of the Salk polio vaccine led to the use of an adequate "placebo." The Hawthorne experiment demonstrated that the design of a proposed "treatment *versus* control" may turn out to be largely a test of *any* treatment *versus lack* of treatment.[11] Many of the initial

successes reported about mental therapy, which later turn into vain hopes, may be due to the hopeful effects of *any* new treatment in contrast with the background of neglect. Shaw, in "The Surprises of Attention and Neglect," writes: "Not until attention has been effectually substituted for neglect as a general rule, will the statistics begin to show the merits of the particular methods of attention adopted."

There is an old joke about the man who drank too much on four different occasions, respectively, of scotch and soda, bourbon and soda, rum and soda, and wine and soda. Because he suffered painful effects on all four occasions, he ascribed, with scientific logic, the common effect to the common cause: "I'll never touch soda again!" Now, to a man (say, from Outer Space) ignorant of the common alcoholic content of the four "treatments" and of the relative physiological effects of alcohol and carbonated water, the subject is not fit for joking, but for further scientific investigation.

Thus, the advantages of experiments over surveys in permitting better control are only relative, not absolute.[12] The design of proper experimental controls is not automatic; it is an art requiring scientific knowledge, foresight in planning the experiment, and hindsight in interpreting the results. Nevertheless, the distinction in control between experiments and surveys is real and considerable; and to emphasize this distinction we refer here to "ideal" experiments in which the control of the random variables is complete.

---

[10] Kish, *op. cit.*

[11] F. J. Roethlisberger and W. J. Dickson, *Management and the Worker*, Cambridge: Harvard University Press, 1939. Troubles with experimental controls misled even the great Pavlov into believing *temporarily*

that he had proof of the inheritance of an acquired ability to learn: "In an informal statement made at the time of the Thirteenth International Physiological Congress, Boston, August, 1929, Pavlov explained that in checking up these experiments it was found that the apparent improvement in the ability to learn, on the part of successive generations of mice, was really due to an improvement in the ability to teach, on the part of the experimenter." From B. G. Greenberg, *The Story of Evolution*, New York: Garden City, 1929, p. 327.

[12] Jerome Cornfield, "Statistical Relationships and Proof in Medicine," *American Statistician*, **8** (December, 1954), pp. 19–21.

Second, it is generally difficult to design experiments so as to represent a specified important population. In fact, the questions of sampling, of making the experimental results representative of a specified population, have been largely ignored in experimental design until recently. Both in theory and in practice, experimental research has often neglected the basic truth that causal systems, the distributions of relations—like the distributions of characteristics—exist only within specified universes. The distributions of relationships, as of characteristics, exist only within the framework of specific populations. Probability distributions, like all mathematical models, are abstract systems; their application to the physical world must include the specification of the populations. For example, it is generally accepted that the statement of a value for mean income has meaning only with reference to a specified population; but this is not generally and clearly recognized in the case of regression of assets on income and occupation. Similarly, the *statistical* inferences derived from the experimental testing of several treatments are restricted to the population(s) included in the experimental design.[13] The clarification of the population sampling aspects of experiments is now being tackled vigorously by Wilk and Kempthorne and by Cornfield and Tukey.[14]

---

[13] McGinnis, *op. cit.*, p. 412, points out that usually "it is not true that one can uncover 'general' relationships by examining some arbitrarily selected population. . . . There is no such thing as a completely general relationship which is independent of population, time, and space. The extent to which a relationship is constant among different populations is an empirical question which can be resolved only by examining different populations at different times in different places."

[14] Martin B. Wilk and Oscar Kempthorne, "Some Aspects of the Analysis of Factorial Experiment in a Completely Randomized Design," *Annals of Mathematical Statistics*, **27** (December, 1956), pp. 950–985; and "Fixed, Mixed and Random Models," *Journal of the American Statistical Association*, **50** (December, 1955), pp. 1144–1167. Jerome Cornfield and John W. Tukey, "Average Values of Mean Squares in Factorials," *Annals of Mathematical Statistics*, **27** (December, 1956), pp. 907–949.

Third, for many research aims, especially in the social sciences, contriving the desired "natural setting" for the measurements is not feasible in experimental design. Hence, what social experiments give sometimes are clear answers to questions the meanings of which are vague. That is, the artificially contrived experimental variables *may* have but a tenuous relationship to the variables the researcher would like to investigate.

The second and third weaknesses of experiments point to the advantages of surveys. Not only do probability samples permit clear statistical inferences to defined populations, but the measurements can often be made in the "natural settings" of actual populations. Thus in practical research situations the experimental method, like the survey method, has its distinct problems and drawbacks as well as its advantages. In practice one generally cannot solve simultaneously all of the problems of measurement, representation, and control; rather, one must choose and compromise. In any specific situation one method may be better or more practical than the other; but there is no over-all superiority in all situations for either method. Understanding the advantages and weaknesses of both methods should lead to better choices.

In social research, in preference to both surveys and experiments, frequently some design of controlled investigation is chosen—for reasons of cost or of feasibility or to preserve the "natural setting" of the measurements. Ingenious adaptations of experimental designs have been contrived for these controlled investigations. The statistical framework and analysis of experimental designs are used, but not the randomization of true experiments. These designs are aimed to provide flexibility, efficiency, and, especially, some control over the extraneous variables. They have often been used to improve considerably research with controlled investigations.

These designs are sometimes called "natural experiments." For the sake of clarity, however, it is important to keep clear the distinctions among the methods and to reserve the word "experiment" for designs in which the uncontrolled variables are randomized. This principle is stated clearly by Fisher,[15] and is accepted often in scientific research. Confusion is caused by the use of terms like "ex post facto experiments" to describe surveys or designs of controlled investigations. Sample surveys and controlled investigations have their own justifications, their own virtues; they are not just second-class experiments. I deplore the borrowing of the prestige word "experiment," when it cloaks the use of other methods.

Experiments, surveys, and investigations can all be improved by efforts to overcome their weaknesses. Because the chief weakness of surveys is their low degree of control, researchers should be alert to the collection and use of auxiliary information as controls against confounding variables. They also should take greater advantage of changes introduced into their world by measuring the effects of such changes. They should utilize more often efficient and useful statistics instead of making tabular presentation their only tool.

On the other hand, experiments and controlled investigations can often be improved by efforts to specify their populations more clearly and to make the results more representative of the population. Often more should be done to broaden the area of inference to more important populations. Thus, in many situations the deliberate attempts of the researcher to make his sample more "homogeneous" are misplaced; and if common sense will not dispel the error, reading Fisher may.[16] When he understands this,

the researcher can view the population base of his research in terms of efficiency—in terms of costs and variances. He can often avoid basing his research on a comparison of one sampling unit for each "treatment." If he cannot obtain a proper sample of the entire population, frequently he can secure, say, four units for each treatment, or a score for each.[17]

Suppose, for example, that thorough research on one city and one rural county discloses higher levels of political interest in the former. It is presumptuous (although common practice) to present this result as evidence that urban people in *general* show a higher level. (Unfortunately, I am not beating a dead horse; this nag is pawing daily in the garden of social science.) However, very likely there is a great deal of variation in political interest among different cities, as well as among rural counties; the results of the research will depend heavily on which city and which county the researcher picked

---

[15] Fisher, *op. cit.*, pp. 17–20. "Controlled investigation" may not be the best term for these designs. "Controlled observations" might do, but "observation" has more fundamental meanings.

[16] *Ibid.*, pp. 99–100. Fisher says: "We have seen that the factorial arrangement possesses two advantages over experiments involving only single factors: (i) Greater *efficiency*, in that these factors are evaluated with the same precision by means of only a quarter of the number of observations that would otherwise be necessary; and (ii) Greater *comprehensiveness* in that,

in addition to the 4 effects of single factors, their 11 possible interactions are evaluated. There is a third advantage which, while less obvious than the former two, has an important bearing upon the utility of the experimental results in their practical application. This is that any conclusion, such as that it is advantageous to increase the quantity of a given ingredient, has a wider inductive basis when inferred from an experiment in which the quantities of other ingredients have been varied, than it would have from any amount of experimentation, in which these had been kept strictly constant. The exact standardisation of experimental conditions, which is often thoughtlessly advocated as a panacea, always carries with it the real disadvantage that a highly standardized experiment supplies direct information only in respect of the narrow range of conditions achieved by standardisation. Standardisation, therefore, weakens rather than strengthens our ground for inferring a like result, when, as is invariably the case in practice, these conditions are somewhat varied."

[17] For simplicity the following illustration is a simple contrast between two values of the "explanatory" variable, but the point is more general; and this aspect is similar whether for true experiments or controlled observations. Incidentally, it is poor strategy to "solve" the problem of representation by obtaining a good sample, or complete census, of some small or artificial population. A poor sample of the United States or of Chicago *usually* has more over-all value than the best sample of freshman English classes at X University.

as "typical." The research would have a broader base if a city and a rural county would have been chosen in each of, say, four different situations—as different as possible (as to region, income, industry, for example)—or better still in twenty different situations. A further improvement would result if the stratification and selection of sampling units followed a scientific sample design.

Using more sampling units and spreading them over the breadth of variation in the population has several advantages. First, some measure of the variability of the observed effect may be obtained. From a probability sample, statistical inference to the population can be made. Second, the base of the inference is broadened, as the effect is observed over a variety of situations. Beyond this lies the combination of results from researches over several distinct cultures and periods. Finally, with proper design, the effects of several potentially confounding factors can be tested.

These points are brought out by Keyfitz in an excellent example of controlled investigation (which also uses sampling effectively): "Census enumeration data were used to answer for French farm families of the Province of Quebec the question: Are farm families smaller near cities than far from cities, other things being equal? The sample of 1,056 families was arranged in a 2^6 factorial design which not only controlled 15 extraneous variables (income, education, etc.) but incidentally measured the effect of 5 of these on family size. A significant effect of distance from cities was found, from which is inferred a geographical dimension for the currents of social change."[18] The mean numbers of children per family were found to be 9.5 near and 10.8 far from cities; the difference of 1.3 children has a standard error of 0.28.

Some Misuses of Statistical Tests

Of the many kinds of current misuses this discussion is confined to a few of the most common. There is irony in the circumstance that these are committed usually by the more statistically inclined investigators; they are avoided in research presented in terms of qualitative statements or of simple descriptions.

First, there is "hunting with a shot-gun" for significant differences. Statistical tests are designed for distinguishing results at a predetermined level of improbability (say at $P = .05$) under a specified null hypothesis of random events. A rigorous theory for dealing with individual experiments has been developed by Fisher, the Pearsons, Neyman, Wold, and others. However, the researcher often faces more complicated situations, especially in the analysis of survey results; he is often searching for interesting relationships among a vast number of data. The keen-eyed researcher hunting through the results of one thousand random tosses of perfect coins would discover and display about fifty "significant" results (at the $P = .05$ level).[19] Perhaps the problem has become more acute now that high-speed computers allow hundreds of significance tests to be made. There is no easy answer to this problem. We must be constantly aware of the nature of tests of null hypotheses in searching survey data for interesting results. After finding a result improbable under the null hypothesis the researcher must not accept blindly the hypothesis of "significance" due to a presumed cause. Among the several alternative hypotheses is that of having dis-

---

[18] Nathan Keyfitz, "A Factorial Arrangement of Comparisons of Family Size," *American Journal of Sociology*, **53** (March, 1953), p. 470.

[19] William H. Sewell, "Infant Training and the Personality of the Child," *American Journal of Sociology*, **53** (September, 1952), pp. 150–159. Sewell points to an interesting example: "On the basis of the results of this study, the general null hypothesis that the personality adjustments and traits of children who have undergone varying training experiences do not differ significantly cannot be rejected. Of the 460 chi square tests, only 18 were significant at or beyond the 5 per cent level. Of these, 11 were in the expected direction and 7 were in the opposite direction from that expected on the basis of psychoanalytic writings. . . . Certainly, the results of this study cast serious doubts on the validity of the psychoanalytic claims regarding the importance of the infant disciplines and on the efficacy of prescriptions based on them" (pp. 158–159). Note that by chance alone one would expect 23 "significant" differences at the 5 per cent level. A "hunter" would report either the 11 or the 18 and not the hundreds of "misses."

covered an improbable random event through sheer diligence. Remedy can be found sometimes by a reformulation of the statistical aims of the research so as to fit the available tests. Unfortunately, the classic statistical tests give clear answers only to some simple decision problems; often these bear but faint resemblance to the complex problems faced by the scientist. In response to these needs the mathematical statisticians are beginning to provide some new statistical tests. Among the most useful are the new "multiple comparison" and "multiple range" tests of Tukey, Duncan, Scheffé,[20] and others. With a greater variety of statistical statements available, it will become easier to choose one without doing great violence either to them or to the research aims.

Second, statistical "significance" is often confused with and substituted for substantive significance. There are instances of research results presented in terms of probability values of "statistical significance" alone, without noting the magnitude and importance of the relationships found. These attempts to use the probability levels of significance tests as measures of the strengths of relationships are very common and very mistaken. The function of statistical tests is merely to answer: Is the variation great enough for us to place some confidence in the result; or, contrarily, may the latter be merely a happenstance of the specific sample on which the test was made? This question is interesting, but it is surely *secondary*, auxiliary, to the main question: Does the result show a relationship which is of substantive interest because of its nature and its magnitude? Better still: Is the result consistent with an assumed relationship of substantive interest?

The results of statistical "tests of significance" are functions not only of the magnitude of the relationships studied but also of the numbers of sampling units used (and the

efficiency of design). In small samples significant, that is, meaningful, results may fail to appear "statistically significant." But if the sample is large enough the most insignificant relationships will appear "statistically significant."

Significance should stand for meaning and refer to substantive matter. The statistical tests merely answer the question: Is there a big enough relationship here which *needs* explanation (and is not merely chance fluctuation)? The word *significance* should be attached to another question, a substantive question: Is there a relationship here *worth* explaining (because it is important and meaningful)? As a remedial step I would recommend that statisticians discard the phrase "test of significance," perhaps in favor of the somewhat longer but proper phrase "test against the null hypothesis" or the abbreviation "TANH."

Yates, after praising Fisher's classic *Statistical Methods*, makes the following observations on the use of "tests of significance":

Second, and more important, it has caused scientific research workers to pay undue attention to the results of the tests of significance they perform on their data, particularly data derived from experiments, and too little to the estimates of the magnitude of the effects they are investigating.

Nevertheless the occasions, even in research work, in which quantitative data are collected solely with the object of proving or disproving a given hypothesis are relatively rare. Usually quantitative estimates and fiducial limits are required. Tests of significance are preliminary or ancillary.

The emphasis on tests of significance, and the consideration of the results of each experiment in isolation, have had the unfortunate consequence that scientific workers have often regarded the execution of a test of significance on an experiment as the ultimate objective. Results are significant or not significant and this is the end of it.[21]

---

[20] John W. Tukey, "Comparing Individual Means in the Analysis of Variance," *Biometrics*, **5** (June, 1949), pp. 99–114; David B. Duncan, "Multiple Range and Multiple F Tests," *Biometrics*, **11** (March, 1955), pp. 1–42; Henry Scheffé, "A Method for Judging All Contrasts in the Analysis of Variance," *Biometrika*, **40** (June, 1953), pp. 87–104.

[21] Frank Yates, "The Influence of *Statistical Methods for Research Workers* on the Development of the Science of Statistics," *Journal of the American Statistical Association*, **46** (March, 1951), pp. 32–33.

For presenting research results statistical estimation is more frequently appropriate than tests of significance. The estimates should be provided with some measure of sampling variability. For this purpose confidence intervals are used most widely. In large samples, statements of the standard errors provide useful guides to action. These problems need further development by theoretical statisticians.[22]

The responsibility for the current fashions should be shared by the authors of statistical textbooks and ultimately by the mathematical statisticians. As Tukey puts it:

Statistical methods should be tailored to the real needs of the user. In a number of cases, statisticians have led themselves astray by choosing a problem which they could solve exactly but which was far from the needs of their clients. . . . The broadest class of such cases comes from the choice of significance procedures rather than confidence procedures. It is often much easier to be "exact" about significance procedures than about confidence procedures. By considering only the most null "null hypothesis" many inconvenient possibilities can be avoided.[23]

Third, the tests of null hypotheses of *zero* differences, of no relationships, are frequently weak, perhaps trivial statements of the researcher's aims. In place of the test of zero difference (the nullest of null hypotheses), the researcher should often substitute, say, a test for a difference of a specific size based on some specified model. Better still, in many cases, instead of the tests of significance it would be more to the point to measure the magnitudes of the relationships, attaching proper statements of their sampling variation. The magnitudes of relationships cannot be measured in terms of levels of significance; they can be measured in terms of

the difference of two means, or of the proportion of the total variance "explained," of coefficients of correlations and of regressions, of measures of association, and so on. These views are shared by many, perhaps most, consulting statisticians—although they have not published full statements of their philosophy. Savage expresses himself forcefully: "Null hypotheses of no difference are usually known to be false before the data are collected; when they are, their rejection or acceptance simply reflects the size of the sample and the power of the test, and is not a contribution to science."[24]

Too much of social research is planned and presented in terms of the mere existence of some relationship, such as: individuals high on variate x are also high on variate y. The *exploratory* stage of research may be well served by statements of this order. But these statements are relatively weak and can serve *only* in the primitive stages of research. Contrary to a common misconception, the more advanced stages of research should be phrased in terms of the quantitative aspects of the relationships. Again, to quote Tukey:

There are normal sequences of growth in immediate ends. One natural sequence of immediate ends follows the sequence: (1) Description, (2) Significance statements, (3) Estimation, (4) Confidence statement, (5) Evaluation. . . . There are, of course, other normal sequences of immediate ends, leading mainly through various decision procedures, which are appropriate to development research and to operations research, just as the sequence we have just discussed is appropriate to basic research.[25]

At one extreme, then, we may find that the contrast between two "treatments" of a labor force results in a difference in productivity of 5 per cent. This difference may appear "statistically significant" in a sample of, say, 1000 cases. It may also mean a

[22] D. R. Cox, "Some Problems Connected with Statistical Inference," *Annals of Mathematical Statistics*, **29** (June, 1958), pp. 357–372.

[23] John W. Tukey, "Unsolved Problems of Experimental Statistics," *Journal of the American Statistical Association*, **49** (December, 1954), p. 710. See also D. R. Cox, *op. cit.*, and David B. Duncan, *op. cit.*

[24] Richard J. Savage, "Nonparametric Statistics," *Journal of the American Statistical Association*, **52** (September, 1957), pp. 332–333.

[25] Tukey, *op. cit.*, pp. 712–713.

difference of millions of dollars to the company. However, it "explains" only about one per cent of the total variance in productivity. At the other extreme is the far-away land of completely determinate behavior, where every action and attitude is explainable, with nothing left to chance for explanation.

The aims of most basic research in the social sciences, it seems to me, should be somewhere between the two extremes; but too much of it is presented at the first extreme, at the primitive level. This is a matter of over-all strategy for an entire area of any science. It is difficult to make this judgment off-hand regarding any specific piece of research of this kind: the status of research throughout the entire area should be considered. But the superabundance of research aimed at this primitive level seems to imply that the over-all strategy of research errs in this respect. The construction of scientific theories to cover broader fields—the persistent aim of science—is based on the synthesis of the separate research results in those fields. A coherent synthesis cannot be forged from a collection of relationships of unknown strengths and magnitudes. The necessary conditions for a synthesis include an *evaluation* of the results available in the field, a coherent interrelating of the *magnitudes* found in those results, and the construction of models based on those magnitudes.

6.5 Group Matching as Research Strategy : How Not to Get Significant Results

C. Alan Boneau and H. S. Pennypacker

Judging from published reports, particularly those involving groups of human *S*s, a common strategy in designing an experiment is the matching of groups (but not individual *S*s) on one or more relevant variables so that the means of the groups are equal or nearly so on these relevant variables. Thus in studies where age and intelligence, for example, are related to the variable one is concerned with, control and experimental groups may be selected from the available pool of *S*s so that the means of the groups on the relevant variables are equated. If hypothesis testing is the aim of the research, *this procedure obviates valid probability statements and generally makes it more difficult to achieve significant differences, if they exist, unless appropriate corrections are made.* In other words, the result of group matching by means is usually a conservative and, hence, less powerful test if the matching variables are correlated with the dependent variables.

Reprinted with permission of authors and publisher: Boneau, C. A., and Pennypacker, H. S. Group matching as research strategy: How not to get significant results. *Psychological Reports*, 1961, **8**, 143–147.

Implicit in the reasoning of a research worker who matches groups by means seems to be the argument that he thereby gets a more precise (in some sense) test of his general hypothesis. At least he knows that his resulting differences are not due to a bad matching of his groups on related, relevant variables. However, the effects of matching must be considered in the statistical evaluation of results. We argue that in the absence of any matching procedures *a large portion of the times the null hypothesis is rejected when true is reserved for those cases where the groups are mismatched by the process of random sampling.* This is heart and soul of statistical inference. Attempting to tamper with natural processes, for example, by unwittingly controlling mean differences on the dependent variable, without suitable adjustments, manipulates the randomness out of the sampling and changes probabilities. If one is compelled to match groups, there are effective ways of doing so, ways which increase the precision of the experiment rather than decrease it.

We have examined two or three volumes each of recent vintage (depending upon availability) of the following journals for examples of the matching error: *Journal of Experimental Psychology*, *Journal of Abnormal and Social Psychology*, *Journal of Clinical Psychology*, *Journal of Consulting Psychology*, *Psychological Reports*, and *Perceptual and Motor Skills*. The search yielded 42 studies in which means of groups rather than of scores of individuals were equated on one or more variables. This number does not include a large number of other studies in which such matching may have been employed but for which it was impossible to decide from the procedural details. Interestingly enough, the authors of a majority of our cases felt constrained to justify their obtained non-significant differences and preempted a good bit of valuable journal space attempting to make their findings jibe with previous reports and with inferences derived by profound reasoning. In fact, it appeared that non-significant findings were permitted to co-mingle with sacredly significant brethren within the confines of journal interiors simply because, coming from more carefully controlled studies, they purportedly contradicted previously reported results. We are by no means going to take the absurd position that these non-significant results would all have been significant had the experiment been analyzed or designed differently. Some of them might have been, however. We merely assume that a research worker who sets the .05 level as his criterion for rejection of the null hypothesis would prefer to operate at that level and not unwittingly stack his statistical deck in favor of the null hypothesis so that the actual alpha value is some indeterminate probability, but probably considerably smaller than .05.

It is well known that changing the alpha level also changes the power of a test. If we unwittingly alter the alpha level to something less than, say, .05, we are automatically bringing into play the power function associated with the new alpha level. The new power function will be below that for the .05 level at all points; thus arises the difficulty in rejecting the false null hypothesis.

We shall devote the remainder of this paper to explicating the rationale behind our contention that matching solely on the basis of group means without appropriate adjustment is an invalid procedure in hypothesis testing. We shall also suggest a number of alternative procedures which not only possess the sanctity of validity but have the added merit of being more prone to yield significant results if real differences exist.

The main feature of our argument follows from the fact that group matching on the basis of group means artificially restricts the ordinary random variation of group means. When we make probability statements about the null hypothesis we are assuming an unrestricted random fluctuation of these means. To take one example, consider the t test (for simplicity the test for equal sample size). Assume that sampling is from a normal population with mean μ and standard deviation σ and that the null hypothesis is true. Under these conditions it can be shown mathematically or demonstrated by random sampling experiments that a large random selection of the differences between the means of two such randomly selected groups is distributed normally with a mean of zero. In addition, the standard deviation of the differences between the means is related to the standard deviation of the original population in a prescribed way. It has a value of $\sigma\sqrt{2/n}$. This is to imply that if we select Ss at random in two groups, we will know the relationship between the standard deviation of S-scores and the standard deviation of the hypothetical population of differences between the means. The t test makes use of this relationship and in effect measures the size of a mean difference in terms of an estimate of standard deviation of mean differences derived from the S to S variability within the samples. That is to say, an estimate of the standard deviation of the S-scores can easily be obtained from the within-sample S-scores themselves by the usual formula. If we multiply this value by $\sqrt{2/n}$ we have estimated the value of the ordinary standard error of the difference between the means.

Now let us consider the effect of matching groups on the basis of a relevant variable correlated with the dependent variable. The greater the correlation between matching and dependent variables the greater is the proportion of the variability of the dependent variable mean which is accounted for by the matching variable. As an extreme case, if perfect correlation were to exist between matching and dependent variable, the dependent variable means would exhibit no random fluctuation at all.[1] The group means on the dependent variable would be equal except for the difference attributable to treatment effects. Yet we would be evaluating these effects in terms of the rather large within-group variability generally displayed by human *S*s. We would, that is, unless we took the matching into account in our evaluation.

If the correlation between matching and dependent variable is r, the variance of the difference between the means of the dependent variable will be decreased to an amount only $(1 - r^2)$ times as great as it would have been had there been no matching. As will be remembered, the quantity $(1 - r^2)$ is the proportion of variance unaccounted for by a correlated variable. This same proportion is unaccounted for if we are predicting individual dependent variable scores from the corresponding matching variable scores or are predicting dependent variable means from the matching variable means. Under the null hypothesis, we are predicting that the group means on the dependent variable will be equal except for the unaccounted, for variability. Given, then, that means have been matched on the matching variable, the expected value of the variance of individual means on the dependent variable will be $\sigma^2(1 - r^2)/n$. The variance of the difference between two such means from independent samples will be the sum of the individual vari-

ances $\sigma^2(1 - r^2)/n_1 + \sigma^2(1 - r^2)/n_2$, that is, $\sigma^2(1/n_1 + 1/n_2)(1 - r^2)$ or, in the equal sample size case considered earlier, $2\sigma^2(1 - r^2)/n$. Since $(1 - r^2)$ is a value between 0 and 1 which varies inversely with size of r, we can readily see that a sizable correlation between the matching variable or variables and the dependent variable will decrease the variance of the differences between sample means considerably. This affects the numerator of the t test since it is the variance of the difference between means which is decreased. If one now measures the numerator in terms of the usual uncorrected, and thereby larger, standard error estimate of $\sqrt{2\sigma^2/n}$, in the denominator, the value of t will tend to be smaller than it should be. The difficulty in achieving the level of significance becomes obvious.

Similarly, in the F test of analysis of variance, recourse is made to the implicit relationship between the variability or variance of group means and the variability of S or other random effects. In analysis of variance, one typically derives an estimate of variance from the deviations between group means and another independent estimate of the same variance (providing the null hypothesis of equality of means) from within-group deviations. In this case the danger of artificially restricting the fluctuations of group means is particularly transparent—the variance estimate used as the numerator is going to be too small if restrictions are imposed.

Valid Procedures to Obtain Benefits of Matching

It certainly is true that one would like to rule out the possibility that factors other than his experimentally manipulated or otherwise independent variables are contributing to observed differences. In other terms we can speak of the advantages of increasing the

---

[1] Or very little. That due to errors of measurement we will consider inconsequential for the sake of the argument.

precision of the experiment. A number of techniques are available for achieving both of these ends.[2] All of these techniques employ the device of decreasing the denominator of the t and F tests by an amount commensurate with the restrictions placed upon the numerator.

Consider the technique of matching by group means, the *bete-noir* of our tale thus far. If we knew the value of r (the correlation between the dependent variable and the matching variable or variables), we could multiply the estimate of the variance in the denominator by $(1 - r^2)$. Since the variance of the numerator is $(1 - r^2)$ times what it would have been without matching, we thereby will have decreased the denominator to the appropriate value. The estimate of the standard error of the difference between means, for our example, would then be $\sqrt{2\sigma^2(1 - r^2)/n}$. This would be the correct denominator in the t test and would yield valid probability statements relating to the null hypothesis.

As an alternative procedure we may simply sample at random without matching and allow simple covariance techniques to take out the effects of the relevant covarying variables. Most statistics texts consider these procedures in detail so they will not be elaborated here. Unfortunately, both covariance analysis and the group matching technique considered above depend for their effectiveness and validity upon a linear relationship between matching and dependent variables. If such a relationship does not hold, these procedures are not appropriate unless one resorts to curvilinear techniques which frequently are not part of the response repertoire of the researcher.

Another variety of procedure is that of forming a series of blocks or levels of Ss who are nearly alike on the relevant variables

and then randomly assigning Ss within each block or level to the treatments. It is then possible to take into account the effects of the relevant variables, that is, in effect to match groups on them, and to remove the effects from the error term of analysis of variance by appropriate techniques. See, for example, Lindquist (1953) who discusses appropriate procedures under the classification "treatments-by-levels" designs. If only two groups are being considered, the ordinary t test for matched pairs is appropriate. A sub-variety of this approach useful when only one concomitant variable is being considered is to rank Ss on the basis of the variable, count off as many as there are experimental treatments, and then to randomly assign treatments to these Ss. Continue by counting off another group of Ss, and so on until the supply is exhausted. An experiment set up in this fashion could be analyzed, say, by Lindquist's "treatment-by-subject" procedure with a great deal more justification than that procedure ordinarily possesses. If more than one relevant variable is to be controlled, one may constitute levels on the basis of one variable and with each level rank Ss on the other variable, randomly assigning the ranked individuals to treatments as above. Or one may constitute levels on two or more relevant variables, use covariance procedures on others, and complicate things to his taste or according to the demands of the analysis. In all these cases one must remember that he achieves a valid test of the null hypothesis with a stated probability level only if, as is done in the above-mentioned analysis of variance designs, he removes from the error term the effects of his systematic manipulations.

Summary

The strategy of matching group means on the basis of relevant variables is examined

---

[2] Feldt (1958) discusses the relative precision of various experimental designs which employ a concomitant variable.

as to its effect on the probability of rejecting the null hypothesis. It is concluded that unless the effects of the matching procedures are accounted for in the analysis, the results are invalid. Some valid procedures are suggested to take relevant variables into account.

References

Feldt, L. S. A comparison of the precision of three experimental designs employing a concomitant variable. *Psychometrika*, 1958, **23**, 335–353.

Lindquist, E. F. *Design and analysis of experiments in psychology and education.* Boston Houghton Mifflin, 1953.

6.6 $N = 1$

William F. Dukes

Abstract. Studies focused on the behavior of only one individual are examined. Despite the limitation on generalizing inherent in such studies, selected examples attest to their importance in the history of psychology. Their frequency in modern psychology is noted, together with their distribution across major topics. The usefulness of an N of 1 in research is viewed as extending beyond the single-case studies of clinician and personologists. An N of 1 is seen as also appropriate when, for the function considered, intersubject variability is low, when opportunities for observing a given class of events are limited, and when a supposed universal relationship is questioned and the obtained evidence is negative.

In the search for principles which govern behavior, psychologists generally confine their empirical observations to a relatively small sample of a defined population, using probability theory to help assess the generality of the findings obtained. Because this inductive process commonly entails some knowledge of individual differences in the behavior involved, studies employing only one subject ($N = 1$) seem somewhat anomalous. With no information about intersubject variability in performance, the general applicability of findings is indeterminate.

Reprinted from *Psychological Bulletin*, 1965, **64** 74–79, by permission of the publisher and author. Copyright 1965 by the American Psychological Association.

Although generalizations about behavior rest equally upon adequate sampling of both subjects and situations, questions about sampling most often refer to subjects. Accordingly, the term "$N = 1$" is used throughout the present discussion to designate the *reductio ad absurdum* in the sampling of subjects. It might, however, equally well (perhaps better, in terms of frequency of occurrence) refer to the limiting case in the sampling of situations—for example, the use of one maze in an investigation of learning, or a simple tapping task in a study of motivation. With respect to the two samplings, Brunswik (1956), foremost champion of the representative design of experiments, speculated:

In fact, proper sampling of situations and problems may in the end be more important than proper sampling of subjects, considering the fact that individuals are probably on the whole much more alike than are situations among one another [p. 39].

As a corollary, the term $N = 1$ might also be appropriately applied to the sampling of experimenters. Long recognized as a potential source of variance in interview data (e.g., Cantril, 1944; Katz, 1942), the investigator has recently been viewed as a variable which may also influence laboratory results (e.g., McGuigan, 1963; Rosenthal, 1963).

Except to note these other possible usages of the term $N = 1$, the present paper is not concerned with one-experimenter or one-situation treatments, but is devoted, as indicated previously, to single-subject studies.

Despite the limitation stated in the first paragraph, $N = 1$ studies cannot be dismissed as inconsequential. A brief scanning of general and historical accounts of psychology will dispel any doubts about their importance, revealing, as it does, many instances of pivotal research in which the observations were confined to the behavior of only one person or animal.

Selective Historical Review

Foremost among $N = 1$ studies is Ebbinghaus' (1885) investigation of memory. Called by some authorities "a landmark in the history of psychology ... a model which will repay careful study [McGeoch & Irion, 1952, p. 1]," considered by others "a remedy ... as least as bad as the disease [Bartlett, 1932, p. 3]," Ebbinghaus' work established the pattern for much of the research on verbal learning during the past 80 years. His principal findings, gleaned from many self-administered learning situations consisting of some 2,000 lists of nonsense syllables and 42 stanzas of poetry, are still valid source material for the student of memory. In another well-known pioneering study of learning, Bryan and Harter's (1899) report on plateaus, certain crucial data were obtained from only one subject. Their letter-word-phrase analysis of learning to receive code was based on the record of only one student. Their notion of habit hierarchies derived in part from this analysis is, nevertheless, still useful in explaining why plateaus may occur.

Familiar even to beginning students of perception is Stratton's (1897) account of the confusion from and the adjustment to wearing inverting lenses. In this experiment according to Boring (1942), Stratton, with only himself as subject,

settled both Kepler's problem of erect vision with an inverted image, and Lotze's problem of the role of experience in space perception, by showing that the "absolute" localization of retinal positions—up-down and right-left—are learned and consist of bodily orientation as context to the place of visual excitation [p. 237].

The role of experience was also under scrutiny in the Kelloggs' (1933) project of raising one young chimpanzee, Gua, in their home. (Although observations of their son's behavior were also included in their report, the study is essentially of the $N = 1$ type, since the "experimental group" consisted of one.) This attempt to determine whether early experience may modify behavior traditionally regarded as instinctive was for years a standard reference in discussions of the learning-maturation question.

Focal in the area of motivation is the balloon-swallowing experiment of physiologists Cannon and Washburn (1912) in which kymographic recordings of Washburn's stomach contractions were shown to coincide with his introspective reports of hunger pangs. Their findings were widely incorporated into psychology textbooks as providing an explanation of hunger. Even though in recent years greater importance has been attached to central factors in hunger, Cannon and Washburn's work continues to occupy a prominent place in textbook accounts of food-seeking behavior.

In the literature on emotion, Watson and Rayner's study (1920) of Albert's being conditioned to fear a white rat has been hailed as "one of the most influential papers in the history of American psychology" (Miller, 1960, p. 690). Their experiment, Murphy (1949) observes,

immediately had a profound effect on American psychology; for it appeared to support the whole conception that not only simple motor habits, but important, enduring traits of personality, such as emotional tendencies, may in fact be 'built into' the child by conditioning [p. 261].

Actually the Albert experiment was unfinished because he moved away from the laboratory area before the question of fear removal could be explored. But Jones (1924) provided the natural sequel in Peter, a child who, through a process of active reconditioning, overcame a nonlaboratory-produced fear of white furry objects.

In abnormal psychology few cases have attracted as much attention as Prince's (1905) Miss Beauchamp, for years the model case in accounts of multiple personality. An excerpt from the Beauchamp case was recently included, along with selections from Wundt, James, Pavlov, Watson, and others, in a volume of 36 classics in psychology (Shipley, 1961). Perhaps less familiar to the general student but more significant in the history of psychology is Breuer's case (Breuer & Freud, 1895) of Anna O., the analysis of which is credited with containing "the kernel of a new system of treatment, and indeed a new system of psychology [Murphy, 1949, p. 307]." In the process of examining Anna's hysterical symptoms, the occasions for their appearance, and their origin, Breuer claimed that with the aid of hypnosis these symptoms were "talked away." Breuer's young colleague was Sigmund Freud (1910), who later publicly declared the importance of this case in the genesis of psychoanalysis.

There are other instances, maybe not so spectacular as the preceding, of influential $N = 1$ studies—for example, Yerkes' (1927)

exploration of the gorilla Congo's mental activities; Jacobson's (1931) study of neuromuscular activity and thinking in an amputee; Culler and Mettler's (1934) demonstration of simple conditioning in a decorticate dog; and Burtt's (1932) striking illustration of his son's residual memory of early childhood.

Further documentation of the significant role of $N = 1$ research in psychological history seems unnecessary. A few studies, each in impact like the single pebble which starts an avalanche, have been the impetus for major developments in research and theory. Others, more like missing pieces from nearly finished jigsaw puzzles, have provided timely data on various controversies.

This historical recounting of "successful" cases is, of course, not an exhortation for restricted subject samplings, nor does it imply that their greatness is independent of subsequent related work.

Frequency and Range of Topics

In spite of the dated character of the citations—the latest being 1934—$N = 1$ studies cannot be declared the product of an era unsophisticated in sampling statistics, too infrequent in recent psychology to merit attention. During the past 25 years (1939-1963) a total of 246 $N = 1$ studies, 35 of them in the last 5-year period, have appeared in the following psychological periodicals: the *American Journal of Psychology, Journal of Genetic Psychology, Journal of Abnormal and Social Psychology, Journal of Educational Psychology, Journal of Comparative and Physiological Psychology, Journal of Experimental Psychology, Journal of Applied Psychology, Journal of General Psychology, Journal of Social Psychology, Journal of Personality,* and *Journal of Psychology*. These are the journals, used by Bruner and Allport (1940) in their survey of 50 years of change in American psychology, selected as significant for and devoted to the advancement of psychology as science. (Also used in their survey were the *Psychological Review, Psy-*

Table 1. Total distribution of $N = 1$ studies (1939–1963).

Category	f	Examples
Maturation, development	29	Sequential development of prehension in a macaque (Jensen, 1961); smiling in a human infant (Salzen, 1963)
Motivation	7	Differential reinforcement effects of true, esophagal, and sham feeding in a dog (Hull, Livingston, Rouse, & Barker, 1951)
Emotion	12	Anxiety levels associated with bombing (Glavis, 1946)
Perception, sensory processes	25	Congenital insensitivity to pain in a 19-year-old girl (Cohen, Kipnis, Kunkle, & Kubzansky, 1955); figural aftereffects with a stabilized retinal image (Krauskopf, 1960)
Learning	27	Delayed recall after 50 years (Smith, 1963); imitation in a chimpanzee (Hayes & Hayes, 1952)
Thinking, language	15	"Idealess" behavior in a chimpanzee (Razran, 1961); opposite speech in a schizophrenic patient (Laffal & Ameen, 1959)
Intelligence	14	Well-adjusted congenital hydrocephalic with IQ of 113 (Teska, 1947); intelligence after lobectomy in an epileptic (Hebb, 1939)
Personality	51	Keats' personality from his poetry (McCurdy, 1944); comparison in an adult of P and R techniques (Cattell & Cross, 1952)
Mental health, psychotherapy	66	Multiple personality (Thigpen & Cleckley, 1954); massed practice as therapy for patient with tics (Yates, 1958)
Total	246	

chological Bulletin, and *Psychometrika,* excluded here because they do not ordinarily publish original empirical work.) Although these 246 studies constitute only a small percent of the 1939–1963 journal articles, the absolute number is noteworthy and is sizable enough to discount any notion that $N = 1$ studies are a phenomenon of the past.

When, furthermore, these are distributed, as in Table 1, according to subject matter, they are seen to coextend fairly well with the range of topics in general psychology. As might be expected, a large proportion of them fall into the clinical and personality areas. One cannot, however, explain away $N = 1$ studies as case histories contributed by clinicians and personologists occupied less with establishing generalizations than with exploring the uniqueness of an individual and understanding his total personality. Only about 30% (74) are primarily oriented toward the individual, a figure which includes not only works in the "understanding" tradition, but also those treating the individual as a universe of responses and applying traditionally nomothetic techniques to describe and predict individual behavior (e.g., Cattell & Cross, 1952; Yates, 1958).

In actual practice, of course, the two orientations—toward uniqueness or generality—are more a matter of degree than of mutual exclusion, with the result that in the literature surveyed purely idiographic research is extremely rare. Representative of that approach are Evans' (1950) novel-like account of Miller who "spontaneously" recovered his sight after more than 2 years of blindness, Rosen's (1949) "George X: A self-analysis by an avowed fascist," and McCurdy's (1944) profile of Keats.

Rationale for $N = 1$

The appropriateness of restricting an idiographic study to one individual is obvious from the meaning of the term. If uniqueness is involved, a sample of one exhausts the population. At the other extreme, an N of 1 is also appropriate if complete population generality exists (or can reasonably be assumed to exist). That is, when between-individual variability for the function under scrutiny is known to be negligible or the data from the single subject have a point-for-point congruence with those obtained from dependable collateral sources, results from a second subject may be considered redundant. Some $N = 1$ studies may be regarded as approximations of this ideal case, as for example, Heinemann's (1961) photographic measurement of retinal images and Bartley and Seibel's (1954) study of entoptic stray light, using the flicker method.

A variant on this typicality theme occurs when the researcher, in order to preserve some kind of functional unity and perhaps to dramatize a point, reports in depth one case which exemplifies many. Thus Eisen's (1962) description of the effects of early sensory deprivation is an account of one quondam hard-of-hearing child, and Bettelheim's (1949) paper on rehabilitation a chronicle of one seriously delinquent child.

In other studies an N of 1 is adequate because of the dissonant character of the findings. In contrast to its limited usefulness in *establishing* generalizations from "positive" evidence, an N of 1 when the evidence is "negative" is as useful as an N of 1,000 in *rejecting* an asserted or assumed universal relationship. Thus Krauskopf's (1960) demonstration with one stopped-image subject eliminates motion of the retinal image as necessary for figural aftereffects; and Lenneberg's (1962) case of an 8-year-old boy who lacked the motor skills necessary for speaking but who could understand language makes it "clear that hearing oneself babble is not a necessary factor in the acquisition of understanding . . . [p. 422]." Similarly Teska's (1947) case of a congenital hydrocephalic, $6\frac{1}{2}$ years old, with an IQ of 113, is sufficient evidence to discount the notion that prolonged congenital hydrocephaly results in some degree of feeblemindedness.

While scientists are in the long run more likely to be interested in knowing *what is* than *what is not* and more concerned with how many exist or in what proportion they exist than with the fact that at least one exists, one negative case can make it necessary to revise a traditionally accepted hypothesis.

Still other $N = 1$ investigations simply reflect a limited opportunity to observe. When the search for lawfulness is extended to infrequent "nonlaboratory" behavior, individuals in the population under study may be so sparsely distributed spatially or temporally that the psychologist can observe only one case, a report of which may be useful as a part of a cumulative record. Examples of this include cases of multiple personality (Thigpen & Cleckly, 1954), unilateral color blindness (Graham, Sperling, Hsia, & Coulson, 1961) congenital insensitivity to pain (Cohen et al., 1955), and mental deterioration following carbon monoxide poisoning (Jensen, 1950). Situational complexity as well as subject sparsity may limit the opportunity to observe. When the situation is greatly extended in time, requires expensive or specialized training for the subject, or entails intricate and difficult to administer controls, the investigator may, aware of their exploratory character, restrict his observations to one subject. Projects involving home-raising a chimpanzee (Hayes & Hayes, 1952) or testing after 16 years for retention of material presented during infancy (Burtt, 1941) would seem to illustrate this use of an N of 1.

Not all $N = 1$ studies can be conveniently fitted into this rubric; nor is this necessary. Instead of being oriented either toward the person (uniqueness) or toward a global theory (universality), researchers may sometimes simply focus on a problem. Problem-centered research on only one subject may, by clarifying questions, defining variables, and indicating approaches, make substantial contributions to the study of behavior. Besides answering a specific question, it may (Ebbinghaus' work, 1885, being a classic example) provide important groundwork for the theorists.

Regardless of rationale and despite obvious limitations, the usefulness of $N = 1$ studies in psychological research seems, from the preceding historial and methodological considerations, to be fairly well established. (See Shapiro, 1961, for an affirmation of the value of single-case investigations in fundamental clinical psychological research.) Finally, their status in research is further secured by the statistician's assertion (McNemar, 1940) that:

The statistician who fails to see that important generalizations from research on a single case can ever be acceptable is on a par with the experimentalist who fails to appreciate the fact that some problems can never be solved without resort to numbers [p. 361].

References

Bartlett, F. C. *Remembering.* Cambridge, England: University Press, 1932.

Bartley, S. H., & Seibel, Jean L. A further study of entoptic stray light. *Journal of Psychology,* 1954, **38**, 313–319.

Bettelheim, B. H. A study in rehabilitation. *Journal of Abnormal and Social Psychology,* 1949, **44**, 231–265.

Boring, E. G. *Sensation and perception in the history of experimental psychology.* New York: Appleton-Century, 1942.

Breuer, J., & Freud, S. Case histories. (Orig. publ. 1895; trans. by J. Strachey) In J. Strachey (Ed.), *The standard edition of the complete psychological works of Sigmund Freud.* Vol. 2. London: Hogarth Press, 1955. Pp. 19–181.

Bruner, J. S., & Allport, G. W. Fifty years of change in American psychology. *Psychological Bulletin,* 1950, **37**, 757–776.

Brunswik, E. *Perception and the representative design of psychological experiments.* Berkeley: Univer. California Press, 1956.

Bryan, W. L., & Harter, N. Studies on the telegraphic language. The acquisition of a hierarchy of habits. *Psychological Review,* 1899, **6**, 345–375.

Burtt, H. E. An experimental study of early childhood memory. *Journal of Genetic Psychology,* 1932, **40**, 287–295.

Burtt, H. E. An experimental study of early childhood memory: Final report. *Journal of Genetic Psychology,* 1941, **58**, 435–439.

Cannon, W. B., & Washburn, A. L. An explanation of hunger. *American Journal of Physiology,* 1912, **29**, 441–454.

Cantril, H. *Gauging public opinion.* Princeton: Princeton Univer. Press, 1944.

Cattell, R. B., & Cross, K. P. Comparison of the ergic and self-sentiment structures found in dynamic traits by R- and P- techniques. *Journal of Personality,* 1952, **21**, 250–271.

Cohen, L. D., Kipnis, D., Kunkle, E. C., & Kubzansky, P. E. Observations of a person with congenital insensitivity to pain. *Journal of Abnormal and Social Psychology,* 1955, **51**, 333–338.

Culler, E., & Mettler, F. A. Conditioned behavior in a decorticate dog. *Journal of Comparative Psychology,* 1934, **18**, 291–303.

Ebbinghaus, H. *Über das Gedächtnis.* Leipzig: Duncker & Humblot, 1885.

Eisen, N. H. Some effects of early sensory deprivation on later behavior: The quondam hard-of-hearing child. *Journal of Abnormal and Social Psychology,* 1962, **65**, 338–342.

Evans, Jean Miller. *Journal of Abnormal and Social Psychology,* 1950, **45**, 359–379.

Freud, S. The origin and development of psychoanalysis. *American Journal of Psychology,* 1910, **21**, 181–218.

Glavis, L. R., Jr. Bombing mission number fifteen. *Journal of Abnormal and Social Psychology,* 1946, **41**, 189–198.

Graham, C. H., Sperling, H. G., Hsia, Y., & Coulson, A. H. The determination of some visual functions of a unilaterally color-blind subject. *Journal of Psychology,* 1961, **51**, 3–32.

Hayes, K. J., & Hayes, Catherine. Imitation in a home-raised chimpanzee. *Journal of Comparative and Physiological Psychology,* 1952, **45**, 450–459.

Hebb, D. O. Intelligence in man after large removals of cerebral tissue: Defects following right temporal lobectomy. *Journal of General Psychology,* 1939, **21**, 437–446.

Heinemann, E. G. Photographic measurement of the retinal image. *American Journal of Psychology,* 1961, **74**, 440–445.

Hull, C. L., Livingston, J. R., Rouse, R. O., & Barker, A. N. True, sham, and esophageal feeding as reinforcements. *Journal of Comparative and Physiological Psychology,* 1951, **44**, 236–245.

Jacobson, E. Electrical measurements of neuromuscular states during mental activities: VI. A note on mental activities concerning an amputated limb. *American Journal of Physiology,* 1931, **96**, 122–125.

Jensen, G. D. The development of prehension in a macaque. *Journal of Comparative and Physiological Psychology,* 1961, **54**, 11–12.

Jensen, M. B. Mental deterioration following carbon monoxide poisoning. *Journal of Abnormal and Social Psychology,* 1950, **45**, 146–153.

Jones, Mary C. A laboratory study of fear: The case of Peter. *Journal of Genetic Psychology*, 1924, **31**, 308–315.

Katz, D. Do interviewers bias poll results? *Public Opinion Quarterly*, 1942, **6**, 248–268.

Kellogg, W. N., & Kellogg, Luella. *The ape and the child*. New York: McGraw-Hill, 1933.

Krauskopf, J. Figural after-effects with a stabilized retinal image. *American Journal of Psychology*, 1960, **73**, 294–297.

Laffal, J., & Ameen, L. Hypotheses of opposite speech. *Journal of Abnormal and Social Psychology*, 1959, **58**, 267–269.

Lenneberg, E. H. Understanding language without ability to speak: A case report. *Journal of Abnormal and Social Psychology*, 1962, **65**, 419–425.

McCurdy, H. G. *La belle dame sans merci. Character and Personality*, 1944, **13**, 166–177.

McGeoch, J. A., & Irion, A. L. *The psychology of human learning*. New York: Longmans, Green, 1952.

McGuigan, F. J. The experimenter: A neglected stimulus object. *Psychological Bulletin*, 1963, **60**, 421–428.

McNemar, Q. Sampling in psychological research. *Psychological Bulletin*, 1940, **37**, 331–365.

Miller, D. R. Motivation and affect. In Paul H. Mussen (Ed.), *Handbook of research methods in child development*. New York: Wiley, 1960. Pp. 688–769.

Murphy, G. *Historical introduction to modern psychology*. New York: Harcourt, Brace, 1949.

Prince, M. *The dissociation of a personality*. New York: Longmans, Green, 1905.

Razran, G. Raphael's "idealess" behavior. *Journal of Comparative and Physiological Psychology*, 1961, **54**, 366–367.

Rosen, E. George X: The self-analysis of an avowed fascist. *Journal of Abnormal and Social Psychology*, 1949, **44**, 528–540.

Rosenthal, R. Experimenter attributes as determinants of subjects' responses. *Journal of Projective Techniques*, 1963, **27**, 324–331.

Salzen, E. A. Visual stimuli eliciting the smiling response in the human infant. *Journal of Genetic Psychology*, 1963, **102**, 51–54.

Shapiro, M. B. The single case in fundamental clinical psychological research. *British Journal of Medical Psychology*, 1961, **34**, 255–262.

Shipley, T. (Ed.) *Classics in psychology*. New York: Philosophical Library, 1961.

Smith, M. E. Delayed recall of previously memorized material after fifty years. *Journal of Genetic Psychology*, 1963, **102**, 3–4.

Stratton, G. M. Vision without inversion of the retinal image. *Psychological Review*, 1897, **4**, 341–360, 463–481.

Teska, P. T. The mentality of hydrocephalics and a description of an interesting case. *Journal of Psychology*, 1947, **23**, 197–203.

Thigpen, C. H., & Cleckley, H. A case of multiple personality. *Journal of Abnormal and Social Psychology*, 1954, **49**, 135–151.

Watson, J. B., & Rayner, Rosalie. Conditioned emotional reactions. *Journal of Experimental Psychology*, 1920, **3**, 1–14.

Yates, A. J. The application of learning theory to the treatment of tics. *Journal of Abnormal and Social Psychology*, 1958, **56**, 175–182.

Yerkes, R. M. The mind of a gorilla. *Genetic Psychology Monographs*, 1927, **2**, 1–193.

6.7 Suggestions for Further Reading

Campbell, D. Reforms as experiments. *American Psychologist*, 1969, **24**, 409–429. A penetrating analysis of the problems and pitfalls involved in accurately evaluating social reform projects.

Gottman, J. M., R. M. McFall, and J. T. Barnett. Design and analysis of research using time series. *Psychological Bulletin*, 1969, **72**, 299–306. The time-series design is often used in experiments where control groups are unavailable but dependent observations can be collected over time. This article examines some of the problems in interpreting the results of such experiments and describes a mathematical model for testing hypotheses.

Heilizer, F. Some cautions concerning the use of change scores. *Journal of Clinical Psychology*, 1959, **15**, 447–449. Examines the problems associated with simultaneously using initial scores as a control variable and change scores to measure improvement or deterioration in performance.

Mantel, N. Random numbers and experimental design. *The American Statistician*, 1969, **23**, 32–34. Describes the use of random-number tables in assigning subjects to experimental conditions.

Solomon, R. L., and M. S. Lessac. A control group design for experimental studies of developmental processes. *Psychological Bulletin*, 1968, **70**, 145–150. Describes a four-group experimental design that enables an experimenter to control or partial out the effects of a number of extraneous variables.

Thorndike, R. L. Regression fallacies in the matched groups experiment. *Psychometrika*, 1942, **7**, 85–102. Examines the problem of regression effects when matched groups are drawn from populations that differ with respect to the characteristics being studied.

Wuebben, P. L. Experimental design, measurement, and human subjects: A neglected problem of control. *Sociometry*, 1968, **31**, 89–101. Presents two designs that enable an experimenter to control measurement effects of particular importance in social-science research.

7

Analysis of Variance

Analysis of variance, one of the most widely used statistical procedures in the behavioral sciences, owes its popularity both to the robustness of the F test statistic and to the general flexibility of the technique. Analysis of variance permits an experimenter to test hypotheses with respect to population means, evaluate trends in data, ascertain strength of association among variables, and estimate the proportion of variation of a dependent variable that is associated with one or more independent variables, to cite only a few applications.

Churchill Eisenhart, in the first article in this chapter, details the assumptions that underlie the analysis of variance procedure. He also discusses the algebra of analysis of variance and distinguishes between the fixed effects and random effects models.

7.1 The Assumptions Underlying the Analysis of Variance

Churchill Eisenhart

1. *Introductory Remarks.* The statistical technique known as "analysis of variance," developed more than two decades ago by R. A. Fisher to facilitate the analysis and interpretation of data from field trials and laboratory experiments in agricultural and biological research, today constitutes one of the principal research tools of the biological scientist, and its use is spreading rapidly in the social sciences, the physical sciences, and in engineering. Numerous textbooks (or should I say " manuals " ?) have been published—and, I dare say, many more are being written—that aim to provide their readers with a working knowledge of the steps of analysis-of-variance procedure with a minimum exposure to mathematical formulas and mathematical thinking. Designed expressly for the "non-mathematical reader," whose mathematical equipment is presumed to be a reasonable competence in arithmetic and elementary algebra—mere previous exposure to these subjects is not enough. The method of instruction adopted in these books consists

chiefly in guiding the reader by easy stages through a series of worked examples that are typical of the more common problems amenable to analysis of variance that arise in the scientific or engineering field with which the author of the book concerned is conversant.[1]

These introductions to analysis of variance have been definitely worthwhile in at least three respects: first, they have acquainted a larger audience with the procedures of analysis of variance and its value as a research tool than probably would have been achieved by more mathematical expositions of the subject (even unfavorable reviews of some of these books have focused attention on the analysis of variance itself as a research tool "that needs further looking into"); second, by studying the worked examples provided and by carrying through analogous steps with data of their own, readers of these books have developed an amazing proficiency with the arithmetical steps involved, even in the cases of analyses associated with fairly complicated

Reprinted from *Biometrics*, 1947, **3**, 1–21, by permission of the publisher and author.

An expository address delivered at a joint session of the Biometrics Section of the American Statistical Association and the Institute of Mathematical Statistics, held on December 28, 1946, in conjunction with the 113th Annual Meeting of the American Association for the Advancement of Science, Boston, Massachusetts.

[1] The author has found the discussions and examples of analysis-of-variance procedures given in the following four books especially valuable both for reference and for purposes of instruction: C. H. Goulden, *Methods of Statistical Analysis*; G. W. Snedecor, *Calculation and Interpretation of Analysis of Variance and Covariance*; G. W. Snedecor, *Statistical Methods*; L. H. C. Tippett, *The Methods of Statistics*. See full bibliographical references at the end of this paper.

experimental designs which probably would not have been attempted or, if attempted, almost certainly would not have been analyzed correctly without the aid of these books; and third, since the worked examples given in these books have generally illustrated statistically sound experimental designs which were more efficient than the designs previously used by their readers, these readers have frequently adopted analogous designs in their own research (in order to be able to follow the book when the data are in and are crying for analysis), with a resulting general improvement of research procedure.

The principal deficiency of these books has been their failure to state explicitly the several assumptions underlying the analysis of variance, and to indicate the importance of each from a practical viewpoint. The mathematical treatments of analysis of variance have shared this deficiency to some extent, for, while they have posited the necessary and sufficient conditions[2] for strict validity of the entire set of analysis-of-variance procedures and associated tests of significance, they have not generally indicated in sufficient detail the actual functions of the respective assumptions—(1) which can be dispensed with for certain purposes; (2) which are absolutely necessary, and what are likely to be the consequences if these are not fulfilled; and (3) what can be done " to bring into line," for purposes of analysis, data which in their original form are not amenable to analysis of variance.[3] In this paper I shall go into these matters in some detail. My

assignment is to enumerate the several assumptions underlying the analysis of variance and to point out the practical importance of each. As we shall see, these assumptions are quite simple to state, and the practical significance of each not difficult to grasp.

2. *Two Distinct Classes of Problems Solvable by Analysis of Variance.* Turning now to my assignment, I am obliged at the outset to draw attention to the fact that analysis of variance can be, and is, used to provide solutions to problems of two fundamentally different types. These two distinct classes of problems are:

(2.1) *Class I: Detection and Estimation of Fixed (Constant) Relations Among the Means of Sub-Sets of the Universe of Objects Concerned.* This class includes all of the usual problems of estimating, and testing to determine whether to infer the existence of, *true* differences among " treatment " means, among " variety " means, and, under certain conditions, among " place " means. Included in this class are all the problems of univariate and multivariate regression and of harmonic analysis. With respect to the problems of estimation belonging to this class, analysis of variance is simply a form of the method of least squares: the analysis-of-variance solutions are the least-squares solutions. The cardinal contribution of analysis of variance to the actual procedure is the *analysis-of-variance table* devised by R. A. Fisher, which serves to simplify the arithmetical steps and to bring out more clearly the significance of the results obtained. The analysis-of-variance tests of significance employed in connection with problems of this class are simply extensions to small samples of the theory of least squares developed by Gauss and others—the extension of the theory to small samples being due principally to R. A. Fisher.

(2.2) *Class II: Detection and Estimation of Components of (Random) Variation Associated with a Composite Population.* This class includes all problems of estimating, and testing to determine whether to infer the existence of, components of variance ascribable to random deviation of the characteristics

---

[2] The conditions here referred to are certainly sufficient: they may be necessary in the mathematical sense, but no proof of this is known to the writer. For a variety of reasons he believes them to be " necessary in practice " in the same sense that, if there are exceptions, the circumstances required would be regarded by the practical man as " pathological."

[3] See, for example, the discussions of analysis of variance in H. Cramér, *Mathematical Methods of Statistics*; in M. G. Kendall, *The Advanced Theory of Statistics*, Volume II; and in S. S. Wilks, *Mathematical Statistics*. Somewhat more complete discussions have been given by J. O. Irwin in his paper entitled " Mathematical Theorems Involved in the Analysis of Variance " and in his note " On the Independence of the Constituent Items in the Analysis of Variance."

of individuals of a particular generic type from the mean values of these characteristics in the "population" of all individuals of that generic type, etc. In a sense, *this is the true analysis of variance*, and the estimation of the respective components of the over-all variance of a single observation requires further steps beyond the evaluations of the entries of the analysis-of-variance table itself. Problems of this class have received considerably less attention in the literature of analysis of variance than have problems of Class I.[4]

The failure of most of the literature on analysis of variance to focus attention on the distinction between problems of Class I and problems of Class II is very likely due to two facts: first, the literature of analysis of variance deals largely with tests of significance in contrast to problems of estimation; second, when analysis of variance is used merely to determine whether to infer (a) the existence of fixed differences among the true means of the sub-sets concerned or (b) the existence of a component of variance ascribable to a particular factor, the computational procedure and the mechanics of the statistical tests of significance are the same in either case —the same test criterion (F or z) is evaluated and referred to the same "levels of significance" in either case. On the other hand, in the estimation of the relevant parameters, and in the evaluation of the efficiency or resolving power of a particular experimental design, the distinction between these two classes of problems needs to be taken into account, since in problems of Class I the parameters involved are *means* and the issues of interest are concerned with the interrelations of these means, i.e., with the differences between pairs

of them, with their functional dependence on some independent variable(s), etc.; whereas in problems of Class II the parameters involved are *variances* and their absolute and relative magnitudes are of primary importance. In other words, the mathematical models appropriate to problems of Class I differ from the mathematical models appropriate to problems of Class II, and consequently, so do the questions to be answered by the data.

3. *The Algebra of Analysis of Variance.* It was remarked above that the computational steps leading to an analysis-of-variance table are the same for problems of Class I and Class II. This is due largely to the fact that the decomposition of the (total) sum of squared deviations of the individual observations from the general mean of the observations into two or more "sums of squares" is based in every case upon an algebraic identity (appropriate to the case concerned) that is valid whatever the meanings of the numbers involved. To demonstrate this in complete generality would render the substance of this paper somewhat complicated, and these complexities would, I fear, distract attention from the main theme. Accordingly, I shall restrict myself to consideration of the algebra of the decomposition for the case of rc numbers arranged in a rectangular array of r rows and c columns. In order to be able to identify the various numbers, let us denote by x_{ij} the number occurring in the ith row and jth column of this array. If we border the rectangular array with a column of row means and a row of column means, then we have a situation such as that portrayed in Table 1, where $x_{i.}$ denotes the arithmetic mean of the c values of x in the ith row, $x_{.j}$ denotes the arithmetic mean of the r values of x in the jth column, and $x_{..}$ denotes the arithmetic mean of all the rc values in the array.

It is evident that the following is an algebraic identity *whatever the interpretation of the numbers x_{ij} involved:*

$$(1) \quad (x_{ij} - x_{..}) = (x_{i.} - x_{..}) + (x_{.j} - x_{..}) + (x_{ij} - x_{i.} - x_{.j} + x_{..}).$$

[4] R. A. Fisher gives a brief discussion of estimating and testing for the existence of components of variance in Section 40 of his *Statistical Methods for Research Workers*. Tippett considers such problems in Sections 6.1–6.2, 6.4, 10.11, and 10.3. Snedecor's treatment is somewhat more complete: *Statistical Methods*, Sections 10.6–10.12, 11.4, 11.7–11.8, 11.14, 11.16. The most complete discussions of problems of Class II are H. E. Daniels, "The Estimation of Components of Variance," and S. Lee Crump, "The Estimation of Variance Components in Analysis of Variance."

Table 1.

		Column											
		1	2	3	·	·	·	·	·	j	·	·	c	*Row means*
Row	1	x_{11}	x_{12}	x_{13}	·	·	·	·	·	x_{1j}	·	·	x_{1c}	$x_{1.}$
	2	x_{21}	x_{22}	x_{23}	·	·	·	·	·	x_{2j}	·	·	x_{2c}	$x_{2.}$
	·	·	·	·	·	·	·	·	·	·	·	·	·	·
	·	·	·	·	·	·	·	·	·	·	·	·	·	·
	·	·	·	·	·	·	·	·	·	·	·	·	·	·
	i	x_{i1}	x_{i2}	x_{i3}	·	·	·	·	·	x_{ij}	·	·	x_{ic}	$x_{i.}$
	·	·	·	·	·	·	·	·	·	·	·	·	·	·
	·	·	·	·	·	·	·	·	·	·	·	·	·	·
	r	x_{r1}	x_{r2}	x_{r3}	·	·	·	·	·	x_{rj}	·	·	x_{rc}	$x_{r.}$
	Col. means	$x_{.1}$	$x_{.2}$	$x_{.3}$	·	·	·	·	·	$x_{.j}$	·	·	$x_{.c}$	$x_{..}$

Remembering that by definition the arithmetic mean of m values of a quantity y is (sum of the m values)/m we see that

(2) Arithmetic Mean of y

$$\equiv \bar{y} \equiv \frac{1}{m} S(y) \text{ implies } S(y - \bar{y}) = 0,$$

and

(3) $$S(y - \bar{y})^2 = S(y^2) - \frac{[S(y)]^2}{m},$$

where S denotes summation over all the values of y involved. Squaring both sides of (1) and summing over all rc observations, the algebraic identity

(4) $$\underset{(A)}{S(x_{ij} - x_{..})^2} = \underset{(B)}{S(x_{i.} - x_{..})^2} + \underset{(C)}{S(x_{.j} - x_{..})^2} + \underset{(D)}{S(x_{ij} - x_{i.} - x_{.j} + x_{..})^2}$$

results, where S denotes summation over all the values in the entire array; the cross-products involved sum to zero by virtue of (2), *on account of the fact that* $x_{i.}$, $x_{.j}$, *etc., and* $x_{..}$ *are means.* The (A), (B), (C), and (D) sums of squared quantities in (4) are what are usually referred to in an analysis-of-variance table as the "total," the "between-row-means," the "between-column-means," and "residual" sums of squares, respectively. Since $(x_{i.} - x_{..})^2$ is identically the same for each of the c observations in the ith row, and $(x_{.j} - x_{..})^2$ is the same for each observation in the jth column, it is sometimes convenient to write (4) as

(5) $$\underset{(A)}{\sum_{i=1}^{r} \sum_{j=1}^{c} (x_{ij} - x_{..})^2} = \underset{(B)}{c \sum_{i=1}^{r} (x_{i.} - x_{..})^2} + \underset{(C)}{r \sum_{j=1}^{c} (x_{.j} - x_{..})^2} + \underset{(D)}{\sum_{i=1}^{r} \sum_{j=1}^{c} (x_{ij} - x_{i.} - x_{.j} + x_{..})^2}$$

where $\sum_{i=1}^{r} \sum_{j=1}^{c}$ denotes summation over all the observations in the array, $\sum_{i=1}^{r}$ denotes summation only over i, $i = 1$ to $i = r$, and $\sum_{j=1}^{c}$ denotes summation only over j, for $j = 1$ to $j = c$.

(3.1) *"Practical" Formulas.* With the aid of the identity (3), it is easy to derive the "practical" formulas used for calculation:

(A) $\displaystyle\sum_{i=1}^{r} \sum_{j=1}^{c} (x_{ij} - x_{..})^2$

$$= \sum_{i=1}^{r} \sum_{j=1}^{c} (x_{ij}^2) - \frac{[\sum\sum(x_{ij})]^2}{rc}$$

= Sum of the squares of all observations

$$- \frac{[\text{Sum of all observations}]^2}{\text{Number of observations}},$$

(B) $\displaystyle c\sum_{i=1}^{r} (x_{i.} - x_{..})^2$

$$= \sum_{i=1}^{r} \frac{(cx_1)^2}{c} - \frac{\sum_{i=1}^{r}(cx_{i.})^2}{cr}$$

(6)   = Sum with respect to i of

$$\frac{(i\text{th-row Total})^2}{c}$$

$$- \{\text{Correction Term, given above}\},$$

(C) $\displaystyle r\sum_{j=1}^{c} (x_{.j} - x_{..})^2$

$$= \sum_{j=1}^{c} \frac{(rx_{.j})^2}{r} - \frac{\left[\sum_{j=1}^{c}(rx_{.j})\right]^2}{rc}$$

= Sum with respect to j of

$$\frac{(j\text{th-Column Total})^2}{r} - idem,$$

(D) $\displaystyle\sum_{i=1}^{r} \sum_{j=1}^{c} (x_{ij} - x_{i.} - x_{.j} + x_{..})^2$

$$= (A) - (B) - (C).$$

I repeat: All of the familiar formulas and procedures for evaluating component "sums of squares" that add up to the "total sum of squares" are based on algebraic identities, and are valid as descriptions of properties of the data whatever the interpretation of the numbers involved. Indeed, the fact that the "components" add up to the total is an algebraic

(or, should I say a "geometric") property and means that (and will only happen when) the respective component "sums of squares" are themselves the squares of, or sums of the squares of, linear combinations of the observations that summarize mutually distinct properties of the data, or, as a geometer would say, linear combinations that define mutually orthogonal vectors in the N-dimensional sample space.[5] Similarly, all of the familiar formulas and procedures for evaluating regression coefficients and the sum of squared deviations from the fitted regression, when the fitting is by the method of least squares, are based upon algebra and calculus, and the results obtained are valid as descriptions of properties of the data in hand, whatever the interpretation of the numbers involved.

In summary, when the formulas and procedures of analysis of variance are used merely to summarize properties of the data in hand, no assumptions are needed to validate them. On the other hand, when analysis of variance is used as a method of statistical inference, for inferring properties of the "population" from which the data in hand were drawn, then certain assumptions, about the "population" and the

[5] To see what we mean by "mutually distinct" and by "orthogonal" in practical language, let us note that, if in the case of numbers arranged as in Table 1 we add a single arbitrary constant to each of the numbers in the first column, a different arbitrary constant to each of the numbers in the second column, and so forth through the cth column, then the several row means will be altered by different amounts, which will be determined by the actual constants added, but the several row means will all be altered by the same amount, so that the *difference* between any pair of row means, $(x_{i_1.} - x_{i_2.})$, will be unchanged. Similarly the values of such quantities as $(x_{i.} - x_{..})$ and $(x_{ij} - x_{i.} - x_{.j} + x_{..})$ will be unchanged by this tampering with the columns, so that the "between-row-means" and the "residual" sums of squares will be unchanged also. This is because *differences among row means* (or *differences of row means from the general mean*) and the *residuals* are *orthogonal* to *differences among column means* (or *differences of column means from the general mean*), that is, summarize mutually distinct properties of the actual numbers involved. This little trick of adding arbitrary constants in accordance with a definite pattern is a convenient practical way of checking whether particular combinations of the observations are mutually orthogonal.

sampling procedure by means of which the data were obtained, must be fulfilled if the inferences are to be valid.

4. *The Assumptions Underlying the Use of Analysis of Variance as a Method of Statistical Inference.* As was remarked earlier, analysis of variance can be, and is, used to provide solutions to two fundamentally different types of problems: On the one hand, it can be used to detect the existence of, and to estimate the parameters defining, fixed (constant) relations among the population means. These were referred to as problems of Class I. On the other hand, analysis of variance can be used to detect the existence of, and to estimate, components of variance. These were termed problems of Class II. To formulate with complete generality the mathematical models upon which the solutions of problems of Class I and Class II by analysis of variance are based would render the substance of this paper somewhat complicated from this point on, and would, I fear, divert attention from the really important distinctions between the two different models, and from the differences between the assumptions required in order to be able to draw valid inferences by analysis of variance in the two cases. Therefore, two different models appropriate to data arranged as in Table 1 will be considered in detail and the relation of each assumption to the inferential steps indicated:

(4.1) *Model I, Special Case: Parameters Are Population Means.* Numbers x_{ij} arranged as in Table 1 do not lie within the province of mathematical statistics, nor can any statistical inferences be based upon them, unless it is assumed that they are (observed values of) *random variables* of some sort. Therefore, in order to bring the discussion within the province of statistical inference we must make

Assumption 1 (Random Variables): The numbers x_{ij} are (observed values of) random variables that are distributed about true mean values m_{ij}, $(i = 1, 2, \ldots, r;$ $j = 1, 2, \ldots, c)$, that are fixed *constants*.

In statistical language this assumption states that, if some particular type of experiment leading to numbers arranged as in Table 1 were repeated indefinitely, then the numbers occurring in the ith cell of the jth column would vary at random about an average value equal to m_{ij}, which is, therefore, a parameter that characterizes the expected value of the number x_{ij}. If, for example, the several rows of Table 1 correspond to different "varieties" and the several columns to different "treatments," then m_{ij} is the so-called *true* (or expected) *yield* of the ith "variety" when subjected to the jth "treatment," under certain growing conditions.

Clearly the parameters m_{ij} can be arranged in a table analogous to Table 1, and bordered by the row-wise means $m_{i.}$, $(i = 1, 2, \ldots, r)$, and the column-wise means $m_{.j}$, $(j = 1, 2, \ldots, c)$ of these parameters, to which may be added, in the lower right corner, the mean $m_{..}$ of all rc of these parameters. If one is merely interested in obtaining unbiased estimates of mean differences such as $m_{12} - m_{52}$, e.g., of the mean difference between variety 1 and variety 5 under treatment 2, then Assumption 1 is sufficient, and $x_{12} - x_{52}$ provides the desired estimate. More generally, Assumption 1 implies that an unbiased estimator of any linear function of the m_{ij} with *known* coefficients is provided by the same linear function of the x_{ij}. Furthermore, if the variances of the x_{ij} about their respective means and their intercorrelations are known, then the variance of any linear function of the x_{ij} can be evaluated, and provides a measure of the precision of this linear function of the x_{ij} as an unbiased estimator of the corresponding linear function of the m_{ij}.

On the other hand, when the entries m_{ij} of such a table of *true means* are simple *additive functions* of the corresponding marginal means and the general mean, that is, when

$$(7) \quad m_{ij} = m_{..} + (m_{i.} - m_{..}) + (m_{.j} - m_{..}),$$

for $i = 2, \ldots, r$ and $j = 1, 2, \ldots, c$, then the statistical inferences that may be based upon the x_{ij} are of a much more satisfactory sort. For instance, when (7) is satisfied, the difference between an arbitrary pair of row-wise marginal means, e.g., $m_{1.}$ and $m_{2.}$, is a *comprehensive* measure of the average difference in effectiveness of the factors identified with these rows. When (7) is not satisfied, then $m_{1.} - m_{2.}$ is merely a measure of the average difference between the effects of the corresponding row factors *when the column factors are as in the experiment concerned*. In other words, when additivity, as defined by (7), does not obtain, then it is not possible to define *the* mean difference in effectiveness of any given pair of the row factors, since the actual mean difference in effectiveness of these row factors will depend upon the column factor(s) concerned; and, conversely, the actual mean difference in effectiveness of a pair of column factors will depend upon the row factor(s) concerned. Hence, when additivity does not prevail, we say that there are *interactions* between row factors and column factors. Thus, in the case of varieties and treatments considered above, additivity implies that, under the general experimental conditions of the test, the true mean yield of one variety is greater (or, less) than the true mean yield of another variety by an amount—an additive constant, not a multiplier—that is the same for each of the treatments concerned, and, conversely, the true mean yield with one treatment is greater (or, less) than the true mean yield with another treatment by an amount that does not depend upon the variety concerned; which is exactly what is meant when we say that there are no "interactions" between varietal and treatment effects.

Therefore, in order to dispense with interactions and thus make possible the drawing of general inferences from the x_{ij}, let us make

Assumption 2 (Additivity):[6] the parameters m_{ij} are related to the means $m_{i.}$, $m_{.j}$, and

$m_{..}$ as specified by (7), for $i = 1, 2, \ldots, r$ and $j = 1, 2, \ldots, c$.

When *Assumption 1* and *Assumption 2* are satisfied, then the difference between any pair of row-wise means of the observations x_{ij}, e.g., $x_{2.} - x_{5.}$, is an unbiased estimator of the *general* average difference in effectiveness of the row factors concerned, i.e., of $m_{2.} - m_{5.}$ in this case; and, similarly, the difference between any pair of column-wise means of the observations is an unbiased estimator of the *general* average difference in effectiveness of the column factors concerned. Furthermore, since such estimators are linear functions of the x_{ij}, the variances of these estimators can be evaluated readily when the variances and intercorrelations of the x_{ij} are known. On the other hand, if these variances and intercorrelations are unknown—the usual case in practice—then it is not possible to derive from the observed values of the x_{ij}, $(i = 1, 2, \ldots, r;$ $j = 1, 2, \ldots, c)$, an unbiased estimate of the variance of any single x_{ij}, or of any particular linear combination of them, unless certain additional conditions are fulfilled by the x_{ij}. For instance, if the x_{ij} are mutually uncorrelated,[7] and if variance of the x_{ij} are given by

$$(8) \qquad \text{variance of } x_{ij} = \frac{\sigma^2}{w_{ij}}$$

where the relative "weights," w_{ij}, are *known* constants, $(i = 1, 2, \ldots, r; j = 1, 2, \ldots, c)$, and σ^2 is an *unknown* constant, then an unbiased estimate of σ^2, and thence unbiased estimates

---

$$m_i = c_{i1}\theta_1 + c_{i2}\theta_2 + \cdots + c_{is}\theta_s$$
$$(i = 1, 2, \ldots, N),$$

and non-singularity of the matrix $\|c_{ij}\|$ signifies that from this set of N equations it is possible to select at least one system of s equations that is soluble with respect to the θ's.

This is known as the *general linear hypothesis*. For details, see the papers by F. N. David and J. Neyman, by S. Kolodziejczyk, and by P. C. Tang cited in the list of references.

[6] In its most general form Model I involves N random variables $x_1, x_2, \ldots, x_N$ with mean values m_i, $(i = 1, 2, \ldots, N)$, and it is assumed that the m_i are linear functions of $s < N$ unknown parameters θ_j, $(j = 1, 2, \ldots, s)$, with *known* coefficients, c_{ij}, the matrix of which is non-singular; thus

[7] That is, if the *covariances*

$$\varepsilon\{[x_{ij} - \varepsilon(x_{ij})] [x_{pq} - \varepsilon(x_{pq})]\},$$

where $(i, j) \neq (p, q)$, $(i$ and $p = 1, 2, \ldots, r;$ j and $q = 1, 2, \ldots, c)$, and ε denotes "expected value of," are *all* equal to zero.

of the variances of linear combinations of the x_{ij}, can be derived from the observations x_{ij} by the method of least squares. For details see, for example, the paper by F. N. David and J. Neyman cited in the list of references. They assume the x's to be mutually independent, whereas it is sufficient to assume that they are mutually uncorrelated.

It should be noted here that thus far the only motivation that has been given for the making of *Assumption 2* is the more general nature of the inferences that may be drawn from the observed means $x_{i.}$ and $x_{.j}$, ($i = 1, 2, \ldots, r; j = 1, 2, \ldots, c$), when it is satisfied. We shall now show that, in general, it is not possible to derive from the observations x_{ij} by the *usual analysis-of-variance procedures*, unbiased estimates of the variances of the x_{ij}, and thence of any particular linear combinations of them, unless *Assumption 1*, *Assumption 2*, and *Assumption 3*, given below, are *all* satisfied.

Assumption 3 (Equal Variances and Zero Correlations): The random variables x_{ij} are *homoscedastic* and *mutually uncorrelated*, that is, they have a common variance σ^2 and all covariances among them are zero.

The foregoing pronouncement can be demonstrated readily by considering the analysis-of-variance table shown as Table 2. This represents the situation when *Assumption 1* and *Assumption 3* are both satisfied, but *Assumption 2* is not. We notice that under these conditions each of the "mean squares" customarily evaluated in such cases will have, in general, an expected value larger than σ^2. If, on the other hand, *Assumption 2* is satisfied also, then the "residual" mean square will be an unbiased estimator of σ^2, the variance of any single observation x_{ij}. This situation is portrayed in Table 3. Hence, when *Assumption 1*, *Assumption 2*, and *Assumption 3* are all satisfied, an unbiased estimate of the variance of the difference of two *observed* row means can be evaluated from 2(residual mean square)/c; and an unbiased estimate of the variance of the difference of two *observed* column means, from 2(residual mean square)/r. Furthermore, under these conditions the between-row-means mean square in general will *tend* to exceed the residual mean square, and this tendency will be greater when the true row means, the $m_{i.}$, differ markedly in magnitude than when they differ only slightly. Similarly, the between-column-means mean square in general will tend to exceed the residual mean square by an amount that depends upon the degree of "scatter" of the true column means, the $m_{.j}$, about $m_{..}$, the mean of all the m_{ij}. Thus we have yardsticks for judging whether there exist real differences

Table 2. Analysis of variance (non-additive case).

Variation	Degree of freedom	Sums of squares	Mean square	Expected value of mean square
Between row means	$r - 1$	$s(x_{i.} - x_{..})^2$	$\dfrac{s(x_{i.} - x_{..})^2}{(r - 1)}$	$\dfrac{\sigma^2 + s(m_{i.} - m_{..})^2}{(r - 1)}$
Between column means	$c - 1$	$s(x_{.j} - x_{..})^2$	$\dfrac{s(x_{.j} - x_{..})^2}{(c - 1)}$	$\dfrac{\sigma^2 + s(m_{.j} - m_{..})^2}{(c - 1)}$
Residual	$(r - 1)(c - 1)$	$s(x_{ij} - x_{i.} - x_{.j} + x_{..})^2$	$\dfrac{s(x_{ij} - x_{i.} - x_{.j} + x_{..})^2}{(r - 1)(c - 1)}$	$\dfrac{\sigma^2 + s(m_{ij} - m_{i.} - m_{.j} + m_{..})^2}{(r - 1)(c - 1)}$
Total	$rc - 1$	$s(x_{ij} - x_{..})^2$	$\dfrac{s(x_{ij} - x_{..})^2}{(rc - 1)}$	$\dfrac{\sigma^2 + s(m_{ij} - m_{..})^2}{(rc - 1)}$

Table 3. Analysis of variance (additive case).

Variation	Degree of freedom	Sum of squares	Mean square	Expected mean square
Between row (variety) means	$r - 1$	$s(x_{i.} - x_{..})^2$	$\dfrac{s(x_{i.} - x_{..})^2}{(r-1)}$	$\sigma^2 + \dfrac{s(m_{i.} - m_{..})^2}{(r-1)}$
Between column (cultivation) means	$c - 1$	$s(x_{.j} - x_{..})^2$	$\dfrac{s(x_{.j} - x_{..})^2}{(c-1)}$	$\sigma^2 + \dfrac{s(m_{.j} - m_{..})^2}{(c-1)}$
Residual	$(r-1)(c-1)$	$s(x_{ij} - x_{i.} - x_{.j} + x_{..})^2$	$\dfrac{s(x_{ij} - x_{i.} - x_{.j} + x_{..})^2}{(r-1)(c-1)}$	σ^2
Total	$rc - 1$	$s(x_{ij} - x_{..})^2$	$\dfrac{s(x_{ij} - x_{..})^2}{(rc-1)}$	$\sigma^2 + \dfrac{s(m_{i.} - m_{..})^2 + s(m_{.j} - m_{..})^2}{rc - 1}$

among the true means for the row factors, and for the column factors. Unfortunately, however, our yardsticks have no scales, i.e., probability levels, marked on them, so that with them we cannot conduct exact tests of significance corresponding to previously agreed upon probability levels. In order to be able to do this, the form of the joint distribution of the x_{ij} must be specified. To this we shall return in a moment.

At this juncture let us pause for an instant to note that *it has not been necessary to postulate mutual* INDEPENDENCE *of the x_{ij} in order to achieve Table 3 and the results deducible from it*—for these, *existence of the mean values of the x_{ij}* (Assumption 1), *additivity* (Assumption 2), *and equal variances and zero covariances* (Assumption 3) *are sufficient.*

Also, let us examine the situation where *Assumption 1* and *Assumption 2* are satisfied, but *Assumption 3* is not. In this case the four values of σ^2 that appear in the last column of Table 3 must be replaced, in general, by four *different* quantities, which we may denote by σ_1^2, σ_2^2, σ_3^2, and σ_4^2. In general these will be complex weighted means of the variances and covariances of the x_{ij}, and the neatness of Table 3 is lost.

In summary, if *Assumption 1* is satisfied, but

if either *Assumption 2*, or *Assumption 3*, or both, is (are) not satisfied, then the strict validity of analysis of variance as a method of solution of problems of Class I vanishes out the window.

Finally, even when *Assumption 1*, *Assumption 2*, and *Assumption 3* are satisfied, it is still not possible to conduct exact tests of significance based on the x_{ij} alone, e.g., tests of significance based upon Fisher's *z*- or Snedecor's *F*-distributions. Fortunately, *normality*, in addition to *Assumptions 1–3*, is sufficient for exact tests of significance. Therefore let us make

Assumption 4 (*Normality*): The x_{ij} are jointly distributed in a multivariate normal (Gaussian) distribution.

It may be noted that when *Assumption 4* is satisfied, *Assumption 1* is partially redundant, and serves principally to define the parameters m_{ij}. Furthermore, zero covariances, as postulated in *Assumption 3*, taken in conjunction with normality, postulated in *Assumption 4*, imply mutual independence of the x_{ij}. Thus independence finally sneaks in by the back door, so to speak.

When *Assumptions 1–4* are all satisfied, then all of the usual analysis-of-variance procedures for estimating, and testing to determine whether to infer the existence of, *fixed linear relations*, e.g., non-zero differences, among population *means*, are strictly valid. In particular, an unbiased estimator of any given linear function of the parameters m_{ij} is provided by the identical linear function of the observations x_{ij}, an unbiased estimate of its variance can be derived from the "residual" mean square and exact confidence limits for the value of the given linear function of the parameters can be deduced with the aid of Student's t-distribution. Furthermore, when the row-wise population means, the $m_{i.}$, are all equal, then the quotient ("between-row-means" mean square)/("residual" mean square) will be distributed according to Snedecor's F-distribution for $n_1 = (r - 1)$ and $n_2 = (r - 1)(c - 1)$ degrees of freedom, respectively, which is the basis of the customary *test* of the hypothesis that the $m_{i.}$ are all equal, and the *power* of the test can be evaluated from the tables provided by P. C. Tang, and by Emma Lehmer—see references. An analogous statement can be made with respect to the column-wise population means, the $m_{.j}$.

Therefore, we can summarize the foregoing by the following theorem:

Theorem I: The necessary[8] and sufficient conditions for the strict validity of analysis-of-variance procedures for solving problems of Class I with respect to data arranged as in Table 1 are that

(9)
$$x_{ij} = m_{..} + (m_{i.} - m_{..}) + (m_{.j} + m_{..}) + z_{ij},$$
$$(i = 1, 2, \ldots, r; j = 1, 2, \ldots, c)$$

where the $m_{i.}$, $m_{.j}$, and $m_{..}$ are constants with

(10) $$m_{..} = \sum_{i=1}^{r} m_{i.}/r = \sum_{j=1}^{c} m_{.j}/c$$

and the z_{ij} are normally and independently distributed about zero with a common variance σ^2.

[8] See footnote 2.

(4.2) *Model II: Parameters are Components of Variance.* The preceding discussion of the application of analysis of variance as a method of drawing statistical inferences about the parameters involved in the mathematical model of an experiment leading to numbers arranged as in Table 1 has been concerned entirely with the problems of Class I, where the parameters are *means* and the object of the analysis is to estimate these means or to infer whether certain differences among them are or are not zero. We shall now consider the application of analysis of variance as a method of statistical inference with respect to *components of variance* involved in the mathematical model of an experiment leading to numbers arranged as in Table 1.

For the sake of concreteness, let us suppose for the moment that r animals are drawn at random from the available (large) stock of a given species and that some characteristic of each, say its body temperature, is measured on each of c days randomly located throughout some period of time. Such measurements could be arranged as in Table 1. Furthermore, let us suppose that our ultimate objective is to determine very precisely *the* body temperature characteristic of this species. By *the* body temperature characteristic of this species we mean that value about which the body temperatures of individual animals from the species will vary as a result of *biological variation*, this variability being accentuated, possibly, by day-to-day vicissitudes in the case of each animal. Under these circumstances it will clearly be of interest (a) to ascertain whether there is a component of variation assignable to day-to-day changes in the body temperature of a single animal, and (b) to compare its magnitude with the component of variation assignable to animal-to-animal variability within the species, in order to have a basis for deciding whether in collecting further data a few animals examined on each one of many days, or many animals examined on each one of only a few days, will lead to a more precise estimate of the mean body temperature characteristic of the species.

These questions may be answered by analysis of variance by arranging the data as in Table 1 and making the following assumptions:

Assumption A (Random Variables): The numbers x_{ij} are (observed values of) random variables that are distributed about a common mean value $m_{..}$, $(i = 1, 2, \ldots, r;$ $j = 1, 2, \ldots, c)$, where $m_{..}$ is some fixed constant.

Assumption B (Additivity of Components): The random variables x_{ij} are sums of component random variables, thus

(11)
$$x_{ij} = m_{..} + (m_{i.} - m_{..}) + (m_{.j} - m_{..}) + z_{ij},$$
$$(i = 1, 2, \ldots, r; j = 1, 2, \ldots, c)$$

where the $(m_{i.} - m_{..})$, the $(m_{.j} - m_{..})$, and the z_{ij} are random variables.[9]

It should be noted that *Assumption A* in conjunction with *Assumption B* implies that the mean values of the $(m_{i.} - m_{..})$, of the $(m_{.j} - m_{..})$ and of the z_{ij}, are all zero.

Assumption C (Zero Correlations and Homogeneous Variances): The random variables $(m_{i.} - m_{..})$, $(m_{.j} - m_{..})$, and z_{ij} are distributed with variances σ_r^2, σ_c^2, and σ^2, respectively, and all covariances among them are zero.

By following a line of reasoning similar to that presented in detail in the preceding section for the case of Model I, it is clear that here, in the case of Model II, the principal function of *Assumption A* is to bring the problem within the province of mathematical statistics; of *Assumption B*, to give specific meaning to the concept of "components of variance"; and of *Assumption C*, to dispense with interactions and render each of the "components of variance" assignable to a

distinct "factor." It should be noted, however, that *independence* of the respective component deviations $(m_{i.} - m_{..}, m_{.j} - m_{..},$ and $z_{ij})$ of an x_{ij} from the general population mean $(m_{..})$ is not assumed—it is merely assumed that all covariances among them are zero, i.e., that they are *mutually* uncorrelated.

Collectively, *Assumptions A, B, and C* imply that

(12) $\sigma_{xij}^2 \equiv$ variance of a single observation
$$\equiv \varepsilon(x_{ij} - m_{..})^2 = \sigma^2 + \sigma_r^2 + \sigma_c^2$$

$\sigma_{xi.}^2 \equiv$ variance of a row-wise mean
$$\equiv \varepsilon(x_{i.} - m_{..})^2 = \sigma_r^2 + \frac{\sigma^2}{c}$$

$\sigma_{x.j}^2 \equiv$ variance of a column-wise mean
$$\equiv \varepsilon(x_{.j} - m_{..})^2 = \sigma_c^2 + \frac{\sigma^2}{r}$$

$\sigma_{x..}^2 \equiv$ variance of the general mean
$$\equiv \varepsilon(x_{..} - m_{..})^2 = \frac{\sigma_r^2}{r} + \frac{\sigma_c^2}{c} + \frac{\sigma^2}{rc}.$$

Whence the expected values of the several mean squares of the customary analysis-of-variance table are as shown in the last column of Table 4. In brief, when *Assumptions A, B, and C* are satisfied, the residual mean square is an unbiased estimate of the "residual" variance, σ^2; subtracting the residual mean square from the between-row-means mean square and dividing this difference by c, the number of columns, yields an unbiased estimate of the "row-factor" component of variance, σ_r^2. By a similar procedure an unbiased estimate of "column-factor" component of variance, σ_c^2, can be obtained. It may be noted in passing that *the "naive" estimate of the over-all variance of a single observation, furnished by the "total" mean square, is a biased estimate*, and becomes unbiased only asymptotically as both r and c increase indefinitely.

In summary, when *Assumptions A, B, and C*, or their analogs in more complex cases, are satisfied, the customary analysis-of-variance procedures yield unbiased estimates of the respective variance components. Details of the procedures appropriate to situations

[9] In the example considered above, $(m_{i.} - m_{..})$ represents the deviation of the long-run mean body temperature of the ith animal from the long-run mean body temperature of the species; similarly, $(m_{.j} - m_{..})$ is an adjustment for the jth day, assumed applicable to the body temperature of any animal from the species on that day. The z_{ij} are "catch-alls" and represent errors of measurement, etc.

Table 4. Analysis of variance (additive case, row- and column-factors *random*).

Variation	Degree of freedom	Sum of squares	Mean square	Expected mean square
Between row (animal) means	$r - 1$	$s(x_{i.} - x_{..})^2$	$\dfrac{s(x_{i.} - x_{..})^2}{(r - 1)}$	$\sigma^2 + c\sigma_r^2$
Between column (day) means	$c - 1$	$s(x_{.j} - x_{..})^2$	$\dfrac{s(x_{.j} - x_{..})^2}{(c - 1)}$	$\sigma^2 + r\sigma_c^2$
Residual	$(r - 1)(c - 1)$	$s(x_{ij} - x_{i.} - x_{.j} + x_{..})^2$	$\dfrac{s(x_{ij} - x_{i.} - x_{.j} + x_{..})^2}{(r - 1)(c - 1)}$	σ^2
Total	$rc - 1$	$s(x_{ij} - x_{..})^2$	$\dfrac{s(x_{ij} - x_{..})^2}{(rc - 1)}$	$\sigma^2 + \dfrac{c(r - 1)}{rc - 1}\sigma_r^2 + \dfrac{r(c - 1)}{rc - 1}\sigma_c^2$

differing in various ways from the situation considered here will be found in the papers by S. Lee Crump and by H. E. Daniels cited in the list of references, and in the additional references that they cite.

Whereas *Assumptions A, B, and C,* or their analogs in more complex cases, are necessary[10] and sufficient for the validity of analysis-of-variance procedures for *unbiased estimation* of *components of variance*, it is not possible to conduct exact tests of significance with respect to these components of variance, nor to derive exact confidence limits for them or their ratios, unless the joint distribution of the several *deviations* in relation (11) is specified. Therefore, we shall make

Assumption D: The deviations $(m_{i.} - m_{..})$, $(m_{.j} - m_{..})$, and z_{ij}, $(i = 1, 2, \ldots, r;$ $j = 1, 2, \ldots, c)$, are all normally distributed.

When *Assumption D* is satisfied, *Assumptions A* and *B* are partially redundant, and serve principally to define the "compositions" of the random variables x_{ij}, $(i = 1, 2, \ldots, r; j = 1, 2, \ldots, c)$. Furthermore, zero covariances, as postulated in *Assumption C,* taken in conjunction with normality, postulated in *Assumption D,* imply mutual independence of the deviations $(m_{i.} - m_{..})$, $(m_{.j} - m_{..})$, and z_{ij}, and thence of the x_{ij} with

respect to each other. So, once again, independence gets in by the back door.

When *Assumptions A–D* are all satisfied, then all of the standard analysis-of-variance procedures for estimating, and testing to determine whether to infer the existence of, *components of variance* are strictly valid. These are based on the fact that these assumptions are sufficient to insure that

(a) The quotient (Between-row-means *sum* of squares)/$(\sigma^2 + c\sigma_r^2)$ will be distributed as χ^2 for $(r - 1)$ degrees of freedom,

(b) The quotient (Between-column-means *sum of squares*)/$(\sigma^2 + r\sigma_c^2)$ will be distributed as χ^2 for $(c - 1)$ degrees of freedom,

(c) The quotient (Residual *sum of squares*)/σ^2 will be distributed as χ^2 for $(r - 1)(c - 1)$ degrees of freedom,

(d) The "quotients" referred to in (a), (b), and (c) will be independent in the probability sense, so that

(e) The quantity

$$\left[\frac{\text{(Between-row-means } mean \ square)}{\sigma^2 + c\sigma_r^2}\right] \Bigg/$$

$$\left[\frac{\text{Residual } mean \ square}{\sigma^2}\right]$$

[10] See footnote 2.

will be distributed in Snedecor's F-distribution for $n_1 = (r - 1)$ and $n_2 = (r - 1)(c - 1)$ degrees of freedom, and

(f) The quantity

$$\left[\frac{\text{Between-column-means } mean \ square}{\sigma^2 + r\sigma_c^2} \right] \Bigg/ \left[\frac{\text{Residual } mean \ square}{\sigma^2} \right]$$

will be distributed according to F for $n_1 = (c - 1)$ and $n_2 = (r - 1)(c - 1)$ degrees of freedom.

Thus (c), which obtains also in the case of Model I when *Assumptions 1–4* are satisfied, is the basis of exact tests of hypotheses regarding the value of σ^2, and of the derivation of exact confidence limits for the value of σ^2. Similarly (e) is the basis of exact tests of hypotheses regarding the value of σ_r^2/σ^2, e.g., that $\sigma_r^2 = 0$, and of the derivation of exact confidence limits for σ_r^2/σ^2. An analogous statement holds for (f) in relation to σ_c^2/σ^2.[11]

Unfortunately, aside from testing the hypothesis that $\sigma_r^2 = 0$ or that $\sigma_c^2 = 0$, it is not possible to conduct exact tests of hypotheses regarding the absolute values of σ_r^2 and σ_c^2, nor is it possible to derive exact confidence limits for their absolute values.

5. *Which Model—Model I or Model II?* In practical work a question that often arises is: which model is appropriate in the present instance—Model I or Model II? Basically, of course, the answer is clear as soon as a decision is reached on whether the parameters of interest specify *fixed relations*, or *components of random variation*. The answer

depends in part, however, upon how the observations were obtained, on the extent to which the experimental procedure employed sampled the respective variables at random. This generally provides the clue. For instance, when an experimenter selects two or more treatments, or two or more varieties, for testing, he rarely, if ever, draws them at random from a population of possible treatments or varieties; he selects those that he believes are most promising. Accordingly Model I is generally appropriate where treatment, or variety comparisons are involved. On the other hand, when an experimenter selects a sample of animals from a herd or a species, for a study of the effects of various treatments, he can insure that they are a random sample from the herd, by introducing randomization into the sampling procedure, for example, by using a table of random numbers. But he may consider such a sample to be a random sample from the species, only by making the assumption that the herd itself is a random sample from the species. In such a case, if several herds (from the same species) are involved, Model II would clearly be appropriate with respect to the variation among the animals from each of the respective herds, and might be appropriate with respect to the variation of the herds from one another.

The most difficult decisions are usually associated with *places* and *times*: Are the *fields* on which the tests were conducted a random sample of the county, or of the state, etc.? Are the *years* in which the tests were conducted a random sample of years?

When a particular experiment is being planned, or when the results are in and are being interpreted, the following parallel sets of questions serve to focus attention on the pertinent issues, and have been found helpful in answering the basic question of random versus fixed effects:

(1) Are the conclusions to be confined to the things actually studied (the animals, or the plots); to the immediate sources

of these things (the herds, or the fields); or expanded to apply to more general populations (the species, or the farm-land of the state)?

(2) In complete repetitions of the experiment would the same things be studied again (the same animals, or the same plots); would new samples be drawn from the identical sources (new samples of animals from the same herds, or new experimental arrangements on the same fields); or would new samples be drawn from the more general populations (new samples of animals from new herds, or new experimental arrangements on new fields)?

It is hoped that these queries will not only aid in reducing the reader's "headaches," but will lead him to the correct decisions.

Finally, it needs to be said—as the reader will no doubt discover for himself, when he considers some specific sets of data or some proposed experiments in the light of the above queries—that real-life investigations rarely fall entirely within the domain of Model I, or entirely within the domain of Model II, unless they are planned and conducted so as to achieve one or the other of these objectives, and then they may not be realistic. In consequence, some of the mean squares of the analysis-of-variance tables may be unbiased estimators of linear combinations of variance components; and others, of linear combinations of variance components *and* "mean squares" of *fixed deviations*. H. E. Daniels, in the paper cited in the list of references, has proposed a method of interpreting analysis-of-variance tables of this sort. His method consists essentially of looking at such an analysis-of-variance table through Model-II spectacles, and interpreting the "mean squares" of *fixed deviations* as variance components also. While this approach may be fruitful in situations of the type to which he has applied his method, it cannot be regarded as a general solution since the objectives of problems of Class I and problems of Class II are in general quite distinct. More general methods need to be devised for interpreting "mixed" analysis-of-variance tables, parcularly in regard to tests of significance for individual factors.

References

Baines, A. H. J. *On Economical Design of Statistical Experiments.* (British) Ministry of Supply, Advisory Service on Statistical Method and Quality Control, Technical Report, Series R, No. QC/R/15, July 15, 1944.

Cramér, H. *Mathematical Methods of Statistics.* Princeton University Press, Princeton, New Jersey. 1946.

Crump, S. Lee. "The Estimation of Variance Components in Analysis of Variance," *Biometrics Bulletin*, vol. 2 (1946), pp. 7–11.

Daniels, H. E. "The Estimation of Components of Variance," *Supplement to the Journal of the Royal Statistical Society*, vol. 6 (1939), pp. 186–197.

David, F. N., and Neyman, J. "Extension of the Markoff Theorem on Least Squares," *Statistical Research Memoirs*, vol. 2 (1938), pp. 105–116.

Fisher, R. A. *Statistical Methods for Research Workers*, 1st and later editions. Oliver & Boyd, Ltd., London and Edinburgh, 1925–1944.

Goulden, C. H. *Methods of Statistical Analysis.* John Wiley and Sons, New York. 1939.

Irwin, J. O. "Mathematical Theorems Involved in the Analysis of Variance," *Journal of the Royal Statistical Society*, vol. 94 (1931), pp. 285–300.

Irwin, J. O. "On the Independence of the Constituent Items in the Analysis of Variance," *Supplement to the Journal of the Royal Statistical Society*, vol. 1 (1934), pp. 236–251.

Kendall, M. G. *The Advanced Theory of Statistics*, Volume II. Charles Griffin & Co., Ltd., London. 1946.

Kolodziejczyk, S. "On an Important Class of Statistical Hypotheses," *Biometrika*, vol. 27 (1935), pp. 161–190.

Lehmer, Emma. "Inverse Tables of Probabilities of Errors of the Second Kind," *Annals of Mathematical Statistics*, vol. 15 (1944), pp. 388–398.

Snedecor, G. W. *Calculation and Interpretation of Analysis of Variance and Covariance.* The Collegiate Press, Inc., Ames, Iowa. 1934.

Snedecor, G. W. *Statistical Methods: Applied to Experiments in Agriculture and Biology*, 4th Edition. The Collegiate Press, Inc., Ames, Iowa. 1946.

Statistical Research Group, Columbia University. *Selected Techniques of Statistical Analysis: For Scientific and Industrial Research and Production and Management Engineering.* McGraw-Hill Book Company, Inc., New York. (In press.)

Tang, P. C. "The Power Function of the Analysis of Variance Tests with Tables and Illustrations of Their Use," *Statistical Research Memoirs*, vol. 2 (1938), pp. 126–149.

Tippett, L. H. C. *The Methods of Statistics*, 2nd Edition. Williams and Norgate, Ltd., London. 1937.

Wilks, S. S. *Mathematical Statistics.* Princeton University Press, Princeton, New Jersey. 1943.

Comments: The student's first course in statistics introduces him to only the most elementary analysis of variance (ANOVA) designs, designs that are rarely used in behavioral research. A non-mathematical introduction to the much broader spectrum of ANOVA designs used by experiments is provided in the following article by the editor of this volume. I describe the principal advantages and limitations of the designs as well as the kinds of research situations for which they are appropriate.

The article also outlines my ANOVA design classification system, which is based on three simple experimental designs called building blocks. These building blocks, which can be used to construct even the most complex designs, provide the basis for my ANOVA design nomenclature.

7.2 Classification of ANOVA Designs

Roger E. Kirk

The variety of analysis of variance (ANOVA) designs used in research today is so extensive that selection of an appropriate design can be one of the most puzzling phases of a research project. Almost as puzzling for readers of research reports is the question of which design has been used. The problems facing the experimenter and reader are magnified because (1) there is no standard nomenclature for ANOVA designs, some of them having as many as five different names, and (2) design classification systems were not developed for the purpose of explicating structural relationships among the various ANOVA designs.[1] In response to these needs this paper presents a nomenclature and a classification system for ANOVA designs. The conceptual structure of the nomenclature and classification system is provided by three simple ANOVA designs. These designs—the completely randomized, the randomized block, and the Latin square—are the building blocks with which all complex ANOVA designs can be constructed. Hopefully this conceptual framework will aid the reader in understanding the more complex ANOVA designs that are so widely used in the behavioral sciences.

This article provides a nonmathematical introduction to those ANOVA designs that have the greatest potential usefulness in the behavioral sciences. No attempt is made to describe the computational procedures or statistical assumptions associated with these designs; for this information the reader is referred to standard reference books by Cochran and Cox (1957), Davies (1956), Edwards (1968), Federer (1955), Kirk (1968), Myers (1966), and Winer (1962).

[1] A number of design classification systems have been proposed. See, for example, Cox (1943); Doxtator et al. (1942); and Federer (1955, pp. 6–12). Miniature nomenclatures for a limited number of ANOVA designs have been presented by Lindquist (1953) and Winer (1962).

Three Building-Block Designs

The three basic building-block designs are appropriate for experiments that have one treatment with $k \geq 2$ treatment levels. The terms "treatment" and "independent variable" will be used interchangeably.[2] In the discussion that follows, the capital letter B will be used to identify a treatment. Specific levels of treatment B are designated by the lower-case letter b and a number subscript—for example, b_1, b_2, ..., b_k. A particular but unspecified treatment level is designated by b_j, where j ranges over the values $j = 1, ..., k$. We will let X_i stand for a score and $\overline{X}_j$ stand for a treatment mean, where the subscripts i and j designate a particular subject and mean, respectively.

Throughout this article it will be assumed that all treatment levels about which an experimenter wants to make inferences are included in the experiment. Under this assumption, conclusions drawn from the experiment apply only to the actual k treatment levels in the experiment. If, however, the k treatment levels in an experiment are a random sample from a population of K levels, the conclusions can be generalized to the population of treatment levels. It is important to distinguish between these two situations because the sampling procedure determines (1) the kinds of generalizations about the treatment that can be made and (2) the composition of the F test statistic for evaluating the significance of differences among the treatment levels. For a discussion of these points the reader is referred to Cornfield and Tukey (1956), Eisenhart (1947),[3] Millman and Glass (1967), and Schultz (1955).

Completely Randomized Design. The simplest ANOVA design from the standpoint of assignment of subjects (experimental units) to treatment levels and statistical analysis is a completely randomized design. This design is appropriate for experiments that meet, in addition to the general assumptions of the analysis of variance model,[4] the following conditions:

1. One treatment with $k \geq 2$ treatment levels.
2. Subjects randomly assigned to the treatment levels so that each subject is observed under only one level. If N subjects are available to participate in the experiment, it is customary to assign them to the k treatment levels so that there is an equal number, n, in each level.[5] If N, the total number of subjects, is not a multiple of k, the subjects should be assigned so that the n's are approximately the same.

The designation for a completely randomized design with k treatment levels is a *type CR-k design.* For example, an experiment with three treatment levels is called a type CR-3 design. This nomenclature indicates the number of treatment levels in an experiment as well as the type of design.

Consider a hypothetical experiment to evaluate three procedures for helping cigarette smokers break the smoking habit. The three levels of the independent variable and their designations are: behavioral therapy $= b_1$, hypnosis $= b_2$, and drug therapy $= b_3$. The choice of a dependent variable may be based on theoretical considerations, although in many experiments it is dictated by such practical considerations concerning the proposed measure as its sensitivity to the treatment, reliability, distribution, or ease of collection. In the present example the dependent variable could be the number of subjects who stopped smoking for, say, six months, or the change in their cigarette consumption, or any one of a number of other measures. The ANOVA designs des-

---

[2] Some writers use the term "factor" in place of "treatment."

[3] Eisenhart's article is the first one in this chapter.

[4] These assumptions are described in detail by Churchill Eisenhart (1947) in the first article in this chapter.

[5] An exception to this general rule is made when each of several treatment levels is compared with a control condition. For a discussion of this exception see Cox (1958, pp. 178–179).

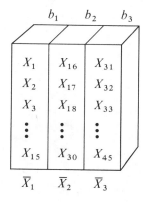

Fig. 1. Layout for type CR-3 design. The letter X_i designates a score for one of the $N = 45$ subjects who are randomly assigned to the three treatment levels. All conditions in the experiment except treatment level b_j should be held constant for the subjects if possible. Those conditions that cannot be held constant—for example, time of day for therapy—should be randomized for the subjects.

cribed in this paper are limited to the assessment of one dependent variable at a time.[6]

Assume that $N = 45$ cigarette smokers who are desirous of stopping are available to participate in the experiment. The subjects are randomly assigned to the $k = 3$ treatment levels with $n = 15$ in each level. The phrase "randomly assigned" is important. It cannot be emphasized too strongly that the validity of the experiment depends on the random assignment of subjects to the treatment levels. This procedure ensures that idiosyncratic characteristics of the subjects will be randomly distributed over the three treatment levels and thus will not selectively bias the outcome of the experiment.[7]

---

[6] If it is necessary to evaluate two or more dependent variables simultaneously, a multivariate analysis of variance design can be used. These designs are beyond the scope of this article; the interested reader is referred to Anderson (1958), Cooley and Lohnes (1962), Fryer (1966), Morrison (1967), and Rao (1952).

[7] In many experiments, factors beyond the investigator's control preclude the random assignment of subjects to treatment levels or the control of important nuisance variables. Campbell and Stanley (1966) refer to such experiments as "quasi-experimental designs." Campbell's paper in Chapter 6 examines potential sources of bias inherent in these designs.

A block diagram of the type CR-3 design for this experiment is shown in Figure 1. A schematic partitioning of the total variation in the experiment is shown in Figure 2. According to Figure 2 the total variation, SS_{total}, can be partitioned into two parts as follows:

1. Sum of squares within groups (SS_{WG}): This variation reflects differences in the dependent measure (X_i) for subjects who receive the same treatment level. It provides an estimate of experimental error (σ_ε^2), those effects *not* attributable to the treatment. The expected value of the within-groups mean square (MS_{WG}) is

$$E(MS_{WG}) = E\left(\frac{SS_{WG}}{kn - k}\right) = \sigma_\varepsilon^2,$$

where $kn - k$ is the degrees of freedom (df) for SS_{WG}.

2. Sum of squares between groups (SS_{BG}): This variation reflects differences among the three treatment means, $\overline{X}_1$, $\overline{X}_2$, and $\overline{X}_3$. $SS_{BG}/(k - 1)$ provides an estimate of treatment effects (β_j) and experimental error (σ_ε^2). If the only treatment levels of interest to the experimenter are those in the experiment, the expected value of the between-groups mean square (MS_{BG}) is

$$E(MS_{BG}) = E\left(\frac{SS_{BG}}{k - 1}\right)$$

$$= \sigma_\varepsilon^2 + n \sum_{j=1}^{k} \beta_j^2/(k - 1),$$

where $k - 1$ is the degrees of freedom (df) for SS_{BG}. The F test statistic and the expected values of the mean squares are, respectively,

$$F = \frac{MS_{BG}}{MS_{WG}}$$

and

$$\frac{E(MS_{BG})}{E(MS_{WG})} = \frac{\sigma_\varepsilon^2 + n \sum_{j=1}^{k} \beta_j^2/(k - 1)}{\sigma_\varepsilon^2}$$

$$\rightarrow \frac{\left(\begin{array}{c}\text{experimental} \\ \text{error}\end{array}\right) + \left(\begin{array}{c}\text{treatment} \\ \text{effects}\end{array}\right)}{\text{experimental error}}.$$

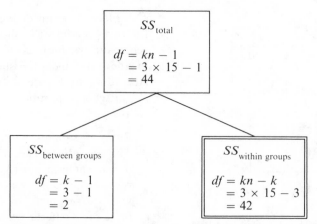

Fig. 2. Schematic partition of total sum of squares (SS_{total}) and degrees of freedom (df) for a type CR-3 design with $n = 15$ subjects randomly assigned to each of $k = 3$ treatment levels. A square within a square identifies the variation used to estimate experimental error.

The F statistic is used to test the null hypothesis that all of the treatment effects, β_j, are equal to zero; that is,

$$H_0 : \beta_j = 0 \text{ for all } j,$$

where β_j is equal to the mean of the jth population minus the grand mean of the $j = 1, \ldots, k$ populations. This hypothesis can be restated in an equivalent but more familiar form as

$$H_0 : \mu_1 = \mu_2 = \cdots = \mu_j = \cdots = \mu_k,$$

where μ_j stands for a population mean. If the obtained F test statistic falls in the upper, say, .05 region of the sampling distribution of F and if we have controlled all nuisance variables, we can conclude either that an improbable event has occurred or that the three population means are not equal and hence the null hypothesis should be rejected. If the latter decision is made, multiple comparison procedures can be used to determine which comparison(s) among population means is not equal to zero.[8]

In this section the nomenclature and main features of a completely randomized design have been described in some detail. With this

as a background let us turn to a somewhat briefer examination of the other two building-block designs.

Randomized Block Design. In behavioral research the experimental units are often people whose aptitudes and past experiences differ markedly. Therefore differences among treatment-level means will reflect not only variation attributable to the different treatment levels but also variation due to individual differences that existed prior to the experiment. Some variation due to individual differences is inevitable, but it is often possible to isolate or partition out a portion of this variation so that it does not appear in estimates of treatment effects and experimental error. One ANOVA design for accomplishing this is a randomized block design. This design, designated by the letters RB-k, is appropriate for experiments that meet, in addition to the general assumptions of the analysis of variance model, the following conditions:

1. The experiment contains one treatment with $k \geq 2$ treatment levels and one nuisance variable (undesired source of variation) with n levels.
2. Subjects are assigned to n blocks (as described below) so that variability within each block is less than the variability

[8] Multiple comparison procedures are discussed in Chapter 8.

among blocks. Each block must contain the same number of experimental units.

3. If each block consists of k matched subjects, the k treatment levels are randomly assigned to the matched subjects within each block. In this case the design requires n sets of k matched subjects, or a total of kn subjects.

4. If each block consists of a single subject who is observed under each of the k treatment levels, the order of presentation of the k levels must be randomized independently for each subject. For this case n subjects are each observed k times.

Variation associated with a nuisance variable is isolated in a type RB-k design by means of a *blocking technique* whereby blocks are formed from subjects who are homogeneous with respect to the nuisance variable. Homogeneity within each block can be achieved through the use of (1) litter mates or subjects with similar heredities, (2) subjects matched on a relevant variable, or (3) observation of each subject under all k treatment levels. If blocks are composed of sets of litter mates, it can be assumed that homogeneity is achieved with respect to genetic characteristics and their behavior is likely to be more homogeneous than the behavior of subjects having different heredities. The second procedure, assigning sets of matched subjects to each block, is an effective means of achieving within-block homogeneity if the variable used to match the subjects correlates positively with the dependent variable. Use of the third procedure, observing each subject under all treatment levels, requires a note of caution. The effects of each treatment level must have dissipated before the subject is observed under a different level. For this reason the kinds of experiments that can be used with repeated measures are limited, although exceptions to this restriction are made when carry-over effects such as learning or fatigue are the principal interest of the investigator.

Let us reconsider the smoking example cited previously. It is reasonable to assume that difficulty in breaking the smoking habit is related to the amount a subject smokes. In this case amount of smoking is an example of a nuisance variable, an undesired source of variation that can affect the dependent variable. Its effects can be experimentally isolated by using the blocking technique, but this requires a knowledge of each subject's cigarette consumption so that subjects with similar smoking habits can be assigned to the same block. In this experiment, one block might contain k subjects who smoke less than one-half pack per day, a second block might contain subjects who smoke one-half to one pack per day, a third block one to one and one-half packs per day, and so on. The subjects within each block should be randomly assigned to the treatment levels.

The layout for a randomized block design is shown in Figure 3 and the partition of the total variation is shown in Figure 5. Figure 5 provides a comparison of the partitioning appropriate for all three building-block designs.

The reader may have wondered how an experimenter benefits by using a type RB-k design in view of the extra work required to match subjects. The answer is simple: use of the blocking technique in a randomized block design usually results in a smaller error term

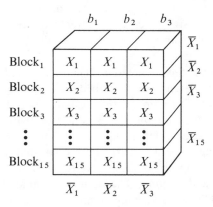

Fig. 3. Layout for type RB-3 design. This design requires $n = 15$ blocks of $k = 3$ homogeneous subjects who are randomly assigned to the treatment levels within each block. As indicated previously, each of the n blocks may, alternatively, contain a single subject who is observed k times. In this case the order of presentation of the k treatment levels must be randomized independently for each subject.

(MS_{residual}) than the error term for a completely randomized design $(MS_{\text{within groups}})$. Thus the experimenter is compensated for the additional experimental effort with a more powerful test of treatment effects. This can be seen from an examination of Figure 5. If MS_{blocks} accounts for an appreciable portion of the total variation, the ratio

$$F = MS_{\text{treatment}}/MS_{\text{residual}}$$

will be larger than

$$F = MS_{\text{between groups}}/MS_{\text{within groups}} .$$

Of course the reduction in MS_{residual} that results from using the blocking technique must more than compensate for the loss of degrees of freedom (by $n - 1$) for this error term.

The randomized block design illustrates an important design principle. If a source of variability can be separated from treatment effects and experimental error, the sensitivity or power of the resulting experiment may be increased. All sources of variation that cannot be estimated remain a part of the uncontrolled sources of variability in the experiment and thus are automatically a part of the experimental error. This helps to explain why investigators generally prefer the more complex multitreatment ANOVA designs described in subsequent sections of this paper. Such designs permit greater experimental control of known sources of variability.

Latin Square Design. The last building-block design to be described is a Latin square design. The name of this design derives from an ancient puzzle that is concerned with the number of different ways Latin letters can be arranged in a square matrix so that each letter appears once and only once in each row and column. The design, designated by the letters LS-k, enables an experimenter to isolate variation attributable to *two* nuisance variables. The levels of one nuisance variable are assigned to rows and the levels of the other to columns of a Latin square; the k treatment levels are assigned to the k^2 cells of the square so that each level appears once in each row

and once in each column. The design is appropriate for experiments that meet, in addition to the general assumptions of the analysis-of-variance model, the following conditions:

1. One treatment with $k \geq 2$ levels and two nuisance variables.
2. Equal numbers of treatment levels and levels of the two nuisance variables.
3. Absence of interactions among the two nuisance variables and the treatment. If this assumption is not fulfilled, one or more tests of significance will be biased.
4. Random assignment of treatment levels to the cells of the square with the restriction that each treatment level must appear only once in a row and once in a column.

Let us return again to the cigarette-smoking example. The randomized block design permitted us to experimentally isolate variation associated with the nuisance variable of cigarette consumption. A Latin square design lets us isolate an additional nuisance variable—say, length of time that a person has smoked. Since we are interested in $k = 3$ kinds of smoking therapies, the design must also have three rows and three columns. Let us assign three levels of the first nuisance variable, cigarette consumption, to the rows of the square and three levels of the second nuisance variable, duration of the smoking habit, to the columns. The layout for this design is shown in Figure 4.

The Latin square shown in Figure 4 has one subject per cell, although it is desirable if k is small to have more than one subject per cell.[9] A comparison of the partitioning of the total sum of squares for this design and the two other building-block designs is shown in Figure 5. The F test statistic for evaluating treatment effects is given by

$$F = \frac{MS_{\text{treatment}}}{MS_{\text{residual}}} .$$

[9] Unless more than one observation per cèll is obtained, squares smaller than 5×5 are not practical because of the small number of degrees of freedom for experimental error.

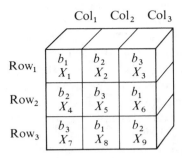

Col$_1$ Col$_2$ Col$_3$

	b_1 X_1	b_2 X_2	b_3 X_3
Row$_1$			
Row$_2$	b_2 X_4	b_3 X_5	b_1 X_6
Row$_3$	b_3 X_7	b_1 X_8	b_2 X_9

Fig. 4. Layout for type LS-3 design. The letter X_i designates a score for one of the $N = 9$ subjects who are assigned to the $k \times n = 9$ cells. Treatment levels (b_j) are randomly assigned to the cells of the square with the restriction that each level must appear only once in a row and once in a column.

The Latin square design is more powerful than either a completely randomized or a randomized block design if variation within rows is small relative to variation among rows and if variation within columns is small relative to variation among columns.

This design permits an investigator to isolate effects associated with two nuisance variables; that is its principle advantage. It has two major disadvantages: (1) the number of treatment levels, rows, and columns must be equal, a balance that may be difficult to achieve, and (2) the design is not appropriate for research situations in which interactions among the treatment, row, and column variables occur.

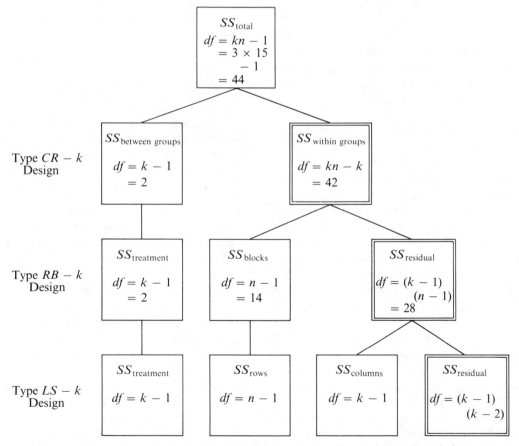

Fig. 5. Schematic partition of total sum of squares for the three building-block designs. A square within a square identifies the experimental error for testing treatment effects. Numerical values for the Latin square degrees of freedom are not given because the total number of observations in this design is only $3 \times 3 = 9$ and hence comparisons with the other designs are not meaningful.

Other Designs for Experiments with One Treatment

Three building-block designs have been described that provide the organizational framework for the classification system and nomenclature presented in this article. All of the other ANOVA designs to be described represent an extension of one or a combination of these designs.

Graeco-Latin Square and Hyper-Graeco-Latin Square Designs. A Graeco-Latin square design is constructed by superimposing one Latin square on a second Latin square orthogonal to the first. Two Latin squares are orthogonal if when they are superimposed every treatment level of one square occurs once and only once with every level of the other square. The building block for this design is a Latin square design. A Graeco-Latin square design, designated by the letters GLS-k, permits an experimenter to evaluate one treatment and at the same time experimentally isolate variation attributable to three nuisance variables.

This design has not proved very useful in the behavioral sciences for several reasons. First, the design is restricted to research situations in which three nuisance variables and a treatment each have the same number of levels. It is generally difficult to achieve this balance for four variables. Second, if interactions among the variables occur, one or more tests of significance are biased. Since interactions are common in behavioral research the usefulness of this design is limited.

A hyper-Graeco-Latin square design is constructed like a Graeco-Latin square design but combines more than two orthogonal Latin squares and is designated by the letters HGLS-k. Each orthogonal square that is added to a hyper-Graeco-Latin square design permits an experimenter to isolate an additional nuisance variable. Although this is an advantage, these designs have the same disadvantages as type GLS-k designs and hence are rarely used in behavioral research.

Incomplete Block Designs. The name "incomplete block design" refers to a large class of designs in which k treatment levels of a single-treatment experiment are assigned to blocks of size q, where $k > q$. Descriptions of most of the designs in this class are beyond the scope of this paper. Two of the potentially more useful designs for the behavioral sciences, a balanced incomplete block design and a Youden square design, will be briefly described.

A balanced incomplete block design, designated by the letters BIB-t, resembles a randomized block design in that it permits an experimenter to isolate variation attributable to one nuisance variable. The design is appropriate for experiments that meet, in addition to the general assumptions of the analysis of variance model, the following conditions:

1. The experiment contains one treatment with $k \geq 2$ levels. The k treatment levels are assigned to blocks of size q, where k is greater than q. The size of the blocks may be limited, for example, by the availability of subjects, equipment, facilities, or time.
2. The number of times any two treatment levels occur together within some block is equal to that of any other pair.

The layout for a type BIB-7 design is shown in Figure 6. This design enables an experimenter to utilize the blocking technique in evaluating $k = 7$ treatment levels even though he cannot have blocks of size seven, a requirement if each treatment level occurs in every block.

A balanced incomplete block design is most often used when an experimenter wants to evaluate a large number of treatment levels. In such situations the use of a randomized block design, the alternative design choice, may not be feasible because of the difficulty in securing a sufficient number of experimental units to form homogeneous blocks.

The statistical analysis of type BIB-k designs is, unfortunately, much more complex than that of type RB-k designs. As is so often the case in the design of experiments, an advantage—in this case small block size—is

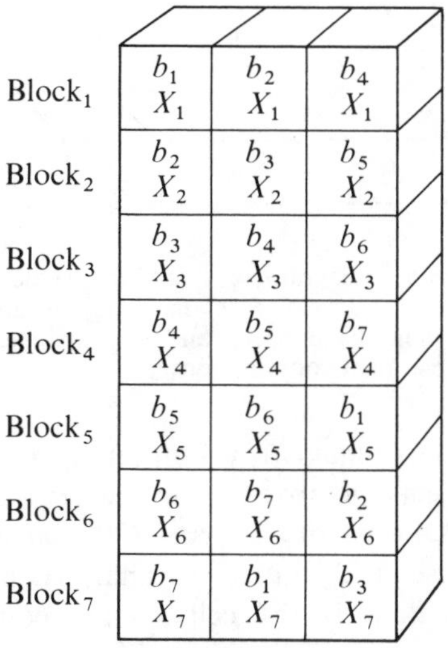

	b_1	b_2	b_4
Block$_1$	X_1	X_1	X_1
Block$_2$	b_2 / X_2	b_3 / X_2	b_5 / X_2
Block$_3$	b_3 / X_3	b_4 / X_3	b_6 / X_3
Block$_4$	b_4 / X_4	b_5 / X_4	b_7 / X_4
Block$_5$	b_5 / X_5	b_6 / X_5	b_1 / X_5
Block$_6$	b_6 / X_6	b_7 / X_6	b_2 / X_6
Block$_7$	b_7 / X_7	b_1 / X_7	b_3 / X_7

Fig. 6. Layout for type BIB-7 design. This design requires $n = 7$ blocks of $k = 3$ homogeneous subjects who are randomly assigned to the treatment levels (b_j) within each block. Alternatively each of the n blocks may contain a single subject who is observed k times. In this case the order of presentation of the k treatment levels must be randomized independently for each subject.

achieved at the price of complications in data analysis.

Another useful incomplete block design is a Youden square design, which combines features of balanced incomplete block designs and Latin square designs. It can be thought of as a special kind of incomplete Latin square that enables an experimenter to isolate variation attributable to two nuisance variables. The Youden square design shares the advantages and disadvantages of the balanced incomplete block design. That is, it enables an experimenter to utilize the blocking technique with blocks of size q, where $q < k$, but involves a relatively complex statistical analysis.

ANOVA Designs for Factorial Experiment

Factorial designs, the subject of this and subsequent sections, differ from the designs covered previously in that two or more treatments can be evaluated simultaneously.

Probably for this reason they are the most widely used designs in the behavioral sciences, a fact that can be readily verified by an examination of recent volumes of behavioral-science journals. Although there are many different kinds of factorial designs, they are all constructed from one or a combination of the three basic building blocks described previously.

It is appropriate at this point to describe the system used to designate treatments in factorial designs. Seven capital letters are used for this purpose, A through G. A particular but unspecified level of a treatment is designated by lower-case letters a through g and a lower-case subscript—for example, $a_i, b_j, \ldots, g_t$. Specific levels of a treatment are designated by number subscripts—for example, $a_1, a_2, \ldots$. In addition to designating treatments, it is also necessary to indicate the number of levels of each treatment. Seven lower-case letters listed under Range of Levels in Table 1 are used for this purpose. The complete treatment designation scheme is shown in Table 1.

Table 1. Treatment designation scheme for factorial designs.

Treatment	Particular level of treatment	Range of levels
A	a_i	$i = 1, \ldots, p$
B	b_j	$j = 1, \ldots, q$
C	c_k	$k = 1, \ldots, r$
D	d_l	$l = 1, \ldots, u$
E	e_o	$o = 1, \ldots, v$
F	f_h	$h = 1, \ldots, w$
G	g_t	$t = 1, \ldots, z$

The letter S is used in the notation scheme to represent a subject or a set of subjects. A specific subject or set of subjects is designated by the letters s_m, where $m = 1, \ldots, n$. The scheme shown in Table 1 uses 24 letters of the alphabet. The remaining two letters, X and Y, are used to designate the magnitude of individual scores. For example, X_{ijkm} represents the magnitude of an observation for subject m in treatment combination a_i, b_j, and c_k.

Completely Randomized Factorial Design. The simplest factorial design, from the standpoint of data analysis and assignment of subjects to treatment levels, is based on a completely randomized building block and hence is called a completely randomized factorial design. A two-treatment design is designated by the letters CRF-*pq* where *p* and *q* refer to the number of levels of treatments *A* and *B*, respectively. The letters CR designate the building block while the letter F indicates that it is a factorial design. A type CRF-2343 design, for example, has two levels of treatment *A*, three levels of *B*, four levels of *C*, and three levels of *D*. If any treatment has ten or more levels, commas are used to separate that level designation from the other designations, as in a type CRF-4,12,32 design. A type CRF-*pq* design is appropriate for experiments that meet, in addition to the assumptions of the completely randomized design, the following conditions:

1. Two or more treatments, with each treatment having two or more levels. If there are *p* levels of treatment *A* and *q* levels of treatment *B*, the experiment consists of *pq* treatment combinations.[10]
2. Random assignment of *n* subjects to the *pq* treatment combinations, with each subject receiving only one combination.

Consider an experiment to evaluate the effects of two treatments *A* and *B* on speed of reading. Let us assume that treatment *A* consists of two levels of room illumination: $a_1 = 5$ foot candles and $a_2 = 30$ foot candles. Treatment *B* consists of three levels of type size: $b_1 = 6$-point type, $b_2 = 12$-point type, and $b_3 = 18$-point type. The designation for this design is a type CRF-23 design, where "2" refers to the two levels of treatment *A* and "3" refers to the three levels of treatment *B*. A block diagram of this design is shown in Figure 7.

[10] When each level of one treatment occurs once with each level of another treatment and vice versa, the two treatments are said to be crossed. This produces *pq* treatment combinations.

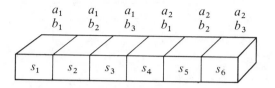

Fig. 7. Block diagram of type CRF-23 design. The letters $s_1, \ldots, s_6$ each represent a random sample of *n* subjects who have been randomly assigned to the $pq = 6$ treatment combinations.

It is customary to assign an equal number of subjects, *n*, to each of the *pq* treatment combinations of a type CRF-*pq* design; otherwise the statistical analysis may be complicated. If the cell *n*'s are equal, the design requires *pq* random samples of *n* subjects each, who are randomly assigned to the *pq* treatment combinations. It is desirable to have more than one observation in each of the *pq* cells, thereby permitting an investigator to compute a within-cell estimate of experimental error. If $n = 1$, an interaction must be used to estimate experimental error under the often tenuous assumption that the interaction is insignificant.

The partition of the total variation for this design is shown in Figure 9. *F* test statistics for evaluating the two treatments and *AB* interaction are given by $F = MS_A/MS_{\text{within cell}}$, $F = MS_B/MS_{\text{within cell}}$, and $F = MS_{AB}/MS_{\text{within cell}}$. In addition to providing for tests of two treatments, a factorial design enables an investigator to determine if the two treatments interact—information that is helpful in interpreting the results of research. If interaction is present it means that one treatment behaves differently under different levels of the other treatment.

Earlier we stated that a type CR-*k* design is the building block for a completely randomized factorial design. A careful examination of the assumptions associated with the mathematical model for a type CRF-*pq* design (Figure 7) and the partition of the total variation (Figure 9) confirms that this design is composed of two type CR-*k* designs in which $k = 2$ and $k = 3$. A design with four treatments—say a type CRF-2343 design—is composed of four type CR-k designs, where $k = 2, 3, 4,$ and 3.

The major advantages of factorial designs are:

1. All subjects are used in evaluating the effects of two or more treatments. The effects of each treatment are evaluated with the same precision as if the entire experiment had been devoted to that treatment alone. Factorial experiments thus permit efficient use of resources.
2. They enable an experimenter to evaluate interaction effects.

The major disadvantages of factorial designs are:

1. If numerous treatments are included in the experiment, the number of subjects required may be prohibitive.
2. A factorial design lacks simplicity in interpretation of results if interaction effects are present. This is not a criticism of the design but simply an acknowledgement of the fact that the interrelationships among variables in the behavioral sciences are often complex.
3. The use of a factorial design commits an investigator to a relatively large experiment. Small exploratory experiments may indicate much more promising lines of investigation than those originally envisioned. Relatively small experiments permit greater freedom in pursuit of serendipity.

Randomized Block Factorial Design. A randomized block factorial design utilizes the main features of the randomized block design, namely the blocking technique, in evaluating two or more treatments and associated interactions. If the design has, say, two treatments, the designation is a type RBF-pq design. This design permits an experimenter to isolate variation attributable to one nuisance variable but requires that each subject or set of matched subjects receive all possible pq treatment combinations in the experiment. A block diagram of this design is shown in Figure 8. As in a type RB-k design, each block may

consist of matched subjects or, alternatively, each block may consist of repeated measures on a single subject. In the latter case the order of administration of the treatment combinations is randomized independently for each subject.

A partition of the total variation for this design is shown in Figure 9. From an examination of Figure 9 it should be apparent that if block effects account for an appreciable portion of the total variation, a type RBF-pq design is more powerful than a type CRF-pq design. Thus the experimenter is usually rewarded for the additional experimental effort required to form homogeneous blocks by obtaining a smaller error term with which to evaluate treatment and interaction effects.

Factorial ANOVA Design with Treatment-Block Confounding

The advantages of experimental designs in which all treatment combinations are assigned to blocks of relatively homogeneous subjects were described in previous sections. Within-block homogeneity can be achieved by using litter mates, matched subjects, or, if the nature of the treatments permits, repeated measures on the same subject. The major problem in using the blocking technique is obtaining a sufficient number of homogeneous subjects to form the blocks. Even a relatively small design such as a type RBF-324 design requires $3 \times 2 \times 4 = 24$ subjects per block. This problem becomes more acute for larger experiments. The use of repeated observa-

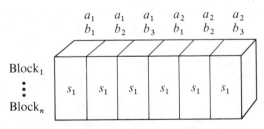

Fig. 8. Block diagram of type RBF-23 design. The letter s_1 represents n blocks of $pq = 6$ homogeneous subjects who are randomly assigned to the six treatment combinations within each block.

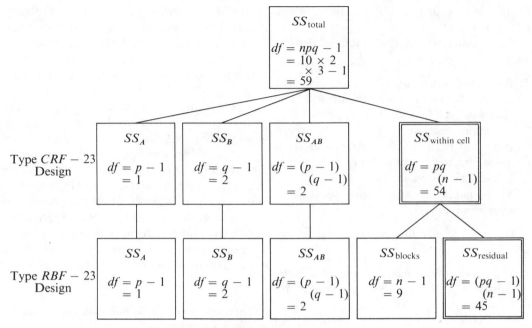

Fig. 9. Schematic partition of total sum of squares for two factorial designs. A square within a square identifies the error term for testing treatment and interaction effects. For purposes of illustration the number of subjects n assigned to each treatment combination of the type CRF-23 design is ten. The number of blocks n in the type RBF-23 design is also ten; hence both designs contain the same number of observations ($npq = 60$).

tions on the same subject is no solution, for there is a practical limit to the amount of participation that can be expected of a subject, and carry-over effects from one treatment combination to another can seriously bias the data.

The factorial design described in this section—a split-plot design—as well as those to be described later have the virtue of reducing the number of treatment combinations that must be assigned to blocks, thus requiring a smaller size block. The reduction in the number of treatment combinations assigned to blocks is accomplished by means of confounding, a procedure whereby treatments are assigned to subjects so that certain effects are indistinguishable from other effects. In a split-plot design, one or more treatments are confounded with between-block variation. Two other forms of confounding exist: interaction-block confounding and treatment-interaction confounding. Designs employing these forms of confounding are described in subsequent sections.

Split-Plot Design. A split-plot design resembles two building-block designs: a completely randomized design and a randomized block design. It is appropriate for experiments that meet, in addition to the general assumptions of the analysis of variance model, the following conditions:

1. The experiment contains two or more treatments with each treatment having two or more levels.
2. The number of treatment combinations exceeds the desired block size.
3. Treatment combinations are assigned to blocks so that only one level of, say, treatment A appears in any one block but all levels of treatment B appear in each of the blocks.

The latter point requires some clarification. Consider the split-plot design shown in Figure 10. The treatment combinations in the first block are ab_{11}, ab_{12}, and ab_{13}. Note that only one level of treatment A appears in this block, but all levels of treatment B appear. A type RBF-23 design would require blocks of size six; the corresponding split-plot design reduces the block size to three.

In a split-plot design, a treatment is referred to as a *between-block* treatment if only one level appears in a block; if all levels appear within a block it is referred to as a *within-block* treatment. The designation for a split-plot design in which treatment A is a between-block treatment and B is a within-block treatment is a type SPF-$p \cdot q$ design. This design nomenclature departs from that used previously in two ways. First the letters SPF appear in the designation instead of the letters of the two building blocks, CR-k and RB-k designs. This is appropriate because the name "split-plot" is a familiar designation for this design, having been used in agricultural research for many years. Second, a dot is used in the nomenclature to distinguish between-block treatments from within-block treatments. The lower case letter(s) preceding the dot indicates the number of levels of a between-block treatment. The letter(s) following the dot indicates the number of levels of a within-block treatment. The pattern underlying this nomenclature can be seen from the following examples:

type SPF-$p \cdot q$ design	(A is a between-block treatment with p levels, B is a within-block treatment with q levels.)
type SPF-$pr \cdot q$ design	(A and C are between-block treatments, B is a within-block treatment.)
type SPF-$p \cdot qr$ design	(A is a between-block treatment, B and C are within-block treatments)
type SPF-$prv \cdot quw$ design	(A, C, and E are between-block treatments, B, D, and F are within-block treatments.)

It is useful to designate which treatments are between-block or within-block treatments because the power associated with tests of the former treatments is characteristically less than that of within-block treatments. In this respect a test of a between-block treatment resembles a test of $MS_{\text{between groups}}$ in a type CR-k design. A similar correspondence exists between a test of a within-block treatment and $MS_{\text{treatment}}$ in a type RB-k design. If an experimenter is interested in evaluating all treatments with equal precision, a design other than a split-plot design should be considered. The logical alternative is a randomized block factorial design or a randomized block confounded factorial design, which is described in the next section.

A partition of the total variation in a split-plot design is shown in Figure 11. Figure 11 illustrates a novel characteristic of this design: there are two error terms, one for evaluating between-block sources of variation and another for within-block variation. As indicated previously, the within-block tests of significance are generally more powerful than between-block tests.

Factorial ANOVA Designs with Interaction Block Confounding

In the previous section a split-plot design with treatment-block confounding was described. When an experimenter uses a split-plot design he sacrifices power in evaluating one or more treatments in order to achieve a small block size. The designs described in this section, confounded factorial designs, involve confounding of one or more interactions with

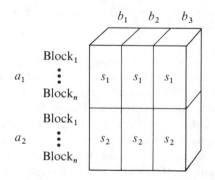

Fig. 10. Block diagram of type SPF-2·3 design. The letters s_1 and s_2 each represent a random sample of n blocks of $q = 3$ homogeneous subjects. The subjects within each block are randomly assigned to one of the q levels of treatment B.

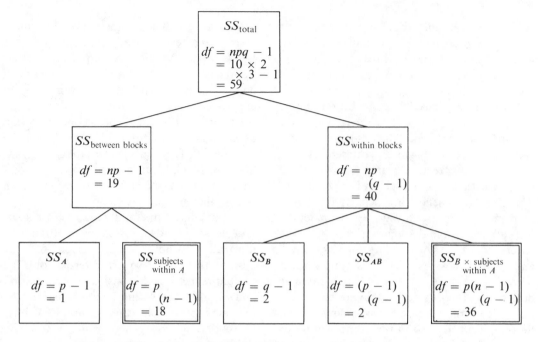

Fig. 11. Schematic partition of total sum of squares for type SPF-2.3 design with $n = 10$ blocks of subjects within each level of A. Squares within squares identify the two error terms for testing between-block and within-block effects.

between-block variation. This type of confounding has an important advantage over the confounding used in a split-plot design. If one of the interactions is known to be negligible, it can be sacrificed (confounded) in order to estimate all treatments and remaining interactions with equal precision. Thus an experimenter can achieve a reduction in block size without sacrificing power in evaluating treatments, which are generally of greater interest than interactions.

Confounded Factorial Designs. A confounded factorial design can be constructed using either randomized block or Latin square building blocks. The letters RBCF or LSCF are used to designate factorial designs in which one or more interactions are completely confounded with between-block variation. If one or more interactions are partially confounded with between-block variation, the design is designated by the letters RBPF. This design provides partial information with respect to the confounded interaction(s).

The assignment of treatment combinations to blocks so as to achieve the desired confounding in type RBCF and RBPF designs is relatively complex. Several schemes have been devised for this purpose (Yates, 1937; Kempthorne, 1952), the most general of which involves modular arithmetic. Confounding of an interaction with between-block variation in a type LSCF design poses no special problems.

Confounded factorial designs based on a randomized block design are restricted to experiments in which the number of levels of each treatment is a prime number. The statistical analysis of these designs is considerably reduced if each treatment has the same number of levels. If we let p stand for the number of levels of each treatment and k for the number of treatments, the designations for confounded factorial designs are RBCF-p^k, RBPF-p^k, and LSCF-p^k.

These designs are appropriate for experiments that meet, in addition to the general assumptions of the analysis of variance model, the following conditions:

1. Two or more treatments with each treatment having p levels where $p \geq 2$. Designs based on a type RB-k building-block design are restricted to the case in which p is a prime number.
2. Treatment combinations assigned to blocks so as to confound between-block variation with one or more interactions. Thus an experimenter must be willing to sacrifice information about an interaction in order to achieve a small block size. The interaction(s) that is confounded with between-block variation should be one believed to be insignificant.
3. Treatment combinations capable of being administered in every possible sequence. This requirement precludes the use, for example, of a treatment whose levels consist of successive periods of time.

A block diagram of a type RBCF-3^2 design is shown in Figure 12. Although confounded factorial designs have the advantage of permitting an experimenter to evaluate all treatments with equal precision, at the same time reducing the number of treatment combinations assigned to each block and hence block size, their layout and statistical analysis are much more complex than that of designs described previously.

Factorial ANOVA Designs with Treatment-Interaction Confounding

One of the characteristics of the factorial designs described previously is that all treatment combinations appear in the experiment.[11] Fractional factorial designs described in this section differ in this respect. These designs include only a fraction—for example, $\frac{1}{3}$—of the treatment combinations of a complete factorial design. Consider a type CRF-333 design which has 27 treatment combinations. This design requires a minimum of 54 subjects

---

[11] Even though not all treatment combinations appear within each block in the split-plot design and confounded factorial designs, the combinations are assigned to blocks so as to ensure that they occur at least once in the experiment.

in order to compute a within-cell error term. By using a $\frac{1}{3}$ fractional replication, the number of treatment combinations can be reduced to 9 and the number of subjects to 18.

Fractional Factorial Designs. Fractional factorial designs represent a fairly recent development in the evolution of experimental designs. The theory of fractional replication was developed for 2^k and 3^k designs by Finney (1945, 1946) and extended by Kempthorne (1947) to designs of the type p^k, where p is any prime number and k designates the number of treatments. Fractional factorial designs have found their greatest use in industrial research. Only limited application of these designs has been made in agricultural research, which historically has provided the impetus for the development of most designs in use today.

A fractional factorial design is appropriate for experiments that meet, in addition to the general assumptions of the analysis of variance model, the following conditions:

1. Two or more treatments, with each treatment having two or more levels. In actual practice a fractional factorial design is rarely used in experiments with less than four or five treatments.

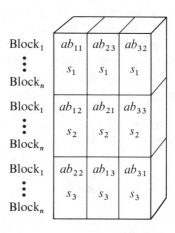

Fig. 12. Block diagram of type RBCF-3^2 design. The letters s_1, s_2, and s_3 represent three random samples of n sets of p homogeneous subjects. The subjects within each set are randomly assigned to the p treatment combinations within each block. In this example the AB interaction is completely confounded with between-block variation.

2. The number of treatment combinations so large as to call for a prohibitively large experiment. Fractional factorial designs are most useful for exploratory research and for situations that permit follow-up experiments to be performed. Thus a large number of treatments can be investigated efficiently in an initial experiment, with subsequent experiments designed to focus on the most promising lines of investigation or to clarify the interpretations of the original analysis.
3. Equal number of levels of each treatment if possible.
4. Some *a priori* reason for believing that most of the higher-order interactions are insignificant.

Fractional factorial designs can be constructed using completely randomized, randomized block, and Latin square building-block designs. The designations for the three designs are CRFF-p^k, RBFF-p^k, and LSFF-p^k, where p stands for the number of levels of each treatment and k stands for the number of treatments. Type CRFF-p^k and RBFF-p^k designs are restricted to experiments in which p is a prime number. A block diagram of a type CRFF-2^4 design is shown in Figure 13.

Fractional factorial designs have much in common with confounded factorial designs. The latter designs through the technique of confounding achieve a reduction in the number of treatment combinations that must be included within a block. A fractional factorial design uses confounding to reduce the number of treatment combinations in the experiment. As is always the case when confounding is used, the reduction is obtained at a price. There is considerable ambiguity in interpreting the outcome of a fractional factorial experiment, since treatments are confounded with interactions. For example, a significant mean square might be attributed to the effects of treatment A or to a $BCDE$ interaction. The two or more designations that can be given to the same mean square are called *aliases*. Treatments are customarily confounded with higher-order interactions that are assumed to equal zero. This helps to minimize but does not eliminate ambiguity in interpreting the outcome of an experiment. It is often possible to carry out a follow-up experiment to help clarify the interpretation of the original experiment, but this requires careful advanced planning.

Although the number of treatment combinations that must be included in an experiment is reduced by using a fractional factorial design, such designs have several serious disadvantages:

1. The interpretation of tests of significance is complicated by the fact that treatments are confounded with interactions.
2. These designs require the assumption of negligible interactions among some or all treatments. This assumption is often unrealistic in the behavioral sciences.
3. The layout and computation procedures for these designs are relatively complex.

ANOVA Designs with Nested Treatments

Factorial experiments are distinguished by the use of crossed treatments in which all possible treatment combinations appear in the experiment. The only exceptions are fractional factorial designs, which include some fraction of the possible treatment combinations.

A different arrangement in multitreatment experiments is possible and involves the *nesting* of one treatment within another treatment. A nested treatment is one in which each level of, say, treatment B occurs with only one level of treatment A instead of occurring with all levels of treatment A. In such an

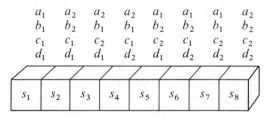

Fig. 13. Block diagram of type CRFF-2^4 design. The letters $s_1, \ldots, s_8$ each represent a random sample of n subjects who have been randomly assigned to the eight treatment combinations.

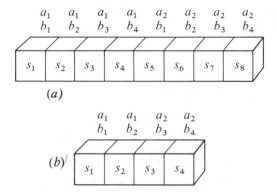

Fig. 14. Comparison of crossed and nested treatments. (a) Type CRF-24 design with crossed treatments. (b) Type CRH-2(4) design with treatment B nested within A.

arrangement, treatment B is said to be nested within treatment A. The difference between crossed and nested treatments is illustrated in Figure 14. Designs in which one or more treatments are nested are called hierarchal designs.

The building block for the design shown in Figure 14b is a completely randomized design —hence the designation type CRH-$p(q)$. The letter in parentheses indicates that treatment B is nested within the p levels of treatment A. Although B is called a nested treatment, in most experiments it is more aptly referred to as a nuisance variable. An example may help to clarify the distinction between crossed and nested treatments and also to indicate the general character of nested treatments. Assume that two types of programmed instructional materials (treatment A) are to be evaluated and that four sixth-grade classes (treatment B) are available for the experiment. We will assume that all children in a particular class must use the same type of programmed material. If under these conditions each type of programmed material is randomly assigned to two classes, we say the classes are nested within treatment A. This hierarchal experiment corresponds to the design illustrated in Figure 14b. Although the variable of classes is of little interest, it must be included in the design because it represents a source of variation that is inherent in the experimental situation.

If an experiment involves three treatments in which B is nested within A and C is nested within B, the design is described as a type CRH-$p(q)(r)$ design. A block diagram for this design is shown in Figure 15.

Experiments may involve three or more treatments, some of which are crossed while others are nested. Such a design, in which B is nested within A but C is crossed with both B and A, is shown in Figure 16. The designation for this design is a completely randomized partial hierarchal design, type CRPH-$p(q)r$ design.

Analysis of Covariance Designs

The designs described in the previous sections have used experimental control to reduce variability due to experimental error and to obtain unbiased estimates of treatment effects. Experimental control can take various forms such as the random assignment of subjects to treatment levels, the stratification of subjects into homogeneous blocks (blocking technique), and the refinement of techniques for measuring a dependent variable. An

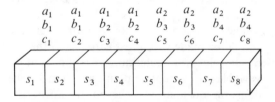

Fig. 15. Block diagram of type CRH-2(4)(8) design. The letters $s_1, \ldots, s_8$ each represent a random sample of n subjects who have been randomly assigned to the eight treatment combinations. Treatment B is nested within A and treatment C is nested within B.

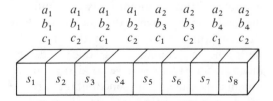

Fig. 16. Block diagram of type CRPH-2(4)2 design. Treatment B is nested within A but treatment C is crossed with both treatments A and B.

alternative approach to reducing experimental error and obtaining unbiased estimates of treatment effects involves the use of statistical control. This approach enables an experimenter to remove potential sources of bias from an experiment, biases that are difficult or impossible to eliminate by experimental control. The statistical control described in this section combines features of regression analysis with analysis of variance and is called analysis of covariance. It involves measuring one or more concomitant variables in addition to the dependent variable. The concomitant variable represents a source of variation that has not been controlled in the experiment and is believed to affect the dependent variable. Through analysis of covariance, the dependent variable can be adjusted so as to remove the effects associated with the concomitant variable.

Analysis of covariance can be used with each of the designs described previously. A layout for a completely randomized analysis of covariance design with one covariate designated by the letters Y_j is shown in Figure 17. The designation for this design is a type CRAC-k design, where the letters AC following the building-block designation indicate that it is an analysis of covariance design.

The selection of a concomitant variable should be made with great care. As in the use of the blocking technique, the concomitant variable should represent an undesired source of variation that is believed to affect the dependent variable. Since the dependent variable is statistically adjusted to remove variation attributable to the concomitant variable, it is essential that the latter source of variation not be affected by the independent variable. If this requirement is not fulfilled, the adjustment made on the dependent variable will remove some of the effects of both the concomitant variable and the independent variable. It is good practice to obtain a measure of the concomitant variable prior to presentation of the independent variable or to arrange the experiment so that one can be absolutely certain that the con-

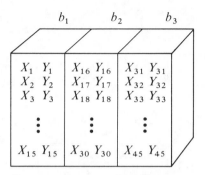

Fig. 17. Layout for type CRAC-3 design. $N = 45$ subjects are randomly assigned to the three treatment levels. X_i and Y_i designate a measure of the dependent variable and the concomitant variable, respectively, for subject i.

comitant variable is unaffected by the independent variable.

The major advantages of analysis of covariance are:

1. These designs enable an experimenter to remove one or more potential sources of bias that are difficult or impossible to eliminate by experimental control.
2. Analysis of covariance provides approximately the same reduction in experimental error as the use of the blocking technique (stratification of subjects in homogeneous blocks). Designs such as type RBAC-k or RBFAC-pq designs incorporate both regression analysis and the blocking technique and can be used to achieve even greater control of nuisance variables.
3. Covariance analysis can be used after the data have been collected if a suitable concomitant variable can be measured. Stratification of subjects into homogeneous blocks must be carried out prior to the collection of data.

The major disadvantages of analysis of covariance are:

1. Computations are more laborious than those for regular analysis-of-variance designs.
2. Analysis of covariance requires a somewhat restrictive set of assumptions that may prove untenable in a particular research application.

Summary of ANOVA Classification System

A variety of ANOVA designs of particular interest to behavioral scientists have been described in terms of their main features, advantages, and limitations. The presentation has been purposely made general and non-mathematical in character in the hope of reaching an audience rarely addressed by the specialist in experimental design.

The principal reason for writing this paper was to present a useful conceptual scheme for understanding the distinctions among various ANOVA designs. Although the literature is replete with such statements as ". . . a two-treatment factorial design was used. . . ," it should be evident that a more precise design nomenclature is required if effective communication between experimenter-author and reader is to be achieved. Such a nomenclature has been described in this article and is summarized in Table 2.

Table 2. Classification of ANOVA designs.

ANOVA design	Nomenclature
I. Designs for experiments with one treatment	
A. Complete block designs	
1. Completely randomized design	CR-k
2. Randomized block design	RB-k
3. Latin square design	LS-k
4. Graeco-Latin square design	GLS-k
5. Hyper-Graeco-Latin square design	HGLS-k
B. Incomplete block designs	
1. Balanced incomplete block design	BIB-t
2. Youden square design	YBIB-t
3. Partially balanced incomplete block design	PBIB-t
II. Factorial experiments	
A. Designs without confounding	
1. Completely randomized factorial design	CRF-pq
2. Randomized block factorial design	RBF-pq
B. Design with treatment-block confounding	
1. Split-plot factorial design	SPF-$p \cdot q$
C. Designs with interaction-block confounding	
1. Randomized block completely confounded factorial design	RBCF-p^k
2. Randomized block partially confounded factorial design	RBPF-p^k
3. Latin square completely confounded factorial design	LSCF-p^k
D. Designs with treatment-interaction confounding	
1. Completely randomized fractional factorial design	CRFF-p^k
2. Randomized block fractional factorial design	RBFF-p^k
3. Latin square fractional factorial design	LSFF-p^k
III. Hierarchal experiments	
A. Design with complete nesting	
1. Completely randomized hierarchal design	CRH-$p(q)$
B. Design with partial nesting	
1. Completely randomized partial hierarchal design	CRPH-$p(q)r$
IV. Analysis of covariance experiments	
A. A covariate adjustment can be used with the designs described above. When this is done the letters *AC* are added to the nomenclature, as in the following examples.	
1. Completely randomized analysis of covariance design	CRAC-k
2. Randomized block analysis of covariance design	RBAC-k
3. Completely randomized factorial analysis of covariance design	CRFAC-pq

References

Anderson, T. W. *Introduction to Multivariate Statistical Analysis.* New York: John Wiley & Sons, Inc., 1958.

Campbell, D. T. Factors relevant to the validity of experiments in social settings. *Psychological Bulletin*, 1957, **54**, 297–312.

Campbell, D. T., and J. C. Stanley. *Experimental and Quasi-Experimental Designs for Research.* Chicago: Rand McNally & Company, 1966.

Cochran, W. G., and G. M. Cox. *Experimental Designs.* New York: John Wiley & Sons, Inc., 1957.

Cooley, W. W., and P. R. Lohnes. *Multivariate Procedures for the Behavioral Sciences.* New York: John Wiley & Sons, Inc., 1962.

Cornfield, J., and J. W. Tukey. Average values of mean squares in factorials. *The Annals of Mathematical Statistics*, 1956, **27**, 907–949.

Cox, D. R. *Planning of Experiments.* New York: John Wiley & Sons, Inc., 1958.

Cox, G. M. Modernized field designs at Rothamsted. *Soil Science Society of America Proceedings*, 1943, **8**, 20–22.

Davies, O. L. (Ed.) *The Design and Analysis of Industrial Experiments.* New York: Hafner Publishing Company, Inc., 1956.

Doxtator, C. W., B. Tolman, C. E. Cormany, H. L. Bush, and V. Jensen. Standardization of experimental methods. *American Society of Sugar Beet Technology Proceedings*, 1942, **3**, 595–599.

Edwards, A. L. *Experimental Design in Psychological Research.* New York: Holt, Rinehart and Winston, Inc., 1968.

Eisenhart, C. The assumptions underlying the analysis of variance. *Biometrics*, 1947, **3**, 1–21.

Federer, W. T. *Experimental Design: Theory and Application.* New York: The Macmillan Company, 1955.

Finney, D. J. The fractional replication of factorial arrangements. *Annals of Eugenics*, 1945, **12**, 291–301.

Finney, D. J. Recent developments in the design of field experiments. III. Fractional replication. *Journal of Agricultural Science*, 1946, **36**, 184–191.

Fryer, H. C. *Concepts and Methods of Experimental Statistics.* Boston: Allyn and Bacon, Inc., 1966.

Kempthorne, O. *The Design and Analysis of Experiments.* New York: John Wiley & Sons, Inc., 1952.

Kempthorne, O. A simple approach to confounding and fractional replication in factorial experiments. *Biometrika*, 1947, **34**, 255–272.

Kirk, R. E. *Experimental Design: Procedures for the Behavioral Sciences.* Belmont, California: Brooks/Cole Publishing Company, 1968.

Lindquist, E. F. *Design and Analysis of Experiments in Psychology and Education.* Boston: Houghton Mifflin Company, 1953.

Millman, J., and G. V. Glass. Rules of thumb for writing the ANOVA table. *Journal of Educational Measurement*, 1967, **4**, 41–51.

Morrison, D. F. *Multivariate Statistical Methods.* New York: McGraw-Hill Book Company, Inc., 1967.

Myers, J. L. *Fundamentals of Experimental Design.* Boston: Allyn and Bacon, 1966.

Rao, C. R. *Advanced Statistical Methods in Biometric Research.* New York: John Wiley & Sons, Inc., 1952.

Schultz, E. F. Rules of thumb for determining expectations of mean squares in analysis of variance. *Biometrics*, 1955, **11**, 123–135.

Winer, B. J. *Statistical Principles in Experimental Design.* New York: McGraw-Hill Book Company, Inc., 1962.

Yates, F. The design and analysis of factorial experiments. *Imperial Bureau of Soil Science Technical Communication No. 35*, Harpenden, England, 1937.

7.3 Beyond Tests of Significance: Estimating Strength of Effects in Selected ANOVA Designs

Graham M. Vaughan and Michael C. Corballis

Abstract. Few psychological reports of experiments using ANOVA designs include estimates of variance components. Such estimates have various uses, including that of assessing the strength of experimental effects, rather than stopping short with the calculation of a significance level. The present paper is devoted to encouraging this additional analysis of data by providing computational formulas appropriate to estimating the strength of effects in basic one-way, two-way, and three-way ANOVA designs. Issues which develop when such estimation is attempted in repeated-measures designs are examined, and limitations to attempting such analysis in respect of specific treatment levels (and contrasts) are indicated.

The publication of *Statistics for Psychologists* (Hays, 1963) heralded the arrival of a popular text but also offered a challenge. Researchers using parametric tests of significance were asked not only to assign a probability value to the tenability of a given hypothesis but to proceed to estimate in addition the degree of association existing between the variables in question. Veteran correlators are well conditioned to such a

point of view, although most need the occasional reminder that a coefficient as high as .70 accounts for just half the variance in common to two variables. The motivation for writing the present paper, however, is that otherwise sophisticated experimenters remain seemingly insensitive to this problem—at least by the criterion of published papers in virtually any psychological journal—and that as long as the popularity of analysis of variance (ANOVA) designs in psychological research continues, a change in attitude is required. A statement of the general principles involved, together with a provision of actual computational formulas for a handful of basic

Reprinted from *Psychological Bulletin*, 1969, **72**, 204–213, by permission of the publisher and authors. Copyright 1969 by the American Psychological Association.

designs, might convince some researchers that they should take that extra step in analyzing their results.

Equations expressing expected mean squares, or E(*MS*)s, as functions of population variance components have been worked out for a wide range of ANOVA designs and are set out in most recent texts in psychological statistics (e.g., Edwards, 1964; Hays, 1963; Winer, 1962). A useful tabular technique for generating the terms and coefficients in these equations has been provided by Scheffé (1959, pp. 284–288). Most often, equations of E(*MS*)s are presented simply as a guide to selecting appropriate error terms in setting up *F* ratios. The present authors agree with Green and Tukey (1960) that there is more to be gained from the analysis of variance than just the chance to apply the conventional *F* test. The purpose here is to show that the same equations can be manipulated to yield estimates of population variance components in terms of calculated mean squares. These estimates are generally both unbiased and consistent, provided that the assumptions underlying the ANOVA models hold. One assumption which is not necessary, however, is that of normality of the underlying distributions (Scheffé, 1959).

Estimation of individual variance components indicates the magnitude, as distinct from the statistical significance, of the variation due to particular effects or interactions. Such estimates may serve a number of useful purposes. A knowledge of the relative contributions of components in a given experiment could guide the researcher in choosing from among a number of specific designs in a subsequent experiment. Again, it may be possible to compare absolute variance estimates between given experiments which employ the same units of measurement.

A further practical exercise is to express each variance estimate as a proportion of total estimated variance, and it is this topic which is the main substance of this paper. In general, the ratio of a single treatment variance to total variance provides an index of the strength of association between that treatment and the dependent variable.[1] Another kind of ratio which a researcher may wish to examine is that between a variance estimate and some subtotal of estimated variance, but this is an exercise which later discussion will show to be less meaningful.

The remainder of the paper deals with the following points: the derivation of convenient computational formulas which express estimated variance components as proportions of total variance in ANOVA designs employing up to three factors, for both the fixed and random model; a treatment of some issues involved in pursuing the same aim in repeated-measures designs; problems in expressing a variance estimate as a proportion of a subtotal of estimated variance, a practice which might follow from the isolation of components due to specific within-treatments contrasts.

Attention is devoted to balanced designs throughout: Equal *n*'s within cells or within groups are assumed, and where there is more than one treatment, each treatment is paired with every other treatment an equal number of times within a common block.

One-Way ANOVA

In the one-way ANOVA design, a general relationship between experimental data and population parameters can be expressed by the basic linear equation

$$X_{ij} = \mu + \alpha_j + \varepsilon_{ij}$$

where X_{ij} is the *i*th score in the *j*th treatment level, μ is the parent population mean, α_j is

---

[1] The reader should be cautioned, however, that *ratios* of variance estimates do not necessarily provide good estimates of the corresponding population ratios. Cramér (1946, p. 255) showed that the ratio of consistent unbiased estimates is itself consistent, though Lord and Novick (1968, pp. 201–203) pointed out that it is likely to be biased. (According to a personal communication from Lord and Novick [February 1969], their assertion on page 202 of a negative bias in the ratio estimate is incorrect; the bias may be either positive or negative.) The risk of bias may not be a serious objection to the use of ratio estimates, particularly if large samples are used, and if one adopts the view that even a biased estimate is better than none.

the differential effect of the jth level of Treatment A, and ε_{ij} is the error component specific to this individual score. From this equation we can proceed to sum the squared deviations of all individual scores from the grand sample mean, and then partition this "total sums of squares" into "sums of squares between" and "sums of squares within." Division of these quantities by the appropriate degrees of freedom yields expected values of respective mean squares, MS_A and MS_w. A good text will distinguish between a "fixed effects" model and a "random effects" model. The first applies when J levels of Treatment A exhaust all possible levels or are deliberately (nonrandomly) selected; the second applies when the levels of A have been randomly selected from a theoretically infinite number of possible levels. In the fixed-effects model, the conclusions drawn at the end of the experiment apply only to the J levels employed, while in the random model the experimenter can generalize to all possible levels. In factorial designs, this distinction becomes important in selecting the error term appropriate to the F ratio, but this does not apply to one-way ANOVA in that MS_w is the correct error term in both the fixed and random model. When, however, the variance component relating to the effect of A is calculated, the distinction between the fixed and random models must always be maintained. We can demonstrate this point by considering the E(MS)s for the two models in a one-way design. Random and fixed factors are designated hereafter as a and A, respectively. The MS_A in Model a is an estimate of $\sigma_\varepsilon^2 + n\sigma_\alpha^2$, where n = number of observations per cell; in Model A it is an estimate of $\sigma_\varepsilon^2 + (n\sum_{j=1}^{J}\alpha_j^2)/(J-1)$. The MS_w in both models is an estimate of σ_ε^2.

First consider the random model a. Our general aim is to set up a ratio of estimates, in this case between the treatment variance component estimate $\hat{\sigma}_\alpha^2$ and the total variance estimate $\hat{\sigma}_\alpha^2 + \hat{\sigma}_\varepsilon^2$. This ratio can be derived in a computationally convenient form as follows: First, we may write

$$n\hat{\sigma}_\alpha^2 = n\hat{\sigma}_\alpha^2 + (\hat{\sigma}_\varepsilon^2 - \hat{\sigma}_\varepsilon^2)$$
$$= (n\hat{\sigma}_\alpha^2 + \hat{\sigma}_\varepsilon^2) - \hat{\sigma}_\varepsilon^2$$
$$= MS_A - MS_w$$

so that the required numerator is

$$\hat{\sigma}_\alpha^2 = \frac{MS_A - MS_w}{n}.$$

The last equation is used to derive the required denominator of the general ratio

$$\hat{\sigma}_\alpha^2 + \hat{\sigma}_\varepsilon^2 = \frac{MS_A - MS_w}{n} + MS_w.$$

It is possible to carry the algebra further to derive a more compact formula which yields a statistic known as the intraclass correlation[2]

$$\frac{\hat{\sigma}_\alpha^2}{\hat{\sigma}_\alpha^2 + \hat{\sigma}_\varepsilon^2} = \frac{MS_A - MS_w}{MS_A + (n-1)MS_w}.$$

The approach here begins with the calculation of individual variance components, which for the random effects one-way ANOVA are shown next to Model a in Table 1. This

Table 1. Estimates of variance components in one-way and two-way designs.

Model	Variance component
A	$\hat{\sigma}_\alpha^2 = (J-1)(MS_A - MS_w)/nJ$ $\hat{\sigma}_\varepsilon^2 = MS_w$
a	$\hat{\sigma}_\alpha^2 = (MS_A - MS_w)/n$ $\hat{\sigma}_\varepsilon^2 = MS_w$
AB	$\hat{\sigma}_\alpha^2 = (J-1)(MS_A - MS_w)/nJK$ $\hat{\sigma}_\beta^2 = (K-1)(MS_B - MS_w)/nJK$ $\hat{\sigma}_{\alpha\beta}^2 = (J-1)(K-1)(MS_{AB} - MS_w)/nJK$ $\hat{\sigma}_\varepsilon^2 = MS_w$
aB	$\hat{\sigma}_\alpha^2 = (MS_A - MS_w)/nK$ $\hat{\sigma}_\beta^2 = (K-1)(MS_B - MS_{AB})/nJK$ $\hat{\sigma}_{\alpha\beta}^2 = (MS_{AB} - MS_w)/n$ $\hat{\sigma}_\varepsilon^2 = MS_w$
ab	$\hat{\sigma}_\alpha^2 = (MS_A - MS_{AB})/nK$ $\hat{\sigma}_\beta^2 = (MS_B - MS_{AB})/nJ$ $\hat{\sigma}_{\alpha\beta}^2 = (MS_{AB} - MS_w)/n$ $\hat{\sigma}_\varepsilon^2 = MS_w$

[2] It is not generally recognized in texts on psychological statistics that this statistic is likely to be a biased estimate of the *population* intraclass correlation, defined by the ratio $\sigma_\alpha^2/(\sigma_\alpha^2 + \sigma_\varepsilon^2)$ (cf. Footnote 1). Calculation of an unbiased estimate involves complexities which need not concern us here (see Olkin & Pratt, 1958).

approach simplifies a comparison of various designs and allows the user to understand more easily what he is actually doing. In all designs, then, it is recommended that the individual variance components be first calculated. Then one can proceed to express any selected component as a proportion of the sum of all components.

Consider now the *fixed* model A. A problem is immediately presented, since the variance component needed as a numerator to estimate the effects of Treatment A as a proportion of total variance has the form $\sum_{j=1}^{J} \alpha_j^2 / J$. Note that the denominator J is appropriate to the concept of a population variance rather than a sample variance, and in the case of the fixed model where all possible levels of A are theoretically present, the required expression needs to be in just this parameter form (cf. Whimbey, Vaughan, & Tatsuoka, 1967). But as indicated earlier, the E(MS) "between" includes an α term, namely $\left(n \sum_{j=1}^{J} \alpha_j^2 \right) / (J - 1)$, whose denominator is $(J - 1)$ rather than J. If the whole term is defined as $n\theta_\alpha^2$, then a relation between σ_α^2 and θ_α^2 can be established thus:

$$\sigma_\alpha^2 = \frac{(J - 1)\theta_\alpha^2}{J}.$$

This adjustment is taken into account in the first grouping of formulas in Table 1. After the calculation of the separate components $\hat{\sigma}_\alpha^2$ and $\hat{\sigma}_\varepsilon^2$, the next step is the expression of the first as a proportion of the sum of both components. The statistic yielded is known as ω^2. The reader should note that this is a ratio of variance estimates, so that the sample value of ω^2 is likely to be a biased estimate of the corresponding population value (cf. Footnotes 1 and 2).

Two-Way ANOVA

As in the one-way designs, the general aim here is to isolate variance components defined in various expected mean squares, and then to express any individual component as a proportion of total variance. The distinction

between fixed and random effects continues to be maintained. The designs considered are balanced, in that it is assumed that J levels of Treatment A are completely crossed with K levels of Treatment B. Three combinations of fixed and random effects, which generate three models, are possible: fixed (A fixed, B fixed); mixed (A random, B fixed); and random (A random, B random). These can be designated more simply: AB, aB, and ab. All models assume a basic linear equation of the form

$$X_{ijk} = \mu + \alpha_j + \beta_k + (\alpha\beta)_{jk} + \varepsilon_{ijk}$$

in which an individual score is expressed in terms of the parent population mean, the effects of the treatments individually, the effect of an interaction between treatments $(\alpha\beta)_{jk}$, and the error component specific to the individual score. As in the case of one-way ANOVA, the present aim is to isolate each variance component and then express this component as a proportion of total variance. For example, in the case of Treatment A, we wish to find

$$\frac{\hat{\sigma}_\alpha^2}{\hat{\sigma}_{tot}^2} = \frac{\hat{\sigma}_\alpha^2}{\hat{\sigma}_\alpha^2 + \hat{\sigma}_\beta^2 + \hat{\sigma}_{\alpha\beta}^2 + \hat{\sigma}_\varepsilon^2}.$$

The means of achieving computational formulas appropriate to the above equation, and to other equations suitable to the components $\hat{\sigma}_\beta^2$ and $\hat{\sigma}_{\alpha\beta}^2$, follow from considering the various E(MS)s. To conserve space, the E(MS)s will not be given. These can be checked elsewhere (e.g., Brownlee, 1965, pp. 508–512; Gaito, 1960; Peng, 1967, pp. 79–85), but the reader is encouraged to use a source (e.g., Edwards, 1964, pp. 95–102) in which a distinction between σ^2 and θ^2 terms is clearly maintained.

Only the random (ab) model consistently incorporates components in "true" variance form, and can be compared with the one-way random model in this respect. The AB and aB models, like the one-way fixed model, require some manipulation of components to ensure that all conform to the notion of a population variance. In the AB model, the α term in MS_A can be defined as

$$\frac{nK \sum_{j=1}^{J} \alpha^2}{J-1} = nK\theta_\alpha^2.$$

As in the one-way fixed model, the symbol θ^2 is used to indicate that the α term does not have the form of a population variance. The β term in MS_B and the $\alpha\beta$ term in MS_{AB} can be expressed as $nJ\theta_\beta^2$ and $n\theta_{\alpha\beta}^2$, respectively. The following manipulations produce the required population variances:

$$\sigma_\alpha^2 = \frac{(J-1)\theta_\alpha^2}{J},$$

$$\sigma_\beta^2 = \frac{(K-1)\theta_\beta^2}{K},$$

$$\sigma_{\alpha\beta}^2 = \frac{(J-1)(K-1)\theta_{\alpha\beta}^2}{JK}.$$

In the *aB* model, in which one random variable is operating, a similar manipulation is required to gain the estimate $\hat{\sigma}_\beta^2$. The other components, $\hat{\sigma}_\alpha^2$ and $\hat{\sigma}_{\alpha\beta}^2$, already exist in variance form. In the use of computational formulas for variance components in two-way models, as given in Table 1, all of the above points are taken into account. Once more, the user is encouraged to calculate separately all variance components for any given model. These can then be summed, and any one component expressed as a proportion of this sum.

Three-Way ANOVA

In this section, crossed designs are assumed, in which there are J levels of Treatment A, K levels of Treatment B, and L levels of Treatment C. Four combinations of fixed and random effects, which generate four models, are possible: fixed (*ABC*), mixed (*aBC*), mixed (*abC*), and random (*abc*). All of these models assume the basic linear equation

$$X_{ijkl} = \mu + \alpha_j + \beta_k + \gamma_l + (\alpha\beta)_{jk} + (\alpha\gamma)_{jl} + (\beta\gamma)_{kl} + (\alpha\beta\gamma)_{jkl} + \varepsilon_{ijkl}$$

in which an individual score is expressed in terms of the parent population mean, the effects of the three treatments individually, the effects of three first-order interactions between treatments, the effect of a second-order interaction between treatments, and the error component specific to the individual score. Once more, single variance components are to be expressed proportional to total variance, but now the latter consists of eight terms. The principles already expounded can be carried forward. The E(MS)s will not be given, nor will the intervening manipulations necessary to convert certain terms in the fixed and mixed models into variance form. Formulas appropriate for the calculation of individual variance components, for all four models, are given in Table 2.

Table 2. Estimates of variance components in three-way designs.

Model	Variance component
ABC	$\hat{\sigma}_\alpha^2 = (J-1)(MS_A - MS_w)/nJKL$
	$\hat{\sigma}_\beta^2 = (K-1)(MS_B - MS_w)/nJKL$
	$\hat{\sigma}_\gamma^2 = (L-1)(MS_C - MS_w)/nJKL$
	$\hat{\sigma}_{\alpha\beta}^2 = (J-1)(K-1)(MS_{AB} - MS_w)/nJKL$
	$\hat{\sigma}_{\alpha\gamma}^2 = (J-1)(L-1)(MS_{AC} - MS_w)/nJKL$
	$\hat{\sigma}_{\beta\gamma}^2 = (K-1)(L-1)(MS_{BC} - MS_w)/nJKL$
	$\hat{\sigma}_{\alpha\beta\gamma}^2 = (J-1)(K-1)$
	$\qquad \times (L-1)(MS_{ABC} - MS_w)/nJKL$
	$\hat{\sigma}_\varepsilon^2 = MS_w$
aBC	$\hat{\sigma}_\alpha^2 = (MS_A - MS_w)/nKL$
	$\hat{\sigma}_\beta^2 = (K-1)(MS_B - MS_{AB})/nJKL$
	$\hat{\sigma}_\gamma^2 = (L-1)(MS_C - MS_{AC})/nJKL$
	$\hat{\sigma}_{\alpha\beta}^2 = (MS_{AB} - MS_w)/nL$
	$\hat{\sigma}_{\alpha\gamma}^2 = (MS_{AC} - MS_w)/nK$
	$\hat{\sigma}_{\beta\gamma}^2 = (K-1)(L-1)(MS_{BC} - MS_{ABC})/nJKL$
	$\hat{\sigma}_{\alpha\beta\gamma}^2 = (MS_{ABC} - MS_w)/n$
	$\hat{\sigma}_\varepsilon^2 = MS_w$
abC	$\hat{\sigma}_\alpha^2 = (MS_A - MS_{AB})/nKL$
	$\hat{\sigma}_\beta^2 = (MS_B - MS_{AB})/nJL$
	$\hat{\sigma}_\gamma^2 = (L-1)$
	$\qquad \times (MS_C - MS_{AC} - MS_{BC} + MS_{ABC})$
	$\qquad\qquad\qquad\qquad\qquad /nJKL$
	$\hat{\sigma}_{\alpha\beta}^2 = (MS_{AB} - MS_w)/nL$
	$\hat{\sigma}_{\alpha\gamma}^2 = (MS_{AC} - MS_{ABC})/nK$
	$\hat{\sigma}_{\beta\gamma}^2 = (MS_{BC} - MS_{ABC})/nJ$
	$\hat{\sigma}_{\alpha\beta\gamma}^2 = (MS_{ABC} - MS_w)/n$
	$\hat{\sigma}_\varepsilon^2 = MS_w$
abc	$\hat{\sigma}_\alpha^2 = (MS_A - MS_{AB} - MS_{AC} + MS_{ABC})$
	$\qquad\qquad\qquad\qquad\qquad /nKL$
	$\hat{\sigma}_\beta^2 = (MS_B - MS_{AB} - MS_{BC} + MS_{ABC})$
	$\qquad\qquad\qquad\qquad\qquad /nJL$
	$\hat{\sigma}_\gamma^2 = (MS_C - MS_{AC} - MS_{BC} + MS_{ABC})$
	$\qquad\qquad\qquad\qquad\qquad /nJK$
	$\hat{\sigma}_{\alpha\beta}^2 = (MS_{AB} - MS_{ABC})/nL$
	$\hat{\sigma}_{\alpha\gamma}^2 = (MS_{AC} - MS_{ABC})/nK$
	$\hat{\sigma}_{\beta\gamma}^2 = (MS_{BC} - MS_{ABC})/nJ$
	$\hat{\sigma}_{\alpha\beta\gamma}^2 = (MS_{ABC} - MS_w)/n$
	$\hat{\sigma}_\varepsilon^2 = MS_w$

Repeated-Measures Designs

In a single-factor experiment with repeated measures, each of N subjects receives J levels of Treatment A. The underlying model in fact allows for "person" (subject) as a second factor taken to be a random variable. The kind of design considered here is that in which each subject has just one score at each level of A, that is, a two-factor experiment with one score per cell ($n = 1$).

A distinction can be drawn between *additive* and *nonadditive* models (Myers, 1966, pp. 153–159; Winer, 1962, pp. 119–120). The nonadditive model assumes the linear equation

$$X_{ij} = \mu + \pi_i + \alpha_j + (\pi\alpha)_{ij} + \varepsilon_{ij}$$

incorporating effects attributable to the person, the treatment, and an interaction between the two. In the additive model, the interaction term $(\pi\alpha)_{ij}$ is assumed zero and dropped. A test to determine whether this assumption can be justified has been devised by Tukey (1949) and is also described by Winer (1962, p. 217) and Myers (1966, p. 166).[3]

Nonadditive Case. Table 3 shows the E(MS)s for nonadditive models $\bar{A}p$ and $\bar{a}p$, where the treatments are fixed and random, respectively; the $\bar{A}$ and $\bar{a}$ terminology denotes repeated measures on Treatment A. The two models can be regarded as special cases of the two-way mixed models, Ab and ab, respectively. The equations in Table 3 pose a problem; for each set there are only three equations for four unknowns, so that one cannot solve uniquely for each and every variance component. However, at least some estimates are possible.

[3] A further assumption usually required of both the additive and nonadditive models is that of equal covariances between population scores on all possible pairs of treatments. If this assumption is not met, the F ratio for testing the treatments main effect is positively biased, and alternative tests of significance are required (see Winer, 1962, p. 123). However, unequal covariances do not alter the variance component estimates, so that the derivations given in this section remain valid even if the covariance assumption does not hold (see Scheffé, 1959, p. 269).

Table 3. E(MS)s for nonadditive one-way repeated-measures designs.

MS	Model	
	$\bar{A}p$	$\bar{a}p$
MS_A	$\sigma_\varepsilon^2 + (N\sum_{j=1}^{J}\alpha_j^2)/(J-1)$ $+ \sigma_{\pi\alpha}^2$	$\sigma_\varepsilon^2 + N\sigma_\alpha^2 + \sigma_{\pi\alpha}^2$
MS_P	$\sigma_\varepsilon^2 + J\sigma_\pi^2$	$\sigma_\varepsilon^2 + J\sigma_\pi^2 + \sigma_{\pi\alpha}^2$
MS_{AP}	$\sigma_\varepsilon^2 + \sigma_{\pi\alpha}^2$	$\sigma_\varepsilon^2 + \sigma_{\pi\alpha}^2$
MS_w	—	—

Note: In the $\bar{A}p$ and $\bar{a}p$ models, N is in fact the number of subjects and could be thought of as N levels of P, this being analogous to K levels of B in the Ab and ab models. Again, in the $\bar{A}p$ and $\bar{a}p$ models, $n=1$, so that there is no MS_w, which means that σ_ε^2 cannot be directly estimated.

When the treatment variable is fixed ($\bar{A}p$ model), one can derive

$$\hat{\sigma}_\alpha^2 = (J-1)\frac{MS_A - MS_{AP}}{NJ}$$

$$\hat{\sigma}_\pi^2 + \frac{\hat{\sigma}_\varepsilon^2}{J} = \frac{MS_P}{J}$$

$$\hat{\sigma}_{\pi\alpha}^2 + \hat{\sigma}_\varepsilon^2 = MS_{AP}.$$

Summing all terms yields an *over*estimate of σ_{tot}^2, with σ_ε^2/J in excess. Hence the following expression is an *under*estimate of the proportion $\sigma_\alpha^2/\sigma_{tot}^2$:

$$\frac{(J-1)(MS_A - MS_{AP})}{(J-1)(MS_A - MS_{AP}) + N\cdot MS_P + NJ\cdot MS_{AP}}.$$

The extent to which this is an underestimate decreases as J, the number of treatment levels, increases.

When the treatment variable is *random* ($\bar{a}p$ model), treatment, person, and total variance can be estimated. The interaction component, however, cannot be separated from the error component. The relevant equations are:

$$\hat{\sigma}_\alpha^2 = \frac{MS_A - MS_{AP}}{N}$$

$$\hat{\sigma}_\pi^2 = \frac{MS_P - MS_{AP}}{J}$$

$$\hat{\sigma}_\varepsilon^2 + \hat{\sigma}_{\pi\alpha}^2 = MS_{AP}.$$

The sum $(\hat{\sigma}_\varepsilon^2 + \hat{\sigma}_{\pi\alpha}^2)$ can be thought of as MS_{residual} and is common to all $E(MS)$s in the $\bar{a}p$ model (See Table 3). In this sense there are three equations and three unknowns, one of the latter being MS_{residual}. Summing the right hand side of all three equations given above yields $\hat{\sigma}_{\text{tot}}^2$; and in this instance, not only $\hat{\sigma}_\alpha^2$ but also $\hat{\sigma}_\pi^2$ can be expressed as a proportion of total variance.

Additive Case. Advantages associated with the additive model are usefully discussed by Myers (1966, pp. 160–169). When the assumption of additivity can be made, for example, by meeting Tukey's criterion, the $\sigma_{\pi\alpha}^2$ term can be dropped. The remaining variance estimates can be calculated, since there are now three equations for three unknowns. The σ_α^2 for the fixed and the random model in turn are estimated as in the nonadditive case. The other components are estimated as follows:

$$\hat{\sigma}_\pi^2 = \frac{MS_P - MS_{AP}}{J}$$

$$\hat{\sigma}_\varepsilon^2 = MS_{AP}.$$

Both of these equations apply whether the treatment variable is fixed or random. The estimates $\hat{\sigma}_\alpha^2$, $\hat{\sigma}_\pi^2$, and $\hat{\sigma}_\varepsilon^2$ can be summed to give $\hat{\sigma}_{\text{tot}}^2$.

Hence, if the additive model can be justified, it is preferred to the nonadditive model in that unbiased estimates of *all* variance can be calculated.

Multifactor Repeated-Measures Designs. There are a large number of possible multifactor repeated-measures designs. Two-way and three-way designs, with repeated measures on one or two factors,[4] include: $A\bar{B}p$, $a\bar{B}p$, $A\bar{b}p$, $a\bar{b}p$; $\bar{A}\bar{B}p$, $\bar{a}\bar{B}p$, $\bar{a}\bar{b}p$;[5] $ABCp$, $aB\bar{C}p$, $ab\bar{C}p$, $AB\bar{c}p$, $aB\bar{c}p$, $ab\bar{c}p$; $A\bar{B}\bar{C}p$,

$a\bar{B}\bar{C}p$, $A\bar{b}\bar{C}p$, $a\bar{b}\bar{C}p$, $A\bar{b}\bar{c}p$, $a\bar{b}\bar{c}p$. The following general principles can be established for designs of this kind:

1. In the *nonadditive* case, there is no design in which an independent estimate of σ_ε^2 is available.

2. However, provided that all factors on which repeated measures are made are random, (e.g., $A\bar{b}p$, $a\bar{b}\bar{c}p$, $A\bar{b}\bar{c}p$, etc.), an MS_{residual} can be formed, and consistent estimates of some variance components, proportional to total variance, are possible. Components which can be isolated always include main effects of treatment, and some interaction effects depending upon the particular design used.

3. (*a*) If *any one* factor on which repeated measures are made is a fixed factor (e.g., $a\bar{B}p$, $ab\bar{C}p$, etc.), estimates of variance components proportional to total variance will be negatively biased. (*b*) The extent of this bias varies from design to design, according to the number of extra fractions of $\hat{\sigma}_\varepsilon^2$ which inflate the total variance. The number of such fractions increases with the number of *fixed* factors on which there are repeated measures. (*c*) The effect of negative bias attributable to any one extra fraction decreases as the number of levels of the relevant fixed factor increases.

4. In the *additive* case, $\hat{\sigma}_\varepsilon^2$ can be obtained in all designs. This permits the calculation in all designs of unbiased estimates of variance components proportional to total variance.

5. In *all* designs, variance components relating to main effects can be isolated. This applies to some interaction effects as well, depending upon the design. This at least allows a comparison *between* treatments to be made (e.g., $\hat{\sigma}_\alpha^2 : \hat{\sigma}_\beta^2 :: 2 : 1$), in lieu of expressing such effects proportional to total variance.

---

[4] General models for some of these sets are given in Winer (1962, pp. 318, 336, 348). Tables of $E(MS)$s for all of the repeated-measures designs specified herein have been prepared by the senior author and are available on request.

[5] Endler (1966) has recently discussed variance component estimation for designs $\bar{A}\bar{B}p$ and $\bar{a}\bar{b}p$, but

certain of his equations require modification (Whimbey, Vaughan, & Tatsuoka, 1967).

Within-Treatments Components[6]

An investigator may wish to probe within a given treatment effect to estimate the contribution made by a specific comparison, or contrast, to the total *treatment* variance (cf. Levin, 1967). For example, the mean score at one treatment level may differ markedly from all the others, so that it may be of interest to estimate the extent to which this deviant score accounts for the total variance between levels. Again, given a significant linear (or quadratic, or cubic) trend among sample means, one may wish to estimate the proportion of treatment variance accounted for by this trend. Questions like these are generally meaningful only when the treatment levels are fixed rather than random. The following discussion is restricted to one-way ANOVA, fixed model, though the principles may be extended to other designs involving fixed variables.

The instances referred to previously represent particular *contrasts*. A contrast ψ among the J levels of some Treatment A may be defined as $\psi = \sum_{j=1}^{J} c_j \mu_j$, where $\sum_{j=1}^{J} c_j = 0$ and μ_j is the population mean under the jth treatment level. The problem is to define and subsequently estimate a variance component of σ_α^2, the total treatment variance, attributable to ψ, a specific contrast within the treatment levels. This is meaningful only if we can regard ψ as one of a set whose total contribution equals σ_α^2. A contrast cannot be considered in isolation. Take, for example, a contrast between just two means, $\mu_j - \mu_k$, with the question in mind of the contribution by this contrast to σ_α^2. The answer depends on whether we consider this contrast in the context of: all possible contrasts (of which there are infinitely many, if $J > 2$), all possible

contrasts between *pairs* of means, all contrasts between adjacent pairs, or some other set. In short, there is no distinctly meaningful answer to an isolated question of this sort.

We should at this point emphasize that we are concerned with estimating a variance contribution attributable to a contrast, and not the contrast itself. A contrast may be simply estimated from the sample means, namely, $\hat{\psi} = \sum_{j=1}^{J} c_j \bar{X}_j$. Such an estimate *can* be considered in isolation and may often be a more useful indicator of the importance of a given contrast than the estimate of its variance contribution. On the other hand, it gives no direct indication of the composition of the total treatment variance.

The most satisfactory subsets of contrasts are those of $(J - 1)$ contrasts which are mutually orthogonal, since in each orthogonal set total treatment variance can be readily broken down into additive components. Before orthogonal contrasts are discussed, however, another set which is not orthogonal, but which seems to have some intuitive appeal (Levin, 1967), is considered.

Components Due to Each α_j. Each α_j itself defines a contrast. The α_j can be defined as $(\mu_j - \mu)$, where μ can be taken as the mean of the μ_j over the J levels. This can be written as a contrast in which $c_j = (J - 1)/J$ and all other $c_i = -1/J$.

Further, it will be recalled that $\sigma_\alpha^2 = \sum_{j=1}^{J} \alpha_j^2/J$. Thus it is clear the σ_α^2 can be regarded as the sum of J additive components α_j^2/J. An estimate of α_j^2/J might therefore be said to indicate how much a given α_j contributes to the total treatment variance.

To find an estimate of α_j^2/J, one can calculate $(\bar{X}_j - \bar{X})^2/J$, where $\bar{X}_j$ is the sample mean under the jth treatment level and $\bar{X}$ is the grand sample mean. It can be shown that this statistic, given the usual assumptions of the ANOVA model, is an unbiased estimate of

$$\frac{(J-1)\sigma_\varepsilon^2}{nJ^2} + \frac{\alpha_j^2}{J}.$$

---

[6] This section involves slightly more complicated algebra than the other sections and may be omitted without disrupting continuity. Further, specification of within-treatments components raises special problems of definition and interpretation, which in some instances can make the calculation of such components less than worthwhile. Nevertheless, development of within-treatments components is a logical extension of breakdown into main effect and interaction components and therefore merits some consideration.

As indicated in Table 1, MS_w is an estimate of σ_ε^2, so that the following estimate of α_j^2/J can be derived:

$$\frac{\hat{\alpha}_j^2}{J} = \frac{(\bar{X}_j - \bar{X})^2}{J} + \frac{J-1}{nJ^2}\, MS_w \,.$$

Expressing this as a proportion of $\hat{\sigma}_\alpha^2$ yields

$$\frac{\hat{\alpha}_j^2}{J\hat{\sigma}_\alpha^2} = \frac{nJ(\bar{X}_j - \bar{X})^2 + (J-1)MS_w}{J(J-1)(MS_A - MS_w)} \,.$$

The last two equations can be regarded as indices of how much a given α_j contributes to the total treatment variance, but they only approximately indicate the contribution of a given *level*. In fact, each α_j depends to some extent on *every* level, because the grand mean μ depends on each μ_j. Moreover, the α_j are not mutually orthogonal contrasts; they are not independent of one another. However, independence increases as J increases, so that it becomes more reasonable to regard $\hat{\alpha}_j^2/J$ as a measure of the contribution of Level j when J is large.

On balance, the qualification concerning lack of independence would generally dictate that treatment variance be divided into components representing orthogonal contrasts.

Components Due to Orthogonal Contrasts. Two contrasts ψ_1 and ψ_2, defined by coefficients c_{1j} and c_{2j}, respectively, are said to be *orthogonal* if $\sum_{j=1}^J c_{1j}c_{2j} = 0$. As we have already noted, there can be no more than $(J-1)$ mutually orthogonal contrasts among J treatment levels, although there are infinitely many such sets (for $J > 2$).

A variance component corresponding to contrast ψ, regarded as one of a set of $(J-1)$ orthogonal contrasts, could be defined as

$$\sigma_\psi^2 = \frac{(\sum_{j=1}^J c_j \mu_j)^2}{J \sum_{j=1}^J c_j^2} \,.$$

It can be shown that the sum of components, defined in this way, over a set of $(J-1)$ orthogonal contrasts is the total treatment variance σ_α^2.

To estimate σ_ψ^2, one may calculate

$$MS_{\text{contrast}} = \frac{n(\sum_{j=1}^J c_j \bar{X}_j)^2}{\sum_{j=1}^J c_j^2} \,,$$

which is an unbiased estimate of $\sigma_\varepsilon^2 + nJ\sigma_\psi^2$. Since MS_w is an estimate of σ_ε^2, one can derive

$$\hat{\sigma}_\psi^2 = \frac{MS_{\text{contrast}} - MS_w}{nJ} \,.$$

Expressing this as a proportion of $\hat{\sigma}_\alpha^2$ gives

$$\frac{\hat{\sigma}_\psi^2}{\hat{\sigma}_\alpha^2} = \frac{MS_{\text{contrast}} - MS_w}{(J-1)(MS_A - MS_w)} \,.$$

These last two expressions indicate the contribution of a contrast ψ to total treatment variance, provided it is recognized that ψ represents one of a set of mutually orthogonal contrasts. For example, the c_j may be chosen from the set of orthogonal polynomials used to define trend coefficients (as in Hays, 1963, p. 551; Winer, 1962, p. 70).

Other Considerations

A few remaining general points can be mentioned:

1. The treatment of variance components in this paper relates only to point estimation. The problem of *interval estimation* of variance components is a complex, even controversial one, and well beyond the scope of this paper. The interested reader is referred to Bulmer (1957), Myers (1966, pp. 290–294), or Scheffé (1959, pp. 229–235).

2. It is possible in practice for a computed estimate of a variance component to assume a *negative quantity*. This seemingly paradoxical result would occur in a one-way design, for example, when the value of MS_w exceeds that of MS_A. In such a case, the most plausible estimate *in that instance* would be zero (cf. Hays, 1963, p. 383). However, replacing a negative estimate by zero introduces a positive bias, and the experimenter is best advised to report the negative value (Scheffé, 1959, p. 229), particularly if the estimate is to be considered in conjunction with estimates from other experiments.

3. The power of the F test increases as a function of the number of degrees of freedom associated with the denominator of a given F ratio. This is sometimes achieved post hoc by *pooling* certain sums of squares and associated df's, on the assumption that some variance components have zero value. The effect of this procedure is to provide several independent estimates of σ_ε^2 which are pooled to produce an averaged estimate of σ_ε^2. The general issue of pooling is controversial (cf. Green & Tukey, 1960). Post hoc pooling complicates the distribution theory underlying variance components, and may introduce bias in the simple procedures outlined in this paper. It is safest not to pool when estimating variance components, even if one does pool for the purpose of testing significance. At the very least, an experimenter ought to specify precisely the rules governing the pooling procedures he uses.

4. Throughout this paper, equal numbers of observations within cells or within groups are assumed. If there are *unequal numbers*, complications arise. For random-effects models, there appear to be no satisfactory solutions when numbers are unequal (Scheffé, 1959, p. 224). The situation for fixed-effects designs is almost as bad. For the one-way fixed model, the expected value of MS_A becomes

$$\sigma_\varepsilon^2 + \frac{\sum_{j=1}^J n_j \alpha_j^2}{J-1}$$

where n_j is the number of observations in the jth group and may differ from group to group. Since MS_w estimates σ_ε^2, we can derive an estimate of the right hand term adjusted if desired so that the denominator is J. However this quantity is not simply related to $\hat{\theta}_\alpha^2$, since it involves a "weighting" factor n_j associated with each level. (However if the n_j are approximately equal one could divide by their mean to obtain a fair approximation to $\hat{\theta}_\alpha^2$.) Further difficulties arise in higher order fixed-effects models, particularly over interaction terms (Scheffé, 1959, pp. 112–119). The experimenter who wishes to avoid unnecessary complication is urged to keep the number of observations per cell constant, if at all possible.

5. Unbiased estimates of variance components, and consistent estimates of ratios involving such components, can be computed in Latin-square designs provided an additive model is valid. The reader can compare the E(MS)s for an additive model and a non-additive model in Myers (1966, pp. 254–255); the number of equations equals the number of unknowns in the additive model only. Winer (1962, pp. 514–577) assumes the additive model in all of the designs which he discusses, so that determinate solutions for the individual variance components could be prepared for the various tables of E(MS)s which he provides.

References

Brownlee, K. A. *Statistical theory and methodology in science and engineering.* New York: Wiley, 1965.

Bulmer, M. G. Approximate confidence limits for components of variance. *Biometrika*, 1957, **44**, 159–169.

Cramér, H. *Mathematical methods of statistics.* Princeton: Princeton University Press, 1946.

Edwards, A. L. *Expected values of discrete random variables and elementary statistics.* New York: Wiley, 1964.

Endler, N. S. Estimating variance components from a mean square for random and mixed effects analysis of variance models. *Perceptual and Motor Skills*, 1966, **22**, 559–570.

Gaito, J. Expected mean squares in analysis of variance techniques. *Psychological Reports*, 1960, **7**, 3–10.

Green, B. F., & Tukey, J. W. Complex analyses of variance: General problems. *Psychometrika*, 1960, **25**, 127–152.

Hays, W. L. *Statistics for psychologists.* New York: Holt, Rinehart & Winston, 1963.

Levin, J. R. Comment: Misinterpreting the significance of "explained variation." *American Psychologist*, 1967, **22**, 675–676.

Lord, F. M., & Novick, M. R. *Statistical theories of mental test scores.* Reading, Mass.: Addison-Wesley, 1968.

Myers, J. L. *Fundamentals of experimental design.* Boston: Allyn & Bacon, 1966.

Olkin, I., & Pratt, J. W. Unbiased estimation of certain correlation coefficients. *Annals of Mathematical Statistics,* 1958, **29**, 201–211.

Peng, K. C. *The design and analysis of scientific experiments.* Reading, Mass.: Addison-Wesley, 1967.

Scheffé, H. *The analysis of variance.* New York: Wiley, 1959.

Tukey, J. W. One degree of freedom for non-additivity. *Biometrics,* 1949, **5**, 232–242.

Whimbey, A., Vaughan, G. M., & Tatsuoka, M. M. Fixed effects versus random effects: Estimating variance components from mean squares. *Perceptual and Motor Skills,* 1967, **25**, 668.

Winer, B. J. *Statistical principles in experimental design.* New York: McGraw-Hill, 1962.

7.4   Suggestions for Further Reading

I. Expectations of Mean Squares and the Choice of an Error Term

Cornfield, J., and J. W. Tukey. Average values of mean squares in factorial. *The Annals of Mathematical Statistics*, 1956, **27**, 907–949.

Gaito, J. Expected mean squares in analysis of variance techniques. *Psychological Reports*, 1960, **7**, 3–10.

Millman, J., and G. V. Glass. Rules of thumb for writing the ANOVA table. *Journal of Educational Measurement*, 1967, **4**, 41–51.

Schultz, E. F. Rules of thumb for determining expectations of mean squares in analysis of variance. *Biometrics*, 1955, **11**, 123–135.

II. Estimating Components of Variance

Crump, S. L. The estimation of variance components in analysis of variance. *Biometrics*, 1946, **2**, 7–11.

Endler, N. S. Estimating variance components from mean squares for random and mixed effects analysis of variance models. *Perceptual and Motor Skills*, 1966, **22**, 559–570. Also see the article by Whimbey, Vaughan, and Tatsuoka for modified equations.

Federer, W. T. Non-negative estimators for components of variance. *Applied Statistics*, 1967, **16**, 171–174.

Whimbey, A., G. M. Vaughan, and M. M. Tatsuoka. Fixed effects vs. random effects: Estimating variance components from mean squares. *Perceptual and Motor Skills*, 1967, **25**, 668.

III. Trend Analysis

Alexander, H. W. A general test for trend. *Psychological Bulletin*, 1946, **43**, 533–557.

Gaito, J. Unequal interval and unequal n in trend analysis. *Psychological Bulletin*, 1965, **63**, 125–127.

Gaito, J., and E. D. Turner. Error terms in trend analysis. *Psychological Bulletin*, 1963, **60**, 464–474.

Grant, D. A. Analysis-of-variance tests in the analysis and comparison of curves. *Psychological Bulletin*, 1956, **53**, 141–154.

IV. Related Topics

Benjamin, L. S. Facts and artifacts in using analysis of covariance to "undo" the law of initial values. *Psychophysiology*, 1967, **4**, 187–206. A penetrating examination of the merits of analysis of covariance in analyzing psychophysiological data when Wilder's Law of Initial Values is assumed.

Cochran, W. G. Analysis of covariance: Its nature and uses. *Biometrics*, 1957, **13**, 261–281. The author discusses five uses of analysis of covariance and gives a readable explication of the nature of the covariance adjustment.

Cochran, W. G. Some consequences when the assumptions for the analysis of variance are not satisfied. *Biometrics*, 1947, **3**, 22–38. This classic paper (1) summarizes the assumptions of analysis of variance, (2) discusses the effects of failure to meet these assumptions, and (3) suggests procedures for determining if for a set of data the assumptions are tenable.

Danford, M. B., H. M. Hughes, and R. C. McNee. On the analysis of repeated-measurements experiments. *Biometrics*, 1960, **16**, 547–565. Some of the statistical problems that may arise when repeated measures are obtained on subjects are discussed and alternative analyses are presented.

Evans, S. H., and E. J. Anastasio. Misuse of analysis of covariance when treatment effect and covariate are confounded. *Psychological Bulletin*, 1968, **69**, 225–234. Examples in the literature are cited in which the independent variable is correlated with the covariate in analysis of covariance. The resulting problems in the interpretation of tests of significance are discussed.

Gaito, J. Statistical models versus empirical intuition: Interactions in analysis-of-variance techniques. *The Journal of General Psychology*, 1964, **70**, 295–303. The implications of carrying out tests of significance when significant interactions are present are discussed.

Lana, R. E., and A. Lubin. The effect of correlation on the repeated measures design. *Educational and Psychological Measurement*, 1963, **23**, 729–739. The authors discuss the problems that can occur whenever an experimenter obtains repeated measures on a set of subjects. To circumvent these problems they suggest using the Geisser-Greenhouse conservative F test or a multivariate analysis of variance design.

Levin, J. R. Misinterpreting the significance of "explained variation." *American Psychologist*, 1967, **22**, 675–676. Many writers have decried the emphasis on tests of significance in analysis of variance to the exclusion of measures of strength of association. However, this article points out a pitfall associated with interpretation in terms of measures of association.

Marascuilo, L. A., and J. R. Levin. Appropriate post hoc comparisons for interaction and nested hypotheses in analysis of variance designs: The elimination of type IV errors. *American Educational Research Journal*, 1970, **7**, 397–421. If an interaction is significant in a factorial design, an experimenter's interest usually turns from tests of main effects to tests of simple main effects. Procedures for evaluating simple main effects and some of the issues involved in such tests are examined in this paper.

Tukey J. W. One degree of freedom for nonadditivity. *Biometrics*, 1949, **5**, 232–242. Describes a test for the presence of row-column non-additivity in a two-way analysis of variance layout. This test is useful in helping an experimenter determine whether an additive or a non-additive model is appropriate for his data.

Tukey, J. W. Query No. 113. *Biometrics*, 1955, **11**, 111–113. A test for non-additivity for a Latin square design is described.

8

Comparisons among Means

As can be expected in a young discipline, the field of statistics has its share of controversial issues. These controversies are chronicled both in the statistical journals and in the unpublished letters of statisticians.

A recent issue that has been of considerable interest to behavioral scientists concerns the use of one-tailed tests in making comparisons among means. The question of whether to use one-tailed or two-tailed tests of significance still continues to puzzle students in introductory statistics courses. Five articles that debate the pros and cons of one- vs. two-tailed tests are presented in the first half of this chapter. After reading these articles, the student should have an understanding of the issues involved and hopefully an appreciation for the way open communication among scholars leads to clarification and resolution of controversy.

The last three articles in this chapter deal with a more complex issue: should the probability of making a type I error be specified for the individual comparison among means, or should the probability of making an error be specified for the statistical hypothesis, the experiment, or some other conceptual unit? The reader will discover that this issue is far from settled.

8.1 Tests of Hypotheses: One-Sided vs. Two-Sided Alternatives

Lyle V. Jones

Psychological literature abounds with experimental studies which utilize statistical tests of the significance of differences between two groups of subjects. Most of these studies present tests based upon either the distribution of Student's t or upon the distribution of χ^2. Since the comparison of an experimental group with a control group of subjects is so fundamental to the experimental method, and since statistical tests of significance are appropriate for testing hypotheses regarding differences between two groups of subjects, it would seem important to correct a common misconception concerning the application of these tests of hypotheses.

One model for a test of significance of mean difference, the more familiar model, is that in which we test the null hypothesis, H_0, against a set of two-sided alternatives, H_1. We might formalize this test,

$$H_0: \mu_1 - \mu_2 = 0$$

$$H_1: \mu_1 - \mu_2 \neq 0,$$

where μ_1 is the mean of the population represented by one sample and μ_2 is the mean of the population represented by a second

sample. Assuming scores X_1, from the first population, and scores X_2, from the second, both to be distributed normally, and assuming the population standard deviations to be equal, we may find

$$t = \frac{\overline{X}_1 - \overline{X}_2}{s\sqrt{\dfrac{1}{N_1} + \dfrac{1}{N_2}}} \ ,$$

where

$$s = \sqrt{\frac{\displaystyle\sum_{i=1}^{N_1} (X_{1i} - \overline{X}_1)^2 + \sum_{j=1}^{N_2} (X_{2j} - \overline{X}_2)^2}{N_1 + N_2 - 2}}$$

and N_1 and N_2 are the numbers of individuals in the samples from the first and second populations.[1] Having stipulated a desired significance level, α, we may enter the t table (1) with $N_1 + N_2 - 2$ degrees of freedom and a p-value equal to α to find a critical value of t, t_c. If the absolute value of the observed t exceeds t_c, we reject H_0 in favor of H_1; otherwise, we accept H_0. In Figure 1 appears a distribution of t showing, graphically, the nature of this decision. This distribution corresponds to the sampling distribution of mean differences, under the null hypothesis. For

This article was prepared while the writer was a National Research Council Fellow. Reprinted from *Psychological Bulletin*, 1952, **49**, 43–46, by permission of the publisher and author. Copyright 1952 by the American Psychological Association.

[1] Of course, if the two samples are not independently selected, we should make use of the correlation between them in the determination of t.

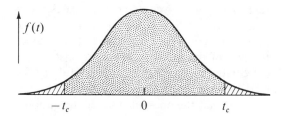

Fig. 1. The two-tailed test model.

any value of t to the right of t_c or to the left of $-t_c$, we reject H_0. The two shaded tails of the distribution, taken together, make up α per cent of the total area under the curve.

The model above, the test of the null hypothesis against two-sided alternatives, is the one used most often by investigators in psychology. Yet in many cases, probably in most cases, it is not the test most appropriate for their experimental problems. More often than not, in psychological research, our hypotheses have a *directional* character. We are interested in whether or not a given diet *improves* maze performance in the rat. We hypothesize that the showing of a particular motion picture to a group of individuals would lead to a *more tolerant* attitude toward certain racial minorities. We wish to test whether or not anxious subjects will respond *more actively* than normal subjects to environmental changes which might be perceived as threatening. In each case, theoretical considerations allow the postulation of the direction of experimental effects. The appropriate experimental test is one which takes this into account, a test of the null hypothesis against a one-sided alternative.

In the one-sided case we test H_0 against H_1, where

$$H_0 : \mu_1 - \mu_2 = 0$$

$$H_1 : \mu_1 - \mu_2 > 0.$$

Under the identical assumptions of the two-sided model we may calculate t as before. Again a significance level, α, is stipulated. The distinction between the one-sided test and the two-sided test arises in the determination of the critical value, t_c. In the present case this critical value is found by entering the t

table with $N_1 + N_2 - 2$ degrees of freedom, as before, but with a p-value equal to 2α. If our observed t is greater than this t_c we reject H_0 in favor of H_1; if t is less than t_c we accept H_0. The t distribution in Figure 2 exemplifies this procedure. A value of t to the right of t_c leads to the rejection of H_0, the acceptance of H_1. While the shaded area under the curve once again represents α per cent of the total area, the shaded portion is restricted, in this case, to one tail of the distribution.

It might be noted that with this formulation of the one-tailed test there is no allowance for the possibility that the true difference, $\mu_1 - \mu_2$, is negative. In the type of problem for which the one-tailed test is suited, such a negative mean difference is no more interesting than a zero difference. In fact, the hypotheses for the one-sided case might be

$$H_0 : \mu_1 - \mu_2 \leqq 0$$

$$H_1 : \mu_1 - \mu_2 > 0.$$

In order to determine a sampling distribution under H_0 we should consider the "worst" of the infinite alternatives under H_0, i.e., that alternative which would make the decision between H_0 and H_1 a most difficult one. Clearly, the decision would be more difficult if the true mean difference were zero than if the true difference were any negative value. Hence we would proceed exactly as in the preceding one-tailed case, utilizing, for our test, the distribution of t based upon the same sampling distribution of mean differences as before. The significance level should be doubled to provide the p-value for entering a table to find a critical t_c, or, if it is desired to ascertain

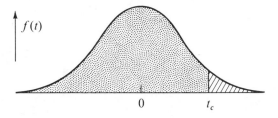

Fig. 2. The one-tailed test model.

the *p*-value corresponding to an observed *t*, the correct value is one-half that given in the typical table of *t*.

While the one-tailed test has been exemplified here as a test of mean difference, based upon the *t* distribution, it is limited in application neither to mean difference problems nor to the *t* statistic. Indeed, wherever an alternative to the null hypothesis is stated in terms of the direction of expected results, the one-tailed test is applicable.

The failure, among psychologists, to utilize the one-tailed statistical test, where it is appropriate, very likely is due to the propagation of the two-tailed model by writers of text books in psychological statistics. It is typical, in such texts, to find little or no attention given to one-tailed tests. Since the test of the null hypothesis against a one-sided alternative is the most powerful test for all directional hypotheses, it is strongly recommended that the one-tailed model be adopted wherever its use is appropriate.

Reference

Fisher, R. A., & Yates, F. *Statistical tables for biological, medical and agricultural research.* Edinburgh: Oliver & Boyd, Ltd., 1938.

8.2 A Brief Note on One-Tailed Tests

C. J. Burke

Concurrent with the recent discussions of one-tailed and two-tailed tests by Hick (4), Jones (5), and Marks (8) there has been a disturbing increase in the use of one-tailed tests in student experimental reports as well as in published and not-yet-published manuscripts. While the popularity of one-tailed tests is undoubtedly attributable in part to the overwillingness of psychologists as a group to make use of the statistical recommendations they have most recently read, there seems to be a certain residual of bad logic, so far as both statistics and psychology are concerned, which merits examination. The writer takes the position already taken by Hick (4) in all important essentials but the argument to be presented differs, at least in emphasis, from that of Hick. It should be noted that some tests, χ^2 and F for example, are naturally single-ended. Nothing said here should be construed so as to apply to them.

Both Jones (5) and Marks (8) seem to the writer to confuse somewhat two quite different notions—that an experimental hypothesis is often directional and that an experimenter may be willing to accept a deviation of any size in the unexpected direction as consonant

Reprinted from *Psychological Bulletin*, 1953, **50**, 384–387, by permission of the publisher and author. Copyright 1953 by the American Psychological Association.

with the null hypothesis. We shall consider two quotations from Jones.

The model above, the test of the null hypothesis against two-sided alternatives, is the one used most often by investigators in psychology. Yet in many cases . . . it is not the test most appropriate for their experimental problems. More often than not, in psychological research, our hypotheses have a *directional* character. . . . theoretical considerations allow the postulation of the direction of experimental effects. The appropriate experimental test is one which takes this into account, a test of the null hypothesis against a one-sided alternative (5, p. 44).

It is a fact that many hypotheses in psychological research, experimentally conceived, are directional for the investigator conducting the experiment, but it does not follow from this that one-sided tests should be used in experimental reports.

To amplify these considerations we point out that there are, in many experiments, two statistical decisions to be made and two different levels of significance may be involved. The first is the decision made by the individual experimenter who frequently plans one experiment from his evaluation of a previous one. We concede that here a one-tailed test is often proper. The second is the decision which determines the place of his findings in the literature of psychology. Here the one-tailed test seems inadmissible. It is the second type of decision with which we are concerned. Marks (8) has in essence repeated from statistical sources a discussion of the Type I and Type II errors which shows that the decisions made in any statistical interpretation depend only upon the underlying populations and the rule of procedure used. Any comparison of alternative rules of procedure must take into account errors of both types, the error of rejecting a hypothesis when it is true and the error of failing to reject it when it is false, but the underlying statistical considerations do not provide automatically a criterion for the selection of one rule over another. Such a criterion is to be sought in the number and kinds of errors the experimenter will tolerate. Roughly, an acceptable criterion is to make the over-all number of errors as small as possible and at the same time to render large and serious errors relatively impossible. Within the class of hypotheses which are considered to be directional it is likely that a one-tailed test might yield a smaller over-all number of errors than a two-tailed test, but there is, under the single-tailed rule, no safeguard whatsoever against occasional large and serious errors when the difference is in the unexpected direction. If one is less willing to commit a large error than to commit a small one, it does not follow from the theory of testing statistical hypotheses that the experimenter's expectation of a given direction for the result necessarily makes the one-tailed test desirable.

To advance this point in our case against the use of the one-tailed test in the public report, we next take up the second quotation from Jones.

It might be noted that with this formulation of the one-tailed test there is no allowance for the possibility that the true difference ... is negative. In the type of problem for which the one-tailed test is suited, such a negative mean difference is no more interesting than a zero difference (5, p. 45).

This statement is perfectly correct.[1] If we consider it carefully we discover its import to be that the investigator should use a one-tailed test when he is willing to accept a difference in the unexpected direction, *no matter how large*, as consonant with the hypothesis of zero difference. This is quite a different matter from using a one-tailed test whenever the direction of the difference is predicted, on some grounds or other, in advance. It is to be doubted whether experimental psychology, in its present state, can afford such lofty indifference toward experimental surprises.

The questions raised by the one-tailed test are to be answered finally by considering the effect of general use of this procedure on the content of psychological literature. The writer cannot agree with Hick (4) that its use makes little difference since there is no practical rule for deciding what "significance really is. In some super-scientific world this point might be well taken, but there is evidence that in our workaday world (where we sometimes read only the concluding sections of reports) it does make a difference whether the investigator has stated that his results were significant. The controversy over the Blodgett effect is a case in point (1, 2, 6, 7, 9, 10, 11, 12).

Remembering that the problem of testing a statistical hypothesis is a statistical problem in which each individual experiment is viewed only as a member of a class of similar experiments and recalling that the properties of any statistical test are determined solely by the procedure followed and by the populations underlying the class, it is pertinent to inquire

---

[1] In his subsequent discussion, Jones spoils the force of this point by confusing hypotheses to be tested with classes of hypotheses to be guarded against as alternatives.

into the effects of widespread adoption of one-tailed tests upon the literature. The writer believes the following statements to be reasonable forecasts.

1. The discovery of new psychological phenomena will be hindered. Our literature abounds with instances in which the outcome of a given experiment has differed reliably and sharply from expectation. These experiments are usually of great interest—new psychological concepts arise from them. Our science is not yet so mature that these can be expected to occur infrequently The most recent instance, known to the writer, of conflicting results from experiments thought to be highly similar was reported by Underwood (13) at the 1952 meetings of the American Psychological Association. From any careful examination of contemporary psychological literature we must conclude that nowhere in the field can we have sufficient a priori confidence in the outcome of any genuinely new experiment to justify the neglect of differences in the unexpected direction.

2. There will be an increase in barren controversy. Fruitless controversies arise from unreliable results. Conclusions at low levels of significance tend to be unreliable, and the adoption of one-tailed tests is equivalent to a general lowering of levels of significance. At a time of severe journal overload this is especially pernicious. There is no substitute in statistical methodology for the carefully designed and controlled experiment in which any important difference between groups will show up at a high enough level of significance to insure a certain reliability in the conclusion.

3. Abuses will be rampant. It is no criticism of the position held on statistical grounds by Jones and Marks to point out that the considerations involved in the choice of a one-tailed test are really rather delicate. A nice instance of what can happen is seen in an experimental report by Gwinn (3). Gwinn reports two experiments which are not markedly different from each other. They turn out in opposite directions, and, by appropriate selection of the position of his "critical tail," Gwinn establishes significance and near significance (1 per cent and 8 per cent levels, approximately) for his results on the basis of one-tailed tests.

The moral can be pointed with advice. We counsel anyone who contemplates a one-tailed test to ask of himself (before the data are gathered): " If my results are in the wrong direction and significant at the one-billionth of 1 per cent level, can I publicly defend the proposition that this is evidence of no difference?" If the answer is affirmative we shall not impugn his accuracy in choosing a one-tailed test. We may, however, question his scientific wisdom.

References

1. Blodgett, H. C. The effect of the introduction of a reward upon maze performance of rats. *Univer. Calif. Publ. Psychol.*, 1929, **4**, 113–134.
2. Blodgett, H. C. Reynolds' repetition of Blodgett's experiment on "latent" learning. *J. exp. Psychol.*, 1946, **36**, 184–186.
3. Gwinn, G. T. Resistance to extinction of learned fear drives. *J. exp. Psychol.*, 1951, **42**, 6–12.
4. Hick, W. E. A note on one-tailed and two-tailed tests. *Psychol. Rev.*, 1952, **59**, 316–318.
5. Jones, L. V. Tests of hypotheses: one-sided vs. two-sided alternatives. *Psychol. Bull.*, 1952, **49**, 43–46.
6. Kendler, H. H. Some comments on Thistlethwaite's perception of latent learning. *Psychol. Bull.*, 1952, **49**, 47–51.
7. Maltzman, I. The Blodgett and Haney types of latent learning experiment: Reply to Thistlethwaite. *Psychol. Bull.*, 1952, **49**, 52–60.
8. Marks, M. R. Two kinds of experiment distinguished in terms of statistical operations. *Psychol. Rev.*, 1951, **58**, 179–184.
9. Meehl, P. E., & MacCorquodale, K. A failure to find the Blodgett effect and some

secondary observations on drive conditioning. *J. comp. physiol. Psychol.*, 1951, **44**, 178–183.

10. Reynolds, B. A repetition of the Blodgett experiment on "latent" learning. *J. exp. Psychol.*, 1945, **35**, 504–516.

11. Thistlethwaite, D. A critical review of latent learning and related experiments. *Psychol. Bull.*, 1951, **48**, 97–129.

12. Thistlethwaite, D. Reply to Kendler and Maltzman. *Psychol. Bull.*, 1952, **49**, 61–71.

13. Underwood, B. J. The learning and reten of serial nonsense lists as a function of distributed practice and intralist similarity. Paper read at Amer. Psychol. Ass., Washington, D. C., September, 1952.

8.3   A Rejoinder on One-Tailed Tests

Lyle V. Jones

In a recent issue of this journal, Burke (1) criticizes earlier discussions of one-tailed and two-tailed tests (3, 4, 5) and suggests need for caution in the application of one-tailed statistical tests to psychological research designs. The writer is in accord with several implications of Burke's note. As is true for most statistical designs, abuses would be reduced markedly were the test model completely specified and justified in terms of the purpose of investigation, before data are viewed by the investigator. To be guided by the data in the specification of hypotheses and statistical tests is a grave breach of the rules of experimental verification (cf. 6).

The argument presented by Burke, however, is more than a plea for careful consideration of the choice of test models for given experimental problems. It is stated that the selection of a one-tailed test model requires that an investigator be willing to "publicly defend the proposition . . . of no difference" if results actually show a large difference in the direction opposite to that predicted (1, p. 387). The proposition appears indefensible, since it demands arguing that a particular observed difference, no matter how large, is only a sampling departure from zero. If accepted,

Burke's argument should lead to universal avoidance of one-tailed tests.

Consider the following experimental problems, selected for simplicity as single variable designs: (*a*) On the basis of a certain behavioral theory, we might predict that an experimental condition imposed upon subjects in the population under study would raise the mean level of performance on a given task. The theory provides a prediction (an alternative hypothesis) that the mean for the experimental group population will exceed the mean for the control group population. The hypothesis under test is that the mean for the experimental population is the same as or less than that for the control population. We should like a statistical test that will yield a decision: either the data are consistent with the hypothesis under test, or we reject the hypothesis in favor of the alternative. (*b*) In a field of applied psychology a new diagnostic technique is developed and is to be adopted if, and only if, we are confident that it is better than the current technique which would be replaced. Assuming the availability of a suitable criterion, the two techniques are applied to comparable samples or to the same sample; interest resides in the extent to which the parametric proportion of successful predictions using the new technique exceeds that using the old. A statistical test is to supply a decision: either the new technique is no

Reprinted from *Psychological Bulletin*, 1954, **51**, 585–586, by permission of the publisher and author. Copyright 1954 by the American Psychological Association.

more adequate than the old, or the new technique is more adequate.

For the class of problems illustrated by these two examples, the hypothesis under test is not simply one of no difference. We wish to test the hypothesis that the algebraic difference between parametric mean performance under experimental and control conditions is zero *or negative* against the alternative hypothesis that the difference is positive. It is meant to stress this formulation of the one-tailed statistical test with the greatest possible emphasis. With this formulation, it is apparent that acceptance of the hypothesis under test does not demand defense of a proposition of no difference; the observed difference, whether negative, zero, or slightly positive, simply does not allow acceptance of the alternative hypothesis at the level of stringency (values of α and β) chosen for the test.

In a footnote, Burke (1, p. 385) criticizes this formulation as it appeared earlier (4, p. 45) on the grounds that it confuses hypotheses to be tested with hypotheses to be guarded against as alternatives. To the contrary, this statement of the problem clarifies the nature of the hypothesis under test. The hypothesis to be tested is[1]

$$H_0 : \mu_e - \mu_c \leqq 0,$$

where μ_e is the population mean for the experimental condition, μ_c is the population mean for the control condition, and the experimental prediction yields an alternative hypothesis,

$$H_1 : \mu_e - \mu_c > 0.$$

Burke's primary argument seems to rest upon the contention that "there is, under the single-tailed rule, no safeguard whatsoever against occasional large and serious errors when the difference is in the unexpected direction" (1, p. 385). Our formulation of the hypothesis under test completely resolves this

---

[1] An equivalent formulation of the one-tailed test model, of which the writer was unaware at the time of his earlier note, is that proposed by Dixon and Massey (2, pp. 100–104).

difficulty, for no error is committed when that hypothesis is accepted on the basis of a large observed difference in the unexpected direction. The event is one of a class of events consistent with the hypothesis tested.

If one were to retain the alternative hypothesis, H_1, above, and to adopt a two-tailed statistical test, accepting H_1 when there were observed large differences between the means *in either direction*, his position would be unenviable. For, following the rules of his test, he would have to reject the hypothesis under test in favor of the alternative, even though an observed difference, $\mu_e - \mu_c$, was a substantial negative value.

The remaining discussion by Burke consists of pragmatic arguments against the adoption of one-tailed tests. The arguments appear valid only under the assumption that every application of the one-tailed test is an abuse of experimental methodology. Certainly, if (*a*) the test model is specified completely (including specification of the significance level to be adopted) before the data are gathered, and if (*b*) the purpose of the test is only to determine whether a particular directional prediction is supported by the data, then the one-tailed test not only is appropriate, but it is in error to use a two-tailed test model.

References

1. Burke, C. J. A brief note on one-tailed tests. *Psychol. Bull.*, 1953, **50**, 384–387.
2. Dixon, W. J., & Massey, F. J., Jr. *Introduction to statistical analysis*. New York: McGraw-Hill, 1951.
3. Hick, W. E. A note on one-tailed and two-tailed tests. *Psychol. Rev.*, 1952, **59**, 316–318.
4. Jones, L. V. Tests of hypotheses: One-sided vs. two-sided alternatives. *Psychol. Bull.*, 1952, **49**, 43–46.
5. Marks, M. R. Two kinds of experiment distinguished in terms of statistical operations. *Psychol. Rev.*, 1951, **58**, 179–184.
6. Marks, M. R. One- and two-tailed tests. *Psychol. Rev.*, 1953, **60**, 207-208.

8.4 Three Criteria for the Use of One-Tailed Tests

Herbert D. Kimmel

Examination of the recent literature on the question of when to use one-tailed tests of significance in psychological research reveals a state of unresolved disagreement. A variety of differing opinions (1, 2, 5, 7, 8, 9, 10, pp. 62–63) have been presented, ranging from Burke's (2) exhortation that psychologists should never report one-tailed tests in the public literature to Jones' (8) statement that we may not only do so, but, in certain instances, we will be in error if we fail to do so.

It is by no means necessary for psychologists to agree on all matters of importance to them. Disagreement regarding methodological considerations, however, especially when they bear on how and when propositions shall be accepted as true or rejected as false, should not be permitted to persist indefinitely. The argument is not settled by noting, as Burke (2) does, that the increased use of one-tailed tests may result in the one-tailers scoring a sociological victory almost before the controversy has begun. Actually, this observation by Burke does not coincide completely with the fact that many responsible investigators have continued to employ two-tailed tests (in situations calling for one-

tailed tests according to Jones' view) long after the opening of the one-tailed avenue.[1]

In attempting to arrive at a set of acceptable criteria for the use of one-tailed tests, it is important to note that the argument is not one of mathematical statistics but primarily one of experimental logic. Burke and Jones would agree that one-tailed tests should be used to test one-tailed hypotheses; their disagreement concerns when one-tailed hypotheses should and should not be made.

Before proceeding to the proposed criteria, it would be of value to consider the difference

Reprinted from *Psychological Bulletin*, 1957, **54**, 351–353, by permission of the publisher and author. Copyright 1957 by the American Psychological Association.

---

[1] An example of an experiment with an explicit directional hypothesis, but employing a two-tailed test, is reported by Davitz (3). This experimenter reasoned that the injection of tetraethylammonium prior to extinction trials would inhibit the punishing effect of the emotional response under study and, consequently, would result in faster extinction in the experimental animals than in a placebo-injected control group. Instead, Davitz found that the experimental group extinguished slower than the control group, the difference in mean number of trials being significant at the 5 per cent level using a two-tail test. A one-tailed hypothesis in this experiment (as would have been urged by Jones) would have made it impossible to evaluate the significance of the obtained difference. A study by Hilgard et al. (6), on the other hand, stated a one-tailed hypothesis in a situation in which a difference in the unpredicted direction could have been predicted with as much justification on the basis of previous work. They obtained a difference in their predicted direction that was significant at the 5 per cent level using a one-tailed test. Their rejection of the null hypothesis on the basis of the difference they obtained is the equivalent of loosening the conventional 5 and 1 per cent standards.

between one- and two-tailed hypotheses from a viewpoint that has not been stressed by previous writers. All concerned agree that a given mean difference in the hypothesized direction is "more significant"[2] under a one-tailed hypothesis (in the correct direction) than under a two-tailed hypothesis. This is due to the fact that there are exactly twice as many chances of committing a type 1 error, with a given mean difference, under a two-tailed hypothesis. The important consideration is that this gain does not accrue without concomitant loss. Even psychology has its law of conservation of energy.

The price that is paid in return for the increased power of one-tailed tests over two-tailed tests stems from the fact that two-tailed null hypotheses are actually more specific than their one-tailed counterparts. A two-tailed null hypothesis can be rejected by a large observed difference in either direction but a one-tailed null hypothesis cannot be rejected by a difference in the unpredicted direction, no matter how large this difference may be. This means that an experimenter using a one-tailed hypothesis *cannot* conclude that an extreme difference in the unpredicted direction is reliably different from zero difference. This limitation cannot be shrugged off by the comment, " We have no interest in a difference in the opposite direction." Scientists are interested in empirical fact regardless of its relationship to their preconceptions.

The meaning of this limitation is exemplified even in applied studies; e.g., those intended to answer the question whether a new product is " better " than the current product. It would be desirable to be able to conclude that the new product is not only " not better " (which is all that failure to reject a one-tailed null hypothesis permits[3]), but, in fact,

" poorer." The decision not to market the proposed new product would follow from either conclusion, it is true, but the additional information available as a result of rejecting a two-tailed null hypothesis from the unexpected side could very well indicate a course of behavior quite different from that indicated by the mere inability to reject a specific one-tailed null hypothesis.

It is hoped that the following criteria will be acceptable to psychological investigators as a group and will be adopted conventionally as a guide. The ultimate consequence of our present state of ambiguity on this matter can only be confusion and subsequent retrogression to a more primitive level of scientific communication and understanding.

Criteria for the Use of One-Tailed Tests

1. Use a one-tailed test when a difference in the unpredicted direction, while possible, would be psychologically meaningless. An example of this situation might be found in the comparison of experimental and control groups on a skilled task for which only the experimental group has received appropriate training. The experiment would have to be designed in such a way as to eliminate all known conditions that could produce opposite results (e.g., not testing immediately after training to avoid fatigue effects, not testing too long after training to avoid memory loss effects, etc.). Since a difference in the unpredicted direction will have been declared beforehand to have no possible meaning (in terms of previous data and present operations) one-tailed hypotheses could not undergo metamorphosis into two-tailed hypotheses to permit testing the significance of differences in the unpredicted direction.

2. Use a one-tailed test when results in the unpredicted direction will, under no conditions, be used to determine a course of behavior different in any way from that determined by no difference at all. This situation is exemplified by the applied study discussed above, in which a new product is compared with one already on the market.

---

[2] That is to say, by chance, the undirectional event is half as probable as the bidirectional; thus its occurrence, being half as likely, is twice as significant.

[3] As Fisher (4) has pointed out, an experimenter never " accepts " the null hypothesis, he merely fails to reject it on the basis of his data. This is one reason why the null hypothesis in a particular experiment should be stated as specifically as possible.

3. Use a one-tailed test when a directional hypothesis is deducible from psychological theory but results in the opposite direction are not deducible from coexisting psychological theory. If results in the opposite direction are explainable in terms of the constructs of existing theory, no matter how divergent from the experimenter's theoretical orientation this theory may be, the statistical hypothesis must be stated in a way that permits evaluation of opposite results. If this criterion were not already implicitly accepted by psychologists, crucial experiments could never be performed.

It should be apparent that the three criteria stated above are actually slightly differing reflections of the same underlying precept. Neither the ethical nor the logical decisions of individual scientists can be prescribed beforehand by any set of standards, no matter how all-pervasive these standards may seem at a given moment. The three criteria proposed above, however, are offered as temporary guideposts until such time as a new set of temporary criteria supersede them. Proponents of one-tailed tests, such as Jones (7, 8), cannot complain that the use of these criteria will reduce the number of one-tailed tests to near zero, without admitting that these tests have been misused in the past. Opponents of one-tailed tests, such as Burke

(1, 2), should welcome this attempt to limit the use of one-tailed tests to those infrequent situations provided for by the proposed criteria.

References

1. Burke, C. J. A brief note on one-tailed tests. *Psychol. Bull.*, 1953, **50**, 384–387.
2. Burke, C. J. Further remarks on one-tailed tests. *Psychol. Bull.*, 1954, **51**, 587–590.
3. Davitz, J. R. Decreased autonomic functioning and extinction of a conditioned emotional response. *J. comp. physiol. Psychol.*, 1953, **46**, 311–313.
4. Fisher, R. A. *The design of experiments.* London: Oliver and Boyd, 1947.
5. Hick, W. E. A note on one-tailed and two-tailed tests. *Psychol. Rev.*, 1952, **59**, 316–318.
6. Hilgard, E. R., Jones, L. V., & Kaplan, S. J. Conditioned discrimination as related to anxiety. *J. exp. Psychol.*, 1951, **42**, 94–99.
7. Jones, L. V. Tests of hypotheses: One-sided vs. two-sided alternatives. *Psychol. Bull.*, 1952, **49**, 43–46.
8. Jones, L. V. A rejoinder on one-tailed tests. *Psychol. Bull.*, 1954, **51**, 585–586.
9. Marks, M. R. Two kinds of experiment distinguished in terms of statistical operations. *Psychol. Rev.*, 1951, **58**, 179–184.
10. McNemar, Q. *Psychological statistics.* New York: Wiley, 1955.

8.5 One-Tailed Tests and "Unexpected" Results

Marvin R. Goldfried

There has been much controversy (Burke: 1953, 1954; Hick, 1953; Jones: 1952, 1954; Marks: 1951, 1953) regarding the use of the one-tailed test of significance. The important question debated is not *if* it should be used, but rather *when* it should be used. Kimmel (1957) has recently attempted to resolve the controversy by suggesting criteria for the use of one-tailed tests. He maintains that one-tailed tests may be used when results in the opposite direction: (*a*) will not be used to determine any new course of behavior, (*b*) will be psychologically meaningless, or (*c*) cannot be deduced by any psychological theory, while an outcome in the expected direction can. It is these last two instances that will be dealt with in this paper.

When an experimenter uses a one-tailed test and finds that his results are in disagreement with his prediction (i.e., if they are in the opposite direction and would have been statistically significant had a two-tailed test been used), he can do one of several things. One course of action is simply to ignore these findings. In practical problems (e.g. deciding whether or not to introduce new machinery

for production), this approach, which is consistent with Kimmel's first criterion, is quite acceptable. With regard to this practice in psychology, on the other hand, Burke has quite correctly pointed out that "It is to be doubted whether experimental psychology, in its present state, can afford such lofty indifference toward experimental surprises" (Burke, 1953, pp. 385–386). Because an outcome is not deducible from any *existing* theory does not mean that it could not be deduced from *future* theories. The implicit assumption in this practice is that no new theoretical approaches will be advanced in the future, and that the task of psychology as a science is to confirm the presently existing theories. A similar criticism might be made of Kimmel's criterion of unpredicted differences being "psychologically meaningless." He defines the "possible meaning" of a difference in the unpredicted direction ". . . in terms of previous data and present conditions" (Kimmel, 1957, p. 352). Whether or not it is "possible" for a given proposition to have meaning, however, depends upon whether or not it is capable of confirmation (Carnap, 1953). It seems that what Kimmel is referring to when he speaks of the "possible meaning" of a given outcome is actually the degree to which a proposition regarding this outcome (i.e., one that states that such an event does not fit into an existing psychological theory) has been confirmed. Thus, since "psychological mean-

The writer is greatly indebted to K. H. Kurtz for his critical evaluation of this paper.

ingfulness" in this sense will change as our knowledge increases, criticisms made of the criterion of theoretical predictability apply here as well.

The experimenter might, on the other hand, wish to take cognizance of his unexpected findings. To do so, however, he must adopt the procedure of changing his original null hypothesis. Instead of testing the hypothesis that $\mu_1 \leq \mu_2$ (one-tailed), he might test the hypothesis that $\mu_1 = \mu_2$ (two-tailed).[1] However, if this practice is adopted, there will be an *increase* in the probability of making a Type I error. This results from the combination of the probability of committing a Type I error when using the original null hypothesis with the probability associated with the new null hypothesis. For example, suppose one has adopted the .05 level of significance when using a one-tailed test; the probability of making a Type I error is thus .05 (the probability associated with the critical region in the given tail). If the null hypothesis is changed to account for the difference which is statistically significant in the opposite direction, what is actually being used now is a two-tailed test. Was this to be used originally (i.e., before the study was conducted), the probability of the experimenter committing a Type I error would be .05 (.025 being associated with the critical point in each direction). Since this null hypothesis has been adopted after the decision had been made to use a one-tailed test (with its associated probability of making this error), the probability of this researcher committing a Type I error is now .075 (i.e., .05 when $H_0: \mu_1 \leq \mu_2$ plus .025 related to the "unexpected" direction when $H_0: \mu_1 = \mu_2$). It should also be noted that even an experimenter who believes he has the option to use such a procedure is operating under the .075 level of significance —whether or not he has occasion to test a difference in the "unexpected" direction. Thus, whether he knows it or not, such an investigator is using a two-tailed test with one tail twice as large as the other.

Another possible course of action would be to repeat the experiment, now using a two-tailed test. Assuming that the errors of measurement are not significantly greater in this replication and this experimenter obtains the same results, he may now conclude that his findings in the previously unexpected direction are significant. Thus, the decision to use a one-tailed test may result in the necessity of repeating the study if results appear in the opposite direction.

The above considerations indicate that the criteria of theoretical predictability and psychological meaninglessness are not as decisive as they may appear to be. Three possible courses of action available when results occur in the unpredicted direction each present difficulties.[2] *Ignoring* differences in the "unexpected" direction leads to the omission of findings which may have important theoretical significance, and thus stifles any fresh theoretical thinking that might have otherwise emerged. If the differences *are* recognized by switching to a two-tailed test within a given study, psychology as a science may unwittingly be led to operate under conclusions with a lower level of statistical significance. The third approach, repeating the experiment and applying a two-tailed test to this new set of data, might be undesirable in terms of the time, expense, etc. that would be involved. The decision to use a one-tailed test should thus be made in light of the difficulties with which the investigator is confronted when the results occur in the "unexpected" direction.

References

Burke, C. J. A brief note on one-tailed tests. *Psychol. Bull.*, 1953, **50**, 384–387.

Burke, C. J. Further remarks on one-tailed tests. *Psychol. Bull.*, 1954, **51**, 587–590.

---

[1] Some experimenters might make the even greater change in their null hypothesis by testing $\mu_1 \geq \mu_2$ (one-tailed, but in the opposite direction).

[2] Whether or not there exist any other possible approaches requires further analysis.

Carnap, R. The two concepts of probability. In H. Feigl & M. Brodbeck (Eds.), *Readings in the philosophy of science.* New York: Appleton-Century-Crofts, 1953. Pp. 438–455.

Hick, W. E. A note on one-tailed and two-tailed tests. *Psychol. Rev.*, 1952, **59**, 316–318.

Jones, L. V. Tests of hypotheses: One-sided vs. two-sided alternatives. *Psychol. Bull.*, 1952, **49**, 43–46.

Jones, L. V. A rejoinder on one-tailed tests. *Psychol. Bull.*, 1954, **51**, 585–586.

Kimmel, H. D. Three criteria for the use of one-tailed tests. *Psychol. Bull.*, 1957, **54**, 351–353.

Marks, M. R. Two kinds of experiment distinguished in terms of statistical operations. *Psychol. Rev.*, 1951, **58**, 179–184.

Marks. M. R. One- and two-tailed tests. *Psychol. Rev.*, 1953, **60**, 207–208.

Editor's comments: If an experiment involves only two treatment levels, it is customary to use a t test statistic to compare the two means. If the experiment contains more than two treatment levels, the number of pairwise comparisons that can be made is given by $k(k-1)/2$ where k refers to the number of levels. Thus, in an experiment with as few as $k = 6$ treatment levels, $6(6-1)/2 = 15$ pairwise comparisons among means can be made. The probability of obtaining at least one significant comparison by chance increases markedly as the number of pairwise comparisons is increased. Hence if an experimenter computes enough t ratios each at α level of significance he will probably reject one or more null hypotheses even though all the null hypotheses are true. This raises the question: should the probability of committing an error be set at α for each individual comparison or should the probability of committing an error be set at α or less for some larger conceptual unit such as the collection of comparisons? This is not an easy question to answer as evidenced by the diversity of practices found in the contemporary behavioral-science literature.

In recent years a number of alternatives to the t test statistic have been developed that enable an experimenter to set the probability of making an error at α or less for the collection of tests. The rationale for using these alternatives to the t test statistic as well as some of the problems that occur when an experimenter wants to compare more than two treatment levels in an experiment are examined in the following three articles.

8.6   Multiple Comparisons in Psychological Research

Thomas A. Ryan

Whenever an experiment involves collecting data from more than two groups or under more than two conditions we become

The writer wishes to express his appreciation to Urie Bronfenbrenner and W. T. Federer for their detailed comments and suggestions upon an earlier draft of this paper.

Abridged from *Psychological Bulletin*, 1959, **56**, 26–47, by permission of the publisher and author. Copyright 1959 by the American Psychological Association.

involved in the problem of multiple comparisons—the problem of comparing each group with every other group or arranging the results in rank order. This becomes a problem when we wish to assign a level of confidence or significance to our conclusions about the relationships among all of the populations involved. Classical methods such as the F test permit us only to reject the overall null hypothesis that all of the means are equal but they do not provide a procedure for comparing specific means with each other.

In the older psychological literature, this problem has been dealt with in a haphazard manner, without recognizing the issues involved. More recently, statistical procedures specifically designed for multiple comparisons have become available and have been discussed briefly in the psychological journals (McHugh & Ellis, 1955; Stanley, 1957). It has not been clear to many psychologists, however, that there are several different methods with different basic assumptions or approaches. There are important questions of logic involved in the use of these methods and these issues have not been clearly faced in the psychological literature. This is partly because many of the papers by statisticians on this subject are in sources which are inaccessible or rarely used by psychologists. In particular, one of the most extensive discussions of the logical problems of multiple comparisons, that of J. W. Tukey, has been available only in a privately circulated paper.[1] Other aspects of the problem have not been dealt with at all, to this writer's knowledge, so that it seems to be time for an attempt to survey the problem systematically.

The emphasis here is upon questions of logic rather than specific methods of computation. For the latter, we shall simply refer to appropriate sources, *after* we have tried to make clear the implications of choosing to use a particular method or set of tables.

Multiple comparisons and other multiple tests. Multiple comparisons are only one instance of the use of multiple statistical tests in a single piece of research. We shall not have space to deal explicitly with the other tests except as we need to distinguish them from the problem of multiple comparisons. One of the sources of confusion in the past has been the failure to distinguish one kind of multiple testing from another.

In order to prevent this kind of confusion from the outset, we may list at least five main

cases in which multiple statistical tests are employed:

1. *Multiple comparisons.* This covers all cases in which results in several different groups are to be compared. Any statistic may be involved—mean, median, frequency, correlation coefficient, etc. For example, we may wish to compare the correlations between intelligence and school grade in Schools A, B, C, D, E, etc. The methods which have been explicitly published and taken up by psychologists have all been concerned, however, with comparisons in terms of the means of groups. Some methods for other statistics, e.g., proportions, are now beginning to appear.

2. *Multiple tests with intercorrelated variables.* The most common instance of this case is the computation of a number of different correlation coefficients with a single batch of Ss. If 10 tests are given there may be 45 intercorrelations and the researcher may wish to state which of these correlations are significant.

3. *Multiple variables in analysis of variance.* A factorial design will permit the computation of several F ratios for the data of an experiment. These F ratios may not be independent if a common error estimate is used for several of them. Whether independent or not, several tests are made in the same experiment, and the implications of this fact need to be analyzed. Similar problems arise if other kinds of analysis are used for what is essentially a factorial design. For example, several nonparametric tests may be made of different rearrangements of the data in a way which is equivalent to analysis of the main effects in analysis of variance.

4. *Replicated tests of a single hypothesis.* In the first three cases mentioned above the statistical tests are concerned with different hypotheses. For example, the different F tests in a factorial design are concerned with different variables. This fourth heading is concerned with cases where the *same* experiment is repeated with different groups of Ss and repeatedly tested for statistical significance.

[1] J. W. Tukey, The Problem of Multiple Comparisons. Privately circulated monograph, 1953.

5. *Overlapping measures relating to a single hypothesis.* Several different ways of measuring the same underlying variable may be available—e.g., different measures of rate of learning—and a significance test is applied to each of the measures separately.

The main purpose of the list is to emphasize that we are concerned only with the first of these headings. Space will not permit us to analyze the other cases, which must be left to later discussions.

General Issues in Multiple Comparisons

A priori vs. a posteriori comparisons. It has been assumed that no special modifications of classical methods are needed where the comparisons to be made are specified in advance of the collection of data (a priori). Most of the recent literature on multiple comparisons has concentrated upon methods for making comparisons suggested by the data (a posteriori, also called post-mortem comparisons). For example, suppose that five conditions of learning are being compared. In advance, the experimenter predicts from his learning theory that Condition A will lead to most rapid learning, Condition B will be second, and so on. Fisher (1947), and others following him, have recommended that the experimenter perform an over-all F test first, then, if this is significant, he may perform ordinary t tests between A and B, B and C, etc. It is pointed out, however, that this method would be incorrect if the comparisons to be tested had not been selected in advance (Fisher, 1947; McHugh & Ellis, 1955; Stanley, 1957). The new methods have been designed for comparisons suggested by an inspection of the data.

We shall contend that the differences between the a priori and the a posteriori situation are slight, or even nonexistent, when everything is taken into account. This is to say that the newer methods are needed for *all* multiple comparisons, and that the classical methods are inappropriate even in a priori

comparisons, except for very special circumstances.

The issue here is similar to that involved in the debate over "one-tailed" vs. "two-tailed" tests of significance for comparing two groups (Burke, 1953; Hick, 1952; Jones, 1952; Marks, 1951). The one-tailed test is appropriate only if the direction of difference is predicted in advance, and if the experimenter is willing to overlook any difference in the opposite direction, no matter how large. Only two conclusions are possible from the data when a one-tailed test is used—either there is a difference in the predicted direction, or the results of the experiment are inconclusive; in effect, the experiment cannot obtain results which are considered as a significant refutation of the prediction. If the experimenter allows for the possibility of a result that contradicts his hypothesis, he must use a two-tailed test, and there is no difference in method of analysis from that used in an empirical experiment where no predictions are made in advance.

In the case of more than two means, the number of possible conclusions is increased. We may have not only confirmation or contradiction of the prediction, but we may also have varying degrees of partial agreement with the prediction. Since it is usually not specified in advance what will be considered as a partial confirmation of the prediction, the situation is reduced essentially to the a posteriori case. Only if the experimenter states in advance all possible conclusions and the rules by which these conclusions will be drawn, would he have an a priori test.

Because of the multiplicity of conclusions which might be drawn, it would appear most feasible to consider the statistical analysis as independent of any predictions of the experimenter. In other words, we consider the statistical analysis as a method of making statements about the state of affairs as revealed by the data. If it turns out that the state of affairs is in complete or partial agreement with the prior prediction, the experiment makes the theory more plausible. If the results are wholly or partially in opposition to the prediction, then the theory needs to be revised.

At this point the position must be stated very dogmatically. After some of the other problems have been dealt with, and a more complete terminology has been developed, we shall be able to give these conclusions further support.

The concept of the error rate. The notions of *significance level* or *confidence level* have been useful ideas so long as we were dealing with a single difference between one pair of means, a single F ratio, a single chi-square value, and so on. The use of these terms becomes confused, however, when we are making simultaneous statements about a number of different comparisons of means, several different F ratios in a single experiment, or the like. The confusion is due to the fact that the concept of significance level may be extended in several different directions when we are considering multiple comparisons or multiple tests. We owe much to J. W. Tukey, who has clarified this point, and who has developed the concept of *error rate* for multiple comparisons (see Footnote 2).

There are several different kinds of error rate involved in the multiple comparison problem (and in other situations involving multiple tests). Some methods of making multiple significance tests fix one of the error rates at a suitably low level, but may allow the other error rates to become absurdly large. The problem becomes that of deciding which error rates should be kept under control, or what compromises may be effected.

Three of the main kinds of error rate are:

1. *Error rate per comparison.* This is the probability that any particular one of the comparisons will be incorrectly considered to be significant. In general this approach is discouraged by statisticians for reasons explained below.

2. *Error rate per experiment.*[2] This is the long-run average number of erroneous state-

ments per experiment. In statistical jargon it is the *expected number* of errors per experiment. Unlike the first error rate, which is a probability, the error rate per experiment could be greater than one. That is, we could set a criterion of "significance" in such a way that we would average three false statements for each experiment.

3. *Error rate experimentwise.* This is the probability that *one or more* erroneous conclusions will be drawn in a given experiment. In other words, *experiments* are divided into two classes: (*a*) those in which all conclusions are correct, and (*b*) those in which some conclusions are incorrect. The error rate experimentwise is the probability that a given experiment belongs in class (*b*).

It may help to understand the distinctions among these error rates if we think of a long series of experiments carried out in a given field, all with the same experimental design. In each experiment a certain number of statements of significance is made—e.g., "Method A is significantly better than Method B"; "Method C is significantly poorer than Method B." To be concrete, suppose that there were 1000 experiments, each with 10 statements of significance, 10,000 statements in all. Of these statements, 90 are actually false, and these false statements are distributed among 70 of the experiments. The different error rates are then as follows:

1. Error rate per comparison: 90/10,000 or .009

---

[2] Tukey's terminology is based upon "families" of comparisons rather than upon the experiment. In the one-dimensional case, these are equivalent terms. That is, the comparison of each mean with each other in the experiment is a "family" of comparisons. If we are concerned with two-variable analysis, however, the experiment may be broken down into two families of comparisons, one for each variable. We could therefore specify an error rate per family and a rate familywise as well as per experiment and experimentwise. Our discussion will be based primarily upon the one-dimensional problem, and it seemed that the issues would be clearer if we emphasized the experiment as a unit. Even where there are several families of comparisons, we shall argue that the experiment should be the basis of analysis of the error rates. Another discussion of experiment-based error rates is found in H. O. Hartley (1955).

2. Error rate per experiment: 90/1000 or .09

3. Error rate experimentwise: 70/1000 or .07

If we look only at the error rate per comparison we would say that the statements of significance were made at better than the ".01 level." Yet the probability is greater than .05 that any given experimental report will contain one or more false claims of significance.

The various error rates are all the same in a simple experiment with a single comparison, but they become more and more divergent as the number of comparisons per experiment increases. Thus, if each of 10 means is compared with each of the others there are 45 comparisons in one experiment. If the "significance level" (Error Rate 1 above) of the test applied to each comparison is .01, we should expect .45 erroneous conclusions *per experiment*. The probability that there will be *one or more* incorrect conclusions in a given experiment (Sense 3) will be somewhere between these two values, usually closer to .45, as will be explained below.

Which of the three values is, then, the "significance level" to be attached to the conclusion from this experiment? This is a point for extensive analysis, but we shall need more concepts before we can do it justice. At this point we shall say only that the basis for the choice between these three rates is still incompletely analyzed. Statistical workers have recognized the problem and have developed their procedures for multiple comparisons primarily on the basis of the third rate of error—the probability that *one or more* erroneous conclusions will be made in a given experiment, the *experimentwise* error rate. The implications of this decision have not, however, been extensively developed, at least to the present writer's knowledge.

Multiple null-hypotheses. The concept of error rate cannot be defined completely without taking account of another important fact. In our distinctions between error rates per comparison, per experiment, and experiment-

wise, the reader may have inferred that the null hypothesis would be that all means are drawn from a single population—the same null hypothesis which is tested in analysis of variance by means of the *F* test. We shall call this the "complete" null hypothesis. This is one possibility which must be considered, but it is not by any means the only one. In our example of 10 means, five might be drawn from one population and five from another, six from one and four from the other, two from each of five different populations, and so on. For each of these different null hypotheses, there is an error rate per comparison, per experiment, and experimentwise, for any given method of testing differences. The question is, therefore, which of these null hypotheses is used to define *the* error rate for our statistical test?

Tukey's answer (see Footnote 2) to the above question is to define the error rate as the *maximum* value it attains under all possible null-hypotheses. Some of the currently proposed methods for multiple comparison are based solely upon the *complete* null hypothesis as the standard, even though the error rate may be higher with some other null hypothesis. Tukey's decision would seem the most reasonable as well as the most cautious approach to this aspect of the problem.

To show how the error rate may be higher for some partial null hypothesis than it is for the complete hypothesis, let us consider a specific method of testing multiple differences based upon traditional approaches. Ten groups are being compared, and we test first with an overall *F* test at the .01 level. Then if the *F* test shows significance, we will test each difference with an ordinary *t* test at the ".01 level." The experimentwise error rate is .01 if we consider only the complete null hypothesis, since no further comparisons will be made if the *F* test does not show significance. The *F* test is specifically designed to produce this error rate under the complete null hypothesis.

Suppose, however, that there are actually five populations, with two groups drawn from each population, and suppose that these populations are widely separated. Then it is almost certain that the F test will be significant, and t tests between pairs drawn from distinct populations will also be almost certainly significant, as they should be. We can still make errors, however, in comparing means in the pairs drawn from identical populations. Since there are five such comparisons, the probability that *one or more* of these will be incorrectly judged to be significant is $(1 - .99^5)$ which is approximately .05. Thus the error rate *experimentwise* is .05 instead of .01 for this particular null hypothesis. The more means there are to be compared by this method, the higher will the experimentwise error rate become, even though the error rate based upon the complete null hypothesis is fixed at .01 for any number of means.

Error rates and a priori comparisons. Now that we have looked at some of the different ways of evaluating error rates, we can deal more concisely with the problem of a priori vs. a posteriori comparisons. As an example, consider a learning experiment in which five conditions are being compared, and suppose that the experimenter has predicted in advance the complete order in which the means should appear. He has, in effect, predicted significant differences for all possible comparisons of the five means, and complete agreement with the theory should produce 10 significant differences. Suppose that he merely computes all 10 t ratios in the standard way, determining their significance by references to the standard "Student" tables, and assume that he uses the .01 levels from these tables. The method which this experimenter has used has an error rate of .10 *per experiment*, and also *experimentwise*, even though all of the tests were computed on the basis of a .01 level for the individual comparisons. In other words, in 10% of experiments analyzed by this method, there will be one or more "significant"[3] differences, even though the complete null hypothesis is true.

[3] In this discussion "significant" in quotes refers to a difference which would be judged to be significant in using the classical tables and based on single comparisons.

Compare this with the case where no predictions were made in advance. The experiment is performed to "see what happens" and, again, all possible t tests are computed. The error is exactly the same as it was when advance predictions were made, if we leave aside the question of "one-tailed" vs. "two-tailed" tests. (If the experimenter in the a priori case wishes to allow for contradictions to his theory which could come out to be "significant" he must use a two-tailed test, just like the experimenter who makes his comparisons after the results are in—a posteriori.)

In other words, the essential factor is the number of comparisons to be made and the error rate to be used, rather than the question of a priori vs. a posteriori comparisons. When ordinary t tests are applied to all comparisons, each of the different kinds of error rate is the same whether predictions were made in advance or not.

The only situation in which advance predictions make a difference would be that in which several groups are studied, but only certain pairs are to be singled out for significance tests. Suppose that in the a priori case, one pair is specified in advance as the only comparison of interest, while in the a posteriori case, only the largest difference is to be studied. The probability that the largest of 10 comparisons will be significant is not the same, of course, as the probability that a pair chosen at random will be significant. The null hypothesis is that the pair chosen in advance by the theory might as well have been chosen at random. Thus a t test applied in the usual way at the .01 level has a probability of .01 of being significant, in the a priori case. When the largest of the 10 differences is chosen, it has a probability of .10 of *appearing* to be significant at the .01 level by classical two-mean tests. The probability that the largest difference will be "significant" is the same as the probability that there will be one or more "significant" pairs among the 10 comparisons. In other words, the *experimentwise* error rate for all comparisons applies to this special case.

The above example is helpful in seeing the

Multiple Comparisons in Psychological Research   297

issues involved in multiple comparisons, but it has little practical application. Tukey suggests that it might occur when all but two of the groups were studied as "camouflage" and only the particular two are of interest to the experimenter. Usually, however, an experimenter who is testing a theory will use five groups for one of two reasons: (1) all are interrelated in the theoretical predictions or (2) some of the groups are predicted from theory while others are unpredictable from the standpoint of theory but the experimenter wishes to find out how they compare with each other and with those which are predicted by the theory. We have already shown that the first case is no different from the completely empirical exploratory study. The second case would be different only if the results were considered as belonging to two separate and unrelated experiments—one group of comparisons being used to test the theory, the other comparisons being considered as part of another empirical exploration.

In all of these examples, we have assumed that the experimenter who is making a priori comparisons will consider each "significant" difference in the predicted direction as supporting his theory, and each "significant" difference in the opposite direction as a contradiction to his theory. He could, of course, have specified other rules for interpreting the results. In actual practice of psychological research, however, he rarely does, and the usual situation is that no rules at all are specified in advance. The decision as to what constitutes "agreement," "partial agreement," and so on, is made only after the results are in and the significance tests are made. This is another strong reason, already mentioned in the preliminary discussion of this problem, for treating all cases of multiple comparison in the same manner, whether there are predictions in advance or not.

Nevertheless, we should investigate to see if carefully specified rules for interpreting the results in relation to the theory would have any effect upon the significance tests. Suppose, for example, that the experimenter says, "If there are at least some significant differences in the predicted direction, and none in

the opposite direction, I shall consider the theory as partially substantiated. If there are any differences which appear to be significant reversals of prediction I shall revise or abandon the theory." The error rate must now include errors of false acceptance of the theory and also errors of false rejection.

Here it is easy to show that there are circumstances in which false acceptance of the theory is almost certain. For example, suppose that Populations A and B are equal and substantially higher in mean than Populations C and D, the latter pair also being equal. Finally, Population E has a considerably lower mean than any other group. The psychological theory has predicted that all groups are different, with mean A the highest, B next, and so on down to E. In other words we are assuming that the actual state of affairs is in partial agreement with the theoretical predictions, but that the theory is wrong in the relation of A to B and of C to D. We suppose in addition that the differences which do exist are so large that significant differences are almost certain in those particular comparisons. In this situation the theory will be rejected only if A is found to be significantly lower than B, or C lower than D. In all other cases the theory will be considered as supported by the experimental results. If t tests are made at the .01 level there is only a .005 probability that Groups A and B will be found in significant contradiction to the theory, and the same value applies to the C-D pair. The probability that one or both will be reversed is approximately .01. Therefore the experimenter has a .99 probability of finding support for his theory and only a .01 probability of contradicting it.

At this point the reader may object that accepting the theory under these circumstances should not be considered as entirely erroneous. After all, the actual state of affairs in the populations is at least partially in agreement with the prediction from theory. Certainly, to accept the theory in this case would not be so bad as to accept the theory when the actual population values are a complete reversal of the predicted levels. The question then becomes: How do we evaluate different degrees of agreement between the actual state of affairs and the theoretical predictions? Clearly this cannot be done on the basis of probability, nor can it be built into a standard significance test. The seriousness of disagreement depends upon the structure of the theory and the nature of the groups being compared. For some theories, the fact that Populations A and B are equal could be a very crucial defect in the theory; in other cases, this might be only a minor point, easily rectified. If the relative importance of all of the possible comparisons were stated in advance, with some kind of numerical weights, it would be possible, although very complicated, to compute probabilities for each outcome and also some kind of a weighted risk function. This would differ from experiment to experiment and would probably be of little practical value.

To summarize, it is argued that comparisons decided upon a priori from some psychological theory should not affect the nature of the significance tests employed for multiple comparisons. Our reasons may be recapitulated as follows:

1. Ordinarily, no statement is made in advance as to what will be considered substantial agreement, partial agreement, partial contradiction, or complete disagreement with the theory. Even if such a statement were made, the probabilities of each of these conclusions being drawn incorrectly would have to be included in the error rate.

2. A theory which predicts the complete order of the results calls for just as many comparisons as the empirical experiment in which no prediction is made. Since the number of comparisons to be made is a crucial factor in the error rate, there is no difference in this respect between a priori and a posteriori comparisons.

3. Some comparisons may be more important to a theory than others. It is not feasible, however, to take account of this fact in devising significance tests or methods of setting confidence limits, since the relative weights would differ from experiment to experiment and would have to be specified quantitatively in advance. It is therefore more practical to examine the results in a common-sense manner and to evaluate qualitatively the degree of support or contradiction offered by the data.

Nonindependence of comparisons. In the textbooks, the student is sometimes warned against a posteriori comparisons, because the different comparisons are not independent of each other. While lack of independence is a factor to be taken into account, it is not at all the main problem. In fact, the error rates per comparison and per experiment are completely unaffected by independence or lack of it. The only important factor in these rates is the number of comparisons to be made. Only the experimentwise error rate is affected by independence. If all of the comparisons are perfectly positively correlated, all are significant or nonsignificant en bloc. Then the experimentwise rate is equal to the rate per comparison. In the case of complete independence, of negatively correlated comparisons, or even of moderate positive correlation, the experimentwise rate is nearly equal to the rate per experiment, when the latter is small. Most cases of multiple comparison fall into the latter category, so that the dependence of the comparisons has but a slight effect.

In the multiple comparison problem, the lack of independence involves the fact that each mean is compared with every other mean, and therefore appears in a number of different significance tests. In many cases also a single error estimate is used for all comparisons. The significance tests are therefore not independent of each other, but this turns out to be less important than was once believed. Another kind of dependency must also be considered. The samples used in determining the various means may also not be independent of each other, notably in the case where the same Ss are used for each experimental condition. Such dependencies are easily taken care of by using two-way analysis of variance with Ss considered as a second variable.

The above conclusions on the relative unimportance of the factor of independence in multiple comparisons do not necessarily apply to other cases of multiple significance tests. The other cases listed in the introduction involve other kinds of dependency and must be analyzed separately.

The Choice of Error Rates

In making multiple comparisons, then, neither specifying the tests in advance nor trying to arrange for independent tests are of much importance, since they have little practical effect upon any of the error rates. It is of much greater practical importance to consider which of the error rates is the best representation of the dependability or "significance" of our results. We may work at the .01 level on a per comparison basis, yet the probability may be almost 1.00 that we have made some erroneous statements of significance in a given experimental report. In our current psychological literature the various bases for error rates are confused and sometimes used interchangeably.

The problem we must consider is the implication of using a particular measure of error rate for clarity and consistency of treatment of our research results. The issue can be made concrete in an example: One experimenter performs a series of four experiments. In the first experiment he compares Groups A and B, in the second, Groups B and C, etc. Each experiment is published separately with a *t* test applied to the difference of means in each case. In each paper he summarizes the results obtained before and in the final paper he compares all five groups, still using simple *t* tests. A second experimenter, not so anxious for rapid and numerous publications waits until all the results are in on all five groups, and performs an analysis of variance on all groups, considering the results as significant only if the *F* value is beyond the .01 point. Both of these kinds of report are quite typical of the psychological literature. The second experimenter has used an experimentwise rate of .01, at least for the complete null hypothesis, but he does not yet have any method of making specific comparisons between groups. The first experimenter has used a .01 level *per comparison*, but his experimentwise rate for the whole series of connected comparisons may be as high as .10, depending on how many of the possible comparisons among the five means are actually made. (To simplify matters, we assume that the first experimenter would have performed all four experiments regardless of the results. If he waited for the results of each experiment before deciding whether to continue the series, matters would be further complicated.)

These examples should make clear that both the per comparison and the experimentwise rates are actually in use in typical researches now in the psychological literature, even when only classical techniques are used. The second example is now the more common approach, and even the first experimenter would probably be more likely to perform an analysis of variance in his last paper to summarize the over-all results. Whether he would be willing to retract earlier conclusions if the final analysis did not prove to be significant is, of course, an embarrassing question.

The current widespread use of analysis of variance would suggest adopting the experimentwise error rate as the standard practice. Current practice is not, however, sufficient justification unless it is based upon careful analysis. We must therefore examine the problem more fully.

Since the rate per comparison is the easiest to use and requires no new methods at all, we may first consider the main argument in its favor. It might be contended that it makes no difference whether specific comparisons are made one at a time by different experimenters, or in groups by a single experimenter. The same amount of data is added to the published literature in either case. Therefore if the simple *t* test is justified in one case it should be justified in the other also. As Tukey (see Footnote 2) states this argument (which he considers fallacious), the man who has studied several means at once has done more work and should be entitled to make more erroneous conclusions.

There is, however, one very strong reason why an experimenter who studies a number of different groups or conditions has less justification for using a per comparison rate than an experimeter who performs a single experiment with two groups. Even if the complete null hypothesis is true, and the experimenter is working with a factor or group of factors completely irrelevant to the behavior he is studying, the more conditions or the more variations of experimental conditions he studies the more chance he has of finding *some* differences which would appear to be significant on a per comparison basis. Thus, he can obtain more "significant" differences by working harder upon irrelevant variables. This is the reason why Tukey considers the point of view of allowing erroneous conclusions in proportion to the amount of work done as an untenable point of view.

The notion of allowing more errors per experiment in proportion to the amount of work done in the experiment would lead to another practice which is contrary to present usage. It would mean that the significance level in the ordinary two-group experimental design could be reduced as the number of cases is increased. In this situation we ordinarily maintain the significance level constant, but we gain through increases in power as the number of observations is increased. In the case of multiple comparisons we do not gain in power in the specific comparisons as the number of groups increases, but there is a compensation in that more information about more different relationships is gained as the number of comparisons increases.

There are even objections to the use of error rate per comparison in a certain type of "experiment" in which only two groups are compared. Consider the following situation: an experimeter is convinced that a certain factor should produce a difference in learning rate. He tries it once and fails to get a significant difference. He is so sure that the experiment should have worked that he reconsiders his experimental technique for possible

errors. He decides that some actually irrelevant feature of the experiment is responsible, changes it, and tries again. Finally after many different revisions of the conditions, all actually irrelevant, he obtains a "significant" difference and publishes the result. We assume that, as an honest scientist, he will mention in his report that several other trials failed, but this will not usually affect his test of significance, and he will usually explain away the earlier, unsuccessful trials as due to errors in technique. Clearly, all of his data should be tested as a single experiment, otherwise obtaining a "significant" difference will depend only upon the experimenter's stubbornness and patience, or upon the number of his research assistants.

Error rate vs. Type II error and power. Several psychologists to whom the above argument has been presented have raised objections to the experimentwise error rate on the ground that it leads to great loss of power. They point out that a t ratio may have to be as high as 4 or 5 for 20 degrees of freedom to be significant at the .01 level instead of 2.85 as it is when significance is measured in the classical way. If this happens, they say, we are obviously going to miss a lot of real differences which might turn out to be important.

While it is perfectly true that the bigger the difference which is required for significance, the less powerful is the test[4] (other things being equal, of course), this fact is irrelevant to the issue involved in the choice of error rates. If the experimenter prefers, he can still use a t value of 2.85 as his criterion of significance even if he reports his results in terms of error rates experimentwise. The difference is that his results will be reported as significant at (say) the .50 level experimentwise instead of the .01 level on a per comparison basis (which is usually not labelled as such).

In other words, the issue is not more or less powerful tests, since the power can be adjusted to any desired level, but simply how we are

---

[4] See Harter (1957) for evaluation of the power of several multiple comparison procedures.

going to evaluate the Type I error. It must be admitted, however, that the tables which are available for establishing error rates on an experimentwise basis tend to limit the experimenter to a fixed value of the error rate (usually .05). This situation can be changed, however, if there is good reason to increase 'he power of the tests.

Duncan's compromise. Duncan (1951, 1955) has argued that there is not only a loss of power in changing from the per comparison to the experimentwise basis, but that this loss of power becomes progressively greater as the number of comparisons increases. Since his method has been used in several recent research papers in psychology, we shall examine his assumptions in detail. Thus if one experiment involves 5 means while another experiment involves 10 means, and both are evaluated by holding the experimentwise error fixed at .05, the experiment with the 10 means is less powerful in the sense that each difference must be larger to be judged significant.

According to Duncan, this state of affairs should be reversed. As more and more conditions are studied it is more and more likely that some real differences exist, and therefore the statistical tests should become more powerful as the number of comparisons increases. This would be the case if we used the error rate per comparison as our basis of establishing significance, but then the probability of Type I error reaches unreasonably high levels. As his compromise, Duncan proposes to base statements of significance upon the rate of error *per independent comparison* or *per degree of freedom*.

The argument for Duncan's method would be that when two different experimenters each perform a simple comparison of two groups we allow them *each* a certain error rate, because they have performed two *independent* comparisons. It is proposed that this allowance be extended to a single experiment involving several comparisons. If 10 means are compared there are 45 comparisons, but only 9 can be made if we are to keep them independent of each other. Table 1 indicates the relationship among the rates per comparison, per degree of freedom and per experiment for multiple means.

Table 1. Error rates per experiment[a] when error rates per comparison and error rates per degree of freedom are controlled.

No. of Means	Error Rate per Comparison Fixed at .01	Error Rate per Degree of Freedom Fixed at .01
2	.01	.01
3	.03	.02
4	.06	.03
5	.10	.04
10	.45	.09
20	1.90	.19
50	12.25	.49

[a] Based on the complete null hypothesis.

The arguments against the per comparison basis of testing also apply, although not so powerfully, against Duncan's compromise procedure. The experimenter still can increase his chances of finding significant differences by multiplying the number of irrelevant conditions which he studies in a given experiment. The probability of erroneous conclusions does not increase as fast as the number of comparisons increases, but it still increases.

While it is true that it becomes less and less likely that all populations have the same mean, as we increase the number of groups, it is not known to what extent these differences are relevant to the problem being studied. We may therefore merely be increasing our probability of detecting differences which are the random result of factors which are not under study in the experiment. In other words, we increase the risk of finding a hodgepodge of "significant differences" which cannot be given a meaningful interpretation.

Duncan's approach is also contrary to common practice, where the *F* test is applied at the same probability level, regardless of the number of groups under study. To be sure, common practice in analysis of variance could also be wrong and could be revised according to Duncan's point of view, but we need to have some stronger arguments for doing so.

Since the degree of conservatism and the inversely related power of the test can be explicitly varied by choosing varying rates of error per experiment or experimentwise, depending upon the type of material being studied and the purposes for which conclusions are being drawn, it does not seem necessary to adopt a rigid compromise between the per comparison and the experiment-based rates. Thus Duncan's special procedure seems unnecessary and may confuse the issues for the user of statistics.

Rates per experiment vs. experimentwise rates. While we cannot say flatly that all significance tests or all confidence limits must be based upon the experiment as a unit, there are, as we have seen, strong reasons to make the experiment the *normal* reference unit at least. In any event, it should always be made clear in an experimental report which approach is being used. If the rate per comparison is chosen, it should require special justification.

Although the two experiment-based error rates are often numerically almost equal, they do represent somewhat different points of view about the nature of the conclusions from an experiment. In one case we control the total number of erroneous statements made in each experiment (rate per experiment). In the other, we consider that *any* erroneous statement spoils the conclusions from that experiment. In other words, the experimentwise rate is based on the assumption that it is just as bad to make one erroneous conclusion as it is to make six in the same experiment.

If we have to make a choice between these two approaches, it will depend upon rather subtle differences in the manner in which the experimental conclusions are to be used. If the total set of conclusions is considered as a *pattern* supporting some theoretical position in such a way that any erroneous statement would destroy the pattern, then the experi-

mentwise basis is clearly the one to use. If each fact can be interpreted independently of the other findings of the experiment, the per experiment basis is more appropriate.

In practice, our interpretation of experimental findings probably does not fall clearly at either of these extremes. One erroneous statement probably will not completely destroy the value of the findings, but, on the other hand, each "fact" must be interpreted in some relation to the other results. We would therefore be in difficulties, if the choice between the two bases were crucial.

In a great many of the common experimental designs, the per experiment basis can be worked out from statistical tables of standard tests already in existence although they must be more extensive than those given in the textbooks. Special tables must be developed for the experimentwise error rates, but a number of these are already available. The principal practical advantage of the per experiment rate lies, therefore, in those cases where special tables are not yet available for experimentwise rates.

Where there is any discrepancy between results computed in the two ways, the per experiment basis is more conservative than the experimentwise basis. We are therefore safe in using the rate per experiment when in doubt, or when the experimentwise rate cannot be calculated, in that the error rate per experiment is always larger than or equal to the experimentwise rate.

An algebraic statement of the relationships among the various error rates may help to show why some of our previous statements about them are true. Let:

$p_{1/1}$ = probability of one erroneous statement in a single trial using a particular critical value in a certain test (for example if t is considered significant when it exceeds 2.75 and the degrees of freedom for error are 30, $p_{1/1} = .01$). This is error rate per comparison.

$p_{1/k}$ = probability of exactly one erroneous statement out of a total of k statements which are made.

$p_{2/k}$ = probability of exactly two erroneous statements out of k, etc.

EP = error rate per experiment.

EW = error rate experimentwise.

Then by definition:

EP = expected number of errors per experiment

$$= p_{1/k} + 2p_{2/k} + 3p_{3/k} + \cdots + kp_{k/k}.$$

$$EW = p_{1/k} + p_{2/k} + p_{3/k} + \cdots + p_{k/k}.$$

It can also be shown that $EP = kp_{1/1}$.

Thus EP is always greater than EW, and the difference between them depends upon the probabilities of more than one erroneous statement. The very simple relationship between EP and $p_{1/1}$ shows why EP can be calculated with standard tables. Suppose, for example, that we are comparing 10 means. There are then $(10)(9)/2 = 45$ comparisons to be made. If each difference were tested with an ordinary t test at the .01 level EP is 45 ($.01) or .45. To reduce EP to the .01 level, we simply reduce $p_{1/1}$ to $.01/45 = .00022$ and find the corresponding value of t. The approximate value of the required t can be obtained from Pearson and Hartley's Table 9, "Probability Integral of the t-Distribution" (Pearson & Hartley, 1954). It turns out to be about 4.1 in the case given in the example above, where there are 30 degrees of freedom. Thus, changing the critical value of t from 2.75 to 4.1 changes our error rate from .01 per comparison to .01 per experiment. (It is assumed that all differences are to be tested against a common critical value of t. Later we shall show that it is possible to obtain a more sensitive significance test by using variable t ratios depending upon the observed order of the means.)

By methods which we shall not discuss here, the EW rate can also be used to find the critical value of t. In the example we are considering, the EW rate turns out to require a value of 4.07 for t. The slight difference is partly due to the gaps in the table of t, so that the 4.1 is only approximate. Table 2 gives some further examples of comparative critical

values of t for experimentwise and per experiment rates of .01.

Fortunately, as the above examples show, it is usually not necessary to make the difficult decision between rates per experiment and experimentwise in terms of the logic of the experimental interpretation. In practice it becomes merely a matter of computational convenience.

Table 2. Critical values of t for testing differences among several means for error rates of .01.

No. of Means	df for error	For $EP = .01$ (approximate)[a]	For $EW = .01$
		Critical Values of t	
5	20	3.9	3.7
	30	3.7	3.6
	60	3.5	3.4
	120	3.4	3.3
10	30	4.1	4.1
	60	4.0	3.9
	120	3.8	3.7
20	60	not covered	4.3
	120	by tables	4.1

[a] To avoid interpolation, values are taken to the next tenth above the critical value.

Significance tests vs. confidence ranges. Several of the methods now available for multiple comparisons give us conclusions in the form of statements of significance—"The difference between A and B is significant, that between B and C is not significant, etc." Others make all comparisons in terms of confidence ranges of the difference—"The difference between A and B is from 2 to 15, the difference between B and C is from −3 to 10, etc."

When there are only two means to be compared, significance statements can be rather easily translated into confidence ranges, and vice versa. In the case of multiple comparisons, however, the relationship is more complex. For example, the confidence range for the difference between B and C above is from −3 to 10, yet a method of testing for significance of differences, with the same error rate, might label the difference between B and C as significant.

This discrepancy is because the most sensitive or powerful significance tests apply a different criterion of significance to different pairs of means, depending on how far apart they are in the total group. Thus the two extreme means must be farther apart for significance than two which are next to each other. The methods of determining confidence ranges have developed a single "allowance" which is applied to each of the differences regardless of where the means are in the total group.

Tukey has argued for the almost universal use of confidence ranges instead of significance statements, basing his case primarily upon the point that confidence ranges contain more information and information which is more useful to future researchers than a statement of significance. Whether or not he is correct in this contention, the fact remains that most of our familiar statistical tools (e.g., the F and chi-square tests) are significance tests. As a result most psychological researchers are more accustomed to thinking in terms of significance. It will therefore require a long period of readjustment if Tukey's point of view is to prevail.

Because it is more in keeping with current practice and ways of thinking in psychology, most of this paper is couched in terms of statements of significance. It is important to realize, however, that there is more difference between the two approaches when we are involved in multiple comparisons than there is in the simple case of two means.

Comparisons and contrasts. All of the discussion so far has been directed at the comparison of one mean with another mean in the group. Sometimes, however, other problems arise. We might, for example, wish to divide the means into two groups of means, by inspection of the data, and to state whether the two groups of means differ significantly from each other. In the literature of this field, the term *contrast* is used for the comparison of any combination of means with another combination. Contrasts include cases where the means are combined with differential weights for different groups.

Some of the procedures now available make it possible to test for significance or place confidence limits on all possible contrasts among the means of a given experiment. Methods which are effective for such broad purposes are not so effective, however, for the case of simple comparisons of one mean with another. . . .

Other cases of multiple tests. As noted at the beginning there are other situations in which a number of statistical tests are made upon one set of experimental results. While we have had space to discuss at length only the problem of multiple comparisons, we must emphasize that the same fundamental issues are involved in the other cases as well.

In all of the cases there is the question of basing the error rate upon the individual comparison (as is frequently done in the literature) or to consider the error rate in relation to the experiment as the unit. Conclusions upon these other cases will not necessarily be the same as for multiple comparisons, since the purpose of the statistical analysis is different in each situation. Each of the cases requires an analysis similar to the one which we have made for the case of multiple comparisons, and, as yet, little has been done on most of them.

As an example of the problems involved, we shall consider briefly just one of the other cases—that of multiple F tests in a factorial experiment (Case 2 on p. 27). Hartley (1955) has described a method for controlling the experimentwise rate of error in a multivariable analysis of variance. By this method it is possible to test each source of variance in such a way that there is a specified probability that there will be one or more incorrect conclusions in the total experiment. Hartley does not go into detail, however, as to the problem of deciding when the experimentwise rate should be used.

The present writer believes that the same arguments which support the experiment-based error rates for multiple comparisons would also apply to multiple F tests. In the multiple comparison situation the experimenter can increase the probability of finding some (erroneously) significant results by studying more and more levels of the variable in the experiment and by basing his significance tests on the single comparison rate. In the same way, one can increase the probability of finding some significant F ratios in an experiment by complicating the experiment with more and more irrelevant variables, while continuing to base the error rate upon the individual F. For example, in a factorial design with five variables there could be as many as 31 F ratios. If each were tested at the standard ".05 level" the probability that some of them would turn out to be significant is almost .80, under the null hypothesis.

Summary

We have considered several basic issues involved in multiple comparisons. Our present position on these problems is as follows:

1. In general the same procedures should be used, whether the direction of differences has been predicted in advance or not. The same procedures which apply to comparisons suggested by the data should be applied when the comparisons have been specified in advance.

2. In general, the experiment should be used as the unit in computing error rates, rather than the individual comparison or test.

3. Following Tukey's lead, the error rate should be determined on the basis of that null hypothesis which maximizes the rate.

4. The error rate *per experiment* is an upper limit for the error rate *experimentwise*, and therefore provides a conservative test which can be used when the experimentwise rate cannot be computed.

5. The choice between the two experiment-based error rates is usually one of convenience, since they differ but little numerically in the cases where both procedures are available.

6. The relative advantages of confidence limits vs. significance tests have not been treated in this discussion, but it is pointed out that the two methods do not lead to parallel conclusions in the case of multiple comparisons. . . .

References

Bechhofer, R. E. A single sample multiple decision procedure for ranking means of normal populations with known variances. *Annals math. Statist.*, 1954, **25**, 16–39.

Bechhofer, R. E., Dunnett, C. E., & Sobel, M. A two-sample multiple decision procedure for ranking means of normal populations with unknown variances. *Biometrika*, 1954, **41**, 170–176.

Bechhofer, R. E., & Sobel, M. A sequential multiple decision procedure for ranking means of normal populations with known variances. *Annals math. Statist.*, 1953, **24**, 136–137 (Abstract).

Burke, C. J. A brief note on one-tailed tests. *Psychol. Bull.* 1953, **50**, 384–387.

Duncan, D. B. A significance test for differences between ranked treatments in an analysis of variance. *Virginia J. Science*, 1951, **2** (New Series), 171–189.

Duncan, D. B. Multiple range and multiple F tests. *Biometrics*, 1955, **11**, 1–42.

Dunnett, C. W. A multiple comparison procedure for comparing several treatments with a control. *J. Amer. Statist. Ass.*, 1955, **50**, 1096–1121.

Fisher, R. A. *Design of experiments.* (4th ed.) Edinburgh: Oliver and Boyd, 1947.

Harter, H. L. Error rates and sample sizes for range tests in multiple comparisons. *Biometrics*, 1957, **13**, 511–536.

Hartley, H. O. Some recent developments in analysis of variance. *Communications on Pure and Applied Mathematics*, 1955, **8**, 47–72.

Hick, W. E. A note on one-tailed and two-tailed tests. *Psychol. Rev.*, 1952, **59**, 316–318.

Jones, L. V. Tests of hypotheses: one-sided vs. two-sided alternatives. *Psychol. Bull.*, 1952, **49**, 43–46.

Keuls, M. The use of studentized range in connection with an analysis of variance. *Euphytica*, 1952, **1**, 112–122.

Kimmel, H. D. Three criteria for the use of one-tailed tests. *Psychol. Bull.*, 1957, **54**, 351–353.

Kozelka, R. M. Approximate upper percentage points for extreme values in multinomial sampling. *Ann. math. Statist.*, 1956, **27**, 507–512.

Marks, M. R. Two kinds of experiment distinguished in terms of statistical operations. *Psychol. Rev.*, 1951, **58**, 179–184.

McHugh, R. B., & Ellis, D. S. The "post-mortem" testing of experimental comparisons. *Psychol. Bull.*, 1955, **52**, 425–428.

Mosteller, F., & Bush, R. R. Selected quantitative techniques. In G. Lindzey (Ed.), *Handbook of social psychology*, Vol. I. Cambridge: Addison Wesley, 1954. Chap. 8.

Newman, D. The distribution of the range in samples from a normal population expressed in terms of an independent estimate of standard deviation. *Biometrika*, 1939, **31**, 20–30.

Pearson, E. S., & Hartley, H. O. *Biometrika tables for statisticians.* Vol. I. Cambridge: Cambridge Univer. Press, 1954.

Scheffé, H. A method for judging all contrasts in the analysis of variance. *Biometrika*, 1953, **40**, 87–104.

Stanley, J. C. Additional "post-mortem" tests of experimental comparisons. *Psychol. Bull.*, 1957, **54**, 128–130.

Tukey, J. W. Comparing individual means in the analysis of variance. *Biometrics*, 1949, **5**, 99–114.

8.7 A Note on the Inconsistency Inherent in the Necessity to Perform Multiple Comparisons

Warner Wilson

Some studies involve only two groups and provide only one difference to be tested for significance. Other studies involve several groups and provide many differences to be tested for significance. A question has arisen in the literature (Duncan, 1955; Ryan, 1959; Tukey, 1949) as to how significance should be determined when a number of tests are to be made in the same experiment. Ryan (1959) has performed a valuable function by pointing out that there are several ways of dealing with this problem.

It would be possible to adopt a strategy that would hold errors constant per comparison, per hypothesis, per experiment, per group, or even per subject. The question is essentially: what is the appropriate unit in which to evaluate research? It is the thesis of this paper that the most defensible decision is to divide our work into separate tests of hypotheses and to hold constant the expected number of errors per hypothesis tested.

The number of groups involved in the test of a single hypothesis may vary depending on the attitude of the experimenter and the nature of the hypothesis. Often an experiment determines the effects of several degrees of a measurable variable: in this case the hypothe-

Reprinted from *Psychological Bulletin*, 1962, **59**, 296–300, by permission of the publisher and author. Copyright 1962 by the American Psychological Association.

sis is usually that there is some relationship between an independent and dependent variable. In this case differences between individual groups may be of little concern. For example, if length of food deprivation is varied at 2-hour intervals from 2 to 24 hours, it is the overall variability between groups that is of interest, not the difference between any particular pair of groups. A failure to find a difference between the 8-hour and 10-hour group would be of little importance. In other cases several groups may be run that do not represent points on a measurable dimension and in such cases the difference between each group and every other group may be viewed as a separate hypothesis. For example, if the results of five different therapies are compared, the significance or nonsignificance of the difference between any two groups would probably be considered important. In this second case there would be more hypotheses but less data relevant to each one. If several variables are studied in a single experiment the significance of the effect of each variable and each interaction may be tested as a separate hypothesis. The practice of holding errors constant per hypothesis tested seems to be by far the most common in the literature: the F test is typically employed when the performance of several groups is subsumed under one hypothesis, and the t test is typically used to test differences between pairs of groups when each pair is construed as bearing on a separate hypothesis. Many, if not most, researchers are not even aware of the various special statistics that have been devised for the purpose of using some unit other than the hypothesis as the basis for error rate.

It is necessary to recognize, however, that all discussions in the literature recommend some unit other than the hypothesis as the basis for determining error rates. Ryan (1959) and Tukey (1953 unpublished), for example, favor the experiment as the preferred unit. The only dissenter to this general approach seems to be Duncan (1955) who favors what is essentially a compromise position. The purpose of this paper is to consider the pros and cons of the per-experiment versus the per-hypothesis approach. An attempt is made to make clear that some inconsistency is involved in either case and that a consequence of this fact is that several of the arguments offered in favor of the per-experiment strategy are in fact offset by parallel, equally logical arguments, in favor of the per-hypothesis strategy. It is pointed out below that while it is impossible to prefer one approach to the other on logical grounds, other considerations actually favor the per-hypothesis approach. Ryan (1959) and Tukey (1953 unpublished) actually speak of a per-comparison (rather than a per-hypothesis) approach as the possible alternative to the per-experiment approach. Although the two may seem to be similar, the per-hypothesis approach is different from the per-comparison in that any number of comparisons may be considered in testing one hypothesis, however, the arguments presented in relation to the per-comparison strategy apply in exactly the same way to the per-hypothesis approach.

As Ryan makes clear, if a per-hypothesis strategy is used, the same number of errors will be expected in 100 small experiments, each of which tests one hypothesis, as will be expected in a large experiment that tests 100 hypotheses (Ryan, 1959, pp. 30–34). Ryan maintains that independence of the tests or lack of it makes no difference: "The error rates per comparison and per experiment are completely unaffected by independence or lack of it" (Ryan, 1959 p. 34). Obviously if the error rate per hypothesis is held constant, the error rate per experiment will vary, depending on the size of the experiment. On the other hand, if the error rate per experiment is held constant, the error rate per comparison will vary, depending again on the size of the experiment. Since inconsistency is involved in either case a choice on purely logical grounds does not seem possible. If the implications of this fact are followed consistently, several of the arguments in favor of the per-experiment solution become meaningless.

Ryan (1959) and Tukey (1953 unpublished) both argue that a per-hypothesis strategy implicitly gives a person license to make relatively more errors per experiment merely because he has been industrious in running many groups. Although this argument seems quite irrelevant to the issue, it is only fair to note the other side of the question. The per-experiment strategy implicitly gives a person license to make relatively more errors per hypothesis, merely because he has been lazy, as evidenced by the running of few groups! It is hard to see how the first argument can be considered more compelling than the second.

Ryan (1959) also argues that a per-hypothesis strategy, by favoring the person who is industrious, as evidenced by the running of many groups, may lead people who run many subjects in a two-group experiment to demand the privilege of using a higher error rate because they too have been industrious, as evidenced by the running of many subjects. While this argument seems a little too artificial to deserve consideration, it is once again easy to point out the parallel counter-argument. The per-experiment solution, by favoring the person who is lazy, as evidenced by the running of few groups, might lead those who run few subjects in a two-group experiment to demand the privilege of using a higher error rate because they too have been lazy! Once again it is hard to argue that the possible consequences of the per-hypothesis approach are worse than the possible consequences of the per-experiment approach.

Some of the comments in the literature (e.g., Ryan, 1959, pp. 35–37) may suggest that the use of a per-hypothesis strategy necessarily results in an inordinate amount of error or at least in more errors than a per-experiment strategy. Such a conclusion would be completely false. It is true that if a per-hypothesis error rate is employed there will be *relatively* more errors *per experiment* in *large* experiments, but it is also true that if a per-experiment error rate is employed there will be *relatively* more errors *per hypothesis* in *small* experiments. The total expected number of errors can be controlled equally well no matter in what unit results of research are measured. Insistence on fewer errors per-experiment would decrease total errors to be sure, but insistence on fewer errors per-hypothesis would decrease total errors equally well. Ryan actually concedes this point at one place, but apparently fails to recognize its implications (Ryan, 1959, pp. 37–38). Unless one wishes to argue that an error does more damage merely because it occurs in a large experiment, it must be concluded once more that there is no logical basis on which to choose between the different strategies.

The writer firmly agrees with those who think a more rigorous control of errors is called for; however, he suggests that the most effective way for workers to achieve this is to hold the expected error rate constant at .001 per hypothesis. Suppose one person publishes at the .05 level per experiment and a second publishes at the .001 level per hypothesis. Assuming that the second person's experiments test less than 50 hypotheses on the average, he will make fewer errors both per experiment and per hypothesis than will the first person. Clearly an experimenter can be as rigorous as he wishes and still use the hypothesis as his research unit.

Another type of consideration relates to the effect that each strategy might have on the behavior of researchers as they design, carry out, and write up experiments. It has been argued (Ryan, 1959, p. 36) that a per-hypothesis type approach encourages investigators to include "irrelevant" variables in their studies merely to increase their chances of obtaining one or more "significant" findings to publish. Surely such motivation is deplorable. However, it is doubtful that many researchers will deliberately resort to such tactics, and surely editors will be reluctant to accept implausible false positives no matter what statistical techniques are used. Furthermore the line between adding irrelevant variables and exploring new possibilities is rather subtle and it is not at all certain that psychology would not profit from some additional blind seeking for relationships. It is necessary to insist on looking at both sides of the picture. What sort of pressures does the per-experiment procedure apply to the researcher? It seems likely that, for better or worse, most experimenters design studies to demonstrate relationships they believe to exist. Their desire is to obtain data that will support their hypotheses and compel others to accept them. Very generally it can be assumed that there is often a choice between testing a number of hypotheses in different experiments by running only the two groups expected to be most extreme versus testing several hypotheses in one experiment by running several groups to determine the effects of each variable.

The latter, more extensive type of study, is greatly to be preferred since it consumes less journal space per hypothesis, it allows for the evaluation of interaction effects, and it gives some idea of the shape of relationships. The per-experiment approach seems to discourage extensive studies because the more extensive the study the less the likelihood of being able to accept any given hypothesis as correct. In other words if a per-experiment strategy is used, the smaller the pieces in which one can publish, the greater his chances of having significant findings to report. When a per-hypothesis strategy is followed this additional encouragement to publish in small pieces is not present. The literature is currently cluttered with small one-shot studies and there is a relative dearth of well conceived, intensive investigations. Certainly all angles should be considered before a strategy is advocated that might intensify this unfortunate tendency. Apparently either the per-experiment or the per-hypothesis strategy might have ill effects on certain researchers, but once again it is hard to see the arguments in favor of the per-experiment approach as more compelling than those favoring the per-hypothesis approach.

In addition it can be pointed out that there are strong advantages to the per-hypothesis solution. The basic question is, what is the most meaningful unit in which to evaluate research? Traditional practice apparently has chosen the hypothesis as the unit and this paper maintains that this is the correct choice. It seems that the hypothesis is psychologically the more logical unit. This writer, at least, would prefer to be confronted with a great array of findings, all of which (statistically speaking) have a comparable probability of being correct, rather than to be confronted with a number of conclusions each of which can be accepted with more or less confidence depending on the size of the experiment from which they were derived.

Another major advantage of the per-hypothesis approach is the fact that it requires no additional learning on the part of researchers. Obviously the more complicated statistics become the more time it will take to learn to use them and the less time will be available for research itself. It seems foolish for researchers to accept additional statistical complications unless there are telling reasons for doing so. It might also be added that it is practically impossible for a statistically naive

researcher to abandon the traditional per-hypothesis techniques because statisticians have not yet agreed upon any other strategy or even on how best to achieve the various alternatives that have been advocated. Duncan (1955) mentions nine different solutions to the problem of multiple comparisons and comments that, "Unfortunately, these tests vary considerably and it is difficult for the user to decide which one to choose for any given problem" (p. 2). One purpose of Duncan's article was to propose still another solution: It has not received general acceptance (Ryan, 1959) and it seems apparent that statisticians have no generally agreed upon alternative to suggest as a possible replacement for the per-hypothesis approach.

It must be concluded that the arguments in favor of the per-hypothesis strategy are more numerous and more compelling than those in favor of the per-experiment solution. Therefore the less effortful per-hypothesis approach should be continued indefinitely unless valid arguments are presented in favor of a different strategy.

References

Duncan, D. B. Multiple range and multiple *F* tests. *Biometrics*, 1955, **11**, 1–42.

Ryan, T. A. Multiple comparisons in psychological research. *Psychol. Bull.*, 1959, **56**, 26–47.

Tukey, J. W. Comparing individual means in the analysis of variance. *Biometrics*, 1949, **5**, 99–114.

8.8 Error Rates for Multiple Comparison Methods: Some Evidence Concerning the Frequency of Erroneous Conclusions

Lewis F. Petrinovich and Curtis D. Hardyck

Abstract. Seven methods currently used for paired comparisons among all group means following analysis of variance were tested for sensitivity to violation of stated requirements. The *t*-test methods and the Duncan multiple range test were found to produce far more Type I errors under null conditions than is generally acceptable. The methods of Scheffé and Tukey produced the fewest Type I errors and are least susceptible to violation of requirements. Type II error rates for all methods were also generated. The appropriate choice of error rate and multiple comparison method is discussed.

Specialized statistical methods for paired comparisons among all groups following an analysis of variance have been used with increasing frequency by psychologists. While it is generally agreed that the procedure of carrying out all possible two-group comparisons by such techniques as the *t* test is inappropriate (due to the rapid increase in Type I errors as the number of comparisons increases), little agreement exists as to choice of alternative methods. Multiple comparison methods developed for comparing several group means use different techniques to control for the frequency of Type I and Type II errors, and disagreement still exists among mathematical statisticians as to the mathematical soundness of certain of the current techniques available (Scheffé, 1959, p. 78).

Still another controversy concerns the appropriate basis for determining the permissible number of Type I and Type II errors. This problem has been discussed by Tukey (1953) and Ryan (1959b), and is evaluated in a later section of this paper.

In general, little has been published on the characteristics and properties of multiple comparison methods. The techniques developed by Fisher (1949), Tukey (1949, 1951, 1953), Keuls (1952), Duncan (1951, 1952, 1955, 1957), and Scheffé (1953, 1959) are

The authors wish to thank Jack Block and Norman Livson for their comments on an earlier version of this paper. The computer programmer for this study was Eleanor Krasnow. The authors would like to thank her for her assistance throughout the course of this project. This research was supported by Grant No. MH-12554 from the National Institutes of Health, United States Public Health Service, and the University of California Medical Center. Preliminary work was accomplished by a grant of free computer time from the Computer Center, University of California, Berkeley.

untested for sensitivity to violations of underlying assumptions; moreover, little is known of their power even when such assumptions are met. Harter (1957) and Wine (1955) have compared the power of several methods by calculating error rates, but only under conditions where all assumptions are met.

It seems doubtful that all multiple comparison methods have the same robustness as does the analysis of variance. The method of Scheffé (1959) should have the same robustness as that displayed by the F statistic since it is based on the F distribution. However the Newman (1939), Keuls (1952), Tukey (1953), and Duncan (1957) methods are based on the q or studentized range (Pearson & Hartley, 1943; Student, 1927)—a statistic which is less powerful overall than the corresponding F statistic (Winer, 1962), and one which has not been studied on other than normal distributions with equal variances. If differences between means are evaluated in terms of the q distribution, it is assumed that all means are based on the same number of observations and have equal standard errors. It is implied that these requirements are not strictly necessary for techniques such as the Newman-Keuls method, but the extent to which these requirements may be disregarded without affecting error rates has not been investigated.

In addition to the question of general robustness, it is important to establish the relative power of the different methods under specified conditions. Not only do the techniques differ in definitions of error rate, but the logic of evaluation also differs. For example, the Newman-Keuls technique compares the range of means in a subset with the appropriate variance; a corresponding F test would compare the variance of the subset means with the error variance. As both Duncan (1955) and Tukey (1953) have pointed out, though these techniques have the same stated "significance levels," the frequencies of Type I errors per experiment are very different.

Textbook authors—at least in the area of psychological statistics—have not been particularly helpful. Authors such as Edwards (1960), Federer (1955), Hays (1963), Mc-Nemar (1952), and Winer (1962) either offer no evaluation as to which method is preferable, or preface their remarks with a cautionary statement to the effect that these methods are still under study and that mathematical statisticians are not entirely in agreement concerning the preferred method. Similarly, disagreement exists as to when these methods may be used. Some discussions state that a significant F ratio over all conditions must be obtained before multiple comparison methods can be used (Hays, 1963; McNemar, 1962); other discussions make no mention of such a requirement (Federer, 1955; Winer, 1962), or deny that it is necessary at all (Edwards, 1960; Ryan, 1959a).

In view of these ambiguities, the following information would be useful for each of the current methods: (*a*) a thorough description of error rates for both total null and real difference conditions among population means when assumptions of normality, equal variances, and equal sample sizes are met; and (*b*) the extent to which error rates are dependent on each of the above assumptions.

Procedure

The procedure in the present study is adapted from an earlier study of the effects of type of measurement scale on the distribution of the t test (Baker, Hardyck, & Petrinovich, 1966). A total of seven types of multiple comparisons were studied. (The last reference for each test lists the location of the computing formula used.)

1. The Duncan multiple range test (Duncan, 1957; Edwards, 1960)

2. The Scheffé test (Scheffé, 1953, 1959)

3. The Tukey test, Type A (Tukey, 1953; Winer, 1962)

4. The Tukey test, Type B (Tukey, 1953; Winer, 1962)

5. The Newman-Keuls test (Newman, 1939; Keuls, 1952; Winer, 1962)

In addition to the above standard measures, two versions of Student's t test were included.

6. The standard t for testing differences between two independent groups. Although the cumulative Type I error rate for individual comparisons is known to be quite high when t is used to test all possible combinations, it was included to provide an empirical base line of error frequency. This technique fails to capitalize on the stability of the overall mean square within treatment groups, but is included to provide a maximum error rate. It is referred to hereafter as t_1.

7. The t test often used in multiple comparisons in which the error term is the square root of the within-groups mean square, with degrees of freedom equal to the within-groups degrees of freedom (Snedecor, 1956). It is referred to hereafter as t_2.

Sampling Method. The random number generator is described by Baker et al. (1966). The basic populations are of two types: a normal distribution with $N = 6,000$, $\mu = 50$, and $\sigma = 15.00$; and an exponential distribution with $N = 6,000$, $\mu = 5.25$, and $\sigma = 8.00$ ($f = 1.283^{(29.8-X)}$). The observation values for both populations range from 1 to 99. Standard score transformations were used to produce populations with means and variances differing from the basic populations. The results reported here are all based on 1,000 "experiments," where an "experiment" is defined as the random selection of samples of size n for k samples, where k may vary from 2 to 10 and n may vary from 5 to 50. Sampling from a designated population was done with replacement after each sample was selected. Thus, the first sample was selected without replacement, the population was then restored to full size and the next sample was selected.

Error Rates. Previous discussions of multiple comparisons (Ryan, 1959b, 1962; Tukey, 1953) have considered the type of error rate appropriate for use with multiple comparisons. Using a conventional t test, where all assumptions of random selection are met, the Type I error rate under true null conditions is .05 if a significance level (α) of .05 is set by the experimenter. However, such a rate cannot be applied to a situation where several means are compared with each other. As Ryan (1959b) has pointed out,

If each of 10 means is compared with each of the others there are 45 comparisons in one experiment. If the "significance level" ... of the test applied to each comparison is .01, we should expect .45 erroneous conclusions *per experiment*. The probability that there will be one or more erroneous conclusions in a given experiment ... will be somewhere between these two values, usually closer to .45 [p. 30].

The error rates used in the present study were defined by Tukey (1953), as cited by Ryan (1959b):

1. *Error rate per comparison.* This is the probability that any particular one of the comparisons will be incorrectly considered to be significant.

2. *Error rate per experiment.* This is the long-run average number of erroneous statements per experiment ... it is the *expected number* of errors per experiment. Unlike the first error rate, which is a probability, the error rate per experiment could be greater than one.

3. *Error rate experimentwise.* This is the probability that *one or more* erroneous conclusions will be drawn in a given experiment. In other words, *experiments* are divided into two classes: (a) those in which all conclusions are correct and (b) those in which some conclusions are incorrect. The error rate experimentwise is the probability that a given experiment belongs in class (b) [p. 29].

The above error rates are studied in the present paper for both Type I and Type II errors to determine both the probability of rejecting the null hypothesis when it is true (Type I—that is, $H_0: \mu_i - \mu_j = 0$ versus $H_1: \mu_i - \mu_j \neq 0$) and the probability of accepting the null when it is false (Type II). Thus, we can determine the power of the various methods to detect real difference both under ideal (all standard statistical assumptions met) and various inappropriate conditions (combinations of nonnormal populations, differing n's, unequal variances).

The following notation is used: $n =$ the number of observations per sample. For comparisons with differing n's, the n for each sample will be listed in order as follows:

Three samples with equal n's of 5 are listed as $n = 5$.

Three samples of differing n's are listed as $n = 5, 10, 15$.

k = the number of samples.

NP = a normally distributed population of 6,000 observations.

EP = an exponentially distributed population of 6,000 observations.

MD = difference between adjoining means expressed in standard deviation units.

For example, in one set of comparisons where $k = 3$, samples of $n = 10$ were drawn from normal populations whose means increased in steps of $.67\sigma$, μ for the lowest population was set at 50.00 and σ at 15.00. For the first sample, $\mu = 50$, for the second sample, $\mu = 60$, and for the third sample, $\mu = 70$. The notation for this when using a normal population is NP, $k = 3$, $n = 10$, $MD = .67\sigma$. A comparison of three samples drawn from populations whose means increased in steps of 2.67σ would be listed as $MD = 2.67\sigma$ and the means would be 50.00, 90.00, and 130.00, respectively.

V = variance.

For comparisons with unequal variances, the standard value of the variance was increased by a factor of four or eight. Thus, for a three-sample comparison, where the variance for Population 1 was 225.00 (15^2), the variance for Population 2 would be 900.00, and the variance for Population 3 would be 3,600. Means and variances are equal, unless otherwise specified. To illustrate: Three samples are drawn from normal populations with $n = 5$ for the first sample, 10 for the second sample, and 15 for the third sample; the mean difference between each successive population is equal to 1σ; the variance of Population 2 is twice that of Population 1, the variance of Population 3 is four times that of Population 1 and twice that of Population 2. The notation for this is NP; $k = 3$; $n = 5, 10, 15$; $MD = 1$; $V = 1, 2, 4$.

Results

Zero-Difference Conditions: True-Null. To begin, the error rates obtained under what Ryan (1959b) has called the "complete-null hypothesis" are considered. In this condition, all samples are drawn from populations with identical means and variances. Any difference which is found to be significant, then, is a Type I error. For our null "experiments," the number of significant F ratios closely approximated the stated α levels under all conditions studied. Similar results have been reported by Norton (1952) and by Box (1953) for F, and by Boneau (1960, 1962) and Baker et al. (1966) for t.

(The overall pattern is, in general, the same for the .01 and the .05 levels of significance. Therefore, in the interests of economy of presentation, only the results for the .05 level are presented unless there is a difference between results for the two levels.)

If we begin with a large sample condition such as NP, $k = 3$, $n = 30$, we find the error rate *per comparison* to be quite low for all procedures except t_1 and t_2, as illustrated in Figure 1.

The error rate preferred by most statisticians to evaluate the effectiveness of multiple comparison methods is the error rate *experimentwise*—the ratio of the number of experiments with at least one error to the total number of experiments performed. Inspection of Figure 2 reveals that the probability of a Type I error *experimentwise* is .138 for t_1, and .110 for t_2.

For the Duncan test the probability of a Type I error *experimentwise* is .098. Tukey A, Tukey B, and the Newman-Keuls method all produce identical probabilities of .054. The extremely conservative Scheffé test results in the fewest Type I errors *experimentwise*.

The results for the error rate *per experiment*, which is the number of incorrect statements divided by the number of experiments, are much the same with t_1, t_2, and Duncan producing the most Type I errors *per experi-*

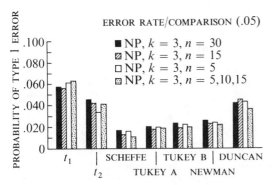

ERROR RATE/COMPARISON (.05)

- ■ NP, $k = 3$, $n = 30$
- ▨ NP, $k = 3$, $n = 15$
- □ NP, $k = 3$, $n = 5$
- ▦ NP, $k = 3$, $n = 5, 10, 15$

Fig. 1. Error rate per comparison for all methods when three samples are compared.

ment (.171, .139, and .130). The Newman-Keuls, Tukey A, and Tukey B are also above the arbitrarily chosen level of .05 by a sufficient margin to cause some concern. Only the Scheffé remains at the .05 level. At the .01 level, only t_1, t_2, and Duncan depart seriously from significance.

Effect of sample size. The above comparisons represent an "idealized" set of data which meet all requirements. If the sample sizes are reduced to $n = 15$ or $n = 5$, leaving all other factors constant, the error rates are as shown in the first three figures for each type of error rate. Reduction in sample size results in no change in error rates from the values established for $n = 30$.

Effects of unequal sample size. If sample sizes differ, there is no appreciable change in error rates. The results for NP; $k = 3$; $n = 5$, 10, 15, are shown in Figures 1, 2, and 3 to illustrate this point.

Effects of nonnormality. No appreciable change in any error rate occurs when the samples are drawn from an exponential population such as EP, $k = 3$, $n = 15$, or EP, $k = 3$, $n = 5$.

Effects of unequal sample size plus unequal variance. These methods not only share in the "robustness" of the F distribution, but also suffer the same limitation. The combination of unequal sample size and unequal variance which produces drastically incorrect values of t (Boneau, 1960) has a similar effect on all multiple comparison methods. For example,

consider the error rates for NP; $k = 3$; $n = 5$, 10, 15; $V = 1$, 2, 4; and NP; $k = 3$; $n = 5$, 10, 15; $V = 4$, 2, 1. The first condition results in fewer significant F values than would be expected by chance under normal theory assuming equal variances, and the second results in many more significant values than would be expected by chance.

Table 1 contains the error rates for all multiple comparison methods examined. In the condition where the larger sample has the larger variance, only t_1 exceeds the .05 error rate *experimentwise* and *per experiment*. The error rate for all other methods was far below the .05 level. In the contrasting condition where the smaller sample has the larger variance, only the Scheffé test has error rates close to the .05 level. In general, the same proscriptions that apply to violations of requirements underlying analysis of variance seem to apply as well to the multiple comparison methods.

Effect of number of samples. Error rate is also a function of the number of samples compared. If we repeat the "ideal conditions," for NP, $k = 6$, $n = 30$, there are some rather drastic changes in t_1, t_2, and Duncan.

For these tests, the error rate *experimentwise* more than triples at the .05 level as a function of doubling the number of samples

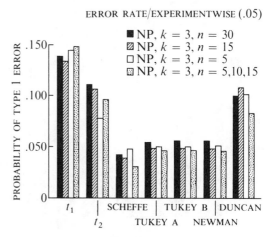

ERROR RATE/EXPERIMENTWISE (.05)

- ■ NP, $k = 3$, $n = 30$
- ▨ NP, $k = 3$, $n = 15$
- □ NP, $k = 3$, $n = 5$
- ▦ NP, $k = 3$, $n = 5, 10, 15$

Fig. 2. Error rate *experimentwise* for all methods when three samples are compared.

Table 1. Type I error rate (.05) for multiple comparison methods.

Population	Error rate[a]	t_1	t_2	Scheffé	Tukey A	Tukey B	Newman-Keuls	Duncan
NP, $k = 3$, $n = 5, 10, 15$	1REW	.119	.038	.016	.012	.012	.012	.030
$V = 1, 2, 4$	1REX	.114	.046	.018	.013	.016	.018	.038
	1RC	.048	.015	.006	.004	.005	.006	.013
NP, $k = 3$, $n = 5, 10, 15$	1REW	.157	.233	.100	.141	.141	.141	.212
$V = 4, 2, 1$	1REX	.216	.341	.141	.192	.213	.235	.319
	1RC	.072	.114	.047	.064	.071	.078	.106

Note.—The percentage of F values significant at $p < .05$ is 2.5 % for the first population and 11.7 % for the second.
[a]1REW = Type I error rate experimentwise; 1REX = Type I error rate per experiment; 1RC = Type I error rate per comparison.

(see Figure 4). This effect cannot be attributed to an increase in the number of significant F tests over the three sample conditions; the percentages of F values significant at the .05 level is 5.4 % for the three-sample condition and 4.8 % for the six-sample condition. The Tukey A, Tukey B, and Newman-Keuls remain almost exactly at the .05 rate. The rates *per experiment* rise quite sharply for t_1, t_2, and Duncan. The error rate per comparison, by contrast, is extremely low, except for t_1 and t_2. Examination of the error rates shown in Figures 4, 5, and 6, for NP, $k = 6$, $n = 15$; and for NP, $k = 6$, $n = 5$, show essentially the same pattern.

When the number of samples is increased to 10, the error rate *experimentwise* reaches .731

for t_1, .353 for Duncan, and .313 for t_2 (see Figure 4). An experimenter using independent t tests (t_1) is much more likely to have an experiment with at least one erroneous conclusion than not! While the error rates for Duncan and t_2 do not reach such dizzying heights, the probability is still far in excess of that tolerated even in the most casual circumstances. (Since earlier comparisons indicated no difference in error rates as a function of sample size or inequalities in sample size, these conditions were not run for the 10-sample condition.)

As has been noted earlier, the Tukey A, Tukey B, and Newman-Keuls maintain an admirable stability at the expected .01 and .05 levels, regardless of the number of samples. Scheffé's test actually seems to produce lower rates of error with 10 samples than

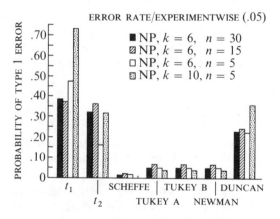

ERROR RATE/EXPERIMENT (.05)

■ NP, $k = 3$, $n = 30$
▨ NP, $k = 3$, $n = 15$
□ NP, $k = 3$, $n = 5$
▩ NP, $k = 3$, $n = 5, 10, 15$

ERROR RATE/EXPERIMENTWISE (.05)

■ NP, $k = 6$, $n = 30$
▨ NP, $k = 6$, $n = 15$
□ NP, $k = 6$, $n = 5$
▩ NP, $k = 10$, $n = 5$

Fig. 3. Error rate per experiment for all methods when three samples are compared.

Fig. 4. Error rate *experimentwise* for all methods when 6 samples and 10 samples are compared.

Table 2. Type II error rates ($\alpha = .05$) for increasing mean differences.

Population	Error rate[a]	MD[b]	t_1	t_2	Scheffé	Tukey A	Tukey B	Newman-Keuls	Duncan
NP, $k = 3, n = 30$	2REW	.6	.545	.543	.781	.740	.652	.548	.547
		1.0	.065	.067	.174	.148	.100	.067	.067
		1.3	.003	.003	.008	.007	.003	.003	.003
		1.6	.000	.000	.000	.000	.000	.000	.000
	2RC	.6	.190	.189	.305	.279	.235	.191	.190
		1.0	.022	.022	.058	.049	.033	.022	.022
		1.3	.001	.001	.003	.002	.001	.001	.001
		1.6	.000	.000	.000	.000	.000	.000	.000
NP, $k = 3, n = 15$	2REW	.6	.875	.874	.974	.973	.930	.875	.874
		1.0	.482	.474	.740	.685	.578	.477	.476
		1.6	.018	.017	.060	.049	.026	.017	.017
		2.0	.000	.000	.004	.003	.001	.000	.000
	2RC	.6	.356	.356	.488	.460	.412	.363	.356
		1.0	.170	.167	.285	.259	.210	.168	.167
		1.6	.006	.006	.020	.016	.009	.006	.006
		2.0	.000	.000	.001	.001	.000	.000	.000
NP, $k = 3, n = 5$	2REW	.6	.979	.976	.995	.993	.992	.976	.976
		1.0	.931	.912	.970	.963	.943	.912	.912
		1.6	.615	.557	.799	.764	.675	.557	.557
		2.0	.389	.337	.605	.558	.446	.337	.337
		2.6	.086	.058	.171	.147	.094	.058	.058
	2RC	.6	.457	.432	.594	.588	.494	.483	.462
		1.0	.419	.392	.541	.509	.456	.401	.392
		1.6	.233	.207	.353	.325	.268	.207	.207
		2.0	.138	.118	.237	.214	.163	.118	.118
		2.6	.029	.019	.061	.051	.031	.019	.019

[a] 2REW = Type II error rate experimentwise; 2RC = Type II error rate per comparison.
[b] Mean difference.

with 3; not even producing .01 errors where a .05 level of significance is established.[1]

Type II Error Rates. When we consider error rates under conditions where real differences exist between the populations, other factors become important. For the large sample condition used earlier (NP, $k = 3$, $n = 30$), we find the error rates shown in Table 2. (The error rates *per experiment* have not been included because the general pattern is quite similar to that for the error rate *experimentwise*.)

All error rates decrease as a function of an increase in the difference between means. All methods follow the same pattern, with t_1, Newman-Keuls, and Duncan displaying a slightly lower error rate than do other methods. The differences between methods almost vanish by the time a mean difference equal to 1.33σ is reached, and they disappear

[1] The error rate *experimentwise* for the Scheffé test is exactly .05 if all (infinitely many) comparisons are tested. Consequently, Scheffé's test will always produce extremely low Type I error rates when comparisons are limited to all possible pairs.

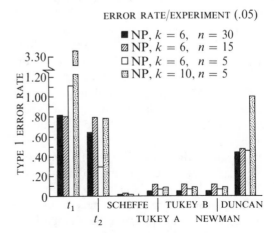

Fig. 5. Error rate per experiment for all methods when 6 samples and 10 samples are compared.

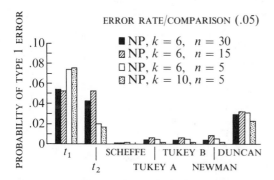

Fig. 6. Error rate per comparison for all methods when 6 samples and 10 samples are compared.

completely by the time the mean difference between each sample reaches 1.67σ. However, note that mean differences between adjoining samples must be at least this large in order to bring the Type II error rate *experimentwise* down to an arbitrarily chosen level of .05.

The rate *per comparison* is also quite high for all methods; adjoining mean differences have to be greater than 1.33σ before this rate drops below .05 for all methods.

If we examine one error rate for NP, $k = 3$, $n = 15$; and NP, $k = 3$, $n = 5$, the effects of sample size become evident (Table 2). Reduction in error rate *experimentwise* does not occur until there is a 1σ difference between successive population means. The drop in power is rather striking—for the NP, $k = 3$, $n = 30$ condition, the .05 error rates *experimentwise* are .000 for all methods when adjoining means are 1.67σ apart. For NP, $k = 3$, $n = 15$ at the same mean difference, the .05 error rates *experimentwise* range from .060 to .017, and in the NP, $k = 3$, $n = 5$ condition, they range from .557 to .709. A difference of at least 1.67σ between adjoining means is necessary to reduce the probability of a Type II error to .06 or lower if $\alpha = .05$ for samples of $n = 15$. For samples of $n = 5$, the Type II error rate at the .05 level of significance is as high as .171 when mean differences are 2.67σ. At least one conclusion is obvious: Unless mean differences are very large, it is pointless to attempt multiple comparisons with samples of less than 10 if one is interested in making few Type II errors.

Effects of unequal sample size. Surprisingly, he probability of a Type II error actually decreases when sample sizes are unequal. The probability of making a Type II error with NP; $k = 3$; $n = 5, 10, 15$ for mean differences ranging from $.33\sigma$ to 2.67σ is intermediate to NP, $k = 3$, $n = 5$; and NP, $k = 3$, $n = 15$. This suggests that it is more important to have a maximum total number of observations than it is to have equal samples if we wish to minimize the number of Type II errors. For example, with NP; $k = 3$; $n = 5$, 10, 15; $MD = 2$, the error rates *experimentwise* range from .046 to .158. The equivalent error rates *experimentwise* for NP, $k = 3$, $n = 5$ are .337 to .605 and for NP, $k = 3$, $n = 15$, they range from .000 to .004. (Error rates for NP, $k = 3$, $n = 10$—not shown—closely approximated the $n = 5, 10, 15$ condition.) Similar results occur with error rates *per experiment* and *per comparison*.

Effects of unequal variance. With unequal variances the general result is an even greater loss of power when the samples are small. The probability of a Type II error *experimentwise* for NP; $k = 3$; $n = 5$; $V = 1, 2, 4$; when $MD = 2.67\sigma$, ranges from .498 to .745 for the different methods! The earlier suggestions regarding the insensitivity of multiple comparison methods if sample size is 10 or less would seem even more applicable when the variances are unequal.

Effects of unequal sample size and unequal variance. When unequal sample sizes are combined with unequal variances, the probability of a Type II error decreases slightly. When the larger sample has the larger variance (NP; $k = 3$; $n = 5, 10, 15$; $V = 1, 2, 4$) the error rates are not noticeably different from NP; $k = 3$; $n = 5$; $V = 1, 2, 4$, for the same mean differences. In contrast, when the smaller sample has the larger variance (NP; $k = 3$; $n = 5, 10, 15$; $V = 4, 2, 1$) the error rates are slightly lower than is the case with NP, $k = 3$, $n = 5$, for the equivalent mean differences. However, in view of the increased probability of Type I errors (Table 1) the increase in power seems to represent a rather dubious gain.

Effects of nonnormal population. If exponential populations are used a somewhat different result is obtained. With EP, $k = 3$, $n = 5$, error rates follow much the same pattern as found with a normally distributed population. This is true until $MD = 1.33\sigma$, at which point error rates suddenly decrease much more rapidly than do the corresponding rates with a normally distributed population. Similarly, with samples of unequal size, error rates for EP; $k = 3$; $n = 5, 10, 15$ decrease more rapidly than do corresponding error rates for NP. This is true until $MD = 1.67\sigma$ at which point error rates for NP and EP are quite close at the .05 level of significance but lower at the .01 level for the EP condition.

This pattern of decreasing Type II error rates with samples drawn from an exponential population also occurs with unequal variances —error rates for both EP; $k = 3$; $n = 5$; $V = 1, 2, 4$, and EP; $k = 3$; $n = 5, 10, 15$; $V = 4, 2, 1$ and the EP and NP conditions initially are quite similar until a sudden sharp drop in error rate occurs when $MD = 1\sigma$ for EP. Following this drop, the error rates for the two conditions continue at approximately the same level. In general, one need not be concerned about the effect of nonnormal populations on multiple comparison techniques. In fact, for many conditions, sampling from an exponential population seems to result in increased power!

Multiple-Null Hypothesis. A final set of conditions was also suggested by Ryan (1959b), who developed the concept of "multiple null hypotheses." Perhaps the best way to outline this issue is with an extensive quotation from Ryan.

The concept of error rate cannot be defined completely without taking account of another important fact. In our distinctions between error rates per comparison, per experiment, and experimentwise ... the null hypothesis would be that all means are drawn from a single population —the same null hypothesis which is tested in analysis of variance by means of the F test. We shall call this the "complete" null hypothesis. This is one possibility which must be considered, but it is not by any means the only one. In our

example of 10 means, five might be drawn from one population and five from another, six from one and four from another, two from each of five different populations, and so on. For each of these different null hypotheses, there is an error rate per comparison, per experiment, and experimentwise, for any given method of testing differences. The question is, therefore, which of these null hypotheses is used to define *the* error rate for our statistical test? ... Suppose ... that there are actually five populations, with two groups drawn from each population, and suppose that these populations are widely separated. Then it is almost certain that the F test will be significant, and t tests between pairs drawn from distinct populations will also be almost certainly significant, as they should be. We can still make errors, however, in comparing means in the pairs drawn from identical populations. Since there are five such comparisons, the probability that *one or more* of these will be incorrectly judged to be significant is $(1 - .99^5)$ which is approximately .05. Thus the error rate *experimentwise* is .05 instead of .01 for this particular null hypothesis. The more means there are to be compared by this method, the higher will the experimentwise error rate become, even though the error rate based upon the complete null hypothesis is fixed at .01 for any number of means [pp. 30–31].

To explore the effects of multiple null hypotheses on the error rate *experimentwise* two sets of conditions were run. The first set was composed of five samples where Populations 1 and 2 had the same mean; Populations 3 and 4 had the same mean, which was 2.67σ above the means of Populations 1 and 2; and the mean for Population 5 was 2.67σ above the means of Populations 3 and 4. The second set was composed of six populations with 1–4 having the same mean and Populations 5–6 having the same mean with 2.67σ separating the means for Populations 1–4 and 5–6. For both sets, the conditions shown in sets 1 and 2 were run.

Set 1

a. NP, $k = 5, n = 5$
b. NP, $k = 5, n = 15$
c. NP; $k = 5$; $n = 5, 15, 5, 15, 5$
d. NP; $k = 5$; $n = 5, 15, 5, 15, 5$;
   $V = 4, 1, 4, 1, 4$

e. NP; $k = 5$; $n = 5, 15, 5, 15, 5$;
    $V = 1, 4, 1, 4, 1$
f. NP, EP, NP, EP, NP, $k = 5$, $n = 5$
g. NP, EP, NP, EP, NP, $k = 5$, $n = 15$
h. NP, EP, NP, EP, NP; $k = 5$; $n = 5$;
    $V = 4, 1, 1, 4, 4$
i. NP, EP, NP, EP, NP; $k = 5$; $n = 5, 15, 15, 5, 5$;
    $V = 4, 1, 4, 1, 8$

Set 2

a. NP, $k = 6$, $n = 5$
b. NP, $k = 6$, $n = 15$
c. NP; $k = 6$; $n = 5, 15, 5, 15, 5, 15$
d. NP; $k = 6$; $n = 5, 15, 5, 15, 5, 15$;
    $V = 4, 1, 4, 1, 4, 1$
e. NP; $k = 6$; $n = 5, 15, 5, 15, 5, 15$;
    $V = 1, 4, 1, 4, 1, 4$
f. NP, EP, NP, EP, NP, EP, $k = 6$, $n = 5$
g. NP, EP, NP, EP, NP, EP, $k = 6$, $n = 15$
h. NP, EP, NP, EP, NP, EP; $k = 6$; $n = 5$;
    $V = 4, 1, 1, 4, 4, 1$
i. NP, EP, NP, EP, NP, EP; $k = 6$; $n = 5, 15, 15, 5, 15, 15$; $V = 4, 1, 4, 1, 8, .5$

These combinations do not represent the most extreme sets that can be devised, but probably do represent the most extreme having any contact with reality. The results are summarized in Table 3.

Two points immediately stand out upon examining Type I and Type II error rates *experimentwise* under these conditions. First, under conditions where real differences exist, violating underlying assumptions by having unequal variances or by combining unequal variances with unequal sample sizes results in drastic increases in the number of Type I errors. The results for Set 2 are particularly interesting here since the comparisons are balanced to provide an almost equal opportunity for Type I errors (seven per set of comparisons) and Type II errors (eight per set of comparisons) to appear. In Set 1, where there are only two possibilities per experiment for Type I errors to occur, the error rates are well within the expected levels for Tukey A and Tukey B (except for Condition d) and for Scheffé. The error rates for t_1, t_2, Duncan, and Newman-Keuls are almost all higher than is generally acceptable. As the conditions examined become more and more extreme, the discrepancies in Type I error for Set 2, as compared to Set 1, become

Table 3. "Maximum value" error rates (.05) for two combinations of groups under nine experimental conditions.

Experimental condition	Error rate	Set 1							Set 2						
		t_1	t_2	Scheffé	Tukey A	Tukey B	Newman-Keuls	Duncan	t_1	t_2	Scheffé	Tukey A	Tukey B	Newman-Keuls	Duncan
a	I[a]	.091	.098	.003	.011	.033	.098	.098	.225	.214	.009	.028	.045	.045	.160
	II[b]	.079	.032	.393	.233	.099	.041	.033	.227	.105	.702	.472	.276	.144	.108
b	I	.092	.047	.005	.011	.031	.097	.097	.237	.242	.013	.031	.051	.096	.184
	II	.000	.000	.000	.000	.000	.000	.000	.000	.000	.000	.000	.000	.000	.000
c	I	.131	.087	.001	.004	.029	.085	.086	.258	.224	.003	.026	.048	.090	.176
	II	.020	.001	.135	.017	.003	.002	.001	.075	.015	.372	.091	.030	.019	.015
d	I	.186	.214	.037	.058	.114	.208	.214	.323	.531	.064	.210	.269	.350	.488
	II	.759	.277	.785	.579	.439	.292	.278	.771	.324	.787	.616	.462	.351	.326
e	I	.108	.013	.000	.000	.000	.012	.012	.259	.057	.001	.001	.004	.016	.040
	II	.077	.281	.976	.818	.579	.367	.285	.238	.488	.998	.932	.800	.656	.500
f	I	.097	.089	.006	.008	.040	.089	.089	.266	.209	.006	.024	.040	.087	.158
	II	.079	.031	.367	.206	.092	.039	.032	.219	.104	.698	.493	.299	.147	.108
g	I	.111	.112	.009	.017	.046	.110	.110	.243	.246	.009	.029	.051	.093	.189
	II	.000	.000	.000	.000	.000	.000	.000	.017	.006	.048	.019	.007	.006	.006
h	I	.131	.064	.001	.005	.015	.061	.064	.329	.246	.020	.056	.091	.145	.222
	II	.617	.420	.864	.776	.634	.491	.423	.701	.441	.922	.825	.695	.549	.444
i	I	.117	.052	.003	.006	.020	.051	.052	.317	.412	.087	.156	.215	.297	.366
	II	.734	.492	.929	.824	.679	.545	.496	.793	.441	.860	.735	.606	.476	.441

[a] Type I error rate experimentwise.
[b] Type II error rate experimentwise.

more and more extreme. Only the Scheffé, Tukey A, and Tukey B approach an acceptable error rate *experimentwise*; the other methods produce *experimentwise* error rates ranging as high as .531 for the nominal .05 level of significance.

The second point is that the increase in error rate for the Newman-Keuls method almost parallels Duncan in frequency of Type I errors for Set 1 conditions. This finding is somewhat surprising in view of the steady error rate displayed earlier by this method.

The probability of making a Type II error is, again, relatively high for Scheffé and almost as high for Tukey A and Tukey B. Therefore, the method of choice would, again, depend on the experimenter's decision regarding which type of error he would most like to minimize.

Discussion

The discussion centers around two issues: (*a*) the choice of the appropriate error rates for use with multiple comparisons, and (*b*) the practical implications of the findings regarding the choice of an appropriate method to estimate the significance of differences among several samples.

Choice of Error Rate. As noted earlier, the error rates used in the present study were defined by Tukey (1953), who argues for the *experimentwise* error rate as the most appropriate. The experiment becomes the unit for computing error rates when this method is used and the experiments are divided into those which have at least one erroneous conclusion and those which have no erroneous conclusions. This approach has been advocated by Ryan (1959b), who also pointed out that error rates remain the same whether comparisons are planned or not; in making all possible comparisons among 10 groups at a significance level of .01, the error rate per experiment is equal to .45. This remains the same whether the comparisons are post hoc or have been elaborately planned and speci-

fied beforehand; exactly the same number of erroneous conclusions will be drawn. The position of Tukey (1953) would still seem the most sensible: A *per comparison* error rate is equivalent to rewarding the industry of the man who has studied several means at once by allowing him to draw more erroneous conclusions. In view of the rather high error rates established in the present study, the argument for the *experimentwise* error rate would, with one possible exception, seem quite compelling.

1. The most compelling reason is the minimization of erroneous conclusions in the published literature. A per comparison approach simply produces much more in the way of fictional results than does the *experimentwise* approach. Of course, the use of this error rate may mean that many real differences are not detected. However, if certain conditions, such as having an *n* of over 10 in every sample, are met in the use of multiple comparison methods, the problem of failing to detect real differences can be minimized.

2. It may also be argued, as many others have done, that it is better to punish truth than to let falsehood gain respectability. Psychologists do not often attempt to replicate the findings of others and journal editors seem reluctant to publish such work on the rare occasions that it is done. If, as seems to be the situation, conclusions must stand or fail on the basis of one experiment, then it is entirely appropriate that we exercise extreme caution in setting the rate at which we admit fiction into our research literature.

3. There is, as we have mentioned earlier, one possible exception to the use of an *experimentwise* error rate. Block (1960), in a discussion of the number of significant findings to be expected by chance, argues that:

it should be recognized ... that in the early stages of problem investigation, research strategy may call for a "shotgun approach" in order to scan empirically for predictive relevance in new and strange variables. The consolidation of findings can come later in the course of a systematic research program. It is most important early in the research sequence not to overlook potential research leads [p. 373].

Under circumstances such as Block discusses, a reasonable argument for use of a per comparison rate could be made, provided that subsequent work returned to a more conservative rate.

Choice of Multiple Comparison Technique. As was noted earlier, the experimenter seeking methods to compare several groups suffers from an embarrassment of riches seldom found in statistics. The present study may be helpful in furnishing some guidelines to prospective users as to choice of methods. Several suggestions would seem to follow from our findings:

1. Certain methods invariably produce a high rate of erroneous conclusions. Use of either independent or pooled-variance *t* tests, or the Duncan method produces Type I error rates *experimentwise* far in excess of currently acceptable standards. Under all conditions, other than the complete null hypothesis, the Newman-Keuls method Type I error rate begins to approximate that of the Duncan technique. These considerations provide a compelling argument against the use of these methods. If these methods are eliminated, we are left with Tukey A, Tukey B, and Scheffé.

2. At this point, some relatively simple criteria can be established for choosing among these methods. If sample size is equal to 15 or more it does not matter greatly which technique is used, since the Type II error rates established in the present study indicate little difference among the three methods. Perhaps the simplest approach would be to argue for use of the Scheffé as the initial test of choice. If differences among groups are found to be significant by the Scheffé method, the possibility of drawing erroneous conclusions is extremely unlikely. If differences are not found to be significant by the Scheffé, the method of choice would seem to be Tukey B, which fixes *experimentwise* error rates at conventional levels and which shows little deviation as a result of the violation of assumptions.[2] The Tukey B is also somewhat more sensitive to real differences between groups.

3. Such conditions as unequal sample size, unequal variances, nonnormal populations, etc., seem to make little difference to the Tukey and Scheffé methods, except as these violations also affect the obtained value of *F*. When the samples are of unequal size, with the smaller groups having the larger variance, an abnormally high percentage of significant *F* values is found. In such circumstances, the rate of Type I errors for the multiple comparison methods is also high—if data are inappropriate for an *F* test, they are also inappropriate for multiple comparisons.

4. If sample size is less than 10, it scarcely seems worthwhile to carry out the computations for multiple comparisons, since the power of any method to detect differences between small groups is extremely low.

In summary, it would seem that the concerns expressed by many textbook authors about the validity of some of the multiple comparison methods seem well substantiated. Certainly, the frequency with which Type I errors may be generated under many of the conditions examined here suggests that methods such as *t* test (regardless of whether planned or unplanned comparisons are made), Duncan, and Newman-Keuls are unsuitable. By contrast, the methods of Scheffé and Tukey seem appropriate to use on any data appropriate for analysis of variance. Although these methods are useful for more detailed analyses of differences among groups, effective detection of such differences is possible only when sufficient effort is made to collect larger samples than usual in many areas of psychology.

References

Baker, B. O., Hardyck, C., & Petrinovich, L. Weak measurements vs. strong statistics: An

[2] Scheffé (1953, 1959) recommends use of Tukey's method where sample sizes are equal and only paired comparisons are made. However, the superiority of Tukey's method is limited to comparisons of the $K(K-1)/2$ type. For all other types of comparisons, Scheffé's method is more efficient.

empirical critique of S. S. Stevens' proscriptions on statistics. *Educational and Psychological Measurement*, 1966, **26**, 291–309.

Block, J. On the number of significant findings to be expected by chance. *Psychometrika*, 1960, **25**, 369–380.

Boneau, C. A. The effects of violations of assumptions underlying the *t* test. *Psychological Bulletin*, 1960, **57**, 49–64.

Boneau, C. A. A comparison of the power of the *U* and *t* tests. *Psychological Review*, 1962, **69**, 246–256.

Box, G. E. P. Non-normality and tests on variances. *Biometrika*, 1953, **40**, 318–335.

Duncan, D. B. A significance test for the differences between ranked treatments in the analysis of variance. *Virginia Journal of Science New Series*, 1951, **2**, 171–189.

Duncan, D. B. On the properties of the multiple comparison test. *Virginia Journal of Science New Series*, 1952, **3**, 49–67.

Duncan, D. B. Multiple range and multiple *F* tests. *Biometrics*, 1955, **11**, 1–42.

Duncan, D. B. Multiple range tests for correlated and heteroscedastic means. *Biometrics*, 1957, **13**, 164–176.

Edwards, A. L. *Experimental design in psychological research*. New York: Holt, Rinehart & Winston, 1960.

Federer, W. T. *Experimental design*. New York: Macmillan, 1955.

Fisher, R. A. *Design of experiments*. (5th ed.) Edinburgh: Oliver & Boyd, 1949.

Harter, H. L. Error rates and sample sizes for range tests in multiple comparisons. *Biometrics*, 1957, **13**, 511–536.

Hays, W. L. *Statistics for psychologists*. New York: Holt, Rinehart & Winston, 1963.

Keuls, M. The use of the "studentized range" in connection with an analysis of variance. *Euphytica*, 1952, **1**, 112–122.

McNemar, Q. *Psychological statistics*. (3rd ed.) New York: Wiley, 1962.

Newman, D. The distribution of range in samples from a normal population, expressed in terms of independent estimate of a standard deviation. *Biometrika*, 1939, **31**, 20–30.

Norton, D. W. An empirical investigation of some effects of non-normality and heterogeneity on the *F*-distribution. Unpublished doctoral dissertation, State University of Iowa, 1952. Cited by E. F. Lindquist, *Design and analysis of experiments in psychology and education*. Boston: Houghton Mifflin, 1953.

Pearson, E. S., & Hartley, H. O. Tables of the probability integral of the studentized range. *Biometrika*, 1943, **33**, 89–99.

Ryan, T. A. Comments on orthogonal components. *Psychological Bulletin*, 1959, **56**, 394–395. (a)

Ryan T. A. Multiple comparisons in psychological research. *Psychological Bulletin*, 1959, **56**, 26–47. (b)

Ryan, T. A. The experiment as the unit for computing rates of error. *Psychological Bulletin*, 1962, **59**, 301–305.

Scheffé, H. A method of judging all contrasts in the analysis of variance. *Biometrika*, 1953, **40**, 87–104.

Scheffé, H. *The analysis of variance*. New York: Wiley, 1959.

Snedecor, G. W. *Statistical methods*. (5th ed.) Ames: Iowa State College Press, 1956.

Student. Errors of routine analysis. *Biometrika*, 1927, **19**, 151–164.

Tukey, J. W. Comparing individual means in the analysis of variance. *Biometrics*, 1949, **5**, 99–114.

Tukey, J. W. Quick and dirty methods in statistics, Part II: Simple analysis for standard designs. *American Society for Quality Control; 5th Annual Proceedings*, 1951, 189–197.

Tukey, J. W. The problem of multiple comparisons. Unpublished manuscript, Princeton University, 1953.

Wine, R. L. A power study of multiple range and multiple *F* tests. Technical Report No. 12, 1955, Virginia Polytechnic Institute.

Winer, B. J. *Statistical principles in experimental design*. New York: McGraw-Hill, 1962.

8.9 Suggestions for Further Reading

I. One- Versus Two-Tailed Tests of Significance*

Five articles setting forth the major issues involved in the choice between one- and two-tailed tests are included in this chapter. Other articles that are an integral part of this controversy are listed below.

Marks, M. R. Two kinds of experiments distinguished in terms of statistical operations. *Psychological Review*, 1951, **58**, 179–184.

Hicks, W. E. A note on one-tailed and two-tailed tests. *Psychological Review*, 1952, **59**, 316–318.

Marks, M. R. One- and two-tailed tests. *Psychological Review*, 1953, **60**, 207–208.

Burke, C. J. Further remarks on one-tailed tests. *Psychological Bulletin*, 1954, **51**, 587–590.

Kaiser, H. F. Directional statistical decisions. *Psychological Review*, 1960, **67**, 160–167.

Eysenck, H. J. The concept of statistical significance and the controversy about one-tailed tests. *Psychological Review*, 1960, **67**, 269–271.

* Articles are listed chronologically.

Bakan, D. The test of significance in psychological research. *Psychological Bulletin*, 1966, **66**, 423–437.

Peizer, D. B. A note on directional inference. *Psychological Bulletin*, 1967, **68**, 448.

II. Multiple Comparisons and the Conceptual Unit for Error Rate*

The following articles supplement the last three articles in this chapter and deal with a wide variety of issues involved in multiple comparisons.

Tukey, J. W. Comparing individual means in the analysis of variance. *Biometrics*, 1949, **5**, 99–114.

McHugh, R. B., and D. S. Ellis. The "postmortem" testing of experimental comparisons. *Psychological Bulletin*, 1955, **52**, 425–428.

Stanley, J. C. Additional "post-mortem" tests of experimental comparisons. *Psychological Bulletin*, 1957, **54**, 128–130.

Harter, H. L. Error rates and sample sizes for range tests in multiple comparisons. *Biometrics*, 1957, **13**, 511–536.

Gaito, J. Multiple comparisons in analysis of variance. *Psychological Bulletin*, 1959, **56**, 392–393.

Ryan, T. A. Comments on orthogonal components. *Psychological Bulletin*, 1959, **56**, 394–396.

Ryan, T. A. Significance tests for multiple comparison of proportions, variances, and other statistics. *Psychological Bulletin*, 1960, **57**, 318–328.

Dunn, O. J. Multiple comparisons among means. *Journal of the American Statistical Association*, 1961, **56**, 52–64.

Ryan, T. A. The experiment as the unit for computing rates of error. *Psychological Bulletin*, 1962, **59**, 301–305.

Balaam, L. N., and W. T. Federer. Error rate bases. *Technometrics*, 1965, **7**, 260–262.

Davis, D. J. Flexibility and power in comparisons among means. *Psychological Bulletin*, 1969, **71**, 441–444.

III. Related Topics

Boneau, C. A. The effects of violations of assumptions underlying the *t* test. *Psychological Bulletin*, 1960, **57**, 49–64. Describes the results of an empirical study of the *t* test statistic for normal, rectangular, and exponential populations. The touted robustness of the *t* test statistic is confirmed.

Grubbs, F. E. Procedures for detecting outlying observations in samples. *Technometrics*, 1969, **11**, 1–21. Experimenters are often hesitant to discard an observation or score that appears to deviate markedly from other members of a sample unless some clear evidence of departure from prescribed research procedures can be invoked to account for the discrepancy. Statistical criteria are suggested in this article for detecting outlying observations.

9
Nonparametric Methods

Recently there has been a resurgence of interest in nonparametric and distribution-free methods. Although these two terms are used interchangeably, this is an oversimplification. In general nonparametric tests are intended to apply to a large class of distributions rather than to all possible distributions, and they usually assume continuity of the population that is sampled. Distribution-free tests, on the other hand, make no assumptions concerning the shape of the population that is sampled.

It is customary to compare the two methods with classical parametric methods. In general neither nonparametric nor distribution-free methods provide tests of hypotheses about population parameters. Parametric methods provide such tests but they involve several restrictive assumptions concerning the shape of the populations that are sampled—namely, normality of the population distributions and equality of the population variances. Thus parametric and nonparametric procedures do not test the same kinds of hypotheses about populations, nor do they involve the same assumptions.

Nonparametric and distribution-free methods also differ from classical parametric methods in other respects. For example, classical methods generally are used to evaluate hypotheses in which the descriptive sample statistics can be defined in terms of arithmetic operations upon observation magnitudes. Nonparametric and distribution-free methods are used more often for evaluating hypotheses in which the descriptive sample statistics are defined in terms of either order-relationships or category-frequencies.

During the past twenty years considerable controversy has developed concerning the relative merits of parametric methods and nonparametric or distribution-free methods. The latter methods have been particularly attractive to experimenters in the behavioral sciences because their assumptions are less restrictive concerning both the shape of the population distributions and the measurement operations used to describe data. Any reluctance to use nonparametric methods has stemmed largely from two considerations: their reputed lack of power-efficiency and the fact that nonparametric analogues for many of the classical tests have not been developed.

The first article in this chapter provides an overview of nonparametric statistics and helps to clarify many issues concerning their use. The author, James V. Bradley, takes exception to the reputed superiority of parametric methods relative to nonparametric methods and provides a fresh perspective from which to view this controversy.

9.1   Nonparametric Statistics

James V. Bradley

Development of Nonparametric Statistics

Most parametric statistics were developed at a time when it was considered natural for a variate to have a bell-shaped, Gaussian distribution, so natural in fact that the bell-shaped distribution was called the normal distribution. Scientific mistakes, once accepted, die hard—and this one has not yet expired. Many variates have distributions that are bell-shaped to a gross approximation, but despite the myths of the past there is no Normal Law of Error that demands explanation for the vagaries of the many variates that don't. Nevertheless, when parametric statistics were developed it seemed natural to adopt as a model the seemingly ubiquitous standard distribution presupposed by the prestigious Normal Law of Error.

The simple fact is that if parametric statistics are to be regarded as generally applicable but highly precise tools, as they were intended to be, then too much must be assumed; normal distributions with identical variances are as rare in the real world as they are common in the minds of Old Guard statisticians.

An uneasy awareness of this discrepancy motivated the development of distribution-free, or nonparametric, statistics. Much of the old format was retained, such as the use

of the Neyman-Pearson decision procedure of testing a narrow null hypothesis against a fairly broad spectrum of alternatives, using either end or both ends of the spectrum as rejection regions, with Type I and Type II errors having the same meaning as before. However, a subtle change of profound importance was made in the problem that probability theory was asked to solve. The question asked by parametric statistics had been "Given the *populations* (implied by H_0 and the assumptions), what is the probability of the test statistic?" Nonparametric statistics changed this to something more like "Given the *combined sample*, what is the probability of the test statistic (if H_0 and the assumptions are true)?" or, in the one-sample case, to "Given the *observation magnitudes*, but not their algebraic signs or sequence in which recorded, etc., what is the probability of the test statistic (if H_0 and the assumptions are true)?" Thus, nonparametric statistics took the set of observation magnitudes as given and then determined (a) the *a priori* probability under H_0 that they would be distributed among the various samples so as to give the test statistic the obtained value or a more extreme value or (b) in the one-sample case, the *a priori* probability under H_0 that the set of magnitudes would have algebraic signs or positions in sequence such as to give the test statistic the obtained value or one more extreme. By taking the magnitudes of the sample observations as given and focusing not upon their values but rather upon their allocation to samples, to sequential positions, or to algebraic signs, the new tests were freed from dependence upon the exact shape of the population distribution of observation magnitudes and therefore from assumptions about it. Furthermore, since in finite-sized samples the number of possible reassignments of observations to samples, sequential positions, or algebraic signs is finite, the probability distribution of the test statistic is finite. Therefore it is discrete and easier to derive and handle than are the continuous probability distributions of parametric statistics, which require calculus.

The magnitudes of n different sample observations vary from sample to sample, but the set of ranks of those magnitudes (in order of increasing size, from 1 to n) does not. Therefore, by substituting ranks for magnitudes one can calculate the distribution of the test statistic, which will be equally applicable for any other sample of the same size. Substituting ranks for magnitudes, therefore, standardizes an otherwise idiosyncratic test and makes it worthwhile to construct and publish tables of probability levels of the test statistic. This step was taken early in the development of nonparametric tests.

The conversion from magnitudes to ranks induced another change in the nature of nonparametric tests. Instead of testing or being sensitive to population statistics such as means or variances that are calculated from observation magnitudes, they test or are sensitive to population statistics such as medians and interquartile ranges that are determined from observation ranks or frequency counts. The transition to ranks also introduced the assumption that no sample observations have the same value. This assumption follows since tied observations should have the same rank and because ranks that are consecutive non-repeating integers are used to calculate the values of the test statistics for tables giving their probability distributions.

Examples. Some typical nonparametric tests can be used to illustrate the foregoing concepts. Perhaps the best-known and most representative nonparametric test is the two-sample Wilcoxon test, which assumes random sampling and no tied observations. It tests the null hypothesis that the two sampled populations have identical distributions against the alternative hypothesis that they do not, and it is sensitive to alternatives in which the two populations have different medians. Suppose that the sample data consist of the randomly drawn observations 17, 22, 29, 35 from an X population and 20, 37, 41, 44, 46, 49, 55 from a Y population. Replacing these observations by their ranks within the pooled sample of X and Y observations, we have

Ranks of X observations   1 3 4 5
Ranks of Y observations   2 6 7 8 9 10 11.

If the null hypothesis of identically distributed populations is true and if sampling was random, then the eleven observation values in the combined sample are in effect a single sample from a single distribution. Chance alone has determined which four observation values receive the arbitrary label "X sample" and which seven the arbitrary label "Y sample." We may therefore regard the X sample as having been randomly drawn from the pooled sample—leaving, of course, the Y sample. The number of different ways of dividing eleven observations or their ranks into two nonoverlapping sets containing four and seven observations or ranks, respectively, is $11!/4!7!$ or 330. Under H_0 each of the resulting 330 sets of "data" had the same *a priori* probability, 1/330, of becoming the actually obtained data. Our test statistic is the sum of the ranks in the smaller sample, which is $1 + 3 + 4 + 5 = 13$. We could have obtained a rank sum of equal or smaller value by the following seven assignments of four of the eleven ranks to the X sample: 1, 2, 3, 4; 1, 2, 3, 5; 1, 2, 4, 5; 1, 3, 4, 5; 1, 2, 3, 6; 1, 2, 4, 6; 1, 2, 3, 7. So the *a priori* probability under H_0 of an X sample rank sum as small as or smaller than that actually obtained is 7/330 or .021. Our result, then, is unlikely if the null hypothesis of identical populations is true. On the other hand, it is more or less what one would expect if the median of the

X population were considerably smaller in value than that of the Y population. So if prior to sampling we had chosen a significance level of .05 for a two-tailed test, we could reject H_0. If we had originally chosen a significance level of .05 (or any value equal to or greater than .021) for a lower-tailed test, we could reject H_0 in favor of the alternative hypothesis that the X and Y populations differ in that the X population's median probably lies below that of the Y population.

A useful one-sample nonparametric test is the sign test for the median. Suppose an experimenter hypothesizes that a certain continuously distributed population has a median of 100, and he wishes to perform a one-tailed test of this H_0 at the .05 level against the alternative hypothesis that the median is greater than 100. He randomly draws a sample of 11 observations from the population in question and then subtracts the hypothesized value of the median from each observation. The data are shown below:

X	134	92	112	109	137	145
	102	71	217	105	137	
$X - 100$	34	-8	12	9	37	45
	2	-29	117	5	37	

The information used by the test is simply that of the eleven differences; two of them have minus signs and the remaining nine are positive. If the hypothesized value of 100 is really the median of the continuously distributed X population, the probability is $\frac{1}{2}$ that a single randomly drawn observation will fall below the median, resulting in a negative value for $X - 100$. The probability that two or fewer observations in a sample of eleven will fall below the median, causing their $X - 100$ difference scores to be minus, is given by the binomial formula

$$P(r \leq 2 \mid n = 11) = \sum_{r=0}^{2} \frac{11!}{r!(11 - r)!} \left(\frac{1}{2}\right)^{11}$$

$$= \frac{67}{2048} = .033 \ .$$

Therefore two or fewer minuses are unlikely to occur if the median is 100 but would be more likely if the median were greater than 100. In that case, instead of 100 being the 50th percentile of the distribution, 100 would be a lower percentile such as the 20th. Consequently, since $.033 < .05$, the experimenter can reject the null hypothesis that $M \leq 100$ in favor of the alternative hypothesis that $M > 100$.

As a final example, suppose we want to test at the .05 level the null hypothesis that prices on the stock exchange are fluctuating randomly against the alternative hypothesis that underlying their gyrations is a nondecreasing upward trend. During the period of interest, the closing prices on the exchange are 803.25, 805.373, 808.125, 806.5, 811.75. Replacing these values by their ranks (for convenience only; it is not actually necessary) we have 1 2 4 3 5. We take as the test statistic the number of "inversions" in the sequence —that is, the number of times a rank is followed by a lower rank. There is one inversion in the sequence: the 4 is followed by the smaller number 3. There are $5! = 120$ different possible sequences in which the five ranks could appear, and they are equally likely under a null hypothesis of strictly random fluctuations. Of the 120 different permutations of ranks, only the following five contain as few or fewer inversions as the sequence actually obtained:

1	2	3	4	5
1	2	3	5	4
1	2	4	3	5
1	3	2	4	5
2	1	3	4	5

So if the null hypothesis is true, the *a priori* probability of obtaining as few inversions as actually obtained in the sample is 5/120 or .042. The obtained results are about what one would expect if the alternative hypothesis of a nondecreasing upward trend is true. Therefore, we may reject H_0 at the .05 level of significance in favor of H_1.

Comparison with Parametric Statistics

When they first appeared, nonparametric statistics were regarded as quick and dirty substitutes for their parametric counterparts. They were widely regarded by practitioners to be distinctly inferior in efficiency but somehow to transcend logic so as to make no assumptions whatever. They were naively believed to be universally applicable despite confounding, missing plots, or malfunctioning apparatus, and to require little more than an *n* of 1 consisting of an observation that could be loosely called a datum. They were, in short, regarded as hopelessly inefficient yet magically impervious to experimental and logical misadventures. Both views are incorrect. Parametric tests are more efficient or equally efficient when their assumptions are met, and the same rules of logical inference apply to both nonparametric and parametric tests.

However, the two types of tests do differ in important respects, the most outstanding of which are that nonparametric statistics, compared with parametric statistics, generally (a) are much simpler, (b) make more modest assumptions, and (c) test hypotheses about different population characteristics. In all three respects the behavioral scientist benefits from the differences between the two types of tests.

Simplicity. Perhaps the most important aspect of nonparametric statistics is the mathematical ease and simplicity with which they are derived. The derivation of parametric tests involves mathematics beyond the training of most behavioral scientists, but the derivation of most nonparametric statistics involves only a knowledge of high school algebra and some elementary probability formulas. The importance of this to the behavioral scientist can hardly be overestimated, since it is extremely demoralizing to a scientist to be forced to use a technique whose basis he does not understand and whose validity and appropriateness he must therefore accept on blind faith. Nonparametric statistics offered the behavioral scientist for the first time a procedure whose rationale he could understand, in place of the cookbook rituals of parametric methods.

In addition to their greater conceptual simplicity, nonparametric statistics are also easier to apply. Often, as with the Wilcoxon test, the test statistic is obtained by simply ranking observations according to size and then adding some of the ranks. Sometimes it is obtained merely by counting the number of observations whose values have a particular algebraic sign. Nonparametric methods based on counting are almost always faster, and those based on ranking are generally faster with modest amounts of data than their parametric counterparts. This greater simplicity in application is a decided economic advantage, since relatively unskilled assistants can be used to conduct the tests. The money or effort saved can be devoted to obtaining more data, thereby increasing precision and tending to offset any shortcomings in statistical efficiency of nonparametric methods.

Assumptions. A test's null hypothesis and assumptions together spell out the conditions under which the test statistic will have its null or "tabled" distribution. If *any* of these conditions is not met, the test statistic's distribution will be altered and this will alter the probability of rejection from α to some other value. Clearly, then, a test tests its assumptions just as truly as it tests its null hypothesis. Indeed, the classical parametric test of population variance is so sensitive to violations of its assumption of normality that it has been seriously proposed as a test for normality!

In order for a test to be an exact test of its null hypothesis, all of its assumptions must be completely met. And in order for a test to be a good approximate test of its null hypothesis, the test must be known to be relatively insensitive to violations of assumptions of the degree existing in the applicational situation. Such knowledge, however, seldom exists outside of well-explored areas, such as quality control, where a vast backlog of data is kept for a highly standardized "experiment" conducted repeatedly under only slightly changing conditions. Except for such situations, it is clear that the more rudimentary and easily met the assumptions of a test are, the better.

Both parametric and nonparametric tests assume randomness in the sampling process. This may require the random drawing of observations from infinite populations, which is often difficult and sometimes not even feasible. If the experiment investigates the effects of treatments applied to subjects or other units, it may suffice simply to draw the various groups of subjects randomly from a common subject pool prior to treatment— that is, one randomly determines which subjects get which treatments. In the latter case, the test's assumption of randomness is completely satisfied, although the extent to which one can generalize his results beyond the common subject pool depends upon the representativeness of the pool. Thus a statistical problem is, in a sense, solved at the expense of introducing a logical problem.

In addition to sampling assumptions there are population assumptions—that is, assumptions about the shape of the population distribution. Some few nonparametric tests (or rather some tests in certain contexts) make no population assumptions that are not automatically met by following the test's prescribed procedure. Most nonparametric tests require either the assumption that there are no tied values among certain sets of observations or the assumption that the members of difference-scores are not equal in value. This requirement that certain scores be unequal is sometimes referred to as the continuity assumption, since there is zero probability of randomly drawing the same value twice in a finite sample from a continuous distribution. But the two assumptions are not really equivalent. Tied observations are obtained even from continuous populations, since the measuring instrument is not infinitely precise and therefore records "close" observations as equal. Finally, a few nonparametric tests assume that the sampled population is symmetric. However, this is about as detailed and

elaborate as the population assumptions of a nonparametric test are likely to be. The exact shape of the population distribution is never specified.

By contrast, most parametric tests assume that all sampled populations have normal distributions and, when there is more than one population, equal variances. The normal distribution is not only continuous and symmetric, as might be assumed by a nonparametric test, but in addition is bell-shaped and extends to infinity in both directions. It therefore represents a very special case of a more general situation to which nonparametric tests are applicable. The parametric assumption of equal population variances together with the normality assumption in tests for means amounts to making them tests for identical normal populations. The corresponding nonparametric tests are, in effect, tests for identical continuous populations. Clearly, then, parametric tests are restricted by their assumptions to a narrow subset of the applicational situations in which nonparametric tests are valid. So nonparametric tests enjoy the advantage of having a much broader scope of application. And since situations meeting parametric assumptions are a mere subset of those meeting nonparametric assumptions, it follows that the assumptions of parametric tests are violated by a much larger class of situations than are those of nonparametric tests. That is, if parametric and nonparametric tests were applied in a large variety of situations, the assumptions of the parametric tests would be violated more frequently and to a greater degree. In this sense, therefore, parametric tests' assumptions are far more susceptible to violation than are those of nonparametric tests.

When the nonparametric assumption of no ties among certain sets of observations is violated, both the fact and the degree of the violation are apparent from the existence and proportion of tied observations in these sets. No analogous clue advises the experimenter of violation of the normality assumption or of the assumption of equal population variances. It is virtually impossible to discern either population normality or population nonnormality from inspection of the empirical distribution of a small sample of data, and small-sample tests for normality or homogeneity are unlikely to detect any but the largest departures. Nonparametric tests, therefore, have the advantage that violations of their population assumptions are more easily detectable than are violations of parametric assumptions.

Closely connected with this feature is the ability of nonparametric tests to assess the effect of violation of the no-ties assumption. The experimenter need only (a) resolve all ties in whatever way minimizes the probability level ρ of the test statistic, calling this minimum ρ-value ρ_m, (b) resolve all ties in the way that maximizes ρ, calling this maximum value ρ_M, (c) assert that $\rho_m \leq \rho \leq \rho_M$, that is, that the true probability level of the test lies somewhere in the interval bounded by ρ_m and ρ_M. If ρ_m and ρ_M are not equal, the test is inexact, but the experimenter has exact knowledge about its possible degrees of inexactitude. If ρ_m and ρ_M both fall on the same side of α (the preselected probability of rejecting a true null hypothesis), there is no problem; if they fall on different sides, the cleanest solution is simply to take more data until the issue is resolved.

There is no analogous way to assess the effect of violation of the nonparametric assumptions of randomness or symmetry. The latter, however, does not pose a real problem, since there are good alternative nonparametric tests for essentially the same hypothesis that do not assume symmetry and that can therefore be used when there is any doubt about the validity of the symmetry assumption. This leaves the randomness assumption as the only unavoidable nonparametric assumption the effect of whose violation cannot be assessed (although, as already pointed out, the purely statistical aspect of the randomness assumption can be easily satisfied when subjects are drawn from a common pool to receive an experimental "treatment"). However, while the effect of

its violation upon probability levels is unknown, the degree to which the randomness assumption is violated should be known to the experimenter, since the assumption concerns the procedure used in sampling and this procedure is under the experimenter's direct control. This knowledge of the extent of the violation should produce some intuition about the probable effect of the violation; however, this is only vague information at best, and the point should not be overstressed. And, of course, the same point can be made for the randomness assumption of parametric tests.

Thus, in many situations nonparametric assumptions are completely satisfied. In other situations, while they produce vexing problems, the experimenter is in a good position to know when they are violated and in a fair position to know the effect of the violation, being able to set bounds upon the true probability level when the no-ties assumption is violated.

The outlook is far dimmer for parametric assumptions. The normality assumption is intrinsically incapable of being exactly met. No variable in the real physical world has a distribution that extends to infinity in both directions as it would have to do, among other things, in order to qualify as a normal distribution. Indeed, for a host of variables such as height, weight, reaction time, or errors, it is impossible for the variable to have values less than zero. (Even the velocity of a gas particle, whose distribution is said to be almost exactly fitted by a normal curve, cannot be *exactly* normal since, if Einstein is correct, the velocity of the particle cannot exceed the speed of light—which is fast but not infinitely so.) The assumption that all sampled populations have the same variance is almost certain to be true when one is testing whether a treatment has changed the mean of a population and when in fact the mean has not changed, since a naturally produced change in variance is almost always accompanied by a change in the mean. The assumption of equal variances is almost certain to be false when one is testing whether a treatment has increased the mean of a population by a nonzero increment of D units and when the mean has in fact changed, since a naturally produced change in the mean is almost always accompanied by a change in variance. In general one would expect the assumption of equal variances to be valid (that is, true if H_0 is true) only when there is a logical implication that if the null hypothesis is true then all sampled populations are identical.

Assumption violation is therefore inevitable in the parametric case. The only question is, what is the degree of the violation and the extent of its effect on probability levels— that is, on the validity of the test. Unless it can be shown that situations involving degrees of violation likely to cause trouble are easily identifiable and therefore avoidable, or that the effect of degrees of violation likely to be encountered in practice is negligible, the parametric statistician will be gambling against dubious odds. But there are all kinds of practices. At one end of the continuum is the mental tester whose tests are weighted during standardization so as to have "normal" distributions with perhaps a mean of 100 and a standard deviation of 10. At the other end is the laboratory psychologist measuring performance on a nonstandard task, in an unexplored research area, using an idiosyncratic tailor-made apparatus. Neither this psychologist nor anyone else knows what is typical of his present practice or what is the worst to be expected.

The writer once investigated the shape of the sampled population (that is, the degree of violation of the normality assumption) in such circumstances and obtained a distribution shaped like an emerging sea serpent. The 2520 observations were spread out over a range of $11\frac{1}{2}$ standard deviations with 90% of the observations concentrated within an interval only one standard deviation in length. By means of a Monte Carlo-type sampling experiment using a large electronic computer, he checked the effect of this degree of nonnormality upon standard parametric tests. One experiment involved Student's t test statistic for samples of 32 observations. The

empirically estimated true probabilities of a Type I error corresponding to nominal one-tailed probabilities of .05, .01, and .001 were .145, .084, and .051, respectively, for the left tail. The latter probability represented an error of 5000 per cent. The corresponding probabilities for the right tail were .012, .0004, and 0, respectively. When the true probability is $\leq \alpha$, discrepancies are often shrugged off as being in the conservative direction so that H_0 will not be rejected more than α of the time. It is seldom recognized that this inevitably means a correspondingly conservative diminution in the power of the test.

Violation of the equal-variances assumption of the t test statistic can have serious consequences. For example, if very large samples of sizes n and $2n$ are randomly drawn from normal populations with the same mean but with standard deviations of σ and 2σ, respectively, it can be proved mathematically that this violation causes the true probabilities of Type I errors corresponding to nominal one-tailed test probabilities of .05, .01, and .001 to be .0220, .0022, and .00008, respectively. Many extravagant claims have been made for the "robustness" of parametric statistics against violation of their assumptions. However, the evidence shows that such claims require such elaborate and special qualifications as to nullify their value as general procedural rules. The fact is that as exact tests parametric tests are invalid and as approximate tests their degree of invalidity can be quite large. Furthermore, to predict their expected degree of invalidity, especially in someone else's area of practice, requires, at best, educated guesswork, and the prediction is vulnerable to the hazards of wishful thinking.

To summarize, the population assumptions of nonparametric tests are relatively modest and are frequently capable of being met. In the case of unavoidable assumptions, their violation has effects that can be easily detected, and consequences that can be readily assessed from the obtained sample data. In contrast, the population assumptions of parametric tests are elaborate; some are incapable of being met, and their violations have effects that are not easily detected and consequences that may be devastating.

Population Characteristics Tested. Parametric tests test hypotheses about population indices, such as means and variances, that are based on the *magnitudes* of the individual units constituting the population. Nonparametric tests test hypotheses about population indices, such as medians and interquartile ranges, that are based on the *positions* of the individual population units when arranged in rank order. When they first appeared, this characteristic of nonparametric tests was considered a flaw. Statisticians had been using means and variances as indices of location and dispersion for so long that they seemed the only proper indices. However, which type of index is preferable depends upon the use to which it is put. The physical sciences may be well served by knowledge of a distribution's first and second moments, but it seems likely that the behavioral sciences will generally find medians and interquartile or other interpercentile ranges more meaningful. A knowledge of the median and interquartile range of the distribution of family incomes in Saudi Arabia would tell a behavioral scientist a great deal about the typical family income. But of what use would it be to know the mean and variance? Indeed, the information might be highly misleading.

Faults and Pseudo-Faults. Perhaps the most grievous fault of nonparametric tests as a group is the absence of a simple test for higher-order interactions. Tests for such interactions exist, but they are cumbersome even for second-order interactions and become increasingly so when applied to interactions of higher and higher order.

Another fault is that since the test statistic has a discrete distribution it is impossible to select any probability of a Type I error one wishes. One might want to use the conventional probability of .05, but this will be impossible (unless one wants to introduce a second, "chance" experiment) if adjacent values of the test statistic have cumulative probabilities of .0461 and .0549. The common practice is to use the probability of .0461 and report it as .05. This becomes extremely misleading at small sample sizes where adjacent cumulative probabilities may differ considerably, having values such as .014 and .072. This fault prevents several tests using different-size samples from being conducted at a common significance level and therefore impairs their comparability. Parametric tests, having continuously distributed test statistics, do not suffer from this shortcoming and can be conducted at any significance level one wishes.

Closely related to the preceding shortcoming is the fact that at very small sample sizes, such as $n < 5$ for one-sample tests, it may be impossible to attain the .05 level of significance no matter how drastically the data belie the null hypothesis. This is due to the fact that the probability of the test statistic taking certain values is the same as the probability of certain permutations or combinations occurring among n things. The number of possibilities is quite limited when n is small, so that the probability of any one of them, even the most unlikely one under H_0, is relatively large. Parametric tests are superior in this regard, since they can attain any significance level for any sample size at which a meaningful value of the test statistic can be calculated. However this may depend on having at least two observations with which to estimate a population variance.

The pseudo-fault of nonparametric tests is their alleged general inefficiency. There is an entire class of nonparametric tests whose members are more efficient for large samples than their best parametric competitors when applied to nonnormal populations and equally efficient when applied to normal populations. Members of another large class of nonparametric tests are more efficient than their best parametric competitors when applied to a certain type of nonnormal population (the type depending on the test) and less efficient when applied to normal populations. So some nonparametric tests are never inferior in large-sample efficiency and

Table 1. Selected distribution-free tests for some common types of application.

Distribution-free test	Type of information used	Counterpart parametric test	Large-sample efficiency relative to counterpart parametric test when all the latter's assumptions are met
For location			
One-sample or two-matched-sample tests			
Sign	Frequencies	Student's *t*	.637
Wilcoxon signed-rank	Ranks	Student's *t*	.955
Fraser normal-scores	Ranks	Student's *t*	1.000
Two-sample tests			
Westenberg-Mood median	Frequencies	Student's *t*	.637
Wilcoxon rank-sum	Ranks	Student's *t*	.955
Terry-Hoeffding normal-scores	Ranks	Student's *t*	1.000
Multi-sample tests			
Brown-Mood median	Frequencies	*F*-test for means	.637
Friedman	Ranks	Matched-observation *F*	.637–.955
Kruskal-Wallis	Ranks	*F*-test for means	.955
Pitman randomization	Values	*F*-test for means	1.000
For dispersion			
Two-sample tests			
Siegel-Tukey rank-sum	Ranks	*F*-test for variances	.608
Klotz normal-scores	Ranks	*F*-test for variances	1.000
For association or correlation			
Blomqvist	Frequencies	*t*-test of *r*	.405
Kendall	Inversions	*t*-test of *r*	.912
For trend			
Cox and Stuart sign	Frequencies	*t*-test of *b*	.827
Mann-Kendall	Inversions	*t*-test of *b*	.985
For poorness of fit			
One-sample tests			
David's empty-cell	Frequencies		
Kolmogorov	Relative frequencies		
Two-sample tests			
Wilks empty-cell	Frequencies		
Smirnov	Relative frequencies		

others are sometimes superior, sometimes inferior. The fact is that the parametric tests' claim to superior efficiency depends largely on the validity of the normality assumption, which cannot in fact be exactly true and which is often seriously false.

Table I performs the dual function of showing the versatility of nonparametric tests, although a much broader range of application exists than is indicated, and of showing the large-sample efficiencies of some of the best nonparametric tests relative to their best parametric competitor when the latter's assumptions are all true. This type of comparison is of course unfair to the

nonparametric statistics since (a) they don't make the parametric tests' assumptions, (b) some of the assumptions are impossible to fulfill, making the efficiencies quoted strictly academic, and (c) nonparametric tests generally have their greatest efficiency in situations other than the assumed conditions. For example, in the class of situations where populations have distributions with identical shapes and variances but possibly different means, the large sample efficiency of Wilcoxon's Rank-Sum test relative to Student's *t*-test can be as high as infinity but no lower than .864. In this case the efficiency of the Terry-Hoeffding Normal-Scores test is always ≥ 1.000, reaching its lower bound of 1.000 only when the populations are normal as assumed by the *t*-test.

Scope. As Table I suggests, at least one and usually several good nonparametric tests exist for virtually all of the standard testing situations. Nor are nonparametric statistics limited to testing; they are also excellent for estimation. For example, one can easily obtain confidence limits for the median or any other population percentile. Nonparametric statistics, in fact, cover all of the standard topics found in an elementary statistics textbook, although as already mentioned they are less satisfactory for investigating complex interactions such as those dealt with by analysis-of-variance techniques.

Summary

Parametric statistical tests (a) have derivations requiring higher mathematics, (b) make elaborate assumptions about the shapes of sampled populations, and (c) test hypotheses about population statistics based on magnitudes such as means and variances.

By contrast, nonparametric statistics (a) have derivations involving elementary probability and simple algebra, (b) make only modest assumptions about the shapes of sampled populations, and (c) test hypotheses about population statistics based on frequencies or ranks such as medians and interquartile ranges.

The behavioral scientist is the beneficiary of the difference, since he (a) usually has a limited knowledge of higher mathematics, (b) seldom knows very much about the precise shape of the populations from which he samples, and (c) is generally much more interested in what is typical than in what is average and more interested in the "range of the ordinary" than in second moments.

Editor's comments: Simplicity of derivation is one of the advantages often cited for nonparametric methods. In the following article, W. J. Dixon and A. M. Mood give the derivation of the sign test, illustrate its use, and describe several modifications of the test for particular applications. Although the article was prepared for professional statisticians, introductory-statistics students will find the main thread of the article surprisingly easy to follow and understand.

9.2   The Statistical Sign Test

W. J. Dixon and A. M. Mood

Abstract. This paper presents and illustrates a simple statistical test for judging whether one of two materials or treatments is better than the other. The data to which the test is applied consist of paired observations on the two materials or treatments. The test is based on the signs of the differences between the pairs of observations.

It is immaterial whether all the pairs of observations are comparable or not. However, when all the pairs are comparable, there are more efficient tests (the *t* test, for example) which take account of the magnitudes as well as the signs of the differences. Even in this case, the simplicity of the sign test makes it a useful tool for a quick preliminary appraisal of the data.

In this paper the results of previously published work on the sign test have been included, together with a table of significance levels and illustrative examples.

Introduction

In experimental investigations, it is often desired to compare two materials or treatments under various sets of conditions. Pairs of observations (one observation for each of the two materials or treatments) are obtained

This paper is an adaptation of a memorandum submitted to the Applied Mathematics Panel by the Statistical Research Group, Princeton University. The Statistical Research Group operated under a contract with the Office of Scientific Research and Development, and was directed by the Applied Mathematics Panel of the National Defense Research Committee.

Reprinted from *American Statistical Association Journal*, 1946, **41**, 557–566, by permission of the publisher and authors.

for each of the separate sets of conditions. For example, in comparing the yield of two hybrid lines of corn, *A* and *B*, one might have a few results from each of several experiments carried out under widely varying conditions. The experiments may have been performed on different soil types, with different fertilizers, and in different years with consequent variations in seasonal effects such as rainfall, temperature, amount of sunshine, and so forth. It is supposed that both lines appeared equally often in each block of each experiment so that the observed yields occur in pairs (one yield for each line) produced under quite similar conditions.

The above example illustrates the circumstances under which the sign test is most useful:

(a) There are pairs of observations on two things being compared.

(b) Each of the two observations of a given pair arose under similar conditions.

(c) The different pairs were observed under different conditions.

This last condition generally makes the *t* test invalid. If this were not the case (that is, if all the pairs of observations were comparable), the *t* test would ordinarily be employed unless there were other reasons, for example, obvious non-normality, for not using it.

Even when the *t* test is the appropriate technique many statisticians like to use the sign test because of its extreme simplicity. One merely counts the number of positive and negative differences and refers to a table of significance values. Frequently the question of significance may be settled at once by the sign test without any need for calculations.

It should be pointed out that, strictly speaking, the methods of this paper are applicable only to the case in which no ties in paired comparisons occur. In practice, however, even when ties would not occur if measurements were sufficiently precise, ties do occur because measurements are often made only to the nearest unit or tenth of a unit for example. Such ties should be included among the observations with half of them being counted as positive and half negative.

Finally, it is assumed that the differences between paired observations are independent, that is, that the outcome of one pair of observations is in no way influenced by the outcome of any other pair.

Procedure

Let *A* and *B* represent two materials or treatments to be compared. Let *x* and *y* represent measurements made on *A* and *B*. Let the number of pairs of observations be *n*. The *n* pairs of observations and their differences may be denoted by:

$$(x_1, y_1), (x_2, y_2), \ldots, (x_n, y_n)$$

and

$$x_1 - y_1, x_2 - y_2, \ldots, x_n - y_n.$$

The sign test is based on the signs of these differences. The letter *r* will be used to denote the number of times the less frequent sign occurs. If some of the differences are zero, half of them will be given a plus sign and half a minus sign.

As an example of the type of data for which the sign test is appropriate, we may consider the following yields of two hybrid lines of corn obtained from several different experiments (Table 1). In this example $n = 28$ and $r = 7$.

If there is no difference in the yielding ability of the two lines, the positive and negative signs should be distributed by the binomial distribution with $p = \frac{1}{2}$. The null hypothesis here is that each difference has a probability distribution (which need not be the same for all differences) with median equal to zero. This null hypothesis will obtain, for instance, if each difference is symmetrically distributed about a mean of zero, although such symmetry is not necessary. The null hypothesis will be rejected when the numbers of positive and negative signs differ significantly from equality.

Table 2 gives the critical values of *r* for the 1, 5, 10, and 25 per cent levels of significance. A discussion of how these values are computed may be found in the appendix. A value of *r* less than or equal to that in the table is significant at the given per cent level.

Thus in the example above where $n = 28$ and $r = 7$, there is significance at the *5%* level, as shown by Table 2. That is, the chances are only 1 in 20 of obtaining a value of *r* equal to or less than 8 when there is no real difference in the yields of the two lines of corn. It is concluded, therefore, at the 5% *level of significance*, that the two lines have different yields.

In general, there are no values of *r* which correspond exactly to the levels of significance 1, 5, 10, 25 per cent. The values given are such that they result in a level of significance as close as possible to, but not exceeding 1, 5, 10, 25 per cent. Thus, the test is a little

Table 1. Yields of two hybrid lines of corn.

Experiment number	Yield of A	B	Sign of x − y	Experiment number	Yield of A	B	Sign of x − y
1	47.8	46.1	+	4	40.8	41.3	−
	48.6	50.1	−		39.8	40.8	−
	47.6	48.2	−		42.2	42.0	+
	43.0	48.6	−		41.4	42.5	−
	42.1	43.4	−	5	38.9	39.1	−
	41.0	42.9	−		39.0	39.4	−
2	28.9	38.6	−		37.5	37.3	+
	29.0	31.1	−	6	36.8	37.5	−
	27.4	28.0	−		35.9	37.3	−
	28.1	27.5	+		33.6	34.0	−
	28.0	28.7	−				
	28.3	28.8	−	7	39.2	40.1	−
	26.4	26.3	+		39.1	42.6	−
	26.8	26.1	+				
3	33.3	32.4	+				
	30.6	31.7	−				

more strict, on the average, than the level of significance which is indicated. For small samples the test is considerably more strict in some cases. For example, the value of r for $n = 12$ for the 10 per cent level of significance actually corresponds to a per cent level less than 5.

The critical values of r in Table 2 for the various levels of significance were computed for the cases where either the +'s or −'s occur a significantly small number of times. Sometimes the interest may be in only one of the signs. For example, in testing two treatments, A and B, A may be identical with B except for certain additions which can only have the effect of improving B. In this case one would be interested only in whether the deficiency of minus signs (for differences in the direction A minus B) were significant or not. In cases of this kind the per cent levels of significance in Table 2 would be divided by two. Thus, 8 minus signs in a sample of 28 would correspond to the 2.5% level of significance.

Size of Sample

Even though there is no real difference, a sample of four or even five with all signs alike will occur by chance more than 5% of the time. Four signs alike will occur by chance 12.5% of the time and five signs alike will occur by chance 6.25% of the time. Therefore, at the 5% level of significance, it is necessary to have at least six pairs of observations even if all signs are alike before any decision can be made. As in most statistical work, more reliable results are obtained from a larger number of observations. One would not ordinarily use the sign test for samples as small as 10 or 15, except for rough or preliminary work.

The question may be raised as to the minimum sample size necessary to detect a given difference in two materials. Suppose that in an indefinitely large number of observations 30% +'s and 70% −'s are to be expected and that we wish the sample to be large enough to detect this difference at the 1% level of significance. Although no sample, however large, will make it absolutely certain that a significant difference will be found, the sample size can be chosen to make the probability of finding a significant result as near to certainty as is desired. In Table 3, this probability has been chosen as 95%; the minimum values of n (sample size) and the corresponding critical values of r to insure a decision 95% of the time are given for various actual percentages p_0 and levels of significance α.

Table 2. Critical values of *r* for the sign test.

n	Per cent level of significance				n	Per cent level of significance			
	1	5	10	25		1	5	10	25
1	—	—	—	—	51	15	18	19	20
2	—	—	—	—	52	16	18	19	21
3	—	—	—	0	53	16	18	20	21
4	—	—	—	0	54	17	19	20	22
5	—	—	0	0	55	17	19	20	22
6	—	0	0	1	56	17	20	21	23
7	—	0	0	1	57	18	20	21	23
8	0	0	1	1	58	18	21	22	24
9	0	1	1	2	59	19	21	22	24
10	0	1	1	2	60	19	21	23	25
11	0	1	2	3	61	20	22	23	25
12	1	2	2	3	62	20	22	24	25
13	1	2	3	3	63	20	23	24	26
14	1	2	3	4	64	21	23	24	26
15	2	3	3	4	65	21	24	25	27
16	2	3	4	5	66	22	24	25	27
17	2	4	4	5	67	22	25	26	28
18	3	4	5	6	68	22	25	26	28
19	3	4	5	6	69	23	25	27	29
20	3	5	5	6	70	23	26	27	29
21	4	5	6	7	71	24	26	28	30
22	4	5	6	7	72	24	27	28	30
23	4	6	7	8	73	25	27	28	31
24	5	6	7	8	74	25	28	29	31
25	5	7	7	9	75	25	28	29	32
26	6	7	8	9	76	26	28	30	32
27	6	7	8	10	77	26	29	30	32
28	6	8	9	10	78	27	29	31	33
29	7	8	9	10	79	27	30	31	33
30	7	9	10	11	80	28	30	32	34
31	7	9	10	11	81	28	31	32	34
32	8	9	10	12	82	28	31	33	35
33	8	10	11	12	83	29	32	33	35
34	9	10	11	13	84	29	32	33	36
35	9	11	12	13	85	30	32	34	36
36	9	11	12	14	86	30	33	34	37
37	10	12	13	14	87	31	33	35	37
38	10	12	13	14	88	31	34	35	38
39	11	12	13	15	89	31	34	36	38
40	11	13	14	15	90	32	35	36	39
41	11	13	14	16	91	32	35	37	39
42	12	14	15	16	92	33	36	37	39
43	12	14	15	17	93	33	36	38	40
44	13	15	16	17	94	34	37	38	40
45	13	15	16	18	95	34	37	38	41
46	13	15	16	18	96	34	37	39	41
47	14	16	17	19	97	35	38	39	42
48	14	16	17	19	98	35	38	40	42
49	15	17	18	19	99	36	39	40	43
50	15	17	18	20	100	36	39	41	43

For $n > 100$, approximate values of *r* may be found by taking the nearest integer less than $\frac{1}{2}n - k\sqrt{n}$, where $k = 1.3, 1, .82, .58$ for the 1, 5, 10, 25 per cent values respectively. A closer approximation to the values of *r* is obtained from $\frac{1}{2}(n-1) - k\sqrt{n+1}$ and the more exact values of *k*, 1.2879, .9800, .8224, .5752.

The sign test merely measures the significance of departures from a 50–50 distribution. If the signs are actually distributed 45–55, then the departure from 50–50 is not likely to be significant unless the sample is quite large. Table 3 shows that if the signs are actually distributed 45–55, then one must take samples of 1,297 pairs in order to get a significant departure from a 50–50 distribution at the 5% level of significance. The number 1,297 is selected to give the desired significance 95% of the time; that is, if a large number of samples of 1,297 each were drawn from a 45–55 distribution, then 95% of those samples could be expected to indicate a significant departure (at the 5% level) from a 50–50 distribution.

Of course, in practice one would not do any testing if he knew in advance the expected distribution of signs (that it was 45–55, for example). The practical significance of Table 3 is of the following nature: In comparing two materials one is interested in determining whether they are of about equal or of different value. Before the investigation is begun, a decision must be made as to how different the materials must be in order to be classed as different. Expressed in another way, how large a difference may be tolerated in the statement that "the two materials are of about equal value ?" This decision, together with Table 3, determines the sample size. If one is interested in detecting a difference so small that the signs may be distributed 45–55, he must be prepared to take a very large sample. If, however, one is interested only in detecting larger differences (for example, differences represented by a 70–30 distribution of signs), a smaller sample will suffice.

In many investigations, the sample size can be left undetermined, and only as much data accumulated as is needed to arrive at a decision. In such cases, the sign test could be used in conjunction with methods of sequential analysis. These methods provide a desired amount of information with the minimum amount of sampling on the average. A complete exposition of the theory and practice of sequential analysis may be found in references 3 and 4.

Modifications of the Sign Test

When the data are homogeneous (measurements are comparable between pairs of observations), the sign test can be used to answer questions of the following kind:

1. Is material A better than B by P per cent?

2. Is material A better than B by Q units ? The first question would be tested by increasing the measurement on B by P per cent and comparing the results with the measurements on A. Thus, let

Table 3. Minimum values of n necessary to find significant differences 95% of the time for various given proportions.

p_0	n				r			
	$\alpha = 1\%$	5%	10%	25%	$\alpha = 1\%$	5%	10%	25%
.45(.55)	1,777	1,297	1,080	780	833	612	512	373
.40(.60)	442	327	267	193	193	145	119	87
.35(.65)	193	143	118	86	78	59	49	37
.30(.70)	106	79	67	47	39	30	26	19
.25(.75)	66	49	42	32	22	17	15	12
.20(.80)	44	35	28	21	13	11	9	7
.15(.85)	32	23	18	14	8	6	5	4
.10(.90)	24	17	13	11	5	4	3	3
.05(.95)	15	12	11	6	2	2	2	1

The italicized values are approximate. The maximum error is about 5 for the value of n, and 2 for the value of r. The values of n and r for 5% were taken from MacStewart (reference 1) who gives a table of values of n and r for a range of confidence coefficients (the above table uses only 95%) and a single value $\alpha = 5\%$.

$(x_1, y_1), (x_2, y_2), (x_3, y_3)$, etc.

be pairs of measurements on A and B, and suppose one wished to test the hypothesis that the measurements, x, on A were 5% higher than the measurements, y, on B. The sign test would simply be applied to the signs of the differences

$$x_1 - 1.05y_1, \ x_2 - 1.05y_2, \ x_3 - 1.05y_3, \text{ etc.}$$

In the case of the second question the sign test would be applied to the differences

$$x_1 - (y_1 + Q), \ x_2 - (y_2 + Q),$$
$$x_3 - (y_3 + Q), \text{ etc.}$$

In either case, if the resulting distribution of signs is not significantly different from 50–50, the data are not inconsistent with a positive answer to the question. Usually there will be a range of values of P (or Q) which will produce a non-significant distribution of signs. If one determines such a range, using the 5% level of significance for example, then that range will be a 95% confidence interval for P (or Q).

Even when the data are not homogeneous, it may be possible to frame questions of the above kind, or it may be possible to change the scales of measurement so that such questions would be meaningful.

Mathematical Appendix

A. *Assumptions.* Let observations on two materials or treatments A and B be denoted by x and y, respectively. It is assumed that for any pair of observations (x_i, y_i) there is a probability $p(0 < p < 1)$ that $x_i > y_i$ $(i = 1, 2, \ldots, n)$; p is assumed to be unknown.[1] It is also assumed that the n pairs of observations (x_i, y_i), $(i = 1, 2, \ldots, n)$ are independent; i.e., the outcome ($+$ or $-$) for (x_j, y_j) is independent of the outcome for (x_i, y_i) $(i \neq j)$.

B. *The Observations.* The purpose of obtaining observations (x_i, y_i) is to make an

---

[1] An additional assumption is that the probability $A_i = B_i$ is 0; thus the probability $B_i > A_i$ is $(1 - p)$.

inference regarding p. The observed quantity upon which an inference is to be based is r, the number of $+$'s or $-$'s (whichever occur in fewer numbers) obtained from n paired observations (x_i, y_i). On the basis of the assumption above it follows that the probability of obtaining exactly r as the minimum number of $+$'s or $-$'s is:

$$\binom{n}{r}[p^r(1-p)^{n-r} + p^{n-r}(1-p)^r]$$

$$r = 0, 1, 2, \ldots, \frac{n-1}{2} \ ; \ n \text{ odd}$$

$$r = 0, 1, 2, \ldots, \frac{n-2}{2} \ ; \ n \text{ even}$$

$$\binom{n}{\frac{1}{2}n} p^{\frac{1}{2}n}(1-p)^{\frac{1}{2}n}$$

$$r = \frac{n}{2} \ ; \ n \text{ even.}$$

C. *The Inference.* In the sign test the hypothesis being tested is that $p = \frac{1}{2}$; in other words that the distributions of the differences $x_i - y_i$ $(i = 1, 2, \ldots, n)$ have zero medians. For the more general tests discussed in Section 5, the hypothesis is that the differences $x_i - f(y_i)$ $(i = 1, 2, \ldots, n)$ have zero medians. The function $f(y)$ may be Py or $Q + y$ (where P and Q are the constants mentioned in Section 5) or any other function appropriate for comparison with x in the problem at hand.

The hypothesis that $p = \frac{1}{2}$ is tested by dividing the possible values of r into two classes and accepting or rejecting the hypothesis according as r falls in one or the other class. The classes are chosen so as to make small (say $\leq \alpha$) the chance of rejecting the hypothesis when it is true and also to make small the chance of accepting the hypothesis when it is untrue. It can be shown that in a certain sense, the best set of rejection values for r is $0, 1, \ldots, R$, where R depends on α and n. R can be determined by solving for $R = $ maximum i in the inequality:

$$\sum_{j=0}^{i} \binom{n}{j} \left(\frac{1}{2}\right)^n = I_{\frac{1}{2}}(n - i, i + 1) \leq \frac{1}{2}\alpha$$

where $I_x(a, b)$ is the incomplete beta function. Table 2 was computed in this way.

D. *Sample Sizes.* When the sample size is small the sign test is likely to reject the hypothesis, $p = \frac{1}{2}$, only if p is near zero or one. If p is near, but not equal to $\frac{1}{2}$, the test is likely to reject the hypothesis, $p = \frac{1}{2}$, only when the sample is large.

The sample size required to reject the hypothesis $p = \frac{1}{2}$ at the α level of significance, $100\lambda\%$ of the time, may be determined by finding the largest i and smallest n which satisfy:

$$\sum_{j=0}^{i} \binom{n}{j} \left(\frac{1}{2}\right)^n \leq \frac{1}{2}\alpha$$

and

$$\sum_{j=0}^{i} \binom{n}{j} p^j (1-p)^{n-j} \geq \lambda \qquad p < \frac{1}{2}.$$

n and i are given in Table 3 for various values of p and α; λ was taken to be .95 in all cases. The tabular values for $1 - p$ are the same as those for p because of the symmetry of the binomial distribution.

E. *Efficiency of the Sign Test.* Let $z = x - y$. Assume z is normally distributed with mean a and variance σ^2. The probability of obtaining a $+$ on a particular z_i is:

$$p = \frac{1}{\sqrt{2\pi}} \int_{-a/\sigma}^{\infty} e^{-(\frac{1}{2})u^2} du.$$

An estimate of p involving only the signs of z_i ($i = 1, 2, \ldots, n$) yields an estimate of (a/σ). Cochran (reference 2) has shown that in large samples the variance of this estimate of (a/σ) is $2\pi pq \, e^{(a/\sigma)^2}/n$. We shall denote a/σ by c.

The efficiency of an estimate based on n independent observations is defined as the limit (as $n \to \infty$) of the ratio of the variance of an efficient estimate to that of the given estimate. An efficient estimate of c is:

$$\frac{\bar{z}}{\sqrt{\dfrac{\sum (z_i - \bar{z})^2}{n-1}}} = \frac{t}{\sqrt{n}}$$

where t is Student's t and $\bar{z} = \Sigma z_i/n$.

The variance of this estimate is $1/(n-2)$; thus the efficiency, E, of the sign test is $e^{-c^2}/2\pi pq$. If $c = 0$, then $p = \frac{1}{2}$ and the efficiency is $2/\pi = 63.7\%$.

The preceding discussion pertains to large values of n; for small values of n, the efficiency is a little better than 63.7%. Computations were made for several smaller values of n, namely, for $n = 18, 30, 44$ pairs of observations at the 10% level of significance. It was found that the sign test using 18 pairs of observations is approximately equivalent to the t-test using 12 pairs of observations; for 30 pairs the equivalent t-test requires between 20 and 21 pairs; and for 44 pairs the equivalent t-test requires between 28 and 29 pairs. Cochran shows that the efficiency of r/n for estimating c decreases as $|c|$ increases.

References

1. MacStewart, W. "A note on the power of the sign test," *Annals of Mathematical Statistics*, Vol. 12 (1941), pp. 236–238.
2. Cochran, W. G. "The efficiencies of the binomial series test of significance of a mean and of a correlation coefficient," *Journal Royal Statistical Society*, Vol. C, Part I (1937), pp. 69–73.
3. Wald, A. "Sequential method of sampling for deciding between two courses of action," *Journal American Statistical Association*, Vol. 40 (1945), pp. 277–306.
4. Statistical Research Group, Columbia University, *Sequential Analysis of Statistical Data: Applications* (1945), Columbia University Press, New York.

Editor's comments: Students in introductory statistics courses unfortunately are rarely introduced to randomization tests, the most powerful of the non-parametric methods. This lack of coverage is understandable since randomization tests often require a prohibitive amount of relatively routine computation. By using approximate randomization tests, so named because they use an approximate sampling distribution instead of the entire sampling distribution, computational labor can be substantially reduced, although a computer is generally required. The power of approximate randomization tests is slightly less than that of randomization tests—a small price to pay for the reduction in computational labor. Eugene S. Edgington describes the nature and uses of approximate randomization tests in the following article.

9.3 Approximate Randomization Tests

Eugene S. Edgington

A. Introduction

In an article on randomization tests, approximate randomization tests were briefly discussed (1). This paper will consider approximate randomization tests in further detail. First, however, it is necessary to correct a statement in the earlier article. In a randomization test, the significance of an obtained statistic is the proportion of the statistics in the sampling distribution that are *as large as* the obtained statistic, *not*, as was erroneously stated, the proportion that exceed the obtained statistic. The correction is required for the computed significance value to be the probability, under the null hypothesis, of obtaining such extreme results.

Reprinted from *Journal of Psychology*, 1969, **72**, 143–149, by permission of the publisher and author.

B. Terminology

The term *entire sampling distribution* will be used to refer to the sampling distribution of a randomization test consisting of all divisions or pairings of measurements that are equally probable under the null hypothesis and random assignment, or to refer to the distribution of statistics associated with the distribution of divisions or pairings. For example, given the null hypothesis of no difference between the effects of Treatments A and B and random assignment of 10 subjects to A and five to B, the entire sampling distribution of divisions consists of every possible division of the 15 obtained measurements that assigns 10 to Treatment A and five to Treatment B. If one were to compute a difference between means for each division in the distribution, the distribution of differences between means thus obtained would be the entire sampling distribution of differences between means.

An *approximate sampling distribution* is a smaller distribution of pairings or divisions of measurements (or a distribution of statistics associated with them) obtained by taking a random sample of all of the pairings or divisions in the entire sampling distribution.

The term *approximate randomization test* refers to a randomization test that uses an approximate sampling distribution instead of the entire sampling distribution in determining the significance of an obtained statistic.

C. Purpose of Approximate Randomization Tests

The advantages of randomization tests over parametric and rank order tests have been widely recognized. The only serious objection that has been raised against the use of randomization tests is that the amount of computation required is excessive. The purpose of approximate randomization tests is to reduce the amount of computation for a randomization test to a practical level, thereby overcoming this objection. The following example will illustrate the reduction in computation that can result from using an approximate randomization test.

Suppose that in an experiment 13 subjects are randomly assigned to 13 different magnitudes of an independent variable and one measurement is obtained from each subject. The null hypothesis is that variation in magnitude of the independent variable has no effect on the dependent variable; i.e., at the time he was measured every subject gave the same response that he would have given at that time to any of the other 12 magnitudes of the independent variable. This hypothesis is tested by computing a product-moment correlation coefficient or some other correlation statistic. Under the null hypothesis and random assignment there are 13! (factorial 13) = 6,227,020,800 equally probable pairings of the 13 magnitudes of the independent

variable with the 13 response magnitudes. Thus, to determine with a randomization test what proportion of the sampling distribution of pairings gives as large a correlation statistic as the obtained value would require the computation of over six billion correlation statistics. Even an electronic computer that paired the measurements and computed a correlation statistic every second would take 197 years of continuous, 24-hours-a-day operation to compute all of the correlation statistics in the entire sampling distribution.

Now consider the application of an approximate randomization test to the same situation. An electronic computer is programmed to use a table of random numbers to simulate taking a random sample, *with replacement*, of 999 pairings from the entire sampling distribution and then to compute a correlation statistic for each of the selected pairings. This provides an approximate sampling distribution of 999 correlation statistics within minutes, in contrast to the 197 years required for the entire sampling distribution.

D. Determination of Significance for Approximate Randomization Tests

Now consider the procedure for determining the significance of results when approximate randomization tests are performed. The approximate sampling distribution is *not* used in exactly the same way as the entire sampling distribution in determining significance. Instead of determining the proportion of the statistics in the approximate sampling distribution that are as large as the obtained statistic, one determines the proportion within the joint distribution consisting of the approximate sampling distribution *plus* the obtained statistic that are as large as the obtained statistic. For example, when the approximate sampling distribution consists of 999 randomly selected statistics, as in the correlation problem, one determines how many of the 1000 correlation statistics (the 999 in the approximate sampling distribution plus the obtained statistic) are as large as the obtained statistic. (Since the obtained statistic is as large as itself, there can be no less than one of the 1000 statistics as large as the obtained statistic.) This number is divided by 1000 to get the probability under the null hypothesis of getting such a large correlation statistic as the one obtained. One can then use this probability to decide whether to reject the null hypothesis. For instance, with a significance level of .05 the null hypothesis is rejected when the computed probability is no greater than .05; and with a significance level of .01, the null hypothesis is rejected when the computed probability is no greater than .01. That is, the null hypothesis is rejected at the .05 level whenever the obtained statistic falls within the upper five percent of the 1000 statistics (including the obtained statistic itself) and at the .01 level whenever the obtained statistic falls within the upper one percent of the 1000 statistics.

It will now be shown that this procedure is valid: i.e., that when the null hypothesis is true the probability of rejecting the null hypothesis is equal to the level of significance. For example, when the null hypothesis is true, for this procedure to be valid the probability of rejecting at the .01 level must be .01.

Random assignment in conjunction with the null hypothesis permits one to regard the obtained statistic as a random sample of one statistic taken from the entire sampling distribution of statistics. When the obtained statistic is added to the 999 randomly selected statistics there are, under the null hypothesis, 1000 statistics randomly selected from the entire sampling distribution. The obtained statistic can be regarded as the first statistic to be drawn. In random sampling there is, of course, no systematic tendency for the first drawn value to be either larger or smaller than values drawn later. So, for 1000 randomly selected statistics the probability that the obtained statistic (the first drawn statistic) will be within the upper half of the statistics is $\frac{1}{2}$, that it will be within the upper quarter is $\frac{1}{4}$, and so on. Thus the probability is .05 that, when the null hypothesis is true, the

obtained statistic will fall within the upper 5 percent of the 1000 statistics, and the probability is .01 that it will fall within the upper one percent of the 1000 statistics. Now, the recommended procedure rejects at the .05 level when the obtained statistic is in the upper five percent of the statistics and at the .01 level when it is in the upper one percent; so the probability under the null hypothesis of rejecting at the .05 level is .05 and at the .01 level is .01. Approximate randomization tests are therefore *valid, not approximately valid*, procedures for testing null hypotheses. Probabilities computed from approximate sampling distributions are not simply approximations to probabilities based on entire sampling distributions, but are probabilities in their own right.

The same argument could have been used to show the validity of the proposed procedure if sampling *without* replacement had been carried out, but the following discussion of the power of the approximate randomization test will be considerably simplified by restricting it to sampling distributions obtained by sampling *with* replacement.

E. Power of Approximate Randomization Tests

The next consideration is the relative power of the approximate randomization test based on an approximate sampling distribution of 999 statistics. First the following statement will be proved:

Statement 1: The probability is .99 that no more than 17 statistics from the upper one percent of the entire sampling distribution will be in the approximate sampling distribution.

Then it will be shown that Statement 1 implies Statement 2:

Statement 2: The probability is .99 that, whenever use of the entire sampling distribution would lead to judging an obtained statistic to be significant at the .01 level, use of the approximate sampling distribution would lead to the assignment of a probability no greater than .018.

For the proof of Statement 1 the entire sampling distribution is to be regarded as consisting of two kinds of statistics: those that are in the upper one percent of the distribution and those that are not. When the entire sampling distribution is randomly sampled with replacement, for a single draw there is a probability, $p = .01$, of drawing a statistic from the upper one percent of the distribution and a probability $q = .99$ of not doing so: that is, of drawing a statistic from the lower 99 percent of the distribution. Replacement after each draw ensures that p and q stay constant over all draws. The binomial distribution of the number of statistics from the upper 1 percent of the entire sampling distribution for $N = 999$, $p = .01$, and $q = .99$ is almost a normal distribution with $\mu = Np = 9.99$ and $\sigma = \sqrt{Npq} = 3.14$. Ninety-nine percent of the area under a normal curve lies below the point that is 2.33 standard deviations above the mean. Consequently the probability is .99 that the number of statistics from the upper one percent of the entire sampling distribution contained in the approximate sampling distribution is no greater than $9.99 + (2.33)$ $(3.14) = 17.31$. This verifies Statement 1, since the discrete frequency 17 is assumed to occupy the interval 16.5 to 17.5 in a normal distribution.

Statement 1 also has been verified by computations by an electronic computer on the exact binomial distribution (not the normal curve approximation).

To see that Statement 1 implies Statement 2, it is necessary to appreciate that Statement 1 is true regardless of the magnitude of the obtained statistic, inasmuch as the selection of the statistics in the approximate sampling distribution is in no way affected by the magnitude of the obtained statistic. Therefore Statement 1 is true even when the obtained statistic is large enough to be judged significant at the .01 level by means of the entire sampling distribution. This fact enables one to derive Statement 2 from Statement 1. An obtained statistic would be judged significant at the .01 level by means of the entire sampling distribution only if it had a value that lies within the upper one percent of the entire sampling distribution of statistics. The only statistics in the approximate sampling distribution that might possibly be as large as the obtained statistic, therefore, would be those from the upper one percent of the entire sampling distribution. This fact, in conjunction with Statement 1 (which has already been verified), leads to this conclusion: When an obtained statistic would have been judged significant at the .01 level by means of the entire sampling distribution, the probability is .99 that there will be no more than 17 values in the approximate sampling distribution that are as large as the obtained statistic. When no more than 17 statistics in the approximate sampling distribution of 999 statistics are from the upper one percent of the entire sampling distribution, no more than 18 (17 in the approximate sampling distribution plus the obtained statistic itself) of the 1000 will be as large as the obtained statistic. Statement 2, therefore, is true.

Statement 2 is correct whether or not the null hypothesis is true: i.e., whether there is a treatment effect or not. Statement 2 is of special interest however when the null hypothesis is false because it shows that an approximate randomization test based on an approximate sampling distribution of 999 statistics is almost as powerful as a randomization test employing the entire sampling distribution.

Analogous computations and arguments in regard to the binomial distribution of the number of statistics from the upper five percent of the entire sampling distribution for $N = 999$, $p = .05$, and $q = .95$ lead to this statement:

Statement 3: The probability is .99 that whenever use of the entire sampling distribution would lead to judging an obtained statistic to be significant at the .05 level, use of the approximate sampling distribution would lead to the assignment of a probability no greater than .066.

Like Statement 2, Statement 3 has also been verified by electronic computer computations on the exact binomial distribution.

One can, of course, follow the same sort of procedure to determine probability statements comparable to Statements 2 and 3 for other levels of significance or for approximate sampling distributions of other sizes. An approximate sampling distribution with more than 999 statistics has more power, but it can be seen that this number provides sufficient power for many purposes.

F. Summary

An approximate randomization test is a randomization test in which the significance of an obtained statistic is determined by using an approximate sampling distribution consisting of a random sample of the statistics in the entire sampling distribution.

The purpose of approximate randomization tests is to reduce the amount of computation for a randomization test to a practical level, thereby overcoming the only serious objection that has been raised against the use of randomization tests.

In an approximate randomization test the significance of an obtained statistic (such as a correlation coefficient or a difference between means) is determined by reference to a distribution composed of the approximate sampling distribution *plus* the obtained statistic. The significance is the proportion of statistics within this distribution that are as large as the obtained statistic. This proportion is the probability under the null hypoth-

esis of getting such a large statistic as was obtained. This procedure is not just a procedure for providing estimates of the probabilities that would be obtained by using the entire sampling distribution, but is valid in its own right: when the null hypothesis is true the probability of rejecting the null hypothesis is equal to the level of significance used.

When the approximate sampling distribution consists of 999 statistics, the power of the approximate randomization test is almost equal to that of the randomization test employing the entire sampling distribution. Specifically, the probability is .99 that whenever use of the entire sampling distribution would lead to judging an obtained statistic to be significant at the .01 level, use of the approximate sampling distribution would lead to the assignment of a probability no greater than .018. Also, the probability is .99 that whenever use of the entire sampling distribution would lead to judging an obtained statistic to be significant at the .05 level, use of the approximate sampling distribution would lead to the assignment of a probability no greater than .066. These probability statements are true for approximate sampling distributions of 999 statistics selected at random, with replacement, from the entire sampling distribution regardless of the size of the entire sampling distribution.

Reference

1. Edgington, E. S. Randomization tests. *J. of Psychol.*, 1964, **57**, 445–449.

9.4 Suggestions for Further Reading

I. On the Use and Misuse of the Chi-Square Test*

The first five articles listed below provide an in-depth study of the Chi-square test. Unfortunately their length prevents their inclusion in this chapter.

Lewis, D., and C. J. Burke. The use and misuse of the Chi-square test. *Psychological Bulletin*, 1949, **46**, 433–489.

Peters, C. C. The misuse of Chi-square—a reply to Lewis and Burke. *Psychological Bulletin*, 1950, **47**, 331–337.

Pastore, N. Some comments on "The use and misuse of the Chi-square test." *Psychological Bulletin*, 1950, **47**, 338–340.

Edwards, A. L. On "The use and misuse of the Chi-square test"—the case of the 2×2 contingency table. *Psychological Bulletin*, 1950, **47**, 341–346.

Lewis, D., and C. J. Burke. Further discussion of the use and misuse of the Chi-square test. *Psychological Bulletin*, 1950, **47**, 347–355.

Cochran, W. G. Some methods for strengthening the common Chi-square tests. *Biometrics*, 1954, **10**, 417–451. Requirements for the use of Chi-square tests are critically examined and suggestions are presented for broadening the scope of the tests.

*Articles are listed chronologically.

II. Other Nonparametric Methods

Bradley, J. V. A Survey of sign tests based on the binomial distribution. *Journal of Quality Technology*, 1969, **1**, 89–101. A lucid examination of the assumptions and applications of the sign test, one of the most useful of the nonparametric methods.

Bresnahan, J. L., and M. M. Shapiro. A general equation and technique for the exact partitioning of Chi-square contingency tables. *Psychological Bulletin*, 1966, **66**, 252–262. Procedures for determining the contribution to the overall Chi-square of a portion of the contingency table are described.

Castellan, N. J. On the partitioning of contingency tables. *Psychological Bulletin*, 1965, **64**, 330–338. Describes procedures for partitioning an $r \times c$ contingency table into $(r - 1)$ $(c - 1)$ independent components which can be evaluated for the presence of interaction.

Marascuilo, L. A., and M. McSweeney. Nonparametric post hoc comparisons for trend. *Psychological Bulletin*, 1967, **67**, 401–412. The use of trend-analysis procedures are described for several of the more common nonparametric analogues of analysis of variance designs.

Moses, L. E. Non-parametric statistics for psychological research. *Psychological Bulletin*, 1952, **49**, 122–143. This article provides a readable introduction to the many and varied classes of nonparametric procedures.

The following two articles describe nonparametric analogues of a factorial design in which the scores in the data matrix are frequencies.

Myers, J. L. Exact probability treatments of factorial designs. *Psychological Bulletin*, 1958, **55**, 59–61.

Sutcliffe, J. P. A general method of analysis of frequency data for multiple classification designs. *Psychological Bulletin*, 1957, **54**, 134–137.

The following two articles describe the computation and application of a nonparametric correlation coefficient. Modifications of certain of the computations suggested by Schaeffer and Levitt are given in the article by Cartwright.

Cartwright, D. S. A note concerning Kendall's tau. *Psychological Bulletin*, 1957, **54**, 423–425.

Schaeffer, M. S., and E. E. Levitt. Concerning Kendall's tau, a non-parametric correlation coefficient. *Psychological Bulletin*, 1956, **53**, 338–346.

Procedures for performing nonparametric multiple comparisons among three or more treatments are described in the following six articles.

Cohen, J. An alternative to Marascuilo's "Large-sample multiple comparisons" for proportions. *Psychological Bulletin*, 1967, **67**, 199–201.

Dunn, O. J. Multiple comparisons using rank sums. *Technometrics*, 1964, **6**, 241–252.

Marascuilo, L. A. Large-sample multiple comparisons. *Psychological Bulletin*, 1968, **65**, 280–290.

Rhyne, A. L., Jr., and R. G. D. Steel. Tables for a treatments versus control multiple comparisons sign test. *Technometrics*, 1965, **7**, 293–306.

Steel, R. G. D. A rank sum test for comparing all pairs of treatments. *Technometrics*, 1960, **2**, 197–208.

Tobach, E., M. Smith, G. Rose, and D. Richter. A table for making rank sum multiple paired comparisons. *Technometrics*, 1967, **9**, 561–567.

10

The Likelihood Function and Bayesian Statistical Methods

The traditional approach to statistical inference was described in Chapter 4. In spite of certain problems associated with this approach, most experimenters have clung to it tenaciously. Two alternative approaches to statistical inference described in the present chapter are the likelihood principle and Bayesian statistical theory. Both approaches emphasize the importance of the likelihood function but use it in different ways. The first article, by D. A. Sprott and J. G. Kalbfleisch, provides a readable introduction to the likelihood function and its use in statistical inference.

10.1   Use of the Likelihood Function in Inference

D. A. Sprott[1] *and J. G. Kalbfleisch*

Abstract. The likelihood function is defined and its use illustrated by a simple coin-tossing experiment. The distinction between the use of the likelihood function and the use of a test of significance is emphasized and illustrated by a simple genetics example. Some examples are given by experiments in psychology where the likelihood function is used to analyse the resulting data; the relative merits of the use of likelihood compared to other more standard methods of analysis are discussed.

The importance of the likelihood function has been emphasized for at least 40 years by Fisher (1922, 1925, 1934, 1946), and more recently specific examples of its use have been given (Barnard, Jenkins & Winston, 1962; Birnbaum, 1962; Fisher, 1955). In fact there is a "likelihood school" of statistics which says that all the information of the sample relative to the population under consideration is contained within the likelihood function and inferences should be based on it (Barnard et al., 1962; Birnbaum, 1962; Fisher, 1955; Sprott, 1961). The Bayesians (Edwards, Lindman, & Savage, 1963) also emphasize the importance of the likelihood function, but the use they make of it is different from that in this paper. Although the use and importance of the likelihood function are now being discussed in the statistical literature, there seems to be very little use or recognition of it in outside fields such as psychology. The purpose of this paper is to describe the likelihood function and its use in statistical inference. Section I defines and discusses the likelihood function and relative likelihood function and illustrates its use in inference by a simple example. Section II compares the likelihood approach with the somewhat similar, but logically distinct significance test. Section III gives an example from psychology of inferences based on the likelihood function from the data that would be difficult or impossible to assess accurately and rigorously in any other way. It is pointed out that perhaps the main drawback to the use of likelihood as a measure of uncertainty is that it is not probability and so is not intuitively obvious to some people. It has however a definite interpretation in terms of probabilities, and it should also be noted that very few methods of inference are capable of making probability statements about hypotheses or unknown parameters. The test of significance and confidence intervals certainly do not.

[1] This research was carried out with a grant from the National Research Council of Canada. The authors would like to thank R. K. Banks, R. V. Thysell, M. D. Vogel-Sprott, E. Tulving, R. G. Stanton, and J. C. Ogilivie for reading the manuscript and for comments, criticisms, and suggestions.

Section I: Likelihood

The use of likelihood is illustrated in the following example: a coin which turns up heads an unknown fraction, θ, of the time is tossed 10 times in order to obtain information about θ. Before the experiment is performed, the probability of observing x heads in 10 tosses for any specified value of θ is given by the binomial distribution

$$f(x|\theta) = \binom{10}{x} \theta^x (1 - \theta)^{10-x}$$

and can thus be calculated numerically for any x. After the experiment a particular value $x = x_0$ has been observed. We can now use $f(x_0|\theta)$ as a function of θ, employing the observed value, x_0, to rank possible values of θ according to their plausibilities. Suppose that $x_0 = 1$ head was observed. For any given θ the probability of the observed result is

$$f(1|\theta) = \binom{10}{1} \theta(1 - \theta)^9. \qquad [1]$$

For example, if $\theta = .1$, the probability of observing one head is

$$f(1|.1) = \binom{10}{1} (.1)(.9)^9 = .387,$$

and if $\theta = .5$, the probability of observing one head is

$$f(1|.5) = \binom{10}{1} (.5)(.5)^9 = .00977.$$

Thus under hypothesis H : $\theta = .1$ the result of the experiment is

$$\frac{.387}{.00977} = 40$$

times more probable than under H : $\theta = .5$. The data favor the hypothesis $\theta = .1$ 40 to 1 over the hypothesis $\theta = .5$. This is expressed by saying the likelihood of $\theta = .1$ is 40 times that of $\theta = .5$, or the likelihood of $\theta = .1$ compared to $\theta = .5$ is 40. Since hypotheses under which the observations are relatively probable will be considered more plausible and hence preferable to those under which the observations are relatively improbable

we can use the function $f(1|\theta)$ to compare the plausibilities of different hypotheses in the light of the observation $x = 1$.

In general, before the experiment, $f(x|\theta)$, considered as a function of the observation x for a specified value of θ, yields the probability of observing x; after the experiment, $f(x|\theta)$, considered as a function of the unknown variable θ, and where x is set equal to its observed value, gives likelihoods of possible values of θ; $f(x|\theta)$ is called the likelihood function of θ and can be denoted by $L(\theta|x)$. Thus, before the experiment probabilities are relevant; after the experiment likelihoods are relevant (Barnard, 1949).

As can be seen from the above example, the likelihood is used to compare the plausibilities of different hypotheses under a given set of observations, so that only likelihood ratios are meaningful. In fact the likelihood ratio

$$\frac{L(\theta_1|x)}{L(\theta_2|x)}, \qquad [2]$$

which is the plausibility of $\theta = \theta_1$ versus $\theta = \theta_2$, is the ratio of the probability of the observations if $\theta = \theta_1$ to their probability if $\theta = \theta_2$. Since only ratios are used, the likelihood may be multiplied by any factor $C(x)$ independent of θ, for the ratio

$$\frac{C(x)L(\theta_1|x)}{C(x)L(\theta_2|x)} = \frac{L(\theta_1|x)}{L(\theta_2|x)}$$

remains unchanged. Thus in the above example the likelihood function for $x = 1$ can be taken to be

$$\theta(1 - \theta)^9. \qquad [3]$$

Since it is desired to rank values of the unknown variable θ according to their plausibilities, special attention will center on the most plausible value or maximum likelihood estimate $\hat{\theta}$ of θ; the likelihoods of all other values of θ can be compared with that of the most plausible value θ by means of the likelihood ratio

$$R(\theta|x) = L(\theta|x)/L(\hat{\theta}|x). \qquad [4]$$

$R(\theta \mid x)$ varies between zero and one and gives the relative plausibility of θ compared to the most likely value $\theta = \hat{\theta}$ in the light of the observations. Any two values of θ can be compared by taking the ratio of their relative likelihoods, for

$$\frac{R(\theta_1 \mid x)}{R(\theta_2 \mid x)} = \frac{L(\theta_1 \mid x)/L(\hat{\theta} \mid x)}{L(\theta_2 \mid x)/L(\hat{\theta} \mid x)} = \frac{L(\theta_1 \mid x)}{L(\theta_2 \mid x)},$$

which by Expression 2 is the plausibility of $\theta = \theta_1$ versus $\theta = \theta_2$. In the first example, where $x = 1$ is observed, the maximum likelihood value $\hat{\theta}$ is $\frac{1}{10}$; inserting this into Expression 3 gives the maximum of the likelihood as

$$L\left(\hat{\theta} = \frac{1}{10} \,\middle|\, 1\right)$$

$$= \frac{1}{10}\left(1 - \frac{1}{10}\right)^9 = \frac{1}{10}\left(\frac{9}{10}\right)^9, \quad [5]$$

and the ratio $L(1 \mid \theta)/L(1 \mid \hat{\theta})$ of Expression 3 and Equation 5 is

$$R(\theta \mid 1) = \frac{\theta(1 - \theta)^9}{\left(\frac{1}{10}\right)\left(\frac{9}{10}\right)^9} = \frac{\theta(1 - \theta)^9 10^{10}}{9^9}. \quad [6]$$

By direct substitution into Equation 6, the values $R(\theta \mid 1)$ can be calculated for various values of θ. Some results are given in Table 1. In Figure 1, $R(\theta \mid 1)$ is plotted as a function of θ. All values of θ between .01 and .35 are fairly plausible. Values of θ smaller than .001 or larger than .5 are suspect, in that there are available hypotheses (e.g., $\theta = .1$) which are 40 times as likely. Values of θ as large as .7 are obviously extremely implausible in the light of the observation $x = 1$ and can be disregarded, as there are hypotheses under which our observation is 1000 times as probable. A discussion of a similar example will be found in Fisher (1955, pp. 66–73).

It can thus be seen that $R(\theta \mid x)$ is a measure of uncertainty or plausibility applicable to an unknown variable θ for given experimental observations x. Like probability, $R(\theta \mid x)$ varies between 0 and 1. Likelihoods are not probabilities however and the distinction is an important one to preserve. Likelihoods do not obey the rules of combination of proba-

bilities. Although this is unimportant when only a single unknown variable, θ, is present, it becomes troublesome when there are many unknown variables, $\theta_1 \cdots \theta_k$.

Section II: Likelihood and Test of Significance

The standard or classical method of examining the foregoing example would be by means of the test of significance. To test the significance of $\theta = .5$, the probability of observing x less than or equal to 1 or greater than or equal to 9 is calculated:

$$\binom{10}{0}\left(\frac{1}{2}\right)^{10} + \binom{10}{1}\left(\frac{1}{2}\right)^{10} + \binom{10}{9}\left(\frac{1}{2}\right)^{10}$$

$$+ \binom{10}{10}\left(\frac{1}{2}\right)^{10} = \frac{11}{512} = .021.$$

Thus $H : \theta = .5$ is significant at the .021 level. It can be noted that the relative likelihood for $\theta = .5$ obtained in Section II for this example is .0255 (Table 1). The test of significance gives an alternative way of examining the evidence in this example. Several points should be noted however:

1. Neither the likelihood nor the test of significance is capable of making any probability statements about hypotheses or unknown variables. Under the latter, an hypoth-

Table 1. Relative likelihoods for coin-tossing example in which $x = 1$ was observed.

θ	$R(\theta/1)$
.0	.0
.001	.0256
.01	.236
.05	.814
.10	1.000
.15	.897
.20	.683
.25	.486
.30	.313
.35	.187
.40	.105
.45	.0535
.50	.0255
.55	.0107
.60	.00406
.65	.00133
.70	.000346

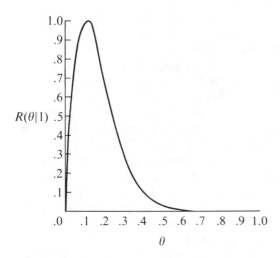

Fig. 1. Relative likelihood for coin-tossing example, $N = 1$ head observed.

esis which is significant exactly at the 5% level does not necessarily have a 5% probability (or any other probability) of being true. Rather, an event as unfavorable as the one observed has a 5% probability of occurring if the hypothesis is true. The likelihood ratio yields statements of relative likelihoods of hypotheses; these are logically different from probabilities.

2. The test of significance makes no logical distinction between certain values not observed and the actual value observed. For instance, in the example the value $x = 0$, which was not observed, is utilized in the same way as $x = 1$, which was observed. Such tests have been criticized on this ground (Fisher, 1955). The likelihood, on the other hand, utilizes only the value of x actually observed, implying that other values of x not observed are irrelevant.

3. The likelihood function can be used only to compare the plausibility of one hypothesis with that of another in the light of the data. In this sense the likelihood shows quantitatively the direction in which the weight of the evidence is pointing. The test of significance shows whether a specified hypothesis is consistent with the data. It should be noted that some hypotheses consistent with the data may still be relatively implausible. Indeed, the fact that an hypothesis is not

significant merely means that no convincing evidence *against* the hypothesis has been demonstrated by that particular test; but there may be much more plausible hypotheses, which the relative likelihood will show. An example can show the relationship between a test of significance and the likelihood.

Example: If there are only two forms, M and N, of a gene at a given locus or place on the chromosome, then the genotypes MM, MN, NN are expected to occur with relative frequencies

$$\theta^2, \ 2\theta(1 - \theta), \ (1 - \theta)^2, \qquad [7]$$

respectively, if mating is at random with respect to these genes, where θ is the relative frequency of the M gene in the population[2] (cf. Li, 1955). Using the observations cited by Li (1955, p. 13) in which the frequency of the MM, MN, and NN blood types in a sample of 1,029 humans was observed to be 342, 500, and 187, respectively, the distinction between the use of a test of significance and the use of the likelihood can be illustrated. (*a*) Use of significance test—The significance test will indicate whether the Statistical Model 7 is consistent with the data at all. The probability of observing 342, 500, and 187 is given by the trinomial distribution with relative frequencies given by Model 7:

$$\Pr(342, 500, 187) = \frac{1,029!}{342! \, 500! \, 187!} \, (\theta^2)^{34}$$

$$\times \ [2\theta(1 - \theta)]^{500}[(1 - \theta)^2]^{187}$$

$$= \frac{1,029! \, 2^{500}}{342! \, 500! \, 187!} \, \theta^{1184}(1 - \theta)^{874}.$$

Thus the likelihood of θ can be taken to be

$$\theta^{1184}(1 - \theta)^{874} \qquad [8]$$

---

[2] If the relative frequency of the M and N genes is θ and $1 - \theta$, respectively, then the probability of having two M genes is θ^2, of having two N genes is $(1 - \theta)^2$, and of having an M and an N gene in either order is $2\theta(1 - \theta)$. This assumes that the acquiring of the M and N genes are independent events like the tossing of a coin.

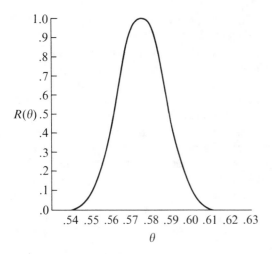

Fig. 2. Relative likelihood function for genetics example.

which attains its maximum at $\theta = \hat{\theta} = .5753$. Using this value for θ, the expected frequencies of MM, MN, NN in a sample of 1,029 are, by the theoretical frequencies in Model 7, $1,029(.5753)^2$, $1,029(2)(.5753)(.4247)$, $1,029(.4247)^2$ which are 340.6, 502.8, and 185.6. Comparing these with the observed frequencies 342, 500, and 187 by means of a chi-square test, and noting that $1df$ has been used to estimate the most likely value of θ (cf. Fisher, 1946), the resulting chi-square is .032 ($df = 1$). Since the probability of observing a chi-square larger than .032 is 85%, there is no evidence that the Model 7 is inconsistent with these data. (*b*) Use of the likelihood—Since the overall model above can be accepted as appropriate, the likelihood can now be used to rate the relative plausibility of values of θ. The maximum of the likelihood can be found by substituting $\theta = \hat{\theta} = .5753$ with Expression 8 to give

$$(.5753)^{1184}(.4247)^{874}.$$

The likelihood ratio obtained by dividing Expression 8 by its maximum in accordance with Equation 4 is

$$R(\theta) = \theta^{1184}(1 - \theta)^{874}/(.5753)^{1184}(.4247)^{874},$$

[9]

and this measures the plausibility of θ relative to the most likely value $\theta = \hat{\theta} = .5753$. In Figure 2, $R(\theta)$ is plotted against θ. Values of θ outside the range (.54, .61) can be disregarded as the data overwhelmingly favor values inside the range.

Had the test of significance in *a* above yielded an improbably large value of χ^2, the entire Statistical Model 7 would have to be rejected as incompatible with the observations. Then the use of the likelihood as above would be irrelevant; it would be meaningless to ask about the relative plausibilities of values of θ in a model that was demonstrably incompatible with the data.

The test of significance can therefore be used to test the validity of the overall model, and, if the model is accepted as appropriate, the likelihood can be used to rate the relative plausibility of the unknown variables in it. The likelihood in fact indicates the direction in which the weight of the evidence is pointing once the hypothesis as a whole is found acceptable.

Section III: An Example from Psychology

The following data arise from an experiment by Banks and Vogel-Sprott (1964). The observations are the number of punished responses in 15 opportunities for response under four different delayed punishment times of 0 seconds, 30 seconds, 60 seconds, and 120 seconds. The results can be presented in a frequency table (Table 2) in which an entry n_{ij} in the body of the table gives the observed frequency of subjects in the *i*th delay-time group who gave *j* responses.

Specific questions under consideration are whether the number of responses depends on the delay-of-punishment time and whether delay times 30, 60, and 120 differ among themselves and from 0 delay. An analysis using number of responses as a continuous normal variate is questionable because the data are discrete and the distribution is U shaped and truncated at 15. A more appropriate procedure would be to use the frequency table (Table 2), which contains counted (multinomial) data. The chi-square test of significance cannot be used because of small frequencies. The table could be condensed and classes combined, but much information would be lost. An extension of Fisher's (1946) exact test of significance for 2×2 tables could be used in principle, but even with a high speed computer this would not be feasible. Thus it is difficult to test whether the above hypotheses are actually inconsistent with the data (see Section II, 3). The use of the likelihood however is computationally simple and can throw some light on which way the experimental evidence is pointing.

Table 2. Frequency of punished responses.

Delay time	\multicolumn{8}{c}{Number of responses}
	1	2	3	4	5	6	7	8
0 seconds	2	3	0	2	0	2	0	0
30 seconds	0	1	1	1	0	0	0	1
60 seconds	0	1	0	1	0	1	1	0
120 seconds	0	1	0	0	0	0	0	0
Total	2	6	1	4	0	3	1	1

	9	10	11	12	13	14	15	Total
0 seconds	0	0	1	0	0	0	0	10
30 seconds	0	0	0	2	0	0	4	10
60 seconds	0	0	0	0	0	1	5	10
120 seconds	0	0	0	0	1	2	6	10
Total	0	0	1	2	1	3	15	40

Suppose p_{ij} is the hypothetical probability that a person makes j responses given delay time i ($i = 1\text{–}4$, $j = 1\text{–}15$). The hypothesis of independence between number of responses and delay time is mathematically

$$H_1: \quad p_{ij} = p_j \text{ independent of } i. \quad [10]$$

The actual frequency of j responses under delay time i is n_{ij} (Table 2). The above data then follow multinomial distributions, and the likelihood function is the product of all $p_{ij}^{n_{ij}}$, which for notational convenience can be written as

$$\prod_{i,j} p_{ij}^{n_{ij}}. \quad [11]$$

This example is somewhat more complex than the preceding examples, as the likelihoods of particular values of p_{ij} are not in question; rather we are concerned with the likelihoods of sets of p_{ij} satisfying Equations 10. In order to see if H_1 is reasonable, we can evaluate the likelihood of the most likely set of values p_{ij} satisfying Equations 10, that is, maximize Expression 11 subject to Equations 10 and evaluate the relative likelihood of the resulting numbers $p_{ij} = \hat{p}_j$. If the most likely values of p_{ij} satisfying Equations 10 are unlikely, then all values satisfying Equations 10 are unlikely, so that H_1 itself is to that extent implausible.

The maximum likelihood values $p_{ij} = \hat{p}_j$ satisfying Equations 10 can be shown to be $\dfrac{n_{.j}}{n}$ where $n_{.j}$ is the total of Column j and n is the overall total. Substituting $p_{ij} = \hat{p}_j = \dfrac{n_{.j}}{n}$ into Expression 11 using the observations n_{ij} in Table 2, the resulting likelihood is

$$\left(\frac{2}{40}\right)^2 \left(\frac{6}{40}\right)^6 \left(\frac{1}{40}\right) \left(\frac{4}{40}\right)^4 \left(\frac{3}{40}\right)^3 \left(\frac{1}{40}\right)$$

$$\times \left(\frac{1}{40}\right) \left(\frac{1}{40}\right) \left(\frac{2}{40}\right)^2 \left(\frac{1}{40}\right) \left(\frac{3}{40}\right)^3 \left(\frac{15}{40}\right)^{15}.$$

The maximum of the likelihood is obtained for $\hat{p}_{ij} = \dfrac{n_{ij}}{n_{i.}}$ where $n_{i.}$ is the total of row i; this can be written, substituting in Expression 11

$$\prod_{i,j} \left(\frac{n_{ij}}{n_{i.}}\right)^{n_{ij}} = \prod_{i,j} n_{ij}^{n_{ij}} \Big/ \prod_{i} n_{i.}^{n_{i.}}. \quad [12]$$

The likelihood ratio or relative likelihood of $\hat{p}_{ij} = \dfrac{n_{.j}}{n_n}$ above is therefore the ratio of these two expressions, which, using the entries in Table 2, is

$$\frac{10^{40} 2^2 6^6 4^4 3^3 2^2 3^3 15^{15}}{40^{40} 2^2 3^3 2^2 2^2 2^2 4^4 5^5 2^2 6^6} = 4.9 \times 10^{-11},$$

ignoring 0's and 1's, since they contribute nothing to the products. The relative likelihood of the most likely set of values p_{ij} satisfying Equations 10 is extremely small so that H_1, even though it might not be contradicted by the data, is a relatively implausible hypothesis certainly not favored by the data. The evidence is pointing strongly away from H_1.

It may be thought that the delay times differ in respect of the rather erratic responses in the middle of the scale, that is, between 6 and 10. If responses 1–5, 6–10, 11–15 are combined, the result is Table 3. The hypothesis H_1, $p_{ij} = p_j$, can be tested in the same way, and the relative likelihood of the most likely values of p_{ij} under H_1 is

$$\frac{10^{40} 13^{13} 5^5 22^{22}}{40^{40} 7^7 2^2 3^3 6^6 2^2 2^2 6^6 9^9} = 2.2 \times 10^{-4}.$$

It can be seen that even under this less sensitive classification the evidence does not favor the hypothesis that delay time produces no effect.

The hypothesis H_2 that delay times 30, 60, and 120 produce the same effect but possibly differ from 0 delay time is more complicated to test. Here the hypothesis is

$$H_2: \quad p_{ij} = p_j; \quad i = 2, 3, 4 \text{ only.}$$

The most likely values of p_{ij} satisfying H_2 can be shown to be

$$\hat{p}_{1j} = \frac{n_{1j}}{n}, \quad \hat{p}_{2j} = \hat{p}_{3j} = \hat{p}_{4j} = \frac{n_{\cdot j} - n_{1j}}{n - n_1} = \hat{p}_j.$$

The likelihood of these values of the p's can be obtained by substituting into Expression 11 using the numerical values in Table 3; the maximum of the likelihood is again found from Equation 12. The likelihood ratio arising from the results in Table 3 is therefore

$$\frac{10^{40} 6^6 3^3 21^{21} 7^7 2^2}{30^{30} 10^{10} 7^7 2^2 3^3 6^6 2^2 2^2 6^6 9^9} = .098.$$

Table 3. Frequency of punished responses.

Delay times	Number of responses			
	1–5	6–10	11–15	Total
0 seconds	7	2	1	10
30 seconds	3	1	6	10
60 seconds	2	2	6	10
120 seconds	1	0	9	10
Total	13	5	22	40

Thus the evidence favors H_2, the equality of delay times 30, 60, and 120, over H_1, the equality of delay times 0, 30, 60, and 120, about

$$446 - 1 = \left(\frac{.098}{2.2 \times 10^4} \right).$$

Also the hypothesis H_2 is itself fairly plausible, so that it is reasonable to conclude that delay times 30, 60, and 120 produce essentially the same effect, the 0 time being responsible for the main differences observed.

The above tests exemplify the use of the likelihood in tables with entries too small to be attacked conveniently in any other way. Of course these tests, like the standard significance tests, are valid only if the hypotheses tested were considered relevant before the experiment was performed; they are not valid if the hypotheses tested were suggested by inspection of the data.

Another Area of Application

Section III dealt with an example in which the use of the likelihood appears to be the most convenient method of attack. A system in which there may be feedback of an unknown kind provides an example where the use of the likelihood function is the only method of attack. For instance, an observational variable y_t may be normally distributed about a mean βx_t with constant variance σ^2, where x_t is another observational variable at time t: If the x_t are independent of all y_t, standard regression analysis applies. If there is feedback of the form $x_t = y_{t-1}$, standard regression theory does not apply and the analysis using significance tests and probability (fiducial) intervals becomes complicated although still possible. If, however, there is feedback of an unknown form, $x_t = g(y_{t-1})$ where g is an unknown function, possibly independent of y_{t-1} (i.e., no feedback), then significance tests and probability intervals cannot even in principle be obtained as the required distributions will not be known. This kind of situation may possibly arise in experiments in which learning takes place. The likelihood function however will not depend on knowing the form of the function $g(y_{t-1})$; all that must be known is the numerical value of $x_t = g(y_{t-1})$. Whatever the form of $g(y_{t-1})$, the likelihood function based on a sample of n pairs of observations: $x_1, y_1, x_2, y_2, \ldots, x_n, y_n$ is

$$\left(\frac{1}{\sigma}\right)^n \exp\left[-\frac{1}{2\sigma^2} \sum (y_t - \beta x_t)^2\right],$$

which attains its maximum at

$$\beta = \hat{\beta} = \frac{\sum x_t y_t}{\sum x_t^2}, \quad \sigma^2 = \hat{\sigma}^2 = \frac{1}{n} \sum (y_t - \hat{\beta} x_t)^2.$$

Thus the relative likelihood can be calculated numerically for any pair (β, σ). For any value $\beta = \beta_0$, one can calculate the most likely value of σ^2 (i.e., σ^2 given β_0); if the relative likelihood of the resulting pair (β_0, σ^2) is small, then the value, β_0, itself is implausible, since no value of σ^2 will yield a pair (β_0, σ^2) with high likelihood. Examples of this kind are discussed in more detail by Barnard et al. (1962).

A final example of the use of the likelihood function is provided by the stochastic learning model of Bush and Mosteller (1955, pp. 238–249) applied to the Solomon-Wynne experiment. Using this model, the probability of a shock on Trial n when it is preceded by avoidances on k trials is

$$q_{n,k} = \alpha_1^k \alpha_2^{n-k},$$

where α_1 is called the avoidance parameter and α_2 the shock parameter. The likelihood function is therefore a product of terms, one for each trial, of the form

$$\alpha_1^k \alpha_2^{n-k} \quad \text{or} \quad 1 - \alpha_1^k \alpha_2^{n-k},$$

the former occurring in the product on shock trials, the latter occurring on avoidance trials. Since, in the Solomon-Wynne experiment, there are 30 dogs each with 24 trials, there are 720 factors in the likelihood function, making it too difficult to write down or compute on a desk calculator. The methods of Bush and Mosteller (1955), based on using suitable subsets of observations, could be used to write

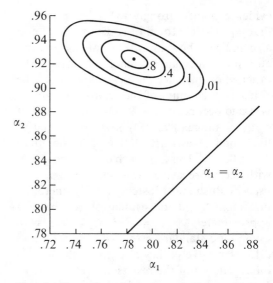

Fig. 3. Contours of constant likelihood for the Solomon-Wynne data.

down an approximate likelihood function amenable to a desk calculator. However, it is relatively simple to program a high speed electronic computer to calculate exact likelihood values, using all of the data to obtain the maximum of the likelihood, and by Equation 4, the relative likelihood. This was done, and the results are shown graphically in Figure 3 in which contours of constant likelihood are plotted in terms of α_1, α_2, and the reference line $\alpha_1 = \alpha_2$ is shown. This graph represents a concise pictorial summary of everything the data have to say about α_1 and α_2. It can be seen from the elongated shape, for instance, that the data throw more light on α_2 than on α_1; that is, whatever be the value of α_2, plausible values of α_1 are in the range (.73, .840), whereas whatever be the value of α_1, the plausible values of α_2 are in the range (.89, .954)—the extremities of the 1% contour—the one range being about twice the size of the other. Also, the likelihood decreases very rapidly for small changes in α_1 and α_2, but about twice as rapidly for α_2. Thus, in terms of relative likelihood it is important for α_1 and α_2 both to be near the maximum. The region of appreciable likelihood is concentrated well above the line $\alpha_1 = \alpha_2$, so that there is little doubt that the evidence points strongly to α_2 being greater than α_1. The contours of constant likelihood also indicate how knowledge of one parameter will affect an inference concerning the other. For instance, if α_1 is known to be .746, the maximum of the relative likelihood is .1 and occurs at $\alpha_2 = .935$, so that values of α_2 in the range (.89, .954) now have a relative likelihood of at least $.01/1 = .1$, rather than .01 as before. This approach can be compared with the more usual large sample approach used by Bush and Mosteller (1955) where the maximum likelihood estimates are obtained approximately as $\hat{\alpha}_1 = .797$, $\hat{\alpha}_2 = .923$ with asymptotic standard errors of about .04 and .015, respectively; they also give a theoretical interpretation of the results.

It can be seen from this and the preceding examples that the use of the likelihood has several advantages:

1. It represents a concise summary of all of the information in the data, and when graphed, it gives a pictorial summary of what the data have to say.

2. Inferences based on the likelihood are exact for samples of all sizes; no assumption or mathematical investigation is required to ascertain if the sample is large enough to justify the method.

3. The rather complicated mathematics of asymptotic or large sample theory is not necessary, giving a relatively simple theoretical approach to inference.

Against these may be set the possible disadvantages of increased numerical computation in addition to the uncertainty being expressed (exactly) in terms of likelihood rather than (approximately in large samples) in terms of probability. Also, the likelihood method is not a test of significance and cannot be used to test the validity of a single hypothesis without reference to alternatives.

References

Banks, R. K., & Vogel-Sprott, M. D. Effect of delayed punishment on an immediately rewarded response in humans, *Journal of Experimental Psychology*, 1965, **70**, in press.

Barnard, G. A. Statistical inference, *Journal of the Royal Statistical Society*, B, 1949, **11**, 116–149.

Barnard, G. A., Jenkins, G. M., & Winsten, C. B. Likelihood inference and time series (with discussions), *Journal of the Royal Statistical Society*, A, 1962, **125**, 321–372.

Birnbaum, A. On the foundations of statistical inference (with discussion), *Journal of the American Statistical Association*, 1962, **47**, 269–326.

Bush, R. R., & Mosteller, F. *Stochastic models for learning*. New York: Wiley, 1955.

Edwards, W., Lindman, H., & Savage, L. J. Bayesian statistical inference for psychological research, *Psychological Review*, 1963, **70**, 193–242.

Fisher, R. A. On the mathematical foundations of theoretical statistics, *Philosophical Transactions of the Royal Society*, A, 1922, **222**, 309–368.

Fisher, R. A. Theory of statistical estimation, *Proceedings of the Cambridge Philosophical Society, Biological Sciences*, 1925, **22**, 700–725.

Fisher, R. A. Two new properties of mathematical likelihood, *Proceedings of the Royal Society*, A, 1934, **144**, 285–307.

Fisher, R. A. *Statistical methods for research workers*. London: Oliver & Boyd, 1946.

Fisher, R. A. *Statistical methods and scientific inference*. London: Oliver & Boyd, 1955.

Li, C. C. *Population genetics*. Chicago: Univer. Chicago Press, 1955.

Sprott, D. A. Similarities between likelihoods and associated distributions a posteriori, *Journal of the Royal Statistical Society*, B, 1961, **23**, 460–468.

Editor's comments: An excellent introduction to the likelihood function was provided in the previous article by Sprott and Kalbfleisch, which both defined this important concept and illustrated its use. That article provides a foundation for the following discussion by Allen Birnbaum of the likelihood function in Bayesian statistical theory and in classical hypothesis-testing methodology.

10.2   Another View on the Foundations of Statistics

Allan Birnbaum

The recently widening discussion of the foundations of statistical inference has been represented in the pages of *The American Statistician* by F. J. Anscombe's "Bayesian Statistics" (Vol. 15, No. 1, Feb. 1961, pp. 21–24) and Irwin D. J. Bross' "Statistical Dogma: A Challenge" (Vol. 15, No. 3, June 1961, pp. 14–15). The purpose of this note is to illustrate a viewpoint somewhat distinct from both the Bayesian and the "standard," "relative frequency" standpoints discussed by those authors.

A traditional standard of work in the natural sciences is accurate reporting of "what was observed, and under what experimental plan and conditions." Such reports are

Research supported by the Office of Naval Research.

Reprinted from *American Statistician*, 1962, **16**, 17–21, by permission of the publisher and author. Later publications by the author which represent the subsequent development of his views include "Concepts of statistical evidence," in *Philosophy, Science, and Method*: *Essays in Honor of Ernest Nagel*, S. Morgenbesser et al., eds., N.Y.: St. Martin's Press, 1969; "Likelihood" in *The International Encyclopedia of the Social Sciences*, 1968; and his letter to the Editor of *Nature*, 1970, **225**, 1033.

an essential part of the empirical sciences, constituting the body of *experimental evidence* available at any stage to support their practical applications and their general laws, theories, and hypotheses. In some circumstances the "experimental plan and conditions" can be represented adequately by a mathematical-statistical model of the experimental situation. Concrete illustration of some relevant concepts is provided by the following highly simplified example: Suppose that some subject-matter of general interest (e.g. a biological tissue or a physical mechanism) is known to be in one of just two possible states (two simple hypotheses, H_1 or H_2) on the basis of general background knowledge, and that information on its actual state will be obtained by using an instrument which gives dichotomous observations, each observation y being either "positive" (denoted by $y = 1$) or "negative" (denoted by $y = 0$). The instrument is known to have equal probabilities of "false positives" and of "false negatives," each equal to .20; that is,

$$\text{Prob}(Y = 1 \,|\, H_1) = \text{Prob}(Y = 0 \,|\, H_2) = .20.$$

Suppose that resources allow just four independent observations (y_1, y_2, y_3, y_4). The mathematical model of this experiment is given by the familiar binomial distributions: Let the number of positives be denoted by

$$x = \sum_{i=1}^{4} y_1.$$

Then we have

$$\text{Prob}(x\,|\,H_1) = \binom{4}{x}(.2)^x(.8)^{4-x}$$

and

$$\text{Prob}(x\,|\,H_2) = \binom{4}{x}(.8)^x(.2)^{4-x}$$

for $x = 0, 1, 2, 3$, and 4.

Consider now the problem of reporting appropriately the results of such an experiment to readers with an intrinsic interest in the material observed, for example in a technical journal. Assuming adequacy of the above model, it would of course be sufficient in one clear sense to describe that model ("binomial, four observations, parameter $p = .2$ under H_1 and $p = .8$ under H_2") and the observed value of x (e.g. "four positive observations" or "$x = 4$"). Denoting this binomial mathematical model of the experiment by E_b, we may say that the *experimental evidence* obtained is reportable in mathematical form by giving E_b and (the numerical observed value) x; or that (E_b, x) is an instance of *statistical evidence*; that is, (E_b, x) is a mathematical model of an instance of experimental evidence, of "what was observed, and under what experimental plan and conditions."

While reports which merely describe or summarize statistical evidence are familiar, it is usual practice in many disciplines to supplement them by applications of statistical techniques, such as significance tests or confidence intervals, which seem appropriate to characterize the nature and meaning of the evidence as it is relevant to the statistical hypotheses or parameters of interest. In the example described, the evidence (E_b, x) would customarily be described as "significant at the $(.2)^4 = .0016$ level," or as "very highly significant." The familiar interpretations are that these phrases indicate strong evidence against H_1 (and favoring H_2 so long as no other alternatives are admitted). The number .0016 (the "*P*-level" or "critical level") is customarily interpreted as a precise index of strength of this evidence, with a status of objective meaningfulness related to its frequency interpretation as a probability: If H_1 were true, such results would have the very small relative frequency .0016 in indefinitely repeated independent experiments of this form. (It is often recommended that the power of tests underlying such interpretations be considered: Here if H_2 were true, the probability of such results (the power of the test which rejects H_1 just when $x = 4$) would be appreciably larger, namely $(.8)^4 = .4096$.) When tests or estimates are used in this general way, they may be called techniques of *evidential interpretation*, intended not merely to describe but to interpret and characterize in suitable objective terms the relevant properties, meaning and strength of statistical evidence as such.

Some of the central issues separating the several theoretical approaches to the foundations of statistics concern the two basic functions of statistics just illustrated, namely the *characterization* and the *interpretation* of statistical evidence as such. These functions, and the problems of determining the theoretical concepts and practical techniques appropriate for them, may be called the functions and the problem-area of *informative* (statistical) *inference*. The preceding paragraph described evidential interpretations of our example (E_b, x) in the customary terms of the "standard," "frequentist" standpoint.

A Bayesian standpoint leads to different interpretations: According to Bayes' principle, the individual making interpretations of experimental results has certain prior probabilities, q_1 and $q_2 = 1 - q_1$, for the respective alternative hypotheses H_1, H_2, which represent appropriately his background information and opinion; and he can specify these numbers q_1, q_2, with adequate precision. The experimental results, combined with these numbers through Bayes' formula, determine numerical posterior probabilities q_1', q_2', for the respective hypotheses; and the latter are considered to represent appropriately his final overall information and opinion. In our example Bayes' formula is equivalent to the simple relation

$$\frac{q_2'}{q_1'} = \frac{q_2}{q_1} \frac{\text{Prob}(x \mid H_2)}{\text{Prob}(x \mid H_1)} = \frac{q_2}{q_1} L(x).$$

Here $L(x) = \text{Prob}(x \mid H_2)/\text{Prob}(x \mid H_1)$ is the likelihood ratio statistic, which appears in various technical roles in the various approaches to statistical theory. If for example $x = 4$, we have $L(4) = (.8)^4/(.2)^4 = 4^4 = 256$. Unless the prior probability q_1 of H_1 is appreciably above $\frac{1}{2}$, the experimental result $x = 4$ will determine a posterior probability q_2' near unity, representing under this approach strong final overall evidence and opinion supporting H_2 against H_1. This illustrates how in the Bayesian treatment strong experimental evidence overshadows and dominates prior opinions when the latter are not extremely strong.

Now each standpoint in statistical theory agrees in principle that it is foolish or worse to set aside background information and competent opinion in scientific and other work involving statistics. The issues concern the place and the form in which such information and opinion should be brought to bear, in reporting, interpreting, and using experimental results. The problem touched on by Bross concerning the possible "dilution" of objective experimental evidence, by combining it with an individual's prior opinions, can be met by Bayesians: Between Bayesian statisticians holding possibly quite different prior probabilities in the same experimental situation, there is no difficulty in recognizing and reporting just the likelihood function, e.g., the value of $L(x)$ above; the numerical value of such a statistic is a complete mathematical characterization of statistical (experimental) evidence as used, interpreted, and applied by Bayesians. In other words, an important

general *consequence* of a Bayesian standpoint is

The likelihood principle (L): Any instance of statistical evidence (e.g. "$(E_b, 4)$" above) is characterized mathematically by just the likelihood function (e.g. "$L = 256$" above), and is otherwise independent of the mathematical form of the experiment and its outcome.

This principle may be taken as the Bayesian answer to the special delimited problem of informative inference; and Bayesians with different prior opinions evidently can in principle supply their different estimates of prior probabilities and proceed independently with their respective theoretical or practical purposes.

If all statisticians and scientists were Bayesians, the technical journals of the empirical sciences would of course continue to report statistical (experimental) evidence as such. But whenever the adequacy of a complete mathematical model of an experiment could be assumed, such a report could always be reduced without loss to a report of just the likelihood function determined by the observations, with no other reference to the structure of the experiment! For example, the report $(E_b, 4)$ in the above example could be replaced by just the report "likelihood ratio $= 256$," and any additional description of the results would be superfluous and irrelevant from this standpoint! A report of the likelihood function (e.g. "$L = 256$") is a *mathematical* characterization of statistical evidence which is appropriate in a Bayesian approach; the other function or problem of informative inference, *interpretation* of statistical evidence as such, requires little attention in this approach since the likelihood function is always to be used in Bayes' formula to obtain final probabilities, on which any questions of interpretation must focus.

Since the likelihood principle is a principal consequence of Bayes' principle, and characterizes the "empirical" part of Bayesian methods, the following questions are of interest to statisticians of various stand-points:

(A) Is the likelihood principle supported by any considerations other than the much-debated principle of Bayes?
(B) When experimental results are reported by just the likelihood function, what kinds of evidential interpretations are possible, and how do these compare with more customary ones?

Some answers to these questions can be obtained as follows by detailed consideration of examples like that above. A similar analysis with greater mathematical generality has been given elsewhere (1), (2).

Conditionality

A number of non-Bayesian statistical theorists concerned with the conceptual and technical problems of informative inference, beginning with R. A. Fisher and including D. R. Cox, J. W. Tukey, D. L. Wallace, and C. Stein (cf. (3) and references therein, and (4)), have been led to study "conditional" properties of interval estimates and of significance tests. Certain conditional probability properties of such techniques seem to be essentially relevant to their reasonableness and adequacy for purposes of informative inference, and the investigation of such properties has seemed significant because of the light it has thrown on relevant basic concepts. The following simple example illustrates this relevance of conditionality concepts: For the same purpose of informative inference considered in the above example, suppose that the experimenter has available three measuring instruments other than the one described above (which was characterized by probabilities .20 of false negatives and of false positives for each observation). Suppose the first instrument has probabilities of false positives and of false negatives each equal to .0039; for the second, such probabilities are each equal to .0588; and for the third, .50. The latter instrument is worthless, since

equally "informative" observations are obtainable by tossing a fair coin and calling heads "positive" and tails "negative." The first instrument is highly informative (very small error-probabilities, even when just a single observation is made), and the second instrument is intermediate in value. Suppose now that only a single observation with one of these instruments will be possible; and suppose that for some reason the instrument which is actually to be used cannot be strictly planned, but is known in advance only in a probability sense: there is a .1536 chance that the worthless instrument will be available, and hence that the experiment will yield no information concerning the material of interest; there is a .4112 chance that the best instrument will be available, and hence that the experiment will yield a highly informative measurement; with the remaining chance, .4352, the second instrument will be used, yielding a moderately informative measurement. This example seems artificial and unrepresentative, but it serves well to illustrate concretely certain concepts and features relevant to typical actual experimental situations. In particular, the alternative instruments available are simple analogues of uncontrollable but observable experimental variables whose values determine experimental precision but do not bias experimental results when taken suitably into account, as is done in standard methods of regression analysis. Moreover, as will be seen below, even such artificial-looking examples are equivalent in a certain mathematical and conceptual sense to more familiar examples, and can help throw light on the latter and on relevant concepts and techniques of informative inference.

If we accept these experimental conditions and the adequacy of the corresponding mathematical model, how should we report an outcome of this experiment, for example in a technical journal for readers interested in the material under study but not interested in irrelevant circumstances encountered by the experimenter? Let us denote by E_2 the sub-experiment consisting of a single observation by the superior instrument; and let us denote by $(E_2, +)$ or $(E_2, -)$, respectively, a positive or negative observation given by this instrument. Similarly, let $(E_1, +)$ and $(E_1, -)$ denote positive and negative results obtained when the second instrument is used. Let E_0 denote use of the third, worthless instrument (observations here need not be described further). Finally, let the overall experiment described be denoted by E_m; E_m may be called a *mixture* of the sub-experiments E_2, E_1, and E_0, since it consists of the specified chances that each of the latter sub-experiments will actually be selected and carried out. E_m has 5 possible outcomes, which may be described respectively (in notation like that used above for E_b) by $(E_m, (E_2, +))$, $(E_m, (E_2, -))$, $(E_m, (E_1, +))$, $(E_m, (E_1, -))$, and (E_m, E_0). For example, the report $(E_m, (E_2, +))$ denotes that the experiment E_m was carried out and resulted in selection of the superior instrument and a positive observation by the latter; (E_m, E_0) denotes that E_m resulted in selection of the worthless instrument. Clearly a report having the appropriate one of these five forms (including a description of the mathematical models denoted by E_m, E_2, E_1, and E_0) would be adequate at least in the sense of not being incomplete. The question of interest is whether such a report is redundant. In particular, does a reader interested in the material studied (but not in irrelevant features of the experimenter's working conditions) have any possible use for the symbol E_m and its interpretation? Would it not suffice for all of his possible purposes to ignore E_m and its interpretation in such reports, and to consider only the appropriate one of the five symbols $(E_2, +)$, $(E_2, -)$, $(E_1, +)$, $(E_1 \ -)$, or E_0, (including a description of the mathematical model for only that instrument actually selected and used), ignoring as irrelevant the instruments which might have been used but in fact were not?

An affirmative answer to the last question will seem appropriate to many on common-sense grounds. The significant general concept illustrated here concerns the possible adoption of a conditional experimental frame of reference, such as E_2, or E_1 or E_0, rather than the overall unconditional experimental frame of reference, like E_m, because the latter seems to contain redundant elements, while the appropriate conditional one seems to be a more appropriately refined frame of reference for evidential interpretations. For example, (E_m, E_0) represents that a measurement was made by a worthless instrument, and in addition informs us that certain other instruments might have been used but in fact were not. But does not the conditional report E_0 (recognizably completely uninformative) tell the whole (uninformative) story as it is *relevant* to the material of interest? Again, $(E_m, (E_2, +))$ tells that the superior instrument (with .0039 error-rates) gave a positive measurement on the material of interest, and in addition informs us that certain other instruments might have been used but in fact were not. But doesn't the abbreviated conditional report $(E_2, +)$ tell the whole *relevant* story? The concept and guiding principle illustrated here, which many statisticians consider appropriate for informative inference, may be formulated as: *The principle of conditionality* (C): When any experiment E_m has the form of a mixture, with components (sub-experiments) E_i, then for purposes of informative inference it is appropriate to interpret any outcome of E_m just as if it were an outcome of the corresponding component experiment E_i, ignoring otherwise the overall structure of E_m and its other components. Adoption of this principle is evidently not compelled by any purely mathematical considerations. Rather it is a principle which may be adopted by those who consider it an appropriate assertion about the nature and structure of experimental evidence (statistical evidence), which is the partly-extra-mathematical subject matter of informative inference. (C) may be called a partial-characterization of evidential meaning or of statistical evidence; it asserts an important property (equivalence between certain instances of statistical evidence), but leaves open remaining problems of more complete characterization of statistical evidence in general.

While many statisticians are inclined to accept (C), it has been seen only recently, and to the surprise of some, that (C) *implies* the likelihood principle (L), by a formal mathematical deduction! This means that the considerable number of non-Bayesian statisticians who find (C) an appropriate general principle for informative inference will, for purely mathematical reasons, be equally able to share an important sector of common methodological ground with Bayesian statisticians, without accepting Bayes' principle itself!

It is possible to illustrate in miniature the way in which (L) can be deduced from (C), using concrete examples which also further illustrates both principles and their relations to more standard concepts and techniques like significance tests and interval estimates. Consider first the simple experiments E_2, E_1, and E_0, which are also the sub-experiments of E_m above. The positive outcome of E_2 determines as the value of the likelihood ratio statistic (which represents the likelihood function here)

$$L = L(E_2, +) = \text{Prob}[+ \,|\, H_2, E_2]/[+ \,|\, H_1, E_2]$$
$$= (1 - .0039)/(.0039) = 256.$$

By the symmetry of E_2, its negative outcome determines the reciprocal value

$$L = L(E_2, -) = \frac{1}{256}.$$

Similarly we find

$$L(E_1, +) = 16, \; L(E_1, -) = \frac{1}{16}, \text{ and } L(E_0) = 1.$$

If only outcomes of one or another of these experiments were under consideration, a report of one of these five numerical values of the likelihood ratio statistic would serve to indicate uniquely a certain outcome of a certain one of these experiments; values of L above unity indicate positive outcomes (supporting H_2 against H_1), with larger values of L representing stronger evidence (corresponding to smaller error-rates). Values of L below unity similarly indicate support of H_1, smaller values representing stronger evidence.

The value $L = 1$ denotes strictly neutral or uninformative evidence. Within the range of all such simple symmetric experiments, there is a one-to-one correspondence between the possible numerical values of the likelihood ratio statistic L, on the one hand, and the strength and direction of statistical evidence as more customarily expressed in terms of error-probabilities α admitting frequency interpretations (significance levels or "P-levels" and power of tests), namely

$$\alpha = 1/(1 + L), \text{ for } L \geq 1,$$
$$\alpha = 1/(1 + L^{-1}) = L/(1 + L), \text{ for } L \leq 1.$$

Such correspondence between the two modes of expressing evidential meanings, likelihood functions and error-probabilities, can be extended to experiments with more general structures, beginning for example with the following observation: The binomial experiment E_b and the mixture experiment E_m described above are *mathematically equivalent*! (They are also "physically equivalent," in the sense of the usual frequency interpretations of probability models!) Hence (assuming adequacy of these models) there are evidently no theoretical or practical grounds on which any experimenter could reasonably base a definite preference between the experiment E_b (four independent observations by the instrument with error-rate .20), and the odd-looking experiment E_m which is like a lottery ticket allowing him a single observation by one of three very different instruments to be chosen randomly! To show the equivalence of E_b and E_m, we arrange the five possible outcomes of each experiment in the following one-to-one correspondence:

E_b yields: $x =$	0	1	2
E_m yields:	$(E_2, -)$	$(E_1, -)$	E_0

E_b yields: $x =$	3	4
E_m yields:	$(E_1, +)$	$(E_2, +)$

Next we observe that under each hypothesis, the respective models E_b and E_m assign identical probabilities to their corresponding points. For example, under H_1 the probability that E_b yields $x = 4$ is (the binomial probability) .0016; a simple calculation shows that .0016 is also the probability under H_1 that E_m yields the corresponding outcome $(E_2, +)$. In this way, the mathematical equivalence of E_b and E_m is directly verified. Hence, there is evidently no reason for giving any outcome of E_b an evidential interpretation different from that which one would give to a corresponding outcome of E_m.

The last statement (which we arrived at on purely mathematical grounds) and the principle of conditionality (C) lead mathematically to (L), insofar as the latter applies to our example E_b, by the following steps:

1. The outcome $x = 4$ of E_b has the same evidential meaning as the outcome $(E_2, +)$ of E_m, on grounds of mathematical equivalence.
2. The outcome $(E_2, +)$ of E_m has the same evidential meaning as the outcome $+$ (positive) of E_2 (considered as a complete experiment, ignoring E_m), according to the principle of conditionality (C).
3. But in the simple symmetric experiment E_2, characterized by error-rates each equal to $\alpha = \frac{1}{257} \doteq .0039$, the positive outcome is labeled by the value $L = 256$ of the likelihood ratio statistic. The latter value is (necessarily) the same value of L which would be obtained by direct calculations in the binomial model.
4. We can discuss similarly each other outcome x of E_b, and thus arrive at the conclusion: The evidential meaning of any outcome x of E_b is the same as that of the (unique) outcome of a simple symmetric experiment which gives the likelihood ratio statistic L the value $L(x)$. We note that the latter evidential interpretations depend on E_b and x only through the observed likelihood function, thus confirming the likelihood principle insofar as it applies to our example.

The fact that (L) can be deduced from the very plausible assertion (C), which many find more acceptable than Bayes' principle, seems to provide both some comfort and some challenge to the general Bayesian standpoint: The important "empirical" part (L) of the Bayesian position is given new support independent of Bayes' principle itself. But this suggests the questions: What are the specific contributions of the qualitative, and the quantitative, parts of Bayesian concepts and techniques to the interpretation and use of statistical evidence? How do these compare with less formalized interpretations, based on direct consideration of the likelihood function and other aspects of the inference situation, without formal use of prior probabilities and Bayes' formula?

These considerations also present some challenge to non-Bayesian statisticians accustomed to use of standard techniques of testing and estimation, in which error-probabilities appear as basic terms of evidential interpretations in a way which does not accord with the likelihood principle. Any techniques or concepts which are incompatible with (L) are also incompatible with the very plausible (C)! The writer has not found any apparent objections to (C) which do not really stem from consideration of a different notion of "conditional" from that formulated above, or else from consideration of purposes other than the modest one of informative inference.

Evidently in our present state of understanding there can be interesting collaboration between Bayesian and non-Bayesian statisticians in exploring the possibilities and limitations of both formal and informal modes of interpreting likelihood functions. Another important area requiring specific study, where such collaboration may also be fruitful, is that of design of experiments when such interpretations are contemplated.

References

1. Birnbaum, Allan, "On the foundations of statistical inference: binary experiments," *Annals of Mathematical Statistics*, **32** (1961), 414–435.

2. Birnbaum, Allan, "On the foundations of statistical inference," to be published, with discussion, in the *Journal of the American Statistical Association*, Vol. 57 (1962). (This paper was read at a discussion session at the December 1961 Annual Meeting of the A.S.A.)

3. Wallace, David L., "Conditional confidence level properties," *Annals of Mathematical Statistics*, **30** (1959), 864–876.

4. Stein, Charles, The Wald Lectures, 1961. Unpublished; presented at the Annual Meeting of the Institute of Mathematical Statistics, Seattle, June 15–17.

Editor's comments: Sequential analysis refers to any form of analysis in which the conclusions drawn depend not only on the data but also on the data-collection stopping rule. Jerome Cornfield in the following article examines the logic of assumptions implicit in sequential analysis and proposes an alternative conception based on the likelihood principle. He shows that acceptance of the likelihood principle leads to rejection of sequential analysis. Although the likelihood principle is a logical outgrowth of Bayes' theorem, Cornfield presents several non-Bayesian arguments for its acceptance that draw upon the classic Neyman-Pearson theory of statistics.

10.3 Sequential Trials, Sequential Analysis and the Likelihood Principle

Jerome Cornfield

1. I shall be concerned in this paper with a single question, the answer to which is of great importance to all those engaged in the sequential collection of data. The question is this: Do the conclusions to be drawn from any set of data depend only on the data or do they depend also on the stopping rule which led to the data? In discussing this question I shall draw heavily upon the theoretical work of others, particularly L. J. Savage, but also F. J. Anscombe, G. A. Barnard, A. Birnbaum, and D. V. Lindley. Biostatisticians have tended to regard the theoretical developments suggested by these names as unduly abstract and perhaps of no great relevance to statistical practice. I shall nevertheless review some of them in the hope that few of us would go

A revision of remarks made at the round table on sequential clinical trials at the 1965 meeting of the American Public Health Association.

Reprinted from *American Statistician*, 1966, **20**, 18–23, by permission of the publisher and author.

so far as to say with Falstaff that because "Honour hath no skill in surgery . . . I'll none of it." These newer developments, abstract or not, are, in my opinion, of great relevance to biostatistical practice and their absorption into thinking, teaching and consultation is becoming overdue.

At the outset it is useful to distinguish between sequential trials and sequential analysis. By a sequential trial I shall mean any form of data collection in which the decision to continue or discontinue further collection depends in some sense on the information previously obtained. By sequential analysis I mean any form of analysis in which the conclusion depends not only on the data, but also on the stopping rule. Although there are many unsettled questions in the construction of particular stopping rules for sequential trials, it is clear that data dependent stopping rules of some type are often appropriate. I shall however concern myself here only with the question of sequential analysis, i.e., whether the conclusion should depend on the stopping rule and not at all on how to choose among the rules.

2. To indicate that the question to be considered is a far from a trivial one I should first like to give a simple example of sequential analysis in which the conclusion is far more dependent on the stopping rule than on the data. Consider the hypothesis that a normal mean has a particular value. This hypothesis might be tested either by observations gathered in a fixed sample size trial or in a Wald three-decision sequential trial [1]. In the former case the most powerful unbiased test would reject the hypothesis at, say, the .05 level whenever the sample mean differed from the hypothesized value by more than 1.96 standard errors, no matter what the value of n or the alternative assumed (for σ known). Proceeding sequentially, however, the number of standard errors by which the sample mean must differ from that hypothesized depends first of all on n, secondly on the value specified as a two-sided alternative, and thirdly, on the power required, given this alternative. Thus let the alternative differ from the hypothesized value by $\pm 0.25\ \sigma$, with power of 0.95. For $n = 1$ the hypothesis could not be rejected at the .05 level unless the observation departed from it by more than 14 standard errors! Even for $n = 10$ the hypothesis could not be rejected with a deviation of less than 5 standard errors [2]. This result cannot be explained by the assumption that σ is known, since if this assumption is relaxed and the sequential t-test [3] is used instead the situation becomes worse. The hypothesis then cannot be rejected at the .05 level until at least 14 observations have been accumulated, no matter what the values of the first 13, i.e., even if they all differ from the hypothesis value by $10^{10^{10}}$. The conclusion is thus markedly, if not pathologically, dependent on the stopping rule, and the question with which we are concerned can hardly be considered a trivial one.

3. To most scientists without previous exposure to statistics, as well as to most intelligent laymen, any dependence of conclusions

on stopping rules, let alone the extreme dependence we have exhibited, seems like a violation of common sense. Those biostatisticians who defend sequential analysis on the other hand would argue that dependence of conclusions on stopping rules is required to preserve the critical level, i.e., the lowest significance level at which the hypothesis can be rejected for given data. If one accepts the importance of preserving the critical level, then clearly conclusions must depend on the stopping rule. But what is not immediately obvious is that the critical level provides an appropriate measure of the amount of evidence in the data for or against the hypothesis. Biostatisticians who accept sequential analysis would argue, if I understand them correctly, that the significance level is an index to or measure of weight of evidence, i.e., that if one is in possession of observations which (in the usual tail area sense) would occur rarely if the hypothesis is true, then one is also in possession of evidence against the hypothesis. The critical level is thus regarded as a universal yardstick. The fundamental postulate, on which sequential analysis is based, thus appears to be, "All hypotheses rejected at the same critical level have equal amounts of evidence against them." I have never heard this postulate, which I shall call the α-postulate, explicitly stated, nor can I point to any statistician that I know for a fact to believe it, although many act as if they do.[1] But yet anyone who denies it seems to me

---

[1] My colleague Dr. Samuel W. Greenhouse points out, however, that numerous first cousins of the postulate have appeared in the literature, as for example,

"We shall describe two different tests T_1 and T_2 associated with critical regions w_1 and w_2 as equivalent when the probabilities $P_1(w_1)$ and $P_1(w_2)$ of making an error of Type I are equal." [4]

The critical level "gives an idea of how strongly the data contradict (or support) the hypothesis." [5]

"When a prediction is made, having a known low degree of probability, such as that a particular throw with four dice shall show four sixes, an event known to have a mathematical probability, in the strict sense, of 1 in 1296, the same reluctance will be felt towards accepting this assertion, and for just the same reason, indeed, that a similar reluctance is shown to accepting a hypothesis rejected at this level of significance . . . In general, tests of significance . . . lead . . . to a rational and well-defined measure of reluctance to the acceptance of the hypotheses they test." [6]

to have denied the only reason known for believing in sequential analysis in the first place. Sequential analysis can be defended, in my opinion, if and only if something like the α-postulate is true.

I propose now to develop two lines of argument: 1. The α-postulate is not as reasonable as it might at first sight appear and 2. There is an alternative formulation of the idea of weight of evidence which is reasonable and which does lead to the conclusion that the stopping rule is irrelevant to the conclusion.

4. The first line of argument consists of citing three situations in which I believe everyone would agree that the critical level has no relation to the idea of weight of evidence, no matter how one chooses to define that elusive concept. (a) For most tests in common use a hypothesis which is rejected at a given significance level, say .01, for a given set of data will be rejected at the .02, .05, and all higher levels for that same set of data. Not all tests of hypotheses have this characteristic and then the α-postulate leads to contradictions. To see this, consider a hypothesis H_1, which is rejected by data S_1 at the .02 and all higher levels but not the .01 level, and hypothesis H_2, which is rejected by the data S_2 at the .01, but not .05 level. If then we consider the .01 level S_2 supplies more evidence against H_2 than S_1 does against H_1, but if we consider the .05 level the contrary is true. The universal applicability of the α-postulate therefore requires that if a most powerful test of a hypothesis against a simple alternative rejects at some significance level, α, for given data, that at any significance level $\alpha' > \alpha$ the hypothesis must also be rejected for the same data. But this is not always so. Lehman [7] has given a counter-example, and others are easily constructed. Whether or not such situations are often encountered in daily statistical practice, their mere existence makes it impossible to believe that the critical level supplies a measure of weight of evidence under all circumstances.

(b) The following example will be recognized by statisticians with consulting experience as a simplified version of a very common situation. An experimenter, having made n observations in the expectation that they would permit the rejection of a particular hypothesis, at some predesignated significance level, say .05, finds that he has not quite attained this critical level. He still believes that the hypothesis is false and asks how many more observations would be required to have reasonable certainty of rejecting the hypothesis if the means observed after n observations are taken as the true values. He also makes it clear that had the original n observations permitted rejection he would simply have published his findings. Under these circumstances it is evident that there is no amount of additional observation, no matter how large, which would permit rejection at the .05 level. If the hypothesis being tested is true, there is a .05 chance of its having been rejected after the first round of observations. To this chance must be added the probability of rejecting after the second round, given failure to reject after the first, and this increases the total chance of erroneous rejection to above .05. In fact as the number of observations in the second round is indefinitely increased the significance approaches .0975 ($= .05 + .95 \times .05$) if the .05 criterion is retained. Thus no amount of additional evidence can be collected which would provide evidence against the hypothesis equivalent to rejection at the $P = .05$ level and adherents of the α-postulate would presumably advise him to turn his attention to other scientific fields. The reasonableness of this advice is perhaps questionable (as is the possibility that it would be accepted). In any event it does not seem possible to argue seriously in the face of this example that all hypotheses rejected at the .05 level have equal amounts of evidence against them.

(c) D. R. Cox [8] has constructed an example which suggests that the most powerful test of the hypothesis that a mean is zero against a particular alternative will sometimes reject the hypothesis when the observed mean is zero. Thus, the observation which would ordinarily be regarded as providing the strongest evidence for the null hypothesis is in this example treated as reason for rejecting it.

These examples are, I think, sufficient to suggest that the seemingly plausible assumption that critical level and weight of evidence are identical concepts is far from firmly established. I realize, of course, that practical people tend to become impatient with counter-examples of this type. Quite properly they regard principles as only approximate guides to practice, and not as prescriptions that must be literally followed even when they lead to absurdities. But if one is unwilling to be guided by the α-postulate in the examples given, why should he be any more willing to accept it when analyzing sequential trials? The biostatistician's responsibility for providing biomedical scientists with a satisfactory explication of inference cannot, in my opinion, be satisfied by applying certain principles when he agrees with their consequences and by disregarding them when he doesn't.

5. I turn now to my second line of argument—which is that there is a reasonable alternative explication of the idea of inference and one which leads to the rejection of sequential analysis. This explication is provided by the likelihood principle—which states that all observations leading to the same likelihood function should lead to the same conclusion.

I shall start by illustrating the idea of a likelihood function. Consider r successes in n independent trials, each with constant probability, p, of leading to a success. If n is a prespecified constant, r is a random variable whose distribution is given by the binomial distribution, namely

$$(1) \qquad \binom{n}{r} p^r (1-p)^{n-r}.$$

If r is a prespecified constant, i.e., if we continue observation until the rth success occurs and then stop, the number of trials, n, is a random variable whose distribution is given by the negative binomial, namely

$$(2) \qquad \binom{n-1}{r-1} p^r (1-p)^{n-r}.$$

These two probabilities refer to different random variables, and even for given r and n will have different numerical values. Thus, for $p = \frac{1}{2}$, $n = 2$, $r = 1$, the first probability is $\frac{1}{2}$, the second $\frac{1}{4}$. It will be observed, however, that each probability can be considered a product of two factors, one of which is a combinatorial term which does not depend on p, and differs for the two distributions, and the other of which depends on p and is the same for both distributions. The factor which depends on p, namely

$$(3) \qquad p^r (1-p)^{n-r}$$

is an instance of a likelihood function. In general, given any probability or joint probability density function it is always possible to factor it into two parts, one of which depends on the unknown parameters, and provides the likelihood function, and the other which does not depend on unknown parameters.

As an additional example, the likelihood function for n independent normal observations, $x_1 \cdots x_n$, with unknown mean and variance, is

$$(4) \qquad \sigma^{-n} \exp - \left[\frac{1}{2\sigma^2} \sum (x_i - \mu)^2 \right],$$

while the second factor, independent of unknown parameters, is

$$(5) \qquad (2\Pi)^{-n/2}.$$

If σ were known, however, then the likelihood function is

$$(6) \qquad \exp - \left[\frac{n}{2\sigma^2} (\bar{x} - \mu)^2 \right]$$

and the second factor is

$$(7) \qquad \sigma^{-n} (2\Pi)^{-n/2} \exp - \left[\frac{\sum (x_i - \bar{x})^2}{2\sigma^2} \right].$$

The motivation for this definition, which goes back to Fisher, is that the likelihood function so defined represents that part of Bayes' formula which is dependent on the data. (See for example p. 326 of [9] and section 6 of Chapter III of [6].) If the entire probability or probability density function, rather than just the likelihood function is inserted in Bayes' formula, the second factor appears as a multiple of both numerator and denominator and cancels out, leaving just the likelihood function. No matter what its motivation, however, the likelihood function is a well-defined mathematical entity, and for the moment this is all we require.

6. Consider now two investigators, one of whom decided to conduct n binomial trials and then stop, no matter what his number of successes, and another, who decided to continue until he had obtained r successes and then stop, no matter how many trials it took. Suppose further that the first investigator obtained exactly r successes, and the second obtained his rth success on exactly the nth trial. Then by (3) they are in possession of the same likelihood function, and if they both accept the likelihood principle they must come to the same conclusion about p, despite the use of quite different stopping rules. If they had adapted some different inferential principle, say that of unbiased estimation, however, the first investigator would have estimated p as r/n and the second as $(r-1)/(n-1)$. The fact that the numerical difference between these two estimates will often be small should not obscure the fact that we are dealing with an important difference in principle and that the conclusions yielded by the likelihood principle differ radically on this point from those yielded by more traditional principles.

For the binomial-negative binomial case the differing stopping rules had no effect upon the likelihood function. Can one be sure that this will always be the case? It is easy to show that this must always be so [10]. The likelihood function does not depend on the stopping rule and if one accepts the likelihood principle one must reject sequential analysis.

7. Why should anyone accept the likelihood principle? I remark first that it is an immediate consequence of Bayes' theorem, since the posterior probability density function of the unknown parameters is proportional to

the product of the likelihood function and the prior probability density function. But acceptance of this justification requires acceptance of prior probabilities and hence a radical revision in the traditional objectivistic outlook towards interpretation of data. A number of non-Bayesian arguments therefore also have been given [11,12]. I should like to sketch out one that owes a good deal to Savage and Lindley [13]. It seems to me to be particularly illuminating because it deduces the principle from concepts with which all of us raised in the Neyman-Pearson tradition are familiar.

We consider a null hypothesis, H_0, a simple alternative, H_1, and k different experimental designs. For each design we can divide the sample space into two parts, one containing points that lead to the acceptance of H_0 and a part that leads to its rejection. Each such division implies particular values for α_i and β_i ($i = 1, 2, \ldots, k$), the errors of the first and second kind. We now ask how to select the "best" critical region for each design. If we accept the α-postulate "best" means that we set each α_i equal to some constant independent of i, and then by application of the Neyman-Pearson lemma (whose truth, of course, is not dependent on the postulate) find the rejection region for each design which minimizes β_i. The sample spaces for the different design need not be the same, but they may have certain points (i.e., possible observations like r successes in n trials) in common. As we have seen, for sample spaces so constructed, it is possible for a common point to be in the rejection region for some designs and outside it for others, i.e., for the inference to depend on the design as well as the data.

As an alternative definition of "best," consider the minimization of a linear function of the two types of errors, $\lambda\alpha_i + \beta_i$, where λ measures the undesirability or cost of an error of the first kind relative to one of the second kind and is the same for all k. We ask for a rejection region for each of k designs which will minimize this function. Consider any sample point, t, and the ratio of the likelihood of the alternative, H_1, given t, to that of the null hypothesis, H_0, given t, and denote the ratio for the ith design by $R_i(t)$. Then it is easy to show that the rejection region for the design must consist of all points for which $R_i(t) > \lambda$. But any common sample point which has the same likelihood function in each design will also have the same value of $R_i(t)$ in each of them. Since λ is the same for all designs, that sample point will then lead to the rejection of the null hypothesis in each of the designs if $R_i(t)$ exceeds λ and its acceptance if it does not. Thus, if instead of minimizing β for a given α, we minimize $\lambda\alpha + \beta$, we must come to the same conclusion for all sample points which have the same likelihood function, no matter what the design.

To extend this argument to designs which differ because of differences in stopping rules, it is sufficient to divide the sample space for each design into three parts, corresponding to rejection of H_0, acceptance of H_0 and suspension of judgment. There are probabilities of falling into each of these three regions, under each hypothesis and appropriate costs. One attempts to define the regions for each design by minimizing a linear function of the error probabilities, where the coefficients in the linear function depend on the costs of the corresponding errors. It is again easy to show that for each design the rejection region for H_0 consists of all points for which $R_i(t) > \lambda_2$, the acceptance region of all points for which $R_i(t) < \lambda_1$ and the region for suspending judgment of all points in which $\lambda_1 \leq R_i(t) \leq \lambda_2$. The constants λ_1 and λ_2 depend on costs and are the same for all k stopping rules.

Thus, the antagonism sometimes pointed to between the likelihood principle and the principle of minimizing errors is seen to depend entirely on a particular formulation of the idea of minimizing errors, namely one dependent on the α-postulate. If a linear function of the errors is minimized instead, one is led directly to the likelihood principle.

One might ask why to minimize a linear rather than some other function of the errors. Savage sketches out a reason for considering the linear function the only appropriate one to minimize. But even without this, it is clear

that the entire basis for sequential analysis depends upon nothing more profound than a preference for minimizing β for a given α rather than minimizing their linear combination. Rarely has so mighty a structure and one so surprising to scientific common sense, rested on so frail a distinction and so delicate a preference.

8. Earlier I remarked that the critical level is taken in current biostatistical practice as a universal yardstick. It will be observed that the likelihood ratio emerges from this argument as such a yardstick instead, at least for the comparison of a given H_0 against a given H_1. The operational sense in which the likelihood ratio supplies a yardstick is illuminated by considering consistent betting behavior. Suppose a statistician who, having observed the outcome of one of the k designs, were willing to bet that H_1 is true at odds of $f(t)$ to 1 or that H_0 is true at odds of 1 to $f(t)$, where $f(t)$ simply denotes the dependence of the odds on the outcome. How should he determine $f(t)$? If he accepts the likelihood principle, $f(t)$ will be constant for all outcomes for which $R(t)$ is constant, and in particular will not depend on the design. If he accepts the α-postulate $f(t)$ will be constant for all outcomes leading to a given critical level. Now it can be demonstrated that when averaging over all k designs one can realize an average gain whether H_0 or H_1 is true by betting against any statistician whose odds depend on the critical level. The strategy is simple. Bet on H_0 for all sample points for which $f(t)/R(t)$ is greater than a given constant and against it when less. If the statistician sets $f(t)/R(t)$ equal to a constant for all t, however, no system of bets can win both when H_0 is true and when H_1 is true. At best one can win when H_0 is true, but will lose when H_1 is true, or vice versa. The result is a direct consequence of deFinetti's demonstration under conditions more general than a simple choice between H_0 and H_1 that a coherent system of betting odds implies and is implied by Bayes' theorem [14].

It is not inappropriate to ask anyone who

denies the relevance of betting odds to scientific inference but accepts critical levels to explain the operational sense in which it is possible for one set of data more strongly to contradict H_0 than another, at the same time that any system of bets based on this measure would lead to loss whether H_0 or H_1 is true.

9. In section 4 doubt was cast upon the α-postulate by considering special situations in which it leads to absurd results. Is a counter-attack possible in which something like this is done for the likelihood principle as well? In particular can one find so absurd a stopping rule, that no matter what the data, one is certain to come to the wrong conclusion if the inference is made without consideration of the stopping rule—as the likelihood principle says it should be. Such a counter-example has been proposed by Armitage [15,16]. It is worth considering, partly on its merits, and partly because it illuminates further the relation between the likelihood principle and Bayes' theorem.

The stopping rule is this: continue observations until a normal mean differs from the hypothesized value by k standard errors, at which point stop. It is certain, using the rule, that one will eventually differ from the hypothesized value by at least k standard errors even when the hypothesis is true. If one looks only at the data, i.e., the likelihood function, one would quite properly reject the hypothesis for reasonably large values of k, whereas if, in the light of the α-postulate, one looks at the stopping rule as well, with its implied α of unity, one would not. Thus, if one disregards the stopping rule and is guided only by the likelihood function one is certain to reject a true hypothesis. In Armitage's words this is a reason for "resisting immediate conversion" to the likelihood principle. Barnard [17] and Birnbaum [18], who accept the likelihood principle, but reject prior probabilities, have, as I understand them, admitted the anomalous nature of this result.

The Bayesian viewpoint of the example is as follows [2]. If one is seriously concerned about the probability that a stopping rule will certainly result in the rejection of a true hypothesis, it must be because some possibility of the truth of the hypothesis is being entertained. In that case it is appropriate to assign a non-zero prior probability to the hypothesis. If this is done, differing from the hypothesized value by k standard errors will not result in the same posterior probability for the hypothesis for all values of n. In fact for fixed k the posterior probability of the hypothesis monotonically approaches unity as n increases, no matter how small the prior probability assigned, so long as it is non-zero, and how large the k, so long as it is finite. Differing by k standard errors does not therefore necessarily provide any evidence against the hypothesis and disregarding the stopping rule does not lead to an absurd conclusion. The Bayesian viewpoint thus indicates that the hypothesis is certain to be erroneously rejected—not because the stopping rule was disregarded—but because the hypothesis was assigned zero prior probability and that such assignment is inconsistent with concern over the possibility that the hypothesis will certainly be rejected when true.

10. The previous remarks have been confined to tests of hypotheses, because this is the only form of sequential analysis for which a general mathematical basis now exists. This should not be interpreted to mean that the difficulties with the α-postulate would disappear if ever a firm mathematical basis for sequential confidence limits were found. The confidence set yielded by a given body of data is the set of all hypotheses not rejected by the data, so that the relation between hypothesis tests and confidence limits is close. In fact the confidence limit equivalent of the α-postulate is, "All statements made with the same confidence coefficient have equal amounts of evidence in their favor." That this may be no more reasonable than the α-postulate is suggested by the very common problem of inference about the ratio of two normal means. The most selective unbiased confidence set for the unknown ratio has the following curious characteristic: for every

sample point there exists an $\alpha > 0$ such that all confidence limits with coefficients $\geq 1 - \alpha$ are plus to minus infinity [19]. But to assert that the unknown ratio lies between plus and minus infinity with confidence coefficient of only $1 - \alpha$ is surely being over-cautious. Even worse, the postulate asserts that there is less evidence for such an infinite interval than there is for a finite interval about a normal mean, but made with confidence coefficient $1 - \alpha'$, where $\alpha' < \alpha$. The α-postulate cannot therefore be considered any more reasonable for confidence limits than it is for hypothesis testing.

It has been proposed by proponents of confidence limits that this clearly undesirable characteristic of the limits on a ratio be avoided by redefining the sample space so as to exclude all sample points that lead to infinite limits for given α. This is equivalent to saying that if the application of a principle to given evidence leads to an absurdity then the evidence must be discarded. It is reminiscent of the heavy smoker, who, worried by the literature relating smoking to lung cancer, decided to give up reading.

References

1. Wald, A., *Sequential Analysis*, (1947), New York, John Wiley and Sons.
2. Cornfield, J., "A Bayesian test of some classical hypotheses—with applications to sequential clinical trials." Submitted for publication.
3. U.S. National Bureau of Standards, *Tables to Facilitate Sequential t-Tests*, (1951), Applied Mathematics Series (NBS-AMS-7) Washington, D. C.: Government Printing Office.
4. Neyman, J. and Pearson, E. S., "The testing of statistical hypotheses in relation to probabilities a priori," *Proceedings of the Cambridge Philosophical Society*, **29** (1933), 492–510.

5. Lehmann, E. L., *Testing Statistical Hypotheses*, (1959), New York, John Wiley and Sons, p. 62.
6. Fisher, R. A. *Statistical Methods and Scientific Inference*, (1956), London, Oliver and Boyd, p. 43.
7. *Op cit.*, Ex. 29, p. 116.
8. Cox, D. R., "Some problems connected with statistical inference," *Annals of Mathematical Statistics*, **29** (1958), 357–372.
9. Fisher, R. A. *Contributions to Mathematical Statistics* (1950), New York, John Wiley and Sons.
10. Anscombe, F. J., Discussion to D. V. Lindley, "Statistical Inference," *Journal of the Royal Statistical Society* (B), **15** (1953), 30–76.
11. Barnard, G. A., "Statistical Inference," *Journal of the Royal Statistical Society* (B), **11** (1949), 115–159.
12. Birnbaum, A., "On the foundations of statistical inference," *Journal of the American Statistical Association*, **57** (1962), 269–326.
13. Savage, L. J., "The foundations of statistics reconsidered," in *Studies in Subjective Probability*, ed. by H. E. Kyburg, Jr. and H. E. Smokler, (1964), New York, John Wiley and Sons.
14. deFinetti, B., "Foresight: its logical laws, its subjective sources," in *Studies in Subjective Probability, op. cit.*
15. Armitage, P., Discussion to Smith, C. A. B., "Consistency in statistical inference and decision," *Journal of the Royal Statistical Society*, (B), **23** (1961), 1–37.
16. Armitage, P., "Sequential medical trials: some comments on F. J. Anscombe's paper," *Journal of the American Statistical Association*, **58**, (1963), 384–387.
17. Barnard, G. A., "Comment on Stein's 'A remark on the likelihood principle'," *Journal of the Royal Statistical Society*, Series A (1962), 569–573.
18. Birnbaum, A., *The Anomalous Concept of Statistical Evidence, Axioms, Interpretations in Elementary Exposition* (invited paper presented to the Joint European Conference of Statistical Societies, Berne, Switzerland, September 14, 1964).
19. Lehmann, *op cit.*, Ex. 11, p. 182.

Editor's comments: Bayesian statistical theory is based on a definition of probability as the measure of the opinions of ideally consistent people. As data become available a Bayesian adherent following Bayes' theorem modifies his prior probability distribution in the light of sample evidence. According to his decision rules the experimenter may decide that he has sufficient information to draw a conclusion, or he may decide that his uncertainty has not been sufficiently reduced and therefore he must continue gathering data until he has a more suitable basis for reaching a conclusion.

In the following article, F. J. Anscombe uses a marketing example to illustrate the basic features of Bayesian statistical theory. This example is also used to compare the classical approach to statistical inference with the Bayesian approach.

10.4   Bayesian Statistics

F. J. Anscombe

During the last few years there has been a revival of interest among statistical theorists in a mode of argument going back to the Reverend Thomas Bayes[1] (1702–61), Presbyterian minister at Tunbridge Wells in England, who wrote an "Essay towards solving a problem in the doctrine of chances," which was published in 1763 after his death. Bayes's

Based on a talk given to the Chicago Chapter of ASA, April 4, 1960. I was much helped in the writing by H. V. Roberts, and subsequently received valuable comments from R. Burges, D. R. Cox, T. Dalenius, W. H. Kruskal, P. Meier, L. J. Savage, J. W. Tukey and D. L. Wallace.

Research was carried out at the Statistical Research Center, University of Chicago, under sponsorship of the Logistics and Mathematical Statistics Branch, Office of Naval Research. Reproduction in whole or in part is permitted for any purpose of the United States Government.

Reprinted from *American Statistician*, 1961, **15**, 21–24, by permission of the publisher and author.

work was incorporated in a great development of probability theory by Laplace and many others, which had general currency right into the early years of this century. Since then there has been an enormous development of theoretical statistics, by R. A. Fisher, J. Neyman, E. S. Pearson, A. Wald and many others, in which the methods and concepts of inference used by Bayes and Laplace have been rejected.

The orthodox statistician, during the last twenty-five years or so, has sought to handle inference problems (problems of deciding what the figures mean and what ought to be done about them) with the utmost "objectivity." He explains his favorite concepts, "significance level," "confidence coefficient," "unbiased estimates," etc., in terms of what he calls "probability," but his notion of probability bears little resemblance to what the man in the street means (rightly) by probability. He is not concerned with probable truth or plausibility, but he defines probability in terms of frequency of occurrence in repeated trials, as in a game of chance. He views his inference problems as matters of routine, and tries to devise procedures that will work well in the long run. Elements of personal judgment are as far as possible to be excluded from statistical calculations. Admittedly, a statistician has to be able to exercise judgment, but he should be discreet about it and at all costs keep it out of the theory. In fact, orthodox statisticians show a great diversity in their practice, and in the explanations they give for their practice; and so the above remarks, and some of the following ones, are no better than crude generalizations. As such, they are, I believe, defensible. (Perhaps it should be explicitly said that Fisher, who contributed so much to the development of the orthodox school, nevertheless holds an unorthodox position not far removed from the Bayesian; and that some other orthodox statisticians, notably Wald, have made much use of formal Bayesian methods, to which no probabilistic significance is attached.)

The revived interest in Bayesian inference starts with another posthumous essay, on "Truth and probability," by F. P. Ramsey[2] (1903–30), who conceived of a theory of consistent behavior by a person faced with uncertainty. Extensive developments were made by B. de Finetti and (from a rather different point of view) by H. Jeffreys. For mathematical statisticians the most thorough study of such a theory is that of L. J. Savage[3,4]. R. Schlaifer[5] has persuasively illustrated the new approach by reference to a variety of business and industrial problems. Anyone curious to obtain some insight into the Bayesian method, without mathematical hardship, cannot do better than browse in Schlaifer's book.

The Bayesian statistician attempts to show how the evidence of observations should modify previously held beliefs in the formation of rational opinions, and how on the basis of such opinions and of value judgments a rational choice can be made between alternative available actions. For him "probability" really means probability. He is concerned with judgments in the face of uncertainty, and he tries to make the process of judgment as explicit and orderly as possible.

It seems likely that during the coming years there will be a change amounting to a revolution in statistical methodology. Methods currently in use will be modified or reinterpreted. Theoretical concepts will be greatly altered. A just appreciation of the situation can only be had by studying the orthodox and the Bayesian approaches to a variety of statistical problems, noting what each one does and how well it does it. Such a study will show, I think, a diversity of relative effectiveness, according to the type of problem considered. When what is at issue is the choice of a decision procedure, to be used impartially in a routine way (as for example a sampling inspection plan operated by a government agency), the analysis made by the orthodox statistician is capable of being good and cogent; the Bayesian statistician will accept this analysis and will add only a little to it. On the other hand, when the problem is one of unique intelligent decision (such as, should this new idea be actively developed, or should it be shelved?), it is not clear that the orthodox statistician's analysis has much cogency; often only the Bayesian can illuminate and assist the workings of common sense.

A Problem in Marketing

The following example is designed to illustrate the difference between the orthodox and Bayesian approaches. It has been shorn of complexities, so that it is not too reminiscent of the real world, but the comparison it suggests is, I believe, fair.

An executive must decide whether or not his company shall place a certain new product on the market. Its possible appeal is to a limited specialized class of consumers, of whom he has a complete list. He has carefully studied the economics of the whole matter. He has determined that if more than 10 per cent of his list of potential customers will in fact be interested in the product, then the company will do well to market it, but that if fewer than 10 per cent of potential customers are interested his company should not market it. The proportion of potential customers who will be interested is the only important feature of the situation which he has not been able to ascertain, and he believes that the profit will be linearly related to this, i.e., the incremental (or marginal) profit will be constant.

Before he makes his decision, the executive has the opportunity to make contact with a random sample of the potential customers and to inquire of them whether they will be interested in the new product if it is made available. Let us suppose that the answers obtained are regarded as perfectly trustworthy, so that the proportion of favorable answers differs only by ordinary sampling fluctuation from the proportion of the whole class of potential customers who will be interested. Let us denote the latter proportion by p, let the sample size be n, which we shall suppose to be only a small fraction of the total number of potential customers on the list, and let the number of favorable answers received in the sampling inquiry be r. (Of these three symbols, p stands for an unknown quantity, but n and r are observed.) It is clear that if r/n turns out to be somewhat higher than 0.1 he will decide that p is probably greater than 0.1 and he ought to market the product, whereas if r/n turns out to be somewhat less than 0.1 he will decide that p is probably less than 0.1 and he ought not to market it. Given the value of n and all the background information, there is presumably some critical (whole) number c, not greatly different from $0.1n$, such that if r turns out to be greater than or equal to c he will judge it wise to market the product, but if r is less than c he will decide against marketing. How can he best choose c?

Orthodox Analysis

Suppose that he consults an orthodox statistician—one who by good fortune is not only orthodox but intelligent too. The statistician will begin by inquiring into the economic situation and will elicit the information outlined above. He will then impart some information as follows. Given any values of n and c, it is possible to calculate the chance that the number of favorable responses in the sample will be greater than or equal to c. Calling this $F(p)$, we have

$$\text{ch}\{r \geq c\} = F(p) = \sum_{i=c}^{n} \binom{n}{i} p^i (1 - p)^{n-i}.$$

(The orthodox statistician will be likely to use the word "probability" rather than "chance," but as his notion of probability is derived from contemplating the phenomena of simple games of chance, flipped coins, rolled dice, spun roulette wheels, etc., and he disbelieves in any other (logical or personal) concept of probability, I here prefer the term "chance," abbreviated "ch.")

Suppose for example that $n = 25$ and c is chosen equal to 1. Then $F(p)$, which is the chance that the new product will be marketed, can be graphed against p, as shown in Fig. 1. Similar curves can be drawn for other values of c, such as 2, 3, 4, 5, as also shown.

Now if in fact $p > 0.1$, the new product ought to be marketed, but the chance is $1 - F(p)$ that $r < c$ and the decision will go against marketing. Incremental profit being assumed constant, the wrong decision will cause a net loss of revenue proportional to $(p - 0.1)$. On the other hand, if $p < 0.1$ the new product ought not to be marketed, but there is the chance $F(p)$ that $r \geq c$ and the decision will be for marketing. This wrong decision will cause a net loss of revenue proportional to $(0.1 - p)$. The expected loss of revenue from wrong decisions is thus proportional to

$$
\begin{aligned}
(p - 0.1)\{1 - F(p)\} \qquad &\text{if } p > 0.1, \\
(0.1 - p)F(p) \qquad &\text{if } p < 0.1.
\end{aligned}
$$

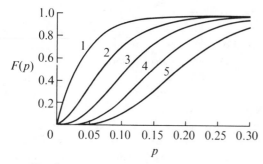

Fig. 1.

Graphs of expected loss are shown in Fig. 2. It will be seen that when $p < 0.1$, the larger c is the smaller are the expected losses, but when $p > 0.1$, the smaller c is the smaller the losses. (Note that the word "loss" is used here exclusively to mean loss of revenue caused by a wrong decision; it does not mean negative profit.)

As a compromise, designed to cope with both of the possibilities that p may be greater than 0.1 and less than 0.1, the statistician will point out the advantage of choosing c equal to 3, as the expected losses have then the least upper bound. This is easily shown to be related to the fact that $F(0.1)$ is close to $\frac{1}{2}$. The statistician may very well suggest a general rule for choosing c, for any given n, namely

RULE A: *Choose c so that as nearly as possible $F(0.1) = 0.5$.*

At the "break-even" value for p, namely 0.1, when it is a matter of indifference which decision is made, there should be equal chances of marketing and not marketing. Then if p is better than the break-even value, the chance of marketing will exceed $\frac{1}{2}$, and if p is worse, the chance will be less than $\frac{1}{2}$. (Rather than Rule A, the statistician may propose directly that the minimax principle be adopted. I have preferred Rule A as being a little simpler to think about. For present purposes the difference is negligible.)

Now all this discussion has been in terms of chances, and the executive may well ask, what have chances got to do with the prob-

lem? He is not, after all, playing roulette. The statistician will tell him that he is basing a decision on a random sample, and that *is* rather like playing roulette. In fact, the statistician's recommendations amount to a policy of play, rather like the betting policy of an inveterate gambler—or better, the betting policy of an insurance company. Chances mean relative frequencies of occurrence in a long series of trials. Expected losses mean average losses over such a long series. If the executive had to make a long sequence of decisions about different products, the break-even value for p and the value of n being always the same, then by following the statistician's policy he could be sure that, whatever values of p might occur, his average loss of revenue from wrong decisions would not be very large (compared with what it might be for other values of c), the worst it could be being as low as possible. Thus in a sense the policy is prudent. Moreover, since decisions are based directly on the sample evidence and not at all on any private hunch the executive may have about the product, the policy has a judicial impartiality—useful if the executive is at loggerheads with his board of directors or under fire from other executives in the company.

If indeed the executive had a series of decisions to make, routine-fashion, and if impartiality were any asset, the statistician's arguments would be much to the point. Most likely, however, the executive will feel that the various decision problems that he encounters are all qualitatively somewhat different from

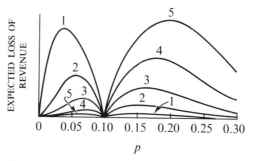

Fig. 2.

each other, that they all need to be considered individually on their merits, and that it is not wise to follow rules of thumb. With luck he is *not* at loggerheads with the board of directors, and does not have to spell out reasons for his decisions; on the contrary, he is expected to act boldly and imaginatively. The statistician's arguments seem no longer very cogent. If n is large, say 1000 or more, r/n will be likely to be so near to p that Rule A seems very reasonable. But what if n is small?

Suppose $n = 2$. It is now impossible to implement Rule A as it stands. If $c = 0$, $F(0.1) = 1$. If $c = 1$, $F(0.1) = 0.19$. If $c = 2$, $F(0.1) = 0.01$. If $c > 2$, $F(0.1) = 0$. These are all the possibilities. At this point the statistician may possibly come out with a suggestion that was on the tip of his tongue when he first mentioned Rule A, namely to ensure that $F(0.1) = \frac{1}{2}$ by following a randomized procedure. The executive should now equip himself with a bowler hat and eight ping-pong balls, of which five have been dyed with red ink and the other three with green ink. Then he should examine the results of the sample inquiry. If one or both of the answers are favorable he should decide to market the product. If neither answer is favorable, he should shake the hat up and draw out one ball without looking, and be guided in the obvious way by its color.

Let us push the matter to the extreme and suppose $n = 0$. How should the executive decide when he has no sample evidence to go on? The randomized version of Rule A now says he should decide by flipping an ordinary coin. This procedure is fair, impartial and eminently "objective." An executive who followed it would not keep his job long.

So while Rule A makes good sense if n is large, it makes nonsense if n is small enough. A good orthodox statistician recognizes this, of course. Rather than push Rule A to the absurd extreme, he will be likely to say that if n is very small or zero no impartial rule can be recommended, and the executive should decide the matter unaided, as his common sense dictates.

More Orthodox Analysis

The above analysis has presented the orthodox point of view in the most favorable possible light. It makes some sense, plenty of sense in its way, but it solves the wrong problem.

Many orthodox statisticians would solve an even wronger problem. Skeptical of economic assessments, disbelieving there can be such a thing as a break-even value for p, blinded by a particular tradition, they would pervert the whole thing into a significance test. They would ask the executive: what is the lowest value for p such that you would much rather decide in favor of marketing—what value for p is too good to miss? He will reply, perhaps, that if $p > 0.15$ he would certainly like to know about it. The statistician will now take 0.15 as an Acceptable Quality Level or Null Hypothesis and dream up an associated Producer's Risk or Error Rate of 5 per cent (or maybe 10 per cent or 1 per cent, there is no knowing what), and he will then enunciate

RULE B: *Choose c so that as nearly as possible $F(0.15) = 0.95$ (or 0.90 or 0.99 or whatever).*

Significance tests have, I believe, an essential and inescapable place in statistics. This isn't it. The less said about Rule B the better.

Bayesian Analysis

The Bayesian statistician will regard it as his function to assist the executive to make up his mind, and in particular to show how the evidence from the sample inquiry should influence the decision. He will begin by asking not only for the economic information we have already considered, but also about the executive's opinion of the value of p before he received the sample results. The executive's knowledge of the market and the customers' needs may well be substantial; it should certainly not be ignored.

In order to keep the calculations as simple as possible, let us suppose that the executive believes that the sale of the new product will depend on a particular feature of the business done by the customers. If the situation is of one sort, then he expects that only about 2 per cent of his list of potential customers will be interested, but if the situation is different he expects that about 15 per cent of his list of potential customers will be interested. That is, he believes that p is either close to 0.02 or close to 0.15. Moreover, in the absence of the sample information, he thinks the first of these possibilities a good deal more likely than the second—he would think it fair to offer odds of $3:1$ in favor of $p = 0.02$. Thus his prior probability distribution is approximately as follows:

$$\text{pr}(p = 0.02) = 0.75,$$
$$\text{pr}(p = 0.15) = 0.25.$$

(I write "pr" instead of "ch," because logical probabilities are referred to, not relative frequencies. More usually in a prior distribution the probability is distributed over more than two possible values of the parameters. This slightly complicates the ensuing calculations, without altering their essential character.)

The executive must decide between two actions, to market or not to market, let us say M and N. We can set out a table of proportional losses from wrong decisions, as follows:

	$p = 0.02$	$p = 0.15$
M	8	0
N	0	5

Thus when $p = 0.02$, it is better not to market. If the decision M is taken, the loss of net revenue, as compared with N, is proportional to $0.1 - p = 0.08$. Similarly when $p = 0.15$, M is the better decision, and the loss if N is taken is proportional to $p - 0.1 = 0.05$. (All that matters here is the ratios of differences in the columns of the table—how much difference it makes whether one decision or another is taken for a given value of p, as compared

with differences for other values of p. We could just as soon work with a table of net revenue, rather than losses; and we can use any units we like, as long as we are consistent about it.)

Now if there is no sample information ($n = 0$), we can calculate the expected loss if M is taken,

$$L(M) = 8 \times \text{pr}(p = 0.02) + 0 \times \text{pr}(p = 0.15)$$
$$= 8 \times 0.75,$$

and the expected loss if N is taken,

$$L(N) = 0 \times \text{pr}(p = 0.02) + 5 \times \text{pr}(p = 0.15)$$
$$= 5 \times 0.25.$$

The ratio of these expected losses is

$$\frac{L(M)}{L(N)} = \frac{8 \times 0.75}{5 \times 0.25} = 4.8.$$

Since this ratio is greater than 1, it will be wise to take action N—and no flipping of coins!

If sample evidence becomes available, the prior odds for 0.02 against 0.15 as the value of p are multiplied by the "likelihood ratio"

$$\frac{(0.02)^r (0.98)^{n-r}}{(0.15)^r (0.85)^{n-r}},$$

and so the ratio of expected losses becomes

$$\frac{L(M)}{L(N)} = \frac{8 \times 0.75 \times (0.02)^r (0.98)^{n-r}}{5 \times 0.25 \times (0.15)^r (0.85)^{n-r}}.$$

The likelihood ratio shows directly how the executive's prior opinion is changed by the sample evidence. The mathematical proposition used here is known as Bayes's theorem.

For example, suppose $n = 2$. Then the likelihood ratio is 1.33 if $r = 0$, 0.15 if $r = 1$, 0.018 if $r = 2$. The ratio of expected losses is 4.8 multiplied by the likelihood ratio, that is,

6.4 if $r = 0$, 0.74 if $r = 1$, 0.085 if $r = 2$.

Clearly he will prefer to take action N if he finds $r = 0$, M if $r = 1$ or 2. Most likely, whether $p = 0.02$ or 0.15, r will be found to be 0, and in that case the sample evidence will have left the decision unchanged—as was to be expected with so little sample information. However, if $p = 0.15$, there is an appreciable chance (about 0.28) that there will be at least one favorable response in the sample, and then decision M will be made.

Suppose now that $n = 25$. We find that the ratio of expected losses is

168 if $r = 0$, 19.5 if $r = 1$, 2.25 if $r = 2$, 0.26 if $r = 3$, etc.,

and the action N is preferred if $r \leq 2$, M if $r \geq 3$. If in fact $p = 0.02$, it is almost certain that the correct decision N will be taken, whereas if $p = 0.15$ the chance is about ¾ that the correct decision M will be made.

The larger n is, the more certain it is that the correct decision will be taken, and the Bayesian procedure approximates the orthodox statistician's Rule A. The decision effectively depends only on the sample information and the loss system. But if n is not very large the prior probabilities substantially influence the decision.

The Bayesian method is based on a theory of consistent behavior which in itself is attractive. Whenever, as here, the results of applying the Bayesian method can be easily compared with intuition and common sense, the agreement is good. One has confidence, therefore, in trusting the method in complex cases where common sense falters.

References

1. Bayes, T. "Essay towards solving a problem in the doctrine of chances," reprinted with bibliographical note by G. A. Barnard, *Biometrika*, **45** (1958), 293–315.
2. Ramsey, F. P. *The Foundations of Mathematics*, London: Routledge and Kegan Paul, 1931.
3. Savage, L. J. *The Foundations of Statistics*, New York: John Wiley, 1954.
4. Savage, L. J. "Subjective probability and statistical practice," to be published in a Methuen Monograph.
5. Schlaifer, R. *Probability and Statistics for Business Decisions: An Introduction to Managerial Economics Under Uncertainty*, New York: McGraw-Hill, 1959.

10.5   Suggestions for Further Reading

Edwards, W., H. Lindman, and L. J. Savage. Bayesian statistical inference for psychological research. *Psychological Review*, 1963, **70**, 193–242. This classic article continues to be one of the most readable references on Bayesian statistical inference for students in the behavioral sciences.

Glossary

Acceptance region A subset of a sample space (or the set of all values of a test statistic) for which a null hypothesis is accepted.

Adjusted sum of squares A sum of squares from which effects associated with a covariate have been removed in analysis of covariance. This sum of squares is also called a reduced sum of squares.

Alias An effect in a fractionally replicated design that cannot be distinguished from another effect.

Alienation, coefficient of A measure of the lack of linear association between two variables; it is designated by k and is equal to $\sqrt{1 - r^2}$ where r is a product-moment correlation coefficient.

Alpha error Error that occurs when an experimenter rejects the null hypothesis when it is true. This error is usually referred to as a Type I error.

Alternative hypothesis The hypothesis that remains tenable when the null hypothesis is rejected. It is designated by H_1.

Analysis of covariance A statistical analysis that combines linear regression and analysis of variance techniques. It permits the statistical control of one or more nuisance variables.

Analysis of variance (ANOVA) A statistical analysis in which the total variability of a set of data is subdivided into components that can be attributed to different sources of variation. Analysis of variance is used primarily to test statistical hypotheses with respect to two or more population means.

A posteriori probability A probability assigned to an event such that information relating directly to the event is taken into account.

A priori probability A probability assigned to an event such that information relating directly to the event is not taken into account.

Arcsin transformation A transformation of a variate, say X, into a variate X' by some relation of the type $X' = \arcsin (X + k)$, where k is chosen for convenience.

Arithmetic mean A measure of location obtained by adding a set of values and dividing their sum by the number of values added. Also called a mean.

Association The degree of dependence or independence that exists between two or more variables measured either quantitatively or qualitatively.

Association, coefficient of A measure of independence for data in a two-by-two contingency table.

Asymptotic normality A distribution is asymptotically normal if it tends to the normal form

387

as a parameter, usually sample size, tends to infinity.

Average   A term not having a precise definition but sometimes used to refer to the arithmetic mean.

Average deviation   See Mean deviation.

Balanced confounding   An arrangement of treatment combinations in a factorial experiment that results in each interaction component of the same order being confounded equally with between-block variation.

Bartlett's test   A test for homogeneity of $k \geq 2$ population variances; specifically a test of the null hypothesis that a set of of independent random samples comes from normal populations having equal variances.

Bayesian statistics   Statistical methods that utilize prior information, either objective or subjective, about parameters.

Bernoulli trial   A simple experiment that can eventuate in only one of two possible outcomes. For example, if a coin is tossed, then one of two events—a head (referred to as success) or a tail (referred to as failure) —must occur. A plot of the two events and their associated probabilities is a *Bernoulli distribution.*

Bias   Any effect that systematically distorts the outcome of an experiment so that the results are not representative of the phenomenon under investigation.

Binomial distribution   A distribution of the number of successes or failures in N independent Bernoulli trials in which p and q, the probabilities associated with two events, remain constant over the N trials. Specifically, binomial distributions constitute a family of theoretical distributions that associate a probability, $p(X)$, with each value of a random variable X according to the rule

$$p(X) = \binom{N}{r} p^r q^{N-r}, \quad 0 \leq X \leq N,$$

with parameters of N (the number of independent trials) and p (the probability of observing exactly r successes in N independent trials).

Bivariate distribution See Joint distribution.

Bivariate normal distribution   A joint distribution of two random variables, say X and Y, where the marginal distribution of X is normal, the regression of Y on X is linear, and the conditional distribution of Y for a fixed value of X is normal with variance independent of the value of X.

Block sum of squares   In analysis of variance, that component of the total sum of squares that can be attributed to differences among blocks. Block differences generally correspond to differences among subjects or experimental units.

Central limit theorem   If a population has a finite variance σ^2 and mean μ, the distribution of sample means, $\overline{X}$'s, from samples of n independent observations approaches a normal distribution with variance σ^2/n and mean μ as the sample n increases. When n is very large, the sampling distribution of $\overline{X}$ is approximately normal.

Characteristic function   If X is a random variable with density function $f(x)$, the expected value of the expression e^{itX} is the characteristic function, $\phi_x(t)$, of X; that is, $\phi_x(t) = E(e^{itX})$. The characteristic function always exists and completely determines the distribution of the random variable X.

Comparison   A linear combination of treatment level sums or means in which the sum of the coefficients, C_J, is equal to zero.

Conditional distribution   A distribution of a random variable or the joint distribution of several random variables in which the values of one or more other random variables are held constant.

Conditional probability   If A and B are two events, the conditional probability denoted by $P(A/B)$ that A will occur given that B has occurred is equal to $P(A \cap B)/P(B)$ where $P(A \cap B)$ is the probability of the joint occurrence of A and B and $P(B)$ is the nonzero probability of B's occurring.

Confidence interval   An interval that, considering all possible samples for estimating a parameter, has some designated probability of including the population parameter.

Confidence limits   Upper and lower boundaries of a confidence interval.

Confounding   A procedure whereby treatments are assigned to subjects so that certain effects cannot be distinguished from other effects. The purpose of confounding is to reduce the number of treatment-level combinations that must be assigned to blocks of subjects.

Conjugate squares   Two Latin squares are conjugate if the rows of one are identical to the columns of the other.

Consistent estimator An estimating procedure is consistent if the estimates it yields tend to approach the parameter more and more closely as the sample size approaches ∞. Such estimates are called consistent estimates.

Consistent test statistic A statistic provides for a consistent test if the probability of rejecting a false null hypothesis approaches one as the sample size approaches ∞.

Contingency coefficient A measure of the strength of association between two variables —variables that are usually qualitative in nature. The maximum value of the coefficient is less than one and depends on the number of rows and columns of the contingency table; the minimum value of the coefficient is zero.

Contingency table A table having two or more rows and two or more columns into which experimental units are classified according to two criteria.

Continuous random variable A random variable, say X, such that in an interval bounded by values a and b every non-zero interval in this range has a non-zero probability. Specifically, a random variable, X, is continuous if for every pair of values a and b where $p(X \leq a) < p(X \leq b)$ it is true that $p(a < X < b) > 0$.

Contrast A linear combination of observations used to estimate some parameter; the coefficients of the linear combination are some set of real numbers subject to the restrictions that their sum must equal zero and at least one coefficient must not equal zero.

Correlation A term denoting relationship or association between two or more variables.

Correlation ratio A measure of the curvilinear relationship between two variables.

Covariance An expected value of the product of the deviations of two random variables from their respective means.

Critical region A subset of a sample space (or the set of all values of a test statistic) for which a null hypothesis is rejected.

Critical value The value of a statistic that corresponds to a given significance level as determined from its sampling distribution. For example if probability $(F > F_{\alpha; v_1 v_2}) = .05$, then $F_{\alpha; v_1, v_2}$ is the critical value of F at the 5 percent level.

Crossed treatments An arrangement in which all possible combinations of levels of two or more treatments occur together in an experiment.

Defining contrast In a fractional factorial analysis of variance design, that source of variation that is used to divide the treatment combinations into fractional replicates.

Degrees of freedom The number of independent observations for a source of variation minus the number of independent parameters estimated in computing the variation.

Density function See Probability density function.

Determination, coefficient of A measure of the proportion of total variation of the dependent variable that is accounted for by the linear relationship with the independent variable. It is equal to the square of the product-moment correlation coefficient.

Discrete random variable A random variable, say X, such that any pair of values a and b each having a non-zero cumulative probability of occurrence there will always exist an interval of values between a and b for which the probability of occurrence is equal to zero. Specifically, a random variable, X, is discrete if for every pair of values a and b where $0 < p(X \leq a) < p(X \leq b)$, there exists another value c between a and b for which $p(a < X \leq c) = 0$ or $p(c \leq X < b) = 0$ or both.

Distribution-free method A method for testing a hypothesis or setting up a confidence interval that does not depend on the form of the underlying distribution.

Distribution function A function whose values $F(x)$ are the probabilities that a random variable X assumes a value less than or equal to x for $\infty < x < \infty$.

Dummy treatment level A hypothetical treatment level that is not actually administered but that is conceptually included in an experiment in order to preserve the symmetry of the design.

Efficient estimator An estimator is asymptotically efficient if the distribution of estimates that it yields tends to normality with the least possible standard deviation of the estimates as the sample size is increased.

Error rate In hypothesis testing, the unconditional probability of erroneously accepting or rejecting a hypothesis.

Error sum of squares In analysis of variance,

that component of the total sum of squares that can be attributed to experimental error.

Estimator   The particular function of observations in a sample that is chosen to estimate a population parameter. For example, the sample mean can serve as an estimator of the population mean. The numerical value obtained is called an estimate.

Expected value   The mean value of a random variable over an indefinite number of samplings. The expected value, $E(X)$, of a discrete random variable X is given by $E(X) = \Sigma Xp(X)$ = mean of X.

Experimental error   A measure that includes all uncontrolled sources of variation affecting a particular score.

Experimental unit   A person, animal, or object to which a treatment is applied.

Fixed-effects model   An experimental design model in which it is assumed that all treatment levels about which inferences are to be made are included in an experiment.

Generalized interaction   In confounded experimental designs, the interaction(s) that is automatically confounded with between-block variation as a result of confounding two other interactions with between-block variation.

Homoscedasticity   In regression analysis, the property that the conditional distributions of Y corresponding to values of the independent variable X all have the same variance.

Hypothesis   *See* Research hypothesis; Statistical hypothesis.

Hypothetical population   A statistical population that has no real existence but is imagined to be generated by repetitions of events of a certain type.

Incomplete block   A block that does not include all treatment levels or combinations of treatment levels contained in an experiment.

Independent events   Two events A and B are independent if the joint probability $p(A \cap B)$ is equal to the probability of A times the probability of B.

Interaction   Two treatments are said to interact if scores obtained under various levels of one treatment behave differently under various levels of the other treatment.

Iteration   A technique of successive approximation in which each step is based on results obtained in the preceding step(s).

Joint density function   An extension of the con-

cept of a density function to two or more random variables.

Joint distribution   An extension of the concept of the distribution of a random variable to two or more random variables. Joint distributions are referred to as *multivariate distributions*, and in the case of two random variables, as *bivariate distributions*.

Joint event   Any event that is in the intersection of two or more events.

Kurtosis   The characteristic of a distribution that is concerned with its peakedness or flatness.

Latin square   A matrix of symbols, usually Latin letters, in which each symbol appears once in each row and each column.

Least squares, method of   A method of curve-fitting that involves minimizing the sum of squares of differences between observed values and the corresponding values predicted by means of a model equation.

Level of significance   Probability of rejecting the null hypothesis when it is true; usually designated by α.

Likelihood ratio test   A general procedure for finding a test statistic with optimum properties for testing any of a broad class of hypotheses.

Linear combination   An expression of the form $a_1 X_1 + a_2 X_2 + \cdots + a_n X_n$ where $(a_1, a_2, \ldots, a_n)$ is any set of n real numbers, not all equal to zero, which are used as weights.

Linear model   A mathematical model in which the random variables and parameters are related by means of linear equations. In analysis of variance, for example, the linear model $X_{ij} = \mu + \beta_j + \varepsilon_{ij}$ states that a random variable, X_{ij}, is equal to the sum of parameters which are, respectively, the grand mean, effect of the jth treatment level, and error effect.

Location, measures of   In general, those statistics that describe the position in a distribution around which individual values cluster. These measures have the property that if a constant is added to each value in the distribution, the measure of location is changed by an amount equal to the constant. Examples are the mean, median, and mode.

Matrix   A rectangular ordered array of elements. An m by n matrix has m rows each containing n elements.

Maximum-likelihood estimator. An estimator that

when substituted for the parameter maximizes the likelihood of the sample result.

Mean See Arithmetic mean.

Mean deviation A measure of dispersion defined as the mean of the sum of absolute deviations from the mean.

Mean square In analysis of variance, a sum of squares, *SS*, divided by its degrees of freedom, *df*.

Median A measure of location defined as that value in a distribution at or below which exactly 50 percent of the cases fall.

Minimax principle A decision rule whereby an experimenter minimizes his maximum-expected loss over all possible true situations.

Minimum-variance estimator The estimator having the smallest possible variance for a given class of estimators.

Mixed model An experimental design model in which some treatments are fixed effects and other treatments are random effects.

Mode A measure of location defined as that value in a distribution having the largest frequency. If two non-adjacent values have the same maximum frequency the distribution is said to be bimodal.

Moment generating function If X is a random variable with a density function $f(x)$, the expected value of the expression e^{tX} is the moment generating function, $M_X(t)$, of X; that is, $M_X(t) = E(e^{tX})$. The distribution of a random variable is completely determined by its moment generating function.

Monte Carlo methods. Methods for obtaining approximate solutions to mathematical and statistical problems by sampling from simulated random processes. Such sampling is usually carried out with the aid of a computer.

Most powerful test The test that has the smallest probability of a Type II error when used to evaluate a particular H_0 against some true alternative value covered by H_1.

Multinomial distribution An extension of the binomial distribution to the case in which the number of possible outcomes of any trial is greater than two.

Multivariate normal distribution A generalization of a univariate normal distribution to the case of p variates, where $p \geq 2$.

Nested treatment A treatment, say B, is nested if each level of treatment B appears in combination with only one level of another treatment.

Nondetermination, coefficient of A measure of the proportion of the total variation of the dependent variable that is not accounted for by the linear relationship with the independent variable. It is usually denoted by k^2 and is equal to $1 - r^2$, where r is the product-moment correlation coefficient.

Nonparametric method A method for testing a hypothesis that does not involve an explicit assertion concerning a parameter.

Normal distribution A graph of the normal density function

$$f(X) = \frac{1}{\sigma\sqrt{2\pi}} e^{-(X-\mu)^2/(2\sigma^2)}$$

that associates a probability density $f(X)$ with each and every possible value of X. The normal distribution, discovered by Abraham DeMoivre, is bell-shaped and extends from $-\infty$ to ∞.

n-tuple An ordered set of n numbers.

Nuisance variable An undesired source of variation in an experiment that may affect the dependent variable.

Null hypothesis A statement concerning one or more parameters that is subjected to statistical test. It is usually designated by H_0.

One-tailed test A test in which the region of rejection, often referred to as the critical region, consists of either the upper or the lower tail of the sampling distribution of the test statistic.

Orthogonal comparisons Two comparisons are said to be orthogonal if the products of their corresponding coefficients sum to zero.

Outlier An observation so far removed from the main body of data as to cast doubt on its correctness. Outliers can result from equipment malfunctions, gross errors in the recording of data, and other mishaps.

Parameter A measure computed from all observations in a population. Parameters are usually designated by Greek letters. For example, the symbols for a population mean and standard deviation are μ and σ, respectively.

Partial correlation A measure of the relationship between two variables where one or more additional variables are statistically held constant.

Percentage point   A level of significance (α) that is expressed in the form of a percentage.

Percentile   A measure of location defined as that value in a distribution at or below which a specified percent of the cases fall. For example, the eightieth percentile is that value in the distribution below which 80 percent of the values lie.

Plot   A set of subjects or materials to which a treatment is administered. The term can be traced to agricultural research where it was used with reference to a plot of land.

Point estimation   Any estimation of a parameter obtained by assigning to it a unique value called a "point estimate." The merits of a method of point estimation are determined on the basis of the resulting point estimate's unbiasedness, consistency, sufficiency, relative frequency, and minimum variance.

Population   A collection of all observations identifiable by a set of rules.

Power efficiency   A measure of the relative efficiency of two test statistics; it is given by $100N_a/N_b$ where N_a and N_b are the sample sizes, respectively, of the more powerful and less powerful tests that are required to detect a true alternative to H_0 for some fixed α and $1 - \beta$ probabilities. The more cases required for test B to attain the same power as test A, the smaller is the power-efficiency of test B relative to test A.

Power function   In hypothesis testing, a function that gives, for the various values of the parameter under consideration, the probabilities of rejecting a false null hypothesis.

Power of a test   Probability of rejecting the null hypothesis when the alternative hypothesis is true. If β is designated as the probability of committing a Type II error, power is equal to $1 - \beta$.

Probability density   An abbreviation for *probability density function;* also denotes a value of such a function.

Probability density function   A function having non-negative values whose integral from a to b where $a \leq b$ gives the probability that a corresponding random variable assumes a value on the interval from a to b.

Random-effects model   An experimental design model in which it is assumed that the treatment levels investigated in an experiment represent a random sample from a population of treatment levels.

Random sample, simple   A sample drawn from a population in such a way that all possible samples of size n have the same probability of being selected.

Random variable   A quantity, say X, which may assume a range of possible values each having an associated probability $p(X)$.

Range   A measure of dispersion defined as the difference between the largest value and the smallest value in a distribution.

Relative efficiency of a statistic   Ratio of experimental error of one statistic to that of another statistic.

Relative information   The ratio of information concerning an effect to the information that would be available concerning the effect if an experimental design involved no confounding.

Replication   The collection of two or more observations under sets of identical experimental conditions.

Research hypothesis   A tentative theory or supposition provisionally adopted to account for certain facts and to guide in the investigation of others. The terms research hypothesis and scientific hypothesis may be used interchangeably.

Sample   A subset of observations from a population.

Sampling distribution   A theoretical probability distribution that describes the functional relation between possible values of a statistic based on N cases drawn at random and the probability associated with each value over all possible samples of size N.

Sample space   A set of all possible distinct outcomes of a simple experiment.

Set   Any well-defined collection of objects; the objects that constitute the set are called its elements.

Significance test   A procedure for choosing, on the basis of sample data and a set of decision rules, between two mutually exclusive and exhaustive statistical hypotheses. The hypothesis that specifies the sampling distribution with which an obtained sample value is compared is the hypothesis that is tested. A significance test is also called a *statistical test*.

Skewness   The characteristic of a distribution that is concerned with its asymmetry.

Standard deviation   A measure of dispersion defined as the square root of the sum of squared deviations from the mean divided by N, the sample size.

Standard error   The standard deviation of the sampling distribution of a statistic.

Standard square   A Latin square in which the first row and first column are ordered alphabetically or numerically.

Statistic   A measure computed from observations in a sample. Statistics are usually designated by Roman letters. For example, the symbols for a sample mean and standard deviation are $\overline{X}$ and S, respectively.

Statistical decision theory   A branch of mathematics concerned with the problem of decision making and the choice of decision rules under uncertain conditions.

Statistical hypothesis   A statement about one or more parameters of a population. Null and alternative hypotheses are two forms of a statistical hypothesis.

Statistical model   A mathematical statement concerning the sampling distribution of random variables; it is used in evaluating the outcome of an experiment or in predicting the outcome of future replications of an experiment.

Statistical test   *See* Significance test.

Sufficient estimator   A statistic that contains all the information concerning a parameter that is available in the data. A statistic is a sufficient or *best* estimator of a parameter if the estimate cannot be improved by utilizing any other aspect of the data not already included in the statistic. For example, the sample mean is a sufficient estimator of the population mean when the distribution is normal.

Test statistic   A statistic whose purpose is to provide a test of some statistical hypothesis. Test statistics such as z, χ^2, t, and F have known sampling distributions that can be employed in determining the probability of an obtained result under the null hypothesis.

Theoretical distribution   The distribution of a random variable; distinguished from a distribution of observed data.

Transformation   A systematic alteration in a set of scores whereby certain characteristics of the set are changed and other characteristics remain unchanged.

Treatment   A set of experimental conditions comprising the independent variable.

Treatment combination   In a factorial experiment, an experimental condition consisting of two or more treatment levels to which a subject is simultaneously exposed.

Treatment level   One of the experimental conditions that comprise a treatment.

Treatment sum of squares   In analysis of variance, that component of the total sum of squares that can be attributed to differences among treatment levels.

Two-tailed test   A test in which the regions of rejection, often referred to as the critical regions, consist of both the upper and the lower tails of the sampling distribution of the test statistic.

Type I error   Error that occurs when the experimenter rejects the null hypothesis when it is true. The probability of committing a Type I error is determined by the level of significance (α) which the experimenter adopts.

Type II error   Error that occurs when the experimenter fails to reject the null hypothesis when it is false. The probability (β) of committing a Type II error is determined by the magnitude of the experimental effect, size of sample, magnitude of random error, and level of significance.

Unbiased estimator   An estimator of a parameter is said to be unbiased if its expected value is equal to the parameter.

Unbiased test statistic   A statistic provides for an unbiased test of the null hypothesis if it makes the probability of rejecting H_0 minimal when H_0 is true.

Uniform distribution   A distribution that assigns in the discrete case the same probability to each value within its domain; in the continuous case, a distribution having a constant probability density over all possible intervals within its domain.

Uniformly most powerful test   A test for which the probability of a Type II error is smaller than that of any other test regardless of the true value covered by the alternative hypothesis.

Univariate distribution   A distribution of one random variable.

Variance-covariance matrix   A square matrix whose elements, a_{ij}'s, are given by the covariance of the ith and jth random variables when $i \neq j$ and by the variance of the ith random variable when $i = j$. The diagonal terms of the matrix are the variances, σ_j^2's, and the off-diagonal terms are the covariances, $\rho\sigma_{jj'}^2$'s. A variance-covariance matrix is also called a "dispersion matrix" and a "covariance matrix."

Variation, measures of Statistics that describe the spread (dispersion) of a distribution.

Examples are the range and the standard deviation.

Vector A matrix having either a single row or a single column.

Name Index

Subject Index

*Italic numbers refer to entries in the glossary.